The Skeptic's Dictionary

A Collection of Strange Beliefs, Amusing Deceptions, and Dangerous Delusions

Robert Todd Carroll

WILEY

John Wiley & Sons, Inc.

Published by John Wiley & Sons, Inc., Hoboken, New Jersey
Published simultaneously in Canada

For general information about our other products and services, please contact our Customer Care Department within the United States at (800) 762-2974, outside the United States at (317) 572-3993 or fax (317) 572-4002.

Wiley also publishes its books in a variety of electronic formats. Some content that appears in print may not be available in electronic books. For more information about Wiley products, visit our web site at www.wiley.com.

Library of Congress Cataloging-in-Publication Data:
Carroll, Robert Todd, date.
    The skeptic's dictionary : a collection of strange beliefs, amusing deceptions, and dangerous delusions / Robert Todd Carroll.
      p. cm.
Includes bibliographical references and index.
    ISBN 0-471-27242-6 (cloth : acid-free paper)
    1. Pseudoscience—Encyclopedias.   I. Title.
    Q172.5.P77C37 2003
    001.9—dc21

                                     2003007878

Printed in the United States of America

10 9 8 7 6 5 4

To Olivia and Flynn
my candles in the dark

Contents

A non-Zoroastrian could think of Zarathustra as simply a madman who led millions of naive followers to adopt a cult of ritual fire worship. But without his "madness" Zarathustra would necessarily have been only another of the millions or billions of human individuals who have lived and then been forgotten.

—John F. Nash Jr.

The most common of all follies is to believe passionately in the palpably not true. It is the chief occupation of mankind.

—H. L. Mencken

Preface

The Skeptic's Dictionary began on the Internet in 1994. It is with some trepidation that I put forth my work in book form. I am not talking about the criticism I am likely to receive for being an atheist, skeptic, debunker, destroyer of hopes, and so on. I am used to that. Nor am I referring to a fear that somebody will actually levitate or prove to have the gift of prophecy or some other paranormal power after I have published my skeptical musings. No, what I fear is that one or more of the 750 sources I have used to write this book will not receive proper credit. I have tried my best to put into quotes whatever I have taken verbatim from another's work. I have tried my best to reference works in the standard fashion of the day. If I have failed to appropriately cite anyone's work, I apologize and I assure you it was inadvertent. Let the publisher know, and should we be blessed to go to a second printing, we will rectify the matter.

Finally, all of the references to URLs of web sites were current at the time of writing.

Acknowledgments

Many, many thanks to Tobias Budke, Leslie Carroll, Ronaldo Cordeiro, Leroy Ellenberger, Jeff Golick, Antonio Ingles, Kim Jeanman, Joe Littrell, Vlado Luknar, Masataka Okubo, Rich Ownbey, Bob Steiner, Ted Weinstein, and the many others who have inspired me, corrected me, guided me, and encouraged me in this mission. Tim Boettcher, Richard Herron, and John Renish deserve special thanks for their generosity in serving as volunteer editors of my web site, www.skepdic.com. Whatever my failings as a writer, they would be much more evident had it not been for their editorial assistance.

I have been motivated and encouraged by the writings of Stephen Barrett, Susan Blackmore, Arthur C. Clarke, Kenneth Feder, Thomas Gilovich, Terence Hines, Ray Hyman, Ivan Kelly, Janja Lalich, Elizabeth Loftus, Joe Nickell, Robert Park, Oliver Sacks, Daniel Schacter, Al Seckel, Michael Shermer, Margaret Thaler Singer, Nicholas Spanos, Victor Stenger, Carol Tavris, and others of like mettle.

But I probably would not have begun this work had it not been for the inspiration I got from reading the books of Stephen Jay Gould, Richard H. Popkin, James Randi, Carl Sagan, and especially the many books of Martin Gardner, whose *Fads and Fallacies in the Name of Science*—I have the 1957 Dover edition—got me started on this mission in the late '60s or early '70s. At the time, I was reading philosophers like Spinoza, Leibniz, and Malebranche, but Gardner introduced me to a world of alternative realities that made the imaginations of these philosophers pale by comparison.

Then there was Freud. His case studies fascinated me, especially his essay on a case of paranoia (1911), a psychological analysis of the memoir of Daniel Paul Schreber (1842–1911), a respected judge and political figure until his psychotic breakdown. Schreber's memoir gives an account of the delusions that landed him in the asylum for several years. What struck me at the time was that I had studied very similar musings in the writings of various mystics such as Plotinus and had even believed notions similar to Schreber's for many years—Virgin birth and impregnation of a human by a divinity, for example. Had Schreber lived in the 2nd century instead of the 19th, would he have taken his place at the same table with Zarathustra instead of with the other patients in the asylum? Had he put forth his fantasies and delusions as metaphysical speculations or scientific insights, would a cult have arisen around his ideas that would have led to an entry in *The Skeptic's Dictionary*? When I consider the list of subjects I still have in my "to investigate further" file, I can only say *Yes, beyond a doubt.*

Introduction

The Skeptic's Dictionary provides definitions, arguments, and essays on subjects supernatural, occult, paranormal, and pseudoscientific. I use the term "occult" to refer to any and all of these subjects. The reader is forewarned that *The Skeptic's Dictionary* does not try to present a balanced account of occult subjects. If anything, this book is a Davidian counterbalance to the Goliath of occult literature. I hope that an occasional missile hits its mark. Unlike David, however, I have little faith, and do not believe Goliath can be slain. Skeptics can give him a few bumps and bruises, but our words will never be lethal. Goliath cannot be taken down by evidence and arguments. However, many of the spectators may be swayed by our performance and recognize Goliath for what he often is: a false messiah. It is especially for the younger spectators that this book is written. I hope to expose Goliath's weaknesses so that the reader will question his strength and doubt his promises.

Another purpose of *The Skeptic's Dictionary* is to provide references to the best skeptical materials on whatever topic is covered. So, for example, if that pesky psychology teacher won't let up about "auras" or "chi" being inexplicable occult phenomena, you can consult your *Skeptic's Dictionary* and become pesky yourself with more than a general skepticism. You may not change your teacher's mind, but you may take away some of his or her power over you.

The Skeptic's Dictionary is aimed at four distinct audiences: the open-minded seeker, who makes no commitment to or disavowal of occult claims; the soft skeptic, who is more prone to doubt than to believe; the hardened skeptic, who has strong disbelief about all things occult; and the believing doubter, who is prone to believe but has some doubts. The one group this book is not aimed at is the "true believer" in the occult. If you have no skepticism in you, this book is not for you.

The open-minded seeker has not had much experience with occult phenomena beyond some religious training but does not dismiss out of hand reports of aura readings, alien abductions, ESP, channeling, ghosts, miracle cures, and so on. The soft skeptic suspends judgment on occult issues and appeals to inexperience, as well as to epistemological skepticism, as reasons for deferring judgment. The hardened skeptic is a disbeliever in all or most occult claims. The believing doubter is attracted to the occult and is a strong believer in one or more (usually more) occult areas but is having some doubts about the validity of occult claims.

My beliefs are clearly that of a hardened skeptic. I don't pretend that I have no experience or knowledge of these matters. For me, the evidence is overwhelming that it is highly probable that any given occult claim is erroneous or fraudulent. Earlier in my life I was a seeker. Looking back, I wish

I had had a book like *The Skeptic's Dictionary,* a book that provides the seeker with arguments and references to the best skeptical literature on occult claims. Though clearly it is my hope that the seeker will become skeptical, I also hope the seeker will investigate these matters before coming to a decision.

The Skeptic's Dictionary will provide the soft skeptic with evidence and arguments, as well as references to more evidence and arguments, on occult issues. In my view, there is sufficient evidence available to convince most reasonable soft skeptics that most occult claims are more probably false than true. However, the soft skeptic recognizes that it does not follow from that fact (if it is a fact) that one should commit oneself to what seems most probable to the rational mind. The soft skeptic often holds that rationality is a value and that the idea that the rational life is the best one for human beings cannot be proven logically, scientifically, or any other way. By way of argument, all one can do is appeal to the consequences of choosing the rational over the irrational life. Also, it seems to be true that belief in the irrational is as appealing to the true believer as belief in the rational is to the hardened skeptic. According to many soft skeptics, whether one chooses a life devoted to rationality or irrationality is a matter of faith. For a good period of my adult life, I was a soft skeptic who believed that my commitment to rationality was as much an act of faith as my earlier commitment to Catholicism had been. For years I remained open to the possibility of all sorts of occult phenomena. My studies and reflections in recent years have led me to the conclusion that there is a preponderance of evidence against the reasonableness of belief in any occult phenomena. I have also concluded that choosing rationality over irrationality is not an act of faith at all. To even pose the question as one requiring thought to answer demonstrates the futility of claiming that everything can be reduced to faith. *One must use reason to argue for faith.* While I do not deny that the consequences of believing in the occult are often beneficial, I do deny that such consequences have anything to do with establishing the reality of occult phenomena. A soft skeptic would have to agree that there is a monumental difference between a believed entity and a real entity. I would agree with the soft skeptic that it is impossible to know anything empirical with absolute certainty. However, I think it is obvious that probabilities serve us well in this life. We have plenty of ways in many, many cases to distinguish among empirical claims that are of differing degrees of probability.

The hardened skeptic doesn't need much more in the way of evidence or argument to be convinced that any given occult claim is probably based on error or fraud. Still, *The Skeptic's Dictionary* has something for the hardened skeptic, too: it will provide ammunition against the incessant arguments of true believers. Most hardened skeptics don't feel it is worth their time to investigate every bizarre idea that comes their way. They dismiss them out of hand. Under most conditions, simply rejecting quackery is intelligent and justified. Often, however, it is better to provide a seeker, soft skeptic, or doubting believer with arguments, both specific and general. But if one's antagonists are true believers, it is probably a waste of time to provide evidence and arguments in response.

Finally, *The Skeptic's Dictionary* will provide the doubting believer with information and sources to consult that will provide, if not a balanced picture, at least a multifaceted one, of a concern about the power of crystals or color therapy or levitation, or other phenomena. It will help the doubter resolve his or her doubts. There may be a few skeptics who can go through all this literature and come out doubting everything, including the skeptical claims, but I think the vast majority will emerge as hardened skeptics. They will not think they must suspend judgment on everything, but will realize that some claims are more probable than others.

As already stated, the one group that this book is not designed for is that of the true believers. My studies have convinced me that arguments or data critical of their beliefs are always considered by true believers to be insignificant, irrelevant, manipulative, deceptive, not authoritative, unscientific, unfair, biased, closed-minded, irrational, and/or diabolical. (It is perhaps worth noting that except for the term "diabolical," these are the same terms some hardened skeptics use to describe the studies and evidence presented by true believers.) Hence, I believe it is highly probable that the only interest a true believer would have in *The Skeptic's Dictionary* would be to condemn and burn it without having read it.

A

acupuncture

A traditional Chinese medical technique for unblocking **chi** by inserting needles at particular points on the body to balance the opposing forces of **yin and yang**. Chi is an **energy** that allegedly permeates all things. It is believed to flow through the body along 14 main pathways called meridians. When yin and yang are in harmony, chi flows freely within the body and a person is healthy. When a person is sick, diseased, or injured, there is an obstruction of chi along one of the meridians. Traditional Chinese medicine has identified some 500 specific points where needles are to be inserted for specific effects.

Acupuncture has been practiced in China for more than 4,000 years. Today, the needles are twirled, heated, or even stimulated with weak electrical current, ultrasound, or certain wavelengths of light. But no matter how it is done, scientific research can never demonstrate that unblocking chi by acupuncture or any other means is effective against any disease. Chi is defined as being undetectable by the methods of empirical science.

A variation of traditional acupuncture is called auriculotherapy, or ear acupuncture. It is a method of diagnosis and treatment based on the unsubstantiated belief that the ear is the map of the bodily organs. For example, a problem with an organ such as the liver is to be treated by sticking a needle into a certain point on the ear that is supposed to be the corresponding point for that organ. (Similar notions about a part of the body being an organ map are held by those who practice **iridology** [the iris is the map of the body] and **reflexology** [the foot is the map of the body].) Staplepuncture, a variation of auriculotherapy, puts staples at key points on the ear hoping to do such things as help people stop smoking.

Traditional Chinese medicine is not based on knowledge of modern physiology, biochemistry, nutrition, anatomy, or any of the known mechanisms of healing. Nor is it based on knowledge of cell chemistry, blood circulation, nerve function, or the existence of hormones or other biochemical substances. There is no correlation between the meridians used in traditional Chinese medicine and the actual layout of the organs and nerves in the human body. Nevertheless, between 10 and 15 million Americans spend approximately $500 million a year on acupuncture for treatment of depression, AIDS, allergies, asthma, arthritis, bladder and kidney problems, constipation, diarrhea, drug addiction, colds, flu, bronchitis, dizziness, smoking, fatigue, gynecologic disorders, headaches, migraines, paralysis, high blood pressure, PMS, sciatica, sexual dysfunction, stress, stroke, tendinitis, and vision problems.

Empirical studies on acupuncture are in their infancy. Such studies ignore notions based on **metaphysics** such as unblocking chi along meridians and seek to find causal connections between sticking needles into traditional acupuncture points and physical effects. Even so, many traditional doctors and hospitals are offering acupuncture as a "complementary" therapy. The University of California at Los

5

Angeles medical school has one of the largest acupuncture training courses in the United States for licensed physicians. The 200-hour program teaches nearly 600 physicians a year. According to the American Academy of Medical Acupuncture, about 4,000 U.S. physicians have training in acupuncture.

In March 1996, the Food and Drug Administration (FDA) classified acupuncture needles as medical devices for general use by trained professionals. Until then, acupuncture needles had been classified as Class III medical devices, meaning their safety and usefulness was so uncertain that they could be used only in approved research projects. Because of that "experimental" status, many insurance companies, as well as Medicare and Medicaid, had refused to cover acupuncture. This new designation has meant both more practice of acupuncture and more research being done using needles. It also means that insurance companies may not be able to avoid covering useless or highly questionable acupuncture treatments for a variety of ailments. Nevertheless, Wayne B. Jonas, director of the Office of Alternative Medicine at the National Institutes of Health in Bethesda, MD, has said that the reclassification of acupuncture needles is "a very wise and logical decision." The Office of Alternative Medicine is very supportive (i.e., willing to spend good amounts of tax dollars) on new studies of the effectiveness of acupuncture.

The most frequently offered defense of acupuncture by its defenders commits the **pragmatic fallacy**. It is argued that *acupuncture works!* What does this mean? It certainly does not mean that sticking needles into one's body opens up blocked chi. At most, it means that it relieves some medical burden. Most often it simply means that some customer is satisfied, that is, feels better at the moment. The National

Council Against Health Fraud issued a position paper on acupuncture (1990, www.ncahf.org/pp/acu.html) that asserts, "Research during the past twenty years has failed to demonstrate that acupuncture is effective against any disease" and that "the perceived effects of acupuncture are probably due to a combination of expectation, suggestion, counter-irritation, operant conditioning, and other psychological mechanisms." In short, most of the perceived beneficial effects of acupuncture are probably due to mood change, the **placebo effect**, and the **regressive fallacy**. Just because the pain went away *after* the acupuncture doesn't mean the treatment was the cause. Much chronic pain comes and goes. An alternative treatment such as acupuncture is sought only when the pain is near its most severe level. Natural regression will lead to the pain becoming less once it has reached its maximum level of severity. Also, much of the support for acupuncture is anecdotal in the form of **testimonial evidence** from satisfied customers. Unfortunately, for every anecdote of someone whose pain was relieved by acupuncture there may well be another anecdote of someone whose pain was not relieved by acupuncture. But nobody is keeping track of the failures (**confirmation bias**).

Nevertheless, it is possible that sticking needles into the body may have some beneficial effects. The most common claim of success by acupuncture advocates is in the area of pain control. Studies have shown that many acupuncture points are more richly supplied with nerve endings than are the surrounding skin areas. There is some research that indicates sticking needles into certain points affects the nervous system and stimulates the body's production of natural pain-killing chemicals such as endorphins and enkephalins, and triggers the release of certain neural hormones including serotonin. Another theory suggests that acupuncture blocks the transmission of pain impulses from parts of the body to the central nervous system.

There are difficulties that face any study of pain. Not only is pain measurement entirely subjective, but traditional acupuncturists evaluate success of treatment almost entirely subjectively, relying on their own observations and reports from patients rather than objective laboratory tests. Furthermore, many individuals who swear by acupuncture or other **alternative health practices** often make several changes in their lives at once, thereby making it difficult to isolate significant causal factors in a **control group study**.

Finally, acupuncture is not without risks. There have been some reports of lung and bladder punctures, some broken needles, and some allergic reactions to needles containing substances other than surgical steel. Acupuncture may be harmful to the fetus in early pregnancy since it may stimulate the production of adrenocorticotropic hormone and oxytocin, which affect labor. There is the possibility of infection from unsterilized needles. And some patients will suffer simply because they avoided a known effective treatment of modern medicine.

Further reading: Barrett and Butler 1992; Barrett and Jarvis 1993; Huston 1995; Raso 1994.

ad hoc hypothesis

A hypothesis created to explain away facts that seem to refute a theory. For example, **psi** researchers have been known to blame the hostile thoughts of onlookers for unconsciously influencing pointer readings on sensitive instruments. The hostile vibes, they say, made it impossible for them to duplicate a positive **ESP** experiment. Being able to duplicate an experiment is essential to confirming its validity. If hostile thoughts can ruin the psi researcher's day, then no experiment on ESP can ever fail. Whatever the results, one can always say they were caused by **paranormal** forces, either the ones being tested or others not being tested.

One key element of the ad hoc hypothesis is that it cannot be independently tested. In the example above, there is no independent way to test for the effect of hostile vibes. Thus, if a hypothesis appears to be ad hoc, one should always ask: Can this be tested independently of the theory it is trying to save? For example, when William Herschel discovered the planet Uranus in 1781 by telescopic observation and its orbit did not fit with predictions made using Newton's laws of planetary motion, it was proposed that another planet must exist further out from the sun than Uranus. This hypothesis could be independently tested. Its size and orbit could be calculated based on how much it perturbed the motion of Uranus. When the math didn't work in accordance with Newton's laws, it was proposed that still another planet awaited discovery. Both of these hypotheses could be independently tested, albeit with some difficulty given the state of knowledge and technology at the time.

The believers in **biorhythms** provide another example of using ad hoc hypotheses. Not only are people who do not fit the predicted patterns of biorhythm theory designated as arrhythmic, but advocates of biorhythm theory claimed that the theory can be used to accurately predict the sex of unborn children. However, W. S. Bainbridge, a professor of sociology at the University of Washington, demonstrated that the chance of predicting the sex of an unborn child using biorhythms was 50:50, the same as flipping a coin. An expert in biorhythms tried unsuccessfully to predict accurately the sexes of the children in Bainbridge's study based on Bainbridge's data. The expert's spouse suggested to Bainbridge an interesting ad hoc hypothesis, namely, that the cases where the theory was wrong probably included many homosexuals with indeterminate sex identities.

Afrocentrism

A **pseudohistorical** political movement that claims that ancient Egypt was dominated by a race of black Africans and that African Americans can trace their roots back to the great civilizations of Egypt. Leading Afrocentrists claim that the ancient Greeks stole their main cultural achievements from black Egyptians and that Jesus, Socrates, and Cleopatra, among others, were black. According to the tenets of Afrocentrism, the Jews created the black African slave trade. None of these claims is supported by the work of traditional historians. The main purpose of Afrocentrism is not so much to achieve historical accuracy as it is to encourage Black Nationalism and ethnic pride as a psychological weapon against the destructive and debilitating effects of universal racism.

Clarence E. Walker (2001), a professor of Black American History at the University of California at Davis, calls Afrocentrism

> a mythology that is racist, reactionary, and essentially therapeutic. It suggests that nothing important has happened in black history since the time of the pharaohs and thus trivializes the history of black Americans. Afrocentrism places an emphasis on Egypt that is, to put it bluntly, absurd.

Walker, an African American, thinks Afrocentrism is harmful because it denies to black Americans the dignity and power that should emerge from a truthful and honest understanding of history.

The leading proponents of Afrocentrism are Professor Molefi Kete Asante of Temple University, Professor Leonard Jeffries of the City University of New York, and Martin Bernal, the author of *Black Athena*. One of the more important Afrocentric texts is the pseudohistorical *Stolen Legacy* (1954), by George G. M. James, who claims that Greek philosophy and the mystery religions of Greece and Rome were stolen from Egyptian black Africans. Many of James's ideas were taken from Marcus Garvey (1887–1940), who thought that white accomplishment is due to teaching children they are superior. If blacks teach their children that they are superior, reasoned Garvey, then they will also accomplish great things.

James's principal sources were Masonic, especially *The Ancient Mysteries and Modern Masonry* (1909), by the Rev. Charles H. Vail. The Masons in turn derived their misconceptions about Egyptian mystery and initiation rites from the 18th-century work of fiction *Sethos, a History or Biography, based on Unpublished Memoirs of Ancient Egypt* (1731), by the Abbé Jean Terrasson, a professor of Greek. Terrasson had no access to Egyptian sources and he would be long dead before Egyptian hieroglyphics would be deciphered, but he knew the Greek and Latin writers well. He constructed an imaginary Egyptian religion based on sources that describe Greek and Latin rites as if they were Egyptian (Lefkowitz 1996). Hence, one of the main sources for Afrocentric Egyptology turns out to be Greece and Rome. The Greeks would have called this irony.

James's pseudohistory is the basis for other Afrocentric pseudohistories such as *Africa: Mother of Western Civilization*, by Yosef A. A. ben-Jochannan, one of James's students, and *Civilization or Barbarism*, by Cheikh Anta Diop of Senegal.

agnosticism

The belief that it is impossible to know whether God exists. It is often put forth as a middle ground between **theism** and **atheism**. Understood this way, agnosticism is skepticism regarding all things theological.

The agnostic holds that human knowledge is limited to the natural world, that the mind is incapable of knowledge of the supernatural. Understood this way, an agnostic could be either a theist or an atheist. The agnostic theist thinks there is some reason for believing in God. The agnostic atheist finds no compelling reason to believe in God.

The term "agnostic" was created by T. H. Huxley (1825–1895), who took his cue from David Hume and Immanuel Kant. Huxley says that he invented the term to describe what he thought made him unique among his fellow thinkers:

> They were quite sure that they had attained a certain "gnosis"—had more or less successfully solved the problem of existence; while I was quite sure I had not, and had a pretty strong conviction that the problem was insoluble.

"Agnostic" came to mind, he says, because the term was "suggestively antithetic to the 'gnostic' of Church history, who professed to know so much about the very things of which I was ignorant." Huxley seems to have agreed with Hume's conclusion at the end of *An Enquiry Concerning Human Understanding:*

> When we run over libraries, persuaded of these principles, what havoc must we make? If we take in our hand any volume; of divinity or school metaphysics, for instance; let us ask, Does it contain any abstract reasoning concerning quantity or number? No. Does it contain any experimental reasoning concerning matter of fact and existence? No. Commit it then to the flames: for it can contain nothing but sophistry and illusion.

In other words, natural theology is, more or less, bunk.

Akashic record

An imagined spiritual realm, supposedly holding a record of all events, actions, thoughts, and feelings that have ever occurred or will ever occur. **Theosophists** believe that the Akasha is an "astral light" containing **occult** records that spiritual beings can perceive by their special "astral senses" and **astral bodies**. Spiritual insight, prophecy, **clairvoyance**, and many other occult notions are allegedly made possible by tapping into the Akasha.

Further reading: Ellwood 1996; Randi 1995.

alchemy

An **occult** art whose practitioners' main goals have been to turn base metals such as lead or copper into precious metals such as gold or silver (the transmutation motif); to create an elixir, potion, or metal that could cure all ills (the panacea motif); and to discover an elixir that would lead to immortality (the transcendence motif). The **philosopher's stone** is the name given to the magical substance that was to accomplish these feats.

Many modern alchemists combine their occult art with **acupuncture**, **astrology**, **hypnosis**, and a wide variety of New Age spiritual quests. Alchemists may have tried out their ideas by devising experiments, but they never separated their methods from the supernatural, the **magickal**, and the superstitious. Perhaps that is why alchemy is still popular, even though it has accomplished practically nothing of lasting value. Alchemists never transmuted metals, never found a panacea, and never discovered the fountain of youth.

Alchemy is based on the belief that there are four basic elements—fire, air, earth, and water—and three essentials: salt, sulfur, and mercury. Great symbolic and occult systems have been built from these seven pillars of alchemy. The foundation of European alchemy, which flourished through the Renaissance, is said to be ancient Chinese and Egyptian occult literature. The Egyptian god Thoth, known as Hermes Trismegistus, allegedly wrote one of the books considered by the alchemists to be most sacred. (Hermes, the thrice-great, was the Greek god who served as a messenger and delivered the souls of the dead to Hades.) The book in question, *Corpus Hermeticum,* began circulating in Florence, Italy, around 1455. The work is full of magic **incantations** and **spells** and is now known to be of European origin.

Some alchemists did make contributions to the advancement of knowledge. For example, Paracelsus (1493–1541) introduced the concept of disease to medicine. Ironically, he rejected the notion that disease is a matter of imbalance or disharmony in the body, a view much favored by modern alchemists. Paracelsus maintained that disease is caused by agents outside of the body that attack it. He recommended various chemicals to fight disease.

Further reading: Trimble 1996.

alien abductions

There is a widespread belief that alien beings have traveled to Earth from other planets and are doing reproductive experiments on earthlings. Despite a lack of credible supportive evidence, a cult has grown up around this belief. According to a Gallup poll done at the end of the twentieth century, about one-third of Americans believe aliens have visited us, an increase of 5% over the previous decade. In the early 1990s, a Roper organization poll found 3% of Americans claiming to have had alien abduction experiences. A 1999 survey by Roper found that 80% of Americans think their government is concealing information on extraterrestrials.

According to the tenets of this cult, aliens crashed at **Roswell**, New Mexico, in 1947. The U.S. government recovered the alien craft and its occupants, and has been secretly meeting with aliens ever since in a place known as **Area 51**. The rise in **UFO** sightings since is due to the increase in alien activity on Earth. The aliens are abducting people in larger numbers and are leaving other signs of their presence in the form of **cattle mutilations** and **crop circles**. Aliens even get credit for the occasional **channeled** book, such as the *Urantia Book*.

Even though the stories of alien abduction do not seem plausible, if there were some physical evidence of alien presence, even the most hardened skeptic would have to take notice. Unfortunately, the only physical evidence that is offered is insubstantial. For example, so-called ground scars allegedly made by UFOs have been offered as proof that the aliens have landed. However, when examined, these sites prove to be quite ordinary and the scars to be little more than fungus or other natural phenomena.

Many abductees point to various scars and scoop marks on their bodies as proof of abduction and experimentation. These marks are not extraordinary in any way and can be accounted for by quite ordinary injuries and experiences.

The most dramatic type of physical evidence are the implants that many abductees claim the aliens have put up their noses or in various other parts of their anatomy. Budd Hopkins, a draftsman by training but an alien abduction researcher by avocation, claims he has examined such an implant and has MRIs (magnetic resonance images) to prove numerous implant claims. When the science TV program *Nova* ("Alien Abductions," first shown on February 27, 1996) put out an offer to abductees to have scientists analyze and evaluate implants, they got no response. Of all the evidence for abduction, the physical evidence is the weakest.

The Barney and Betty Hill story shares top billing with the Roswell story in the lore of cult beliefs about alien visitation and experimentation. The Hills claim to have been abducted by aliens on September 19, 1961. Barney claims the aliens took a sample of his sperm. Betty claims they stuck a needle in her navel. She took people to an alien landing spot, but only she could see the aliens and their craft. The

"Alien."

Hills recalled most of their story under **hypnosis** a few years after the alleged abduction. Barney Hill reported that the aliens had "wraparound eyes," a rather unusual feature. However, twelve days earlier an episode of *The Outer Limits* featured just such an alien being (Kottmeyer 1990). Usually, the aliens are described as small and bald with big crania and small chins, having white, gray, or green skin, and large slanted eyes, pointed ears, or no ears at all. "We can find all the major elements of contemporary UFO abductions in a 1930 comic adventure, *Buck Rogers in the 25th Century*" (Schaeffer 1996).

The main features of the Hills' account of abduction have been repeated many times. There is a period of amnesia that follows the alleged encounter. There is then usually a session of hypnosis, counseling, or psychotherapy during which the subject recalls the abduction and experimentation. The only variation in the abductees' stories is that some claim to have had implants put in them, and many claim to have scars and marks on their bodies put there by aliens.

For example, Whitley Strieber, who has written several books about his abductions, realized aliens had abducted him only after psychotherapy and hypnosis. Strieber claims that he saw aliens set his roof on fire. He says he has traveled to distant planets and back during the night. He wants us to believe that he and his family can see the aliens and their spacecraft, even though others see nothing. Strieber seems to be a very disturbed person and he was certainly in a very agitated psychological state prior to his alleged visitation by aliens. A person in such a heightened state of anxiety is prone to hysteria and especially vulnerable to radically changing behavior or belief patterns. When Strieber was having an anxiety attack he consulted his analyst, Robert Klein, as well as Budd

Hopkins. Under hypnosis, Strieber recalled the horrible aliens and their visitations.

Hopkins demonstrated his investigative incompetence on an "Alien Abductions" episode of *Nova*. The camera followed him to Florida, where he cheerfully helped a visibly unstable mother inculcate in her children the belief that they had been abducted by aliens. In between more sessions with more of Hopkins's subjects, the viewer heard him repeatedly give plugs for his books and witnessed a total absence of skepticism regarding the very bizarre claims he was eliciting.

Dr. Elizabeth Loftus, an expert on **false memory**, was asked by *Nova* to evaluate Hopkins's method of counseling the children. She noted that Hopkins did much encouraging of his subjects to remember more details, and gave many verbal rewards when new details were brought forth. Loftus characterized the procedure as risky, because we do not know what effect this counseling will have on the children. It seems we can safely predict one effect: They will grow up thinking they've been abducted by aliens. This belief will be so embedded in their **memory** that it will be difficult to get them to consider that the experience was planted by their mother and cultivated by alien enthusiasts such as Hopkins.

Another enthusiast is Harvard psychiatrist Dr. John Mack, who has written books about patients who claim to have been abducted by aliens. Hopkins has referred many of Mack's patients to him. Mack claims that his psychiatric patients are not mentally ill and that he can think of no other explanation for their stories than that they are true. Dr. Mack also appeared on the *Nova* "Alien Abductions" program. He claimed that his patients are otherwise normal people who have nothing to gain by making up their incredible stories.

It is often thought by intelligent people that if a person's motives can be trusted, then his or her testimony can be trusted, too. It is true we are justified in being skeptical of a person's testimony if she has something to gain by the testimony (such as fame or fortune), but it is not true that we should trust every testimony given by a person who has nothing to gain by giving the testimony. The fact that a person is kind, decent, and otherwise normal except for a single bizarre belief and has nothing to gain by lying does not make him or her immune to error in the interpretation of perceptions to justify that bizarre belief.

People who believe they have been abducted by aliens may not be insane, but they are certainly fantasy-prone. Being fantasy-prone is not an abnormality, if abnormality is defined in terms of minority belief or behavior. The vast majority of humans are fantasy-prone, otherwise they would not believe in God, **ghosts**, **angels**, or **Satan**. A person can function normally in a million and one ways and hold the most irrational beliefs imaginable, as long as the irrational beliefs are culturally accepted delusions.

Alien abductees seem analogous to medieval nuns who believed they'd been seduced by devils. They also seem like the ancient Greeks who believed they had sex with gods in the form of animals. The abductees' counselors and therapists are analogous to priests who do not challenge delusional beliefs, but encourage and nurture them. The delusions of the ancients and the medievals are not couched in terms of aliens and spacecraft; these latter are our century's creations. We can laugh at the idea of gods taking on the form of swans to seduce beautiful women, or of devils impregnating nuns, because they do not fit with our cultural prejudices and delusions. The ancients and medievals would have laughed at anyone claiming to have been picked up by aliens from another planet for sex or reproductive surgery. The only reason anyone takes the abductees seriously today is that their delusions do not blatantly conflict with our cultural beliefs that intergalactic space travel is a real possibility and that it is highly probable that ours is not the only inhabited planet in the universe. In other times, no one would have been able to take these claims seriously.

Dr. Mack noted that his patients gain a lot of attention by being abductees. The same might be said of Dr. Mack and Mr. Hopkins. Both have much to gain in fame and fortune by encouraging their clients to come up with more details of their abductions. Mack received a $200,000 advance for his first book on alien abductions. He also benefits by publicizing and soliciting funds for his Center for Psychology and Social Change and his Program for Extraordinary Experience Research.

Another contributor to the mythology of alien abductions is Robert Bigelow, a wealthy Las Vegas businessman who likes to use his money to support **paranormal** research (see Charles **Tart**) and who partially financed the Roper survey on alien abductions. The survey did not directly ask its 5,947 respondents whether aliens had abducted them. Instead, it asked them if they had undergone any of the following experiences:

- Waking up paralyzed with a sense of a strange person or presence or something else in the room.
- Experiencing a period of time of an hour or more in which you were apparently lost, but you could not remember why, or where you had been.
- Seeing unusual lights or balls of light in a room without knowing what was causing them or where they came from.

- Finding puzzling scars on your body and neither you nor anyone else remembering how you received them or where you got them.
- Feeling that you were actually flying through the air although you didn't know why or how.

Saying yes to four of the five "symptoms" was taken as evidence of alien abduction. A 62-page report, with an introduction by John Mack, was mailed to some 100,000 psychiatrists, psychologists, and other mental health professionals. The implication was that aliens have abducted some 4 million Americans or some 100,000,000 earthlings. As Carl Sagan wryly commented: "It's surprising more of the neighbors haven't noticed" (Sagan 1995). The timing of the mailing was impeccable: shortly before the 1992 CBS-TV miniseries based on Strieber's *Intruders*.

It is possible that abductees describe similar experiences because they've had similar hallucinations due to similar brain states (Persinger 1987). These states may be associated with **sleep paralysis** or other forms of sleep disturbances, including mild brain seizures. Sleep paralysis occurs in the **hypnagogic** or **hypnopompic state**. The description abductees give of their experience—being unable to move or speak, feeling some sort of presence, feeling fear and an inability to cry out—is a list of the symptoms of sleep paralysis. Sleep paralysis may account for not only many alien abduction delusions, but also other delusions involving paranormal or supernatural experiences (Blackmore 1998). Using electrodes to stimulate specific parts of the brain, Michael Persinger has duplicated key aspects associated with the alien abduction experience, the mystical experience, and **out-of-body experiences**.

Of course, it is possible that aliens have visited us. There may well be life elsewhere in the universe, and some of that life may be intelligent. There is a high mathematical probability that among the trillions of stars in the billions of galaxies there are millions of planets in age and proximity to a star analogous to our Sun. The chances seem very good that on some of those planets life has evolved. It is highly probable that natural selection governed the evolution of that life (Dawkins 1988). However, it is not inevitable that the results of that evolution would yield intelligence, much less intelligence equal or superior to ours. It is possible we are unique (Pinker 1997: 150).

We should not forget that the closest star (besides our Sun) is so far away from Earth that travel between the two would probably take more than a human lifetime. The fact that it takes our Sun about 200 million years to revolve once around the Milky Way gives one a glimpse of the perspective we have to take of interstellar travel. We are 500 light-seconds from the sun. The next nearest star, Alpha Centauri, is about 4 light-years away. That might sound close, but it is actually something like 24 trillion miles away. Even traveling at 1 million miles per hour, it would take more than 2,500 years to get there. To get there in 25 years would require traveling at more than 100 million miles an hour for the entire trip. Our fastest spacecraft, *Voyager,* travels at about 40,000 miles an hour and would take 70,000 years to get to Alpha Centauri.

Furthermore, any signal from any planet in the universe broadcast in any direction is very unlikely to be in the path of another inhabited planet. It would be folly to explore space for intelligent life without knowing exactly where to go. Yet waiting for a signal might require a wait longer than any life on any planet might last. Finally, if we do get a signal, the waves carrying that signal left hundreds or thousands of years earlier, and by the time we tracked down its source, the sending planet may no longer be habitable or even exist.

Thus, while it is possible that there is intelligent life in the universe, traveling between solar systems in search of that life poses some serious obstacles. Such travelers would be gone for a very long time. We would need to keep people alive for hundreds or thousands of years. We would need equipment that can last for hundreds or thousands of years and be repaired or replaced in the depths of space. Or, of course, we would need a technology and materials that can far exceed the speed of light, and a whole new theory of reality to go with them. These are not impossible conditions, perhaps, but they seem to be significant enough barriers to make interstellar and intergalactic space travel highly improbable. It is difficult to imagine beings capable of overcoming these barriers coming here to abduct our people, rape and experiment on them, mutilate our cattle, create artwork in our wheat fields, and deliver such commonplace messages as "The goal of human self-realization should be spiritual, not material."

See also **flying saucers** and **Men in Black**.

Further reading: Baker 1987–88; Dudley 1999; Frazier 1997; Klass 1988; Loftus 1994; Matheson 1998; Persinger 1983; Schaeffer 1986.

allopathy

A term used to refer to conventional medicine by American **chiropractors**, **homeopaths**, **naturopaths**, **osteopaths**, and other advocates of **alternative health practices**. The *Random House Dictionary of the English Language* (unabridged edition) defines allopathy as "the method of treating disease by the use of agents that produce effects different from those of the disease treated (opposed to homeopathy)."

The word was invented by the homeopath Samuel Hahnemann as a term for those who are other than homeopaths.

alphabiotics

An **alternative health practice** based on the unverifiable notion that all disease is the result of an imbalance and lack of Life Energy. Health depends on "aligning" and "balancing" this alleged **energy**.

Alphabiotics is the brainchild of Dr. V. B. Chrane, who started practicing it in the 1920s near Abilene, Texas. It was "established as a unique new profession by Dr. Virgil Chrane Jr. on December 28, 1971," according to Virgil Chrane, Jr., himself. The practice is still flourishing with Virgil Jr. and his son, Dr. Michael Chrane.

alpha waves

Oscillating electrical voltages in the brain. Alpha waves oscillate in the range of 7.5–13 cycles per second. Because alpha waves occur in relaxed states such as meditation and under **hypnosis**, they have been mistakenly identified as desirable. Alpha waves also occur under unpleasant conditions and when one is not relaxed. They are not a measure of peace and serenity, nor are they indicative of an **altered state of consciousness**. Alpha waves are indicative of lack of visual processing and lack of focus: the less visual processing and the more unfocused, generally the stronger the alpha waves. If you close your eyes and don't do any deep thinking or concentrating on vivid imagery, your alpha waves will usually be quite strong.

There is no evidence that "when asleep, the brain goes into a 'repair and rebuild' mode under alpha wave energy," as an ad for Calorad, a protein supplement, claims. Nor is there evidence that the brain is more insightful, creative, or productive while producing alpha waves. Some think that increasing alpha waves can enhance the immune system and can lead to self-healing or the prevention of illness. This belief seems to be based on the mistaken inference that since alpha waves increase while meditating, they are indicative of lack of stress, which can only be good for you. Increasing alpha waves is no guarantee either that one is reducing stress or that one is enhancing one's immune system.

See also **naturopathy** and **Silva Mind Control**.

Further reading: Beyerstein 1985, 1996a.

altered state of consciousness (ASC)

A state of consciousness that differs significantly from baseline or normal consciousness often identified with a *brain state* that differs significantly from the brain state at baseline or normal consciousness. However, it is not the brain state itself that constitutes an ASC. The brain state is an objective matter, but it should not be equated with an EEG or MRI reading. Otherwise, we would end up counting such things as sneezing, coughing, sleeping, being in a coma, thinking of the color red, and being dead as ASCs. Brain state readings reveal brain activity or inactivity, but are not a good measure of ASCs. **Alpha waves**, for example, have been identified with an ASC, but they usually measure lack of visual processing and lack of focus. Alpha waves occur in athletes who reach what they call "the Zone" and in some video-game players who seem to be on "auto-pilot."

The baseline brain state might be best defined by the presence of two important *subjective* characteristics: the psychological sense of a self at the center of one's perception and a sense that this self is identified with one's body. States of consciousness where one loses the sense of identity with one's body or with one's perceptions are definitely ASCs. Such states may be spontaneously achieved, instigated by such things as trauma, sleep disturbance, sensory deprivation or sensory overload, neurochemical imbalance, epileptic seizure, or fever. They may also be induced by social behavior, such as frenzied dancing or chanting. Finally, they may be induced by electrically stimulating parts of the brain or by ingesting psychotropic drugs.

Many think the **hypnotic** state is an ASC. It certainly often resembles one, but it is doubtful that it is truly an ASC. A hypnotized person closely resembles certain amnesiacs who can be primed by being shown certain words. Later they have no conscious recollection of having been shown the words, but they give evidence of implicit **memory** of the words. It is doubtful that amnesia should be considered an ASC.

There is little evidence that ASCs can transport one into a transcendent realm of higher consciousness or truth, as **parapsychologists** Raymond **Moody** and Charles **Tart** maintain, but there is ample evidence that some ASCs bring about extremely pleasant feelings and can profoundly affect personality. Some religious experiences, for example, are described as providing a very pleasant sense of divine presence and of the oneness, interrelatedness, and significance of all things. Drugs such as LSD and mescaline can induce similar feelings. Some patients suffering from temporal lobe epilepsy think of their disease as temporal lobe "ecstasy," since it leaves them with a feeling of being united with God (Ramachandran and Blakeslee 1998). Also, by electrically stimulating the temporal lobes, Michael Persinger has been able to duplicate the sense of presence, the sense of leaving the body, and other feelings associated with mysticism and **alien abduction** (Persinger 1987). Dr. Olaf Blanke of Geneva University Hospital in Switzerland found that electrically stimulating the right angular gyrus (located at the juncture of the temporal and parietal lobes) triggers out-of-body experiences. (In a related matter, Dr. Stuart Meloy, an anesthesiologist and pain specialist in Winston-Salem, North Carolina, was testing his pain-relieving invention on a patient when he accidentally discovered that by electrically stimulating a woman's spinal column he could induce orgasm.)

Are the brain states that elicit the feelings of mysticism in the religious ecstatic, the epileptic, the one on an "acid" trip, and the one with electrodes attached to his cranium caused by God? Perhaps, but if so there is no way of finding this out. Most likely, however, the mechanisms that trigger these feelings are completely natural. They may be a pleasant side effect of some evolutionary adaptation, but as yet we do not know why such brain states are triggered. And while it is an extremely interesting discovery that religious experiences can be induced by disease, electrodes, and by drugs, it hardly seems a compelling reason for believing in God, although it might be a compelling reason for taking drugs, for not seeking treatment, or for using a transcranial electromagnetic stimulator and hoping for Orgasmatron-like results achieved by the Woody Allen character in *Sleeper*. Most religions identify the ideal state as an ASC: losing one's body and one's self, uniting with some sort of divine being, and feeling ecstatic pleasure. In this sense, to seek an ASC is to seek to kill your sense of self while enjoying the ultimate orgasm.

Further reading: Beyerstein 1996a; Blackmore 1993; Newberg et al. 2001; Sacks 1974, 1984, 1985, 1995; Spanos 1996.

"alternative" health practices (AHPs)

Health or medical practices are called "alternative" if they are based on untested, untraditional, or unscientific principles, methods, treatments, or knowledge. (Such practices are not truly alternatives to conventional treatments, and hence I prefer to put the term "alternative" in quotes when writing about alternative medicine. However, the quotes are a distraction and will be omitted henceforth.) Alternative medicine (AM) is often based on spiritual beliefs and is frequently antiscientific. Because truly alternative medical practices would be ones that are known to be equally or nearly equally effective as the ones they replace, most alternative medical practices are not truly alternative. Thus, many clinics that offer conventional and alternative treatments prefer the term "complementary" or "integrative" health practices. Critics refer to the treatments as **quackery**.

It is estimated that AM is a $15 billion a year business. Traditionally, most insurance companies have not covered alternative medicine, but that is changing rapidly as the demand for AM increases and insurance companies figure out that such coverage is very profitable.

The National Institutes of Health's National Center for Complementary and Alternative Medicine (NCCAM) has supported a number of research studies of unorthodox cures, including the use of **shark cartilage** to treat cancer and the effectiveness of bee pollen in treating allergies. NCCAM has also supported studies on spirituality and patients with AIDS, **prayer** and cancer patients, as well as many unsubstantiated but popular claims among alternative practitioners regarding the effectiveness of numerous herbs and botanicals. NCCAM also strongly supports studies on alternative therapies for the reduction of pain, including **acupuncture**, **chiropractic**, and **magnet therapy**. Alternative practitioners have long complained about lack of funding as the main reason they rarely do scientific studies. Perhaps NCCAM support will put an end to this complaint and to the criticism of skeptics that alternative practitioners prefer religious **faith**, superstition, and magical thinking to science.

On the other hand, many questionable products touted as cure-alls or as cures for serious illnesses such as cancer or heart disease are promoted with scientific gobbledygook and misrepresentation or falsification of scientific studies. Jodie Bernstein, director of the FTC's Bureau of Consumer Protection, offers the following list of signs of quackery:

- The product is advertised as a quick and effective cure-all for a wide range of ailments.
- The promoters use words such as *scientific breakthrough, miraculous cure, exclusive product, secret ingredient,* or *ancient remedy.*
- The text is written in "medicalese": impressive-sounding terminology to disguise a lack of good science.
- The promoter claims the government, the medical profession, or research scientists have conspired to suppress the product.
- The advertisement includes undocumented case histories claiming amazing results.
- The product is advertised as available from only one source.

The general rule is, if it sounds too good to be true, it probably is.

The *New England Journal of Medicine* reported on a study in January 1993 that showed that about one-third of American adults sought some sort of unorthodox therapy during the preceding year. Why is AM so popular? There are many reasons.

1. *Drugs and surgery are not part of AM.* Fear of surgery and apprehension regarding the side effects of drugs alienate many people from conventional medicine. AM is attractive because it does not offer these frightening types of treatments. Furthermore, conventional medicine often harms patients. AM treatments are usually inherently less risky and less likely to cause direct harm. The harm to AM patients comes not from positive intervention but from not getting treatment (drugs or surgery) that could improve their health and increase their life span.

2. *Conventional medicine often fails to discover the cause of an illness or to relieve pain.* This is true of AM as well. But conventional practitioners are not as likely to express hopefulness when their medicine fails. Alternative practitioners often encourage their patients to be hopeful even when the situation is hopeless.

3. *When conventional medicine does discover the cause of an illness, it often fails to offer treatment that is guaranteed to be successful.* Again, AM offers hope when conventional medicine can't offer a safe and sure cure. For example, a television news anchor, Pat Davis, rejected chemotherapy for her breast cancer in favor of Gerson Therapy. She followed a rigorous 13-hour-a-day regimen of diet (green vegetables and green juices), exercise, and coffee enemas (four a day) developed by Dr. Max Gerson. Davis's mother has had breast cancer twice, undergoing chemotherapy and a mastectomy. Davis knew the dangers of chemotherapy and the effects of breast surgery. She refused to accept that there were no alternatives. Gerson therapy gave her hope. When it was clear that the Gerson treatment was ineffective, Davis agreed to undergo chemother-

apy. She died four months later on March 20, 1999, at the age of 39, after two and a half years of fighting her cancer. Her mother was still alive in 2002. Could chemotherapy have saved her had she sought the treatment earlier? Maybe. The odds may have been against her, but the hope offered by scientific medicine was at least a real hope. The hope offered by Gerson is a false hope through and through.

4. *AHPs often use "natural" remedies.* Many people believe that what is **natural** is necessarily better and safer than what is artificial (such as pharmaceuticals). Just because something is natural does not mean that it is good, safe, or healthy. There are many natural substances that are dangerous and harmful. There are also many natural products that are ineffective and of little or no value to one's health and well-being.

5. *AHPs are often less expensive than conventional medicine.* This fact has made alternative treatments attractive to health maintenance organizations (HMOs) and to insurance companies, both of whom are coming to realize that it is cheaper and thus more profitable to offer alternative treatments. If alternative therapies were truly alternatives, it would make no sense to pay more for the same quality treatment. However, most so-called alternative therapies are not truly alternatives; they are not equally effective treatments. Thus, the fact that they are cheaper is of little significance.

6. *AHPs are often sanctioned by state governments, which license and regulate alternative practices and even protect alternative practitioners from attacks by the medical establishment.* Chiropractors, for example, won a major restraint-of-trade lawsuit against the American Medical Association (AMA) in 1987. A federal judge permanently barred the AMA from "hindering

the practice of chiropractic." Being government licensed, regulated, and protected is seen as legitimizing an AHP. Actually, much of the licensing and regulation is aimed at protecting the public from frauds and quacks.

7. *Many alternative practitioners falsely claim that doctors of conventional medicine treat diseases first and people second.* Alternative practitioners claim they are "holistic" and treat the mind, body, and soul of the patient. Many people are attracted to the spiritual claims made by AM practitioners and to the hope they give by convincing the patient that attitude is more important than the facts of their illness.

8. *Conventional physicians often work out of large hospitals or HMOs and see hundreds or thousands of patients for their specialized needs.* Alternative therapists, on the other hand, often work out of their homes or small offices or clinics, and typically see many fewer patients than a conventional physician. Their patients are often attracted to their personalities and world views, rather than their knowledge and experience with the disease. Those who seek help from a conventional physician usually do not care what his or her personal religious, metaphysical, or spiritual beliefs are. For example, a person with diabetes who goes to an endocrinologist probably will not be interested in his or her physician's belief in **chi** or any other spiritual notions. Whether the doctor believes in God or the **soul** is irrelevant. If the doctor is kind and personable, that is all to the better. A cold and indifferent alternative practitioner would not have much business. A cold and indifferent traditional doctor may have patients standing in line for treatment if he or she is an excellent physician.

Many people apparently do not understand that conventional medicine has the same shortcomings as all other forms of human knowledge: It is fallible. It also is correctable. Systems of thought that are fundamentally metaphysical in nature are not testable and can therefore never be proven incorrect. Hence, once they get established they tend to become dogmatically adhered to and rarely change. The only way to change dogma is to become a heretic and set up your own counter-dogma. When scientific therapies prove to be unnecessary, ineffective, or harmful, they are eventually abandoned.

Alternative practices and treatments are often based on faith and belief in spiritual entities such as chi, and lend themselves to **ad hoc hypotheses** to explain away failure or ineffectiveness. In scientific medicine there will be disagreement and controversy, error and argument, testing and more testing, and so on. Fallible human beings will make imperfect decisions. But scientific medicine will grow, it will progress, it will change dramatically. On the other hand, **homeopathy**, **iridology**, **reflexology**, **therapeutic touch**, and other therapies will not change in any fundamental ways over the years. Their practitioners do not challenge each other, as scientific medicine requires. Instead, alternative practitioners generally do little more than reinforce each other.

9. *Alternative therapies appeal to magical thinking.* Ideas with little scientific backing, such as those of **sympathetic magic**, are popular among alternative practitioners and their clients. Conventional medicine is rejected by some simply because it is not magical. While conventional medicine may sometimes seem to work **miracles**, the miracles of modern medicine are based on **science**, not faith.

10. *The main reason people seek alternative health care is because they think it "works."* That is, they feel better, healthier, more

vital, and so on after the treatment. Those who say alternative medicine "works" usually mean little more than that they are satisfied customers. For many AM practitioners, having satisfied customers is all the proof they need that they are true healers. In many cases, however, a person's condition would have improved had he or she done nothing at all. But since the improvement came *after* the treatment, it is believed that the improvement must have been caused *by* the treatment (the **post hoc fallacy** and the **regressive fallacy**). In many cases, the successful treatment may be due to the **placebo effect**. In some cases, treatment by conventional medicine causes more harm than good, and the improvement one feels is due to stopping the traditional treatment rather than to starting the alternative one. In many cases, the cure was actually due to the conventional medicine taken *along with* the alternative therapy, but the credit is given to the "alternative." Also, many so-called cures are not really cures at all in any objective sense. The patient may have been misdiagnosed in the first place, so no cure actually took place. Also, a patient subjectively reports that he or she "feels better" and the change in mood is taken as proof that the therapy is working. Psychological effects of therapies are not identical to objective improvements. A person may feel much worse but actually be getting much better. Conversely, a person may feel much better but actually be getting much worse.

11. *Many advocates of alternative therapies refuse to admit failure.* When comedian Pat Paulsen died while receiving alternative cancer therapy in Tijuana, Mexico, his daughter did not accept that the therapy was useless. Rather, she believed that the only reason her father died was that he had not sought the alternative therapy sooner. Such faith is common among those who are desperate and vulnerable, common traits among those who seek alternative therapies.

Further reading: Barrett and Jarvis 1993; Gardner 1957, 1991; Randi 1989a; Raso 1994, 1995; Trafford 1995.

amulets

Ornaments, gems, and so on worn as **charms** against evil. Amulets are often inscribed with magical **incantations**.

Amway

See **multilevel marketing**.

ancient astronauts

This term designates the speculative notion that aliens are responsible for the most ancient civilizations on earth. The most notorious proponent of this idea is Erich von Däniken, author of several popular books on the subject. His *Chariots of the Gods?: Unsolved Mysteries of the Past,* for example, is a sweeping attack on the memories and abilities of ancient peoples. Von Däniken claims that the myths, arts, social organizations, and so on of ancient cultures were introduced by astronauts from another world. He questions not just the capacity for **memory** in ancient peoples, but the capacity for culture and civilization itself. Visitors from outer space must have taught art and science to our ancestors.

Where is the proof for von Däniken's claims? Some of it is fraudulent. For example, he produced photographs of pottery that he claimed had been found in an archaeological dig. The pottery depicts flying saucers and, according to von Däniken, dated from Biblical times. However, investigators from *Nova* (the award-winning public-television science program) found the potter who had made the allegedly ancient pots. They confronted von Däniken with evidence of his fraud. His reply was that his deception was justified because some people would believe only if they saw proof ("The Case of the Ancient Astronauts," first aired on March 8, 1978, and was done in conjunction with BBC's *Horizon*).

However, most of von Däniken's evidence is in the form of specious and fallacious arguments. His data consist mainly of archaeological sites and ancient myths. He begins with the ancient astronaut assumption and then **shoehorns** the data to fit the assumption. For example, in **Nazca**, Peru, he explains giant animal drawings in the desert as an ancient alien airport. The likelihood that these drawings related to the natives' religion, science, or daily life is not considered.

There have been many critics of von Däniken's notions, but Ronald Story stands out as the most thorough. Most critics of von Däniken's theory point out that prehistoric peoples were not the helpless, incompetent, forgetful savages he makes them out to be. (They must have at least been intelligent enough to understand the language and teachings of their celestial instructors—no small feat!) It is true that we still do not know how the ancients accomplished some of their more astounding physical and technological feats. We still wonder how the ancient Egyptians raised giant obelisks in the desert and how Stone Age men and women moved huge cut stones and placed them in position in dolmens and passage graves. We are amazed by the giant carved heads on Easter Island and wonder why they were done, who did them, and why the natives abandoned the place. We may someday have the answers to our questions, but they are most likely to come from scientific investigation, not pseudoscientific speculation. For example, observing contemporary stone age peoples in Papua New Guinea, where huge stones are still found on top of tombs, has taught us how the ancients may have accomplished the same thing with little more than ropes of organic material, wooden levers and shovels, a little ingenuity, and a good deal of human strength. *Nova*'s "Secrets of Lost Empires" used historical information and speculation requiring no intervention from gods or aliens in its attempt to explain the accomplishments of ancient engineers. L. Sprague De Camp's *The Ancient Engineers* (1977) also provides earthly explanations for the accomplishments of the ancients that are much more plausible than von Däniken's wild speculations.

We have no reason to believe our ancient ancestors' memories were so much worse than our own that they could not remember these alien visitations well enough to preserve an accurate account of them. There is little evidence to support the notion that ancient myths and religious stories are the distorted and imperfect recollection of ancient astronauts recorded by ancient priests. The evidence to the contrary—that prehistoric or "primitive" peoples were (and are) quite intelligent and resourceful—is overwhelming.

Of course, it is possible that visitors from outer space did land on earth a few thousand years ago and communicate with our ancestors. But it seems more likely that prehistoric peoples themselves were responsible for their own art, technology, and culture. Why concoct an explanation such as von Däniken's? To do so may seem to increase the mystery and romance of one's theory, but it also makes it less reasonable, especially when one's theory seems inconsistent with what we already know about the world. The ancient astronaut hypothesis is unnecessary. **Occam's razor** should be applied and the hypothesis rejected.

See also **Dogon**, Zecharia **Sitchin**, and **UFOs**.

Further reading: Bullard 1996; Feder 2002; Story 1976, 1980.

anecdotal evidence

See **testimonial evidence**.

angel

A bodiless, immortal **spirit**, limited in knowledge and power. Religions such as Judaism, Christianity, and Islam believe

Ceramic cherub guarding skeptical books.

God created angels to worship Him. Not all God's angels acted angelically, however. Some angels, led by **Satan**, rebelled against a life of submission, and were cast out of Heaven. These bad angels were sent to Hell and are known as devils.

Not all angels are created equal. Angels have different functions. Some do nothing but worship their Lord. Others are sent to deliver messages to creatures on earth. Some are sent as protectors of earthlings. Still others are sent to do battle with devils.

Even though angels are spirits and devoid of a physical nature, believers in angels have had no problem depicting and describing them. Angels, say their advocates, are invisible but can take the form of visible things. Angels are usually depicted with wings and looking like human adults or children. The wings may be related to their work as messengers from God, who lives in the sky. The anthropomorphizing is understandable. Depiction enhances belief. A bodiless creature cannot be depicted. A depiction of a creature of less than human stature would be undignified and unworthy of celestial creatures. Nevertheless, it is puzzling how a bodiless creature thinks and feels. As Hobbes noted 400 years ago, to talk of a spirit as a nonbodily body seems to be akin to talking of a round square.

Since angels are invisible but capable of taking on visible forms, it is understandable that there have been many sightings. Literally anything could be an angel and any experience could be an angel experience. The existence of angels cannot be disproved. The down side of this tidy picture is that angels cannot be proved to exist, either. Everything that could be an angel could be something else. Every experience that could be due to an angel could be due to something else. Belief in angels, angel sightings, and angel experiences is entirely a matter of **faith**.

Even if they exist only in the imagination, however, angels can be very useful. They can serve as monitors of behavior and protectors of children. Much entertainment in books, films, and television programs is based on the concept of the guardian angel, often transformed into a superhuman master of **occult** powers.

Traditional religionists are not the only ones who love angels. Angel figurines seem to be popular with people from all walks of life. And New Age mythmakers have made an industry out of angels. There are many popular books connecting angels with everything from guidance in daily life to talking to the dead to **psychic** healing.

angel therapy

A New Age **psychotherapy** based on the notion that communicating with **angels** will facilitate healing because the angels will guide the patient in the right direction.

Doreen Virtue, Ph.D., is the author of several books on healing with the angels and seems to be the leading proponent of angel therapy. She claims to be a "natural **clairvoyant**" and to have a doctorate in Counseling Psychology from somewhere.

animal quackers

People who apply **quackery** to animals, such as holistic **massage therapy** for dogs and horses; **reiki** and **therapeutic touch** for pets; and **acupuncture**, **aromatherapy**, Ayurvedic medicine, and **homeopathy** for animals of all sorts. The queen of all animal quackers has to be pet **psychic** Sonya Fitzpatrick, who wisely disclaims any responsibility for the accuracy of the content on her web site or television show, *The Pet Psychic*. The king of the animal quackers has to be Rupert Sheldrake, who

thinks he's proved that some pets are psychic.

See also **morphic resonance**.

anomalous cognition

A term coined by Science Applications International Corporation (SAIC) to refer to **ESP**, including **clairvoyance**, **precognition**, **remote viewing**, and **telepathy**.

SAIC, the largest employee-owned research and engineering company in the nation, claims their terminology is neutral. It also sounds more scientific and looks better on grant applications.

anomalous luminous phenomena (ALP)

Lights of various sizes that are generated by stresses and strains within the earth's crust preceding earthquakes, according to Michael Persinger, Ph.D. He developed the tectonic strain theory (TST) as an explanation for what is going on when people observe **UFOs**.

According to Persinger,

[ALPs] display odd movements, emit unusual colors or sounds and occasionally deposit physical residues. When these phenomena closely approach a human observer, exotic forces and perceptions are frequently reported. Most ALPs [sic] display life times in the order of minutes and appear to show spatial dimensions in the order of meters. Despite their remarkably similar descriptions over time and across cultures, the transience and localized occurrence of these phenomena have limited their systematic investigation. [www .laurentian.ca/neurosci/tectonic.htm]

Persinger claims that TST is meant not to debunk UFO claims but to provide a means of predicting earthquakes. "Persinger has apparently done a computer analysis of about 3,000 UFO sightings and has found that many of them occurred weeks or months before the start of earth tremors" (LeBoeuf: www.the-spa.com/thirteen/ufo's/eth.htm).

Persinger believes that the energy from tectonic fractures causes some observers to hallucinate or lose consciousness. Those who share the experience confirm and reinforce each other's UFO observations.

anomalous perturbation

A term coined by Science Applications International Corporation (SAIC) to refer to **psychokinesis**.

anomaly

An event that does not fit a standard rule or known law of nature. The word derives from the Greek (*a nomos,* no law or rule). For example, a frog when dropped should move through the air toward the ground, according to the law of gravity. If the frog were to remain suspended in mid-air, such **levitation** would be an anomaly. If it were discovered, however, that the frog was being suspended in mid-air by electromagnetic devices, the anomaly would dissolve.

In a loose sense, anything weird, abnormal, strange, odd, statistically unexpected, or difficult to classify is called an anomaly, such as frogs raining from the sky.

In science, an anomaly is an event that cannot be explained by currently accepted scientific theories. Sometimes the new phenomenon leads to new rules or theories, for example, the discovery of x-rays and radiation.

See also Charles **Fort** and the **Piltdown Hoax**.

Further reading: Kuhn 1996; Reed 1988; Zusne and Jones 1990.

anoxia

Cerebral anoxia is the lack of oxygen to the brain. If severe, it can cause irreversible brain damage. Less severe cases can cause sensory distortions and hallucinations. Some researchers have cited cerebral anoxia as the cause of **near-death experiences** (Blackmore 1993). However, many "patients have had a near-death experience even though it was determined that their brain was not deprived of oxygen" (Wynn and Wiggins 2001: 79).

Further reading: Lutz and Nilsson 1998.

anthropometry

The study of human body measurements for use in anthropological classification and comparison.

In the 19th and early 20th centuries, anthropometry was a **pseudoscience** that classified potential criminals by facial characteristics. For example, Cesare Lombroso's *Criminal Anthropology* (1895) claimed that murderers have prominent jaws and pickpockets have long hands and scanty beards. The work of Eugene Vidocq, which identifies criminals by facial characteristics, is still used nearly a century after its introduction in France.

The most infamous use of anthropometry was by the Nazis, whose Bureau for Enlightenment on Population Policy and Racial Welfare recommended the classification of Aryans and non-Aryans on the basis of measurements of the skull and other physical features. **Craniometric** certification was required by law. The Nazis set up certification institutes to further their racial policies. Not measuring up meant denial of permission to marry or work. For many it meant the death camps.

Today, anthropometry has many practical uses, most of them benign. For example, it is used to assess nutritional status, to monitor the growth of children, and to assist in the design of office furniture.

See also **metoposcopy**, **phrenology**, and **physiognomy**.

Further reading: Gould 1993a.

anthroposophy

The Austrian-born Rudolf Steiner (1861–1925) was the head of the German Theosophical Society from 1902 to 1912. Two theosophical notions particularly attracted him: a special spiritual consciousness provides direct access to higher spiritual truths, and being mired in the material world hinders spiritual evolution. He left **theosophy** and formed the Anthroposophical Society because the theosophists did not treat Jesus or Christianity as special. Steiner had no problem, however, in accepting such Hindu notions as **karma** and **reincarnation**. By 1922 he had established what he called the Christian Community, with its own liturgy and rituals for anthroposophists. Both the Anthroposophical Society and the Christ-

Rudolph Steiner (1861–1925).

ian Community still exist, though they are separate entities.

Steiner was a true polymath. He dabbled in agriculture, architecture, art, chemistry, drama, literature, math, medicine, philosophy, physics, and religion, among other subjects. His doctoral dissertation at the University of Rostock was on Fichte's theory of knowledge. It wasn't until Steiner was nearly 40 that he became deeply interested in the **occult**. He was the author of many books and lectures, such as *How to Know Higher Worlds* (1904), *Investigations in Occultism* (1920), *Occult Science: An Outline* (1913), and *The Philosophy of Spiritual Activity* (1894). He was also much attracted to Goethe's mystical ideas and worked as an editor of Goethe's works for several years. Much of what Steiner wrote seems like a rehash of Hegel. He thought Marx had it wrong; that spiritual forces, not material ones, drive history. Steiner even spoke of the tension between the search for community and the experience of individuality, which he believed are not really contradictions but represent polarities rooted in human nature.

Although Steiner broke from the Theosophical Society, he did not abandon the eclectic mysticism of the theosophists. He thought of anthroposophy as a spiritual science. Convinced that reality is essentially spiritual, he wanted to train people to overcome the material world and learn to comprehend the spiritual world by the higher, spiritual self. He taught that there is a kind of spiritual perception that works independently of the body and the bodily senses. Apparently, it was this special spiritual sense that provided him with information about the occult.

According to Steiner, people have existed on Earth since the creation of the planet. Humans, he taught, began as spirit forms and progressed through various stages to reach today's form. Humanity is currently living in the Post-Atlantis Period, which began with the gradual sinking of **Atlantis** in 7227 B.C. The Post-Atlantis Period is divided into seven epochs, the current one being the European-American Epoch, which will last until the year 3573. After that, humans will regain the **clairvoyant** powers they allegedly possessed prior to the time of the ancient Greeks (Boston 1996a).

Steiner's most lasting and significant influence, however, has been in the field of education. In 1913, Steiner built his Goetheanum, a "school of spiritual science," at Dornach, near Basel, Switzerland. The Goetheanum was a forerunner of the Steiner or Waldorf schools. (The term "Waldorf" comes from the school Steiner opened for the children of workers at the Waldorf-Astoria cigarette factory in Stuttgart, Germany, in 1919. The owner of the factory invited Steiner to give a series of lectures to his factory workers and was so impressed that he asked Steiner to set up the school.) The first U.S. Waldorf School opened in New York City in 1928. Today, there are more than 600 Waldorf schools in over 32 countries with approximately 120,000 students. About 125 Waldorf schools are said to be currently operating in North America. There is even a nonaccredited Rudolf Steiner College offering degrees in Anthroposophical Studies or in Waldorf Education.

Steiner designed the curriculum of his schools around notions derived by spiritual insight rather than empirical study. He believed we are each comprised of body, spirit, and **soul**. He believed that children pass through three seven-year stages and that education should be appropriate to the spirit for each stage. Birth to age seven, he claimed, is a period for the spirit to adjust to being in the material world. At this stage, children best learn through imitation. Academic content is held to a min-imum during these years. Children are told fairy tales, but do no reading until about the second grade. They learn about the alphabet and writing in the first grade.

According to Steiner, the second stage of growth is characterized by imagination and fantasy. Children learn best from ages 7 to 14 by acceptance and emulation of authority. The children have a single teacher during this period and the school becomes a "family" with the teacher as the authoritative "parent."

The third stage, from 14 to 21, is when the **astral body** is drawn into the physical body, causing puberty. These anthroposophical ideas are not part of the standard Waldorf school curriculum, but apparently are believed by those in charge of the curriculum. Waldorf schools tend to be spiritually oriented and are based on a generally Christian perspective, but they leave religious training to parents.

Even so, because they are not taught fundamentalist Christianity from the Bible, Waldorf schools are often attacked for encouraging **paganism** or even **Satanism**. This may be because they emphasize the relation of human beings to nature and natural rhythms, including an emphasis on festivals, myths, ancient cultures, and various celebrations.

Some of the ideas of the Waldorf School are not Steiner's, but are in harmony with the master's spiritual insights. For example, television viewing is discouraged because of its content and because it discourages the growth of the imagination. This idea is undoubtedly attractive to parents, since it is very difficult to find anything of positive value for young children on television.

Waldorf schools also discourage computer use by young children. The benefits of computer use by children have yet to be demonstrated, though it seems to be widely believed and accepted by educators, who spend billions each year on the latest computer equipment for students who often can barely read or think critically, and have minimal social and oral skills. Waldorf schools, on the other hand, are as committed to the arts as public schools are to technology. What the public school consider frills, Waldorf schools consider essential, for example, weaving, knitting, playing a musical instrument, woodcarving, and painting.

One of the more unusual parts of the curriculum involves something Steiner called "eurythmy," an art of movement that tries to make visible what he believed were the inner forms and gestures of language and music. According to the Waldorf FAQ (www.fortnet.org/rsws/waldorf/faq .html),

> it often puzzles parents new to Waldorf education, [but] children respond to its simple rhythms and exercises which help them strengthen and harmonize their body and their life forces; later, the older students work out elaborate eurythmic representations of poetry, drama and music, thereby gaining a deeper perception of the compositions and writings. Eurythmy enhances coordination and strengthens the ability to listen. When children experience themselves like an orchestra and have to keep a clear relationship in space with each other, a social strengthening also results.

Perhaps the most interesting consequence of Steiner's spiritual views was his attempt to instruct the mentally and physically handicapped. Steiner believed that it is the *spirit* that comprehends knowledge and the spirit is the same in all of us, regardless of our mental or physical differences.

Most critics of Steiner find him to have been a truly remarkable man, most decent and admirable. Unlike many other gurus, Steiner seems to have been a truly moral man who didn't try to seduce his followers and who remained faithful to his wife.

His moral stature has been challenged by charges of racism, however. Steiner believed in reincarnation and that souls pass through stages, including racial stages, with African races being lower than Asian races and European races being the highest form.

There is no question that he made contributions in many fields, but as a philosopher, scientist, and artist he rarely rises above mediocrity and is singularly unoriginal. His spiritual ideas seem less than credible. Some of his ideas on education, however, are worth considering. He was correct to note that there is a grave danger in developing the imagination and understanding of young people if schools are dependent on the government. State-funded education is likely to lead to emphasis on a curriculum that serves the state, that is, one mainly driven by particular economic and social policies. Education is driven not by the needs of children but by the material and political needs of society. The competition that drives most public education benefits society and those in power more than it benefits most individuals. An education where cooperation and love, rather than competition and resentment, mark the essential relationship among students might be more beneficial to the students' intellectual, moral, and creative well-being.

Steiner was ahead of his time in understanding sexism. He recognized that the social position of women is largely determined by stereotyping, rather than by the particular talents of individual women. On the other hand, it is likely that some of anthroposophy's weirder notions about such things as **astral bodies** and **Atlantis** will get passed on in a Waldorf education, even if Steiner's philosophical theories are not part of the curriculum for children.

apophenia

The spontaneous perception of connections and meaningfulness of unrelated phenomena. K. Conrad coined the term in 1958 (Brugger 2001).

Peter Brugger of the Department of Neurology, University Hospital Zurich, gives examples of apophenia from August Strindberg's *Occult Diary,* the playwright's own account of his psychotic break:

> He saw "two insignia of **witches**, the goat's horn and the besom" in a rock and wondered "what demon it was who had put [them] . . . just there and in my way on this particular morning."
>
> He sees sticks on the ground and sees them as forming Greek letters which he interprets to be the abbreviation of a man's name and feels he now knows that this man is the one who is persecuting him.
>
> He sees sticks on the bottom of a chest and is sure they form a **pentagram**.
>
> He sees tiny hands in prayer when he looks at a walnut under a microscope and it "filled me with horror."
>
> His crumpled pillow looks "like a marble head in the style of Michaelangelo." Strindberg comments that "these occurrences could not be regarded as accidental, for on some days the pillow presented the appearance of horrible monsters, of gothic gargoyles, of dragons, and one night . . . the Evil One himself." (Brugger 2001: 195–213)

According to Brugger, "The propensity to see connections between seemingly unrelated objects or ideas most closely links psychosis to creativity . . . apophenia and creativity may even be seen as two sides of the same coin." Brugger notes that one psychoanalyst thought he had support for Freud's penis envy theory because more females than males failed to return their pencils after a test. Another spent nine pages in a prestigious journal describing how sidewalk cracks are vaginas and feet are penises, and the old saw about not stepping on cracks is actually a warning to stay away from the female sex organ.

Brugger's research indicates that high levels of dopamine affect the propensity to find meaning, patterns, and significance where there is none, and that this propensity is related to a tendency to believe in the **paranormal**.

In statistics, apophenia is called a Type I error, seeing patterns where none exist. It is highly probable that the apparent significance of many unusual experiences and phenomena are due to apophenia.

See also Carl **Jung** and **pareidolia**.

Further reading: Belz-Merk: www.igpp .de/english/counsel/project.htm; Leonard and Brugger 1998.

applied kinesiology

An alternative therapy created by George Goodheart, D.C. According to the International College of Applied Kinesiology, the therapy "is based on **chiropractic** principles and requires manual manipulation of the spine, extremities and cranial bones as the structural basis of its procedures." However, Goodheart and his followers unite chiropractic with traditional Chinese medicine (among other things); not only do they accept the notion of **chi** and the meridians of **acupuncture**, they posit a universal intelligence of a spiritual nature running through the nervous system. They believe that muscles reflect the flow of chi and that by measuring muscle resistance one can determine the health of bodily organs and nutritional deficiencies. These are empirical claims and have been tested and shown to be false (Hyman 1999; Kenney et al. 1988). Other claims made by practitioners are supported mainly by anecdotes supplied by advocates. No reputable scientific journal has ever published a paper supporting the validity of applied kinesiology, according to Janice Lyons, R.N. (www.hcrc.org/contrib/lyons/kinesiol.html).

Applied kinesiology should not be confused with kinesiology proper: the scientific study of the principles of mechanics and anatomy in relation to human movement.

apport

An object allegedly materialized during a **séance**. Believers see apports as signs or gifts from **spirits**. Skeptics see them as evidence of **conjuring**.

Good magicians can seem to produce objects out of nowhere. They can also make objects seem to disappear. When a **medium** does this it is referred to as a **deport**.

See also **teleportation**.

Further reading: Keene 1997.

Area 51

A part of an off-limits military base near Groom Dry Lake in Nevada where aliens are being hidden, according to some **UFO** followers. The state of Nevada, showing either extreme insensitivity or humor, recently designated a barren 98-mile stretch of Route 375, which runs near Area 51, as the Extraterrestrial Highway.

Since you can be shot if you try to trespass on the military base where Area 51 is located, UFO tourists must view the sacred ground from distant vantage points. Many do this, hoping for a glimpse of a UFO landing. According to a common UFO legend, our government has a treaty with the aliens that allows them to fly into this area

at will, as long as we can experiment on them and try to duplicate their aircraft.

Skeptics don't doubt that something secret is going on in Area 51. What is going on may be more sinister than building secret aircraft or developing new weapons. Leslie Stahl of CBS's *Sixty Minutes* suggested that Area 51 might be an illegal dumping ground for toxic substances. Several former workers at Area 51 and widows of former workers have filed lawsuits against the government for injuries or death resulting from illegal hazardous waste practices. So far the government has been protected from such suits because of "national security." In fact, the government does not even acknowledge the existence of the base known as Area 51. Such denials, according to UFO followers, support their belief in a conspiracy to hide the truth about aliens in Area 51, including the belief that everything has been moved to Area 52.

See also **alien abductions**, **Roswell**, and **UFOs**.

Further reading: Klass 1997.

argument from design

One of the "proofs" for the existence of God. In its basic form, this argument infers from the intelligent order and created beauty of the universe that there is an intelligent designer and creator of the universe. The argument has been criticized for **begging the question**: By assuming that the universe has an intelligent order and created beauty, one assumes that it is the work of a designer. The argument also suppresses evidence: For all its beauty and grandeur, the universe is also full of obvious defects: babies born without brains, good people suffering monstrous tortures such as neurofibromatosis, evil people basking in the sun and enjoying power and reputation, volcanoes erupting, earthquakes rattling the planet, hurricanes and tornadoes blindly wiping out thousands of lives a day, and comets smashing into planets. Is it unfair to call these *defects*, what are blithely referred to by theologians as nonmoral or physical evils? To say that these defects only seem so to us, because we are ignorant of God's plan and vision and cannot know their purpose, is self-refuting. If we can't know what's good and what's not, we can't know whether the design, if any, is good or bad, intelligent or incompetent.

One of the argument's more famous variations involves an analogy with a watch. William Paley (1743–1805), the Archdeacon of Carlisle, writes in his *Natural Theology* (1802):

> In crossing a heath, suppose I pitched my foot against a stone and were asked how the stone came to be there, I might possibly answer that for anything I knew to the contrary it had lain there forever; nor would it, perhaps, be very easy to show the absurdity of this answer. But suppose I had found a watch upon the ground, and it should be inquired how the watch happened to be in that place, I should hardly think of the answer which I had before given, that for anything I knew the watch might have always been there.

The reason, he says, that he couldn't conceive of the watch having been there forever is because it is evident that the parts of the watch were put together for a purpose. It is inevitable that "the watch must have had a maker," whereas the stone apparently has no purpose revealed by the complex arrangement of its parts.

Clarence Darrow noted that some stones would be just as puzzling and as marvelous as a watch. Some stones are complex and could easily have been designed by someone for some purpose. Paley's point was that a watch could be seen to be analogous to the universe. The design of the watch implies an intelligent designer; likewise for the design of the universe: "[E]very manifestation of design which existed in the watch, exists in the works of nature, with the difference on the side of nature of being greater and more, and that in a degree which exceeds all computation." The implication is that the works of nature must have a designer of supreme intelligence to have contrived such a magnificent mechanism as the universe.

According to Darrow, however, this "implication" is actually an assumption. The design of the universe is manifest in its order, but to say that something shows order one must have some norm or pattern by which to determine whether the matter concerned shows any order. The pattern is the universe itself. We have observed this universe and its operation and we call it order. To say that the universe is patterned on order is to say that the universe is patterned on the universe. It can mean nothing else (Darrow 1932).

David Hume (1711–1776) took up the design analogy a few years before Paley. In his *Dialogues Concerning Natural Religion*, Philo—one of the characters in the dialogue—says the world

> plainly resembles more an animal or a vegetable than it does a watch or knitting-loom. Its cause, therefore, it is more probable, resembles the cause of the former. The cause of the former is generation or vegetation. The cause, therefore, of the world we may infer to be something similar or analogous to generation or vegetation.

Finally, the watch analogy would be more convincing of divine purpose if, while observing it in his imaginary scenario, Paley's watch suddenly and for no reason shot a lightning bolt through his forehead. That would be more in harmony with the universe most of us have come to know and love.

Another common form of the argument from design lists facts about nature that, if they were different, would mean that our planet, or life on our planet, would not exist. We wouldn't be here, it is noted,

- if the sun were just slightly farther away or half as powerful
- if the axis of the earth were slightly different
- if the moon were larger or closer or farther away
- if gravity weren't such a weak force
- if DNA didn't replicate and genetic mutation did not happen
- if molecules were larger or smaller
- if there were sixty planets in our solar system
- if the speed of light were half what it is
- if the rotation of the earth were one-tenth of what it is

Furthermore, look at all the signs of design: salmon, eels, birds, butterflies, and whales are able to migrate and find the same breeding and feeding grounds year after year. Finally, ecological systems so balanced, so orderly, so harmonious—give proof that there must be a caring designer.

One cannot deny the facts. If things were different, then things would be different. But they aren't different, so what is the point of this argument? The sun will be unable to support life on this planet some day. It is already unable to support life on several other planets. What does this fact prove about design? The axis of the earth has been different and will be different again. There was a time when no life existed on this planet and there will be another time when no life exists here. At one time this planet did not exist. There will come another time when it ceases to exist. There are countless planets that exist that do not have the conditions necessary for life. What do these facts prove about design? Nothing.

Another form of the argument focuses on the notion that the odds are a billion billion to one that all these circumstances accidentally coincided to make life on earth possible. Yet, it's happened. So, the odds are 100% that it *can* happen.

> Suppose you put ten pennies, marked from one to ten, into your pocket and give them a good shuffle. Now try to take them out in sequence from one to ten, putting back the coin each time and shaking them all again. Mathematically we know that your chance of first drawing number one is one in ten; of drawing one and two in succession, one in 100; of drawing one, two and three in succession, one in 1000, and so on; your chance of drawing them all, from number one to number ten in succession, would reach the unbelievable figure of one in ten billion. (Cressy Morrison, "Seven Reasons a Scientist Believes in God," *Reader's Digest*, January 1948)

Morrison begs the question. The earth with life on it is here. The odds are 1:1 of its existing. In any case, if one had 20 billion years to pull 10 numbered pennies out of a pocket, the odds of drawing out the coins in sequence at least once seem decent.

> [R]arity by itself shouldn't necessarily be evidence of anything. When one is dealt a bridge hand of thirteen cards, the probability of being dealt that particular hand is less than one in 600 billion. Still, it would be absurd for someone to be dealt a hand, examine it carefully, calculate that the probability of getting it is less than one in 600 billion, and then conclude that he must not have been dealt that very hand because it is so very improbable. (Paulos 1990)

Are there naturalistic and mechanistic explanations for ecological systems and what is called "animal wisdom"? Of course. Does this prove they were not designed? Of course not. Nor does their existence prove design.

See also **intelligent design**.

Further reading: Dawkins 1988, 1996; Stenger 1995a, 1995b.

argument to ignorance (*argumentum ad ignorantiam*)

The argument to ignorance is a logical fallacy claiming something is true because it has not been proved false, or that something is false because it has not been proved true. A claim's truth or falsity depends on supporting or refuting evidence for the claim, not the lack of support for a contrary or contradictory claim. (*Contrary* claims can both be false, unlike *contradictory* claims.)

The fact that it cannot be proved that the universe *is not* designed by an intelligent creator does not prove that it is. Nor does the fact that it cannot be proved that the universe *is* designed by an intelligent creator prove that it isn't.

The argument to ignorance seems to be more seductive when it can play upon **wishful thinking**. People who want to believe in immortality, for example, may be more prone to think that the lack of proof to the contrary of their desired belief is somehow relevant to supporting it.

Further reading: Browne and Keeley 1997; Carroll 2000; Damer 2001; Kahane 1997; Moore and Parker 2000.

aromatherapy

A term coined by French chemist René Maurice Gattefossé in the 1920s to describe the practice of using essential oils taken from plants, flowers, roots, seeds, and so on in healing. The term is a bit misleading since the *aromas* of oils, whether natural or synthetic, are generally not themselves therapeutic. Aromas are used to identify the oils, to determine adulteration, and to stir the memory, but not to directly bring about a cure or healing. It is the "essence" of the oil—its chemical properties—that gives it whatever therapeutic value the oil might have. Vapors are used in *some* cases of aromatherapy, and, of course, an aroma can affect mood, and therefore one's sense of well-being, by arousing a memory. In most cases, however, the oil is rubbed onto the skin or ingested in a tea or other liquid. Some aromatherapists even consider cooking with herbs a type of aromatherapy.

The healing power of essential oils is the main attraction in aromatherapy. It is also the main question for the skeptic. There is very little evidence for most of the claims made by aromatherapists regarding the various healing properties of oils. Most of the support for the healing power of substances such as tea tree oil is in the form of anecdotes such as the following:

> In the plane on my way to India [from Europe] a few years ago, my index finger began throbbing violently. A rose thorn had lodged in it two days before, as I pruned my roses. It was now turning septic. I straight away applied tea tree oil undiluted to the finger. By the time I arrived in Bangalore, the swelling had almost gone and the throbbing had stopped. (Ryman 1993)

This kind of **post hoc** reasoning abounds in the literature of **alternative health care**. What would be more convincing would be some **control group studies** such as the following:

> Professor *SoandSo* of the Department of Microbiology at the University of Washington has published a paper in [*well-*

regarded scientific journal] that demonstrates that tea tree oil kills many bacteria present in common infections, including some staphylococci and streptococci.

When references are made, they are made not to scientific studies but to anecdotes or beliefs of other aromatherapists; for example, "Marguerite Maury prescribed rose for frigidity, ascribing aphrodisiac properties to it. She also considered rose a great tonic for women who were suffering from depression" (ibid.: 205). Such testimonials are not met with skepticism or even curiosity as to what evidence there is for them. They are just passed on as articles of faith.

On the other hand, many of the claims of aromatherapists are not testable at all, as in how certain oils will affect their "subtle body," bring balance to their **chakra**, restore harmony to their **energy** flow, return them to their center, or contribute to spiritual growth. One aromatherapist claims that incense "cleanses the air of negative energies." Another asserts that benzoin resinoid will "drive out evil spirits" (McCutcheon 1996). Such claims are part of New Age mythology and can't engender any meaningful scientific discussion or debate.

When aromatherapists get into professional debates about empirical matters, it is generally over such issues as whether natural oils are superior to synthetic ones, though even here references to scientific studies of the issue are sought in vain.

There have been some controlled studies on aromatherapy, but they are not favorable. For example, one study compared the effects on intensive care patients of aromatherapy using lavender, **massage therapy**, and rest. The study concluded that rest was best (Dunn et al. 1995).

What aromatherapy lacks is a nose for sniffing out nonsense.

See also **Bach's flower therapy**.

Further reading: Barrett 2001b; Barrett and Butler 1992; Barrett and Jarvis 1993; Raso 1994.

astral body

One of seven bodies each of us has, according to Madame Blavatsky, the queen of **Theosophy**. The astral body is the seat of feeling and desire and has an **aura**. How the physical body and the other alleged bodies interact is unknown, but it is said to be by **occult** forces. The astral body is said to be capable of leaving the other bodies for an **out-of-body experience** known as **astral projection**.

astral projection

A type of **out-of-body experience** (OBE) in which the **astral body** leaves its other six bodies and journeys far and wide to anywhere in the universe. The notion that we have seven bodies (one for each of the seven planes of reality) is a teaching of **theosophist** Madame Blavatsky. On its trips, the astral body perceives other astral bodies rather than their physical, etheric, emotional, spiritual, or other bodies. In an ordinary OBE, such as **remote viewing** or the **near-death experience** (NDE), there is a separation of a person's consciousness from his or her body. In the NDE, there may be the experience of hovering above and perceiving one's body and environs. One might hear conversations of surgeons or rescue workers as they tinker with one's body. In astral projection, it is the astral body, not the **soul** or consciousness, that leaves the body. The astral body is the seat of feeling and desire, and is generally described as being connected to the physical body during astral projection by an infinitely elastic and very fine silver cord, a kind of cosmic umbilical cord or Ariadne's thread.

There is scant evidence to support the

claim that anyone can project his or her mind, soul, psyche, spirit, astral body, etheric body, or any other entity to somewhere else on this or any other planet. The main evidence for this claim is in the form of **testimonials** of those who claim to have experienced being out of their bodies when they may have been out of their minds.

Further reading: Grim 1990; Sagan 1979.

astrology

In its traditional form, astrology is a type of **divination** based on the theory that the positions and movements of celestial bodies (stars, planets, sun, and moon) at the time of birth profoundly influence a person's life. In its psychological form, astrology is a type of New Age therapy used for self-understanding and personality analysis (**astrotherapy**). Ivan Kelly, who has written many articles critical of astrology, thinks that astrology

> has no relevance to understanding ourselves, or our place in the cosmos. Modern advocates of astrology cannot account for the underlying basis of astrological associations with terrestrial affairs, have no plausible explanation for its claims, and have not contributed anything of cognitive value to any field of the social sciences.

Even so, astrology is believed by millions of people and it has survived for thousands of years. The ancient Chaldeans and Assyrians engaged in astrological divination some 3,000 years ago. By 450 B.C.E. the Babylonians had developed the 12-sign zodiac, but it was the Greeks—from the time of Alexander the Great to their conquest by the Romans—who provided most of the fundamental elements of modern astrology. The spread of astrological practice was checked by the rise of Christianity,

"Hermes Zodiac." Drawing from Athenasius Kircher, *OEdipus Aegyptiacus* (1653).

which emphasized divine intervention and free will. During the Renaissance, astrology regained popularity, in part due to rekindled interest in science and astronomy. Christian theologians, however, warred against astrology, and in 1585 Pope Sixtus V condemned it. At the same time, the work of Kepler and others undermined astrology's tenets. Its popularity and longevity are, of course, irrelevant to the truth of astrology in any of its forms.

The most popular form of traditional Western astrology is sun sign astrology, the kind found in the horoscopes of many daily newspapers. A horoscope is an astrological forecast. The term is also used to describe a map of the zodiac at the time of one's birth. The zodiac is divided into 12 zones of the sky, each named after the constellation that originally fell within its zone (Taurus, Leo, etc.). The apparent paths of the sun, the moon, and the major planets all fall within the zodiac. Because of the precession of the equinoxes, the equinox and solstice points have each moved westward about 30 degrees in the last 2,000 years. Thus, the zodiacal constellations named in ancient times no longer correspond to the segments of the zodiac represented by their signs. In short, had you been born at the same time on the same day of the year 2,000 years ago, you would have been born under a different sign.

Traditional Western astrology may be divided into tropical and sidereal. (Astrologers in non-Western traditions use different systems.) The tropical, or solar, year is measured relative to the sun and is the time between successive vernal equinoxes (365 days, 5 hr, 48 min, 46 sec of mean solar time). The sidereal year is the time required for the earth to complete an orbit of the sun relative to the stars (365 days, 6 hr, 9 min, 9.5 sec of mean solar time). The sidereal year is longer than the tropical year because of the precession of the earth's axis of rotation.

Sidereal astrology uses the actual constellation in which the sun is located at the moment of birth as its basis; tropical astrology uses a 30-degree sector of the zodiac as its basis. Tropical astrology assigns its readings based on the time of the year, while generally ignoring the positions of the sun and constellations relative to each other. Sidereal astrology is used by a minority of astrologers and bases its readings on the constellations near the sun at the time of birth.

According to some astrologers, the data support the hypothesis that there is a causal connection between heavenly bodies and human events. Appeals are made to significant correlations between astrological signs and such things as athleticism. However, even a statistically significant correlation between x and y is not a sufficient condition for reasonable belief in a causal connection, much less for the belief that x causes y. Correlation does not prove causality; nevertheless, it is extremely attractive to defenders of astrology. For example: "Among 3,458 soldiers, Jupiter is

to be found 703 times, either rising or culminating when they were born. Chance predicts this should be 572. The odds here: one million to one" (Gauquelin 1975). Let's assume that the statistical data show significant correlations between various planets rising, falling, and culminating, and various character traits. It would be more surprising if, of all the billions and billions of celestial motions conceivable, there weren't a great many that could be significantly correlated with dozens of events or individual personality traits.

Defenders of astrology are fond of noting that "the length of a woman's menstrual cycle corresponds to the phases of the moon" and "the gravitational fields of the sun and moon are strong enough to cause the rising and falling of tides on Earth." If the moon can affect the tides, then surely the moon can affect a person. But what is the analog to the tides in a person? We are reminded that humans begin life in an amniotic sea and the human body is 70 percent water. If oysters open and close their shells in accordance with the tides, which flow in accordance with the electromagnetic and gravitational forces of the sun and moon, and humans are full of water, then isn't it obvious that the moon must influence humans as well? It may be obvious to some, but the evidence for these **lunar effects** is lacking.

Astrologers emphasize the importance of the positions of the sun, moon, and planets at the *time of birth*. However, the birthing process isn't instantaneous. There is no single *moment* that a person is born. The fact that some official somewhere writes down a time of birth is irrelevant. Do they pick the moment the water breaks? The moment the first dilation occurs? When the first hair or toenail peeks through? When the last toenail or hair passes the last millimeter of the vagina? When the umbilical cord is cut? When the first breath is taken? Or does birth occur at the moment a physician or nurse looks at a clock to note the time of birth?

Why are the *initial* conditions more important than all subsequent conditions for one's personality and traits? Why is the moment of *birth* chosen as the significant moment rather than the moment of *conception*? Why aren't other initial conditions such as one's mother's health, the delivery place conditions, forceps, bright lights, dim room, back seat of a car, and so on more important than whether Mars is ascending, descending, culminating, or fulminating? Why isn't the planet Earth—the closest large object to us in our solar system—considered a major influence on who we are and what we become? Other than the sun and the moon and an occasional passing comet or asteroid, most planetary objects are so distant from us that any influences they might have on anything on our planet are likely to be wiped out by the influences of other things here on earth.

No one would claim that in order to grasp the effect of the moon on the tides or potatoes one must understand initial conditions before the Big Bang, or the positions of the stars and planets at the time the potato was harvested. If you want to know what tomorrow's low tide will be, you do not need to know where the moon was when the first ocean or river was formed, or whether the ocean came first and then the moon, or vice-versa. Initial conditions are less important than present conditions to understanding current effects on rivers and vegetables. If this is true for the tides and plants, why wouldn't it be true for people?

Finally, there are those who defend astrology by pointing out how accurate professional horoscopes are. Astrology "works," it is said, but what does that mean? Basically, to say astrology works

means that there are a lot of satisfied customers, and one can **shoehorn** any event to fit a chart. It does not mean that astrology is accurate in predicting human behavior or events to a degree significantly greater than mere chance. There are many satisfied customers who believe that their horoscope accurately describes them and that their astrologer has given them good advice. Such evidence does not prove astrology so much as it demonstrates the **Forer effect** and **confirmation bias**. Good astrologers give good advice, but that does not validate astrology. There have been several studies that have shown that people will use **selective thinking** to make any chart they are given fit their preconceived notions about themselves and their charts. Many of the claims made about signs and personalities are vague and would fit many people under many different signs. Even professional astrologers, most of whom have nothing but disdain for sun sign astrology, can't pick out a correct horoscope reading at better than a chance rate. Yet astrology continues to maintain its popularity, despite the fact that there is scarcely a shred of scientific evidence in its favor. Even the former First Lady of the United States, Nancy Reagan, and her husband, Ronald, consulted an astrologer while he was the leader of the free world, demonstrating once again that astrologers have more influence than the stars do.

See also **cosmobiology**.

Further reading: Culver and Ianna 1988; Dean, Mather, and Kelly 1996; Jerome 1977; Kelly 1997, 1998; Kelly, Dean, and Saklofske 1990; Martens and Trachet 1998; Randi 1982a.

astrotherapy

The use of **astrology** by counselors and therapists as a guide to self-actualization, self-transcendence, and the transformation of personality. Astrotherapists believe astrology can aid in psychological healing and growth.

According to defenders of astrotherapy, most critics of astrology misunderstand how human destiny is actually linked to the heavens. Frederick G. Levine, author of *The Psychic Sourcebook* (1988), claims that modern astrologers are more "holistic" than their ancient counterparts. The contemporary astrologer doesn't believe in anything so crude as direct causal connection between the heavenly bodies and a person's destiny. He or she believes in the interrelatedness of all things.

> [T]here are larger patterns of energy that govern all interactions in the universe and . . . these patterns or cycles are reflected in the movements of stars and planets in the same way they are reflected in the movements of people and cultures. Thus it is not that planetary motions cause events on earth, but simply that those motions are indicators of universal patterns.

To back up his claim, Levine cites Linda Hill, "a New York astrological consultant." Says Ms. Hill, "I don't think anyone knows exactly why it works; it just works. Carl **Jung** used the term synchronicity. It's simply a synchronization. . . . We are somehow synchronized to the celestial patterns that were present at our birth."

Dane Rudhyar is seen as the father of astrotherapy. In the 1930s he applied Jungian psychological concepts to astrology. He liked Jung's notion that the psyche seeks psychic wholeness or "individuation," a process Rudhyar believed is evident in the horoscope.

Rudhyar's work is carried on today by Glen Perry, who boasts a Ph.D. in clinical psychology from the Saybrook Institute in San Francisco, a regionally accredited graduate school "dedicated to fostering the full expression of the human spirit and humanistic values in society." In astrotherapy, says Perry, "astrology is used to foster empathy for the client's internal world and existing symptoms, and promote positive personality growth and fulfillment." He thinks astrology is both a theory of personality and a diagnostic tool, yet he provides neither arguments nor evidence to support this notion. Here is an example of how he uses astrology:

> Saturn opposed Venus in the natal chart indicates not simply "misfortune in love," but the potential to love deeply, enduringly, and responsibly along with the patience and determination to overcome obstacles. While realization of this potential may require a certain amount of hardship and suffering, to predict only hardship and suffering with no understanding of the potential gains involved is shortsighted at best and damaging at worst.

How Perry knows this is not made clear. Other claims, equally profound, do not require argument or evidence because they are vacuous: "the horoscope symbolizes the kind of adult that the individual may become." Still other claims are nearly unintelligible: "What the individual experiences as a problematic situation or relationship can be seen in the chart as an aspect of his or her own psyche. In this way, the horoscope indicates what functions have been denied and projected, and through what circumstances (houses) they will likely be encountered." "Simply put," says Perry, "the goal is to help the client realize the potentials that are symbolized by the horoscope." What systematic analysis and methodological tools he used to arrive at this notion are not mentioned, much less how one could go about verifying the specific symbolizations of any given horoscope. He does, however, seem to rely heavily upon questionable psychological concepts promoted by Jung and **Freud**.

Another astrotherapist, Brad Kochunas, makes it clear that one of the chief virtues of applying astrology to the *inner life* rather than to outward patterns of behavior is that it takes astrology out of the realm of the scientific, where it has not fared too well when it has been thoroughly examined. Kochunas calls this concern with the inner life "the imaginal perspective" and says it

> is not concerned with whether something is true or not but rather with its usefulness for the task at hand. Questions of truth or falsity belong to the realm of the rational and are irrelevant to the value of the imaginal. It is the functional validity and not the factual validity which is primary for the imaginal perspective. Does something work for a person? Is it useful in the sense of providing depth, meaning, value, or purpose to an individual or community? If so, then there is little call for its cultural degradation, it has power. [Kochunas: www.mountainastrologer.com/kochunas .html]

Kochunas, unlike Perry, firmly locates astrotherapy in mythology and proudly proclaims it to be outside of the realm of science and without concern for empirical truth or falsity. His message seems to be very simple and straightforward: If you can find satisfied customers, you have a valid myth.

Ivan Kelly, author of several articles critical of astrology, also has a simple message: "Astrology is part of our past, but astrologers have given no plausible reason why it should have a role in our future."

See also New Age **psychotherapies**.

Further reading: Dean et al. 1996; Kelly 1997, 1998.

atheism

Traditionally defined as *disbelief in the existence of God*. As such, atheism involves *active rejection* of belief in God's existence.

However, since there are many concepts of God and these concepts are usually rooted in some culture or tradition, atheism might be defined as the belief that a particular word used to refer to a particular god is a word that has no reference. Thus, there are as many different kinds of atheism as there are names of gods.

Some atheists may know of many gods and reject belief in the existence of all of them. Such a person might be called a *polyatheist*. But most people who consider themselves atheists probably mean that they do not believe in the existence of the local god. For example, most people who call themselves atheists in a culture where the Judeo-Christian-Islamic God (JCIG) dominates would mean, at the very least, that they deny that there is an omnipotent, omniscient, providential, personal creator of the universe. On the other hand, people who believe in the JCIG would consider such denial tantamount to atheism. Spinoza, for example, defined God as being identical to nature and as a substance with infinite attributes. Many Jews and Christians considered him an atheist because he rejected both the traditional JCIG and personal immortality. Hobbes was also considered an atheist because he believed that all substances are material and that God must therefore be material, not spiritual. Yet neither Spinoza nor Hobbes called themselves atheists.

Epicurus did not call himself an atheist, either, but he rejected the concept of the gods popular in ancient Greece. The gods are perfect, he said. Therefore, they cannot be the imperfect beings depicted by Hesiod, Homer, and others. Their gods have human flaws, including jealousy. Perfect beings would not be troubled by anything, including the behavior of humans. Hence, the notion that the gods will reward or punish us is absurd. To be perfect is to be unperturbed. The concept of perfection, therefore, requires that the gods be indifferent to human behavior. Some have rejected belief in the Christian God for similar reasons. The idea of a perfect being creating the universe is self-contradictory. How can perfection be improved upon? To create is to indicate a lack, an imperfection. If that objection can be answered, another arises: If God is all-good and all-powerful, evil should not exist. Therefore, either God is all-good but allows evil because God is not all-powerful, or God is all-powerful but allows evil because God is not all-good. Such an argument clearly does not deny the existence of *all* gods.

Others have rejected the Christian God because they believe that the concept of *worship*, essential to most Christians, contradicts the concept of *omnipotence* (Rachels 1989). Still others reject a belief in the JCIG because they consider the Scriptures supporting that belief to be unbelievable. Some theologians have tried to prove through reason alone that this God exists. Rejection of such proofs, however, is not atheism.

Some Christians consider Buddhists to be atheists, apparently for the same reason they consider Spinoza or Plato to be atheists: Anyone who rejects the Omnipotent and All-Good Providential Personal Creator rejects God. Yet rejecting the JCIG is not to reject all gods. Nor is rejecting God the same as rejecting belief in an ultimate ground or principle of being and goodness that explains why there is something rather than nothing and why everything is as it is. Nor is rejection of the JCIG the same as rejecting belief in a realm of beings such as devas or spirits that are not limited by mortality and other human or animal frailties.

Finally, atheists do not deny that people have "mystical" or "religious" experiences, where they feel God's presence or a sense of the oneness and significance of everything in the universe. Nor do atheists deny that many people experience God's presence in their everyday lives. Atheists deny that the brain states that results in such feelings and experiences have supernatural or extraterrestrial causes.

How widespread is atheism? A worldwide survey in 2000 by the Gallup polling agency found that 8% do not think there is any spirit, personal God, or life force. Another 17% are not sure. However, more than half the world's population and more than 90% of the world's scientists do not believe in a personal God, and hence would be considered atheists by many Christians.

See also **agnosticism**, **miracles**, and **theism**.

Further reading: Berman 1996; Freud 1927; Hobbes 1651; Hume 1748; Johnson 1981; Martin 1990; Mill 1859; Newberg et al. 2001; Persinger 1987; Rachels 1989; Russell 1977; Smith 1979; Smith 1952; Spinoza 1670.

Atlantis

A legendary island in the Atlantic, west of Gibraltar, that sunk beneath the sea during a violent eruption of earthquakes and floods some 9,000 years before Plato wrote about it in his *Timaeus* and *Critias*. In a discussion of utopian societies, Plato claims that Egyptian priests told Solon about Atlantis. Plato was not describing a real place any more than his allegory of the cave describes a real cave. The purpose of Atlantis is to express a moral message in a discussion of ideal societies, a favorite theme of his. The fact that nobody in Greece for 9,000 years had mentioned a battle between Athens and Atlantis should serve as a clue that Plato was not talking about a real place or battle. Nevertheless, Plato is often cited as the primary source for the reality of a place on earth called Atlantis. Here is what the Egyptian priest allegedly told Solon:

> Many great and wonderful deeds are recorded of your state in our histories. But one of them exceeds all the rest in greatness and valour. For these histories tell of a mighty power which unprovoked made an expedition against the whole of Europe and Asia, and to which your city put an end. This power came forth out of the Atlantic Ocean, for in those days the Atlantic was navigable; and there was an island situated in front of the straits which are by you called the Pillars of Heracles; the island was larger than Libya and Asia put together, and was the way to other islands, and from these you might pass to the whole of the opposite continent which surrounded the true ocean; for this sea which is within the Straits of Heracles is only a harbour, having a narrow entrance, but that other is a real sea, and the surrounding land may be most truly called a boundless continent.
>
> Now in this island of Atlantis there was a great and wonderful empire which had rule over the whole island and several others, and over parts of the continent, and, furthermore, the men of Atlantis had subjected the parts of Libya within the columns of Heracles as far as Egypt, and of Europe as far as Tyrrhenia. This vast power, gathered into one, endeavoured to subdue at a blow our country and yours and the whole of the region within the straits; and then, Solon, your country shone forth, in the excellence of her virtue and strength, among all mankind. She was pre-eminent in courage and military skill, and was the leader of the Hellenes. And when the rest fell off from her, being compelled to stand alone, after having undergone the very extremity of danger, she defeated and triumphed over the invaders, and preserved from slavery those who were not yet subjugated, and generously liberated all the rest of us who dwell within the pillars. (*Timaeus*, Jowett translation)

The story is reminiscent of what Athens did against the Persians in the early fifth century B.C.E., but the battle with Atlantis allegedly took place in the eighth or ninth millennium B.C.E. It would not take much of a historical scholar to know that Athens in 9,000 B.C.E. was either uninhabited or occupied by very primitive people. This fact would not have concerned Plato's readers because they would have understood that he was not giving them a historical account of a real city. To assume, as many believers in Atlantis do, that there is a parallel between Homer's *Iliad* and *Odyssey* and Plato's *Critias* and *Timaeus* is simply absurd. Those who think that just as Schliemann found Troy so too will we someday crack Plato's code and find Atlantis are drawing an analogy where they should be drawing the curtains. Plato's purpose was not to pass on stories, but to create stories to teach moral lessons.

Different seekers have located Atlantis in the mid-Atlantic, Cuba, the Andes, and dozens of other places. Some have equated ancient Thera with Atlantis. Thera is a volcanic Greek island in the Aegean Sea that was devastated by a volcanic eruption in 1625 B.C.E. Until then it had been associated with the Minoan civilization on Crete.

To many, however, Atlantis is not just a lost continent. It is a lost world. The Atlanteans were extraterrestrials who destroyed themselves with nuclear bombs or some other extraordinarily powerful device. Atlantis was a place of advanced civilization and technology. Lewis Spence, a Scottish mythologist who used "inspiration" instead of scientific methods, attributes Cro-Magnon cave paintings in Europe to displaced Atlanteans (Feder 2002). Helena Blavatsky, founder of the Theosophical Society in the late 19th century, originated the notion that the Atlanteans had invented airplanes and explosives and grew extraterrestrial wheat. The theosophists also invented Mu, a lost continent in the Pacific Ocean. Psychic healer Edgar **Cayce** claimed to have had **psychic** knowledge of Atlantean texts, which assisted him in his prophecies and cures. J. Z. Knight claims that **Ramtha**, the spirit she **channels**, is from Atlantis.

The serious investigator of the myth of Atlantis must read Ignatius Donnelly's *Atlantis: The Antediluvian World* (1882). Donnelly could have been an inspiration for von Däniken, **Sitchin**, and **Velikovsky**, since he assumes that Plato's myth is true history. Much of the popularity of the myth of Atlantis, however, must go to popular writers such J. V. Luce (*The End of Atlantis*, 1970) and Charles Berlitz, the man who popularized the **Bermuda Triangle** and the discovery of **Noah's Ark**. His *Doomsday, 1999 A.D.* (1981) comes complete with maps of Atlantis. Graham Hancock is doing much to keep alive this tradition of "alternative" and "speculative" history and archaeology, which seeks a single source for ancient civilizations.

These "alternative" archaeologists have credited the Atlanteans with teaching the Egyptians and the Meso-Americans how to build pyramids and how to write, arguing similarly to von Däniken that ancient civilizations burst on the scene in a variety of different places on earth and have a common source. Atlanteans or aliens—either way the case can be made for a common source for ancient civilizations only if one selectively ignores the gradual and lengthy development of those societies. One must also ignore that the writing of the Egyptians is no clue to the writing of the Mayans, or vice-versa, and that the purpose of their pyramids was quite different. The Meso-Americans rarely buried anyone in their pyramids; they were primarily for religious rituals and sacrifices. The Egyptians used pyramids exclusively for tombs or monuments over tombs. Why would the aliens or Atlanteans not teach the same writing techniques to the two cultures? And why teach step building in Meso-America, a technique not favored by the Egyptians? If you ignore the failures of the early pyramid builders and ignore their obvious development over time, including the development of underground tombs with several chambers, then you might be able to persuade uncritical minds that Giza couldn't have occurred without alien intervention (Feder 2002).

Finally, if the Atlanteans were such technological geniuses who shared their wisdom with the world, why did Plato depict them as arrogant warmongers? To paraphrase Whitehead, the belief in Atlantis, the ancient and great civilization, is another footnote to Plato.

Further reading: Asimov, Greenberg, and Waugh 1988; De Camp 1975; R. Ellis 1998; Gardner 1957.

aura

A colored outline, or set of contiguous outlines, allegedly emanating from the surface of an object. Auras are not to be confused with the aureoles or halos of **saints**, which are devices of Christian iconography used to depict the radiance of light associated with divine infusion. In the New Age, even the lowly amoeba has an aura, as does the mosquito and every lump of goat dung. The aura supposedly reflects an occult **energy** field or life force that permeates all things. Human auras allegedly emerge from the **chakras**. Under ordinary circumstances, auras are only visible to certain people with special **psychic** power. However, with a little bit of training, or with a special set of Aura Goggles with "pinacyanole bromide" filters, anyone can see auras. You may also use **Kirlian photography** to capture auras on film. You may not be psychic if you see auras, however. You may have a migraine, a certain form of epilepsy, a visual system disorder, or a brain disorder.

You may also see auras by doing certain exercises. Most aura training exercises involve staring at an object placed against a white background in a dimly lit room. What one sees is due to retinal fatigue and other natural perceptual processes, not the unleashing of hidden psychic powers. Something similar happens when you stare at certain colored or black and white patterns. Vision is not the verbatim recording

of the outside world. When looking at a colored object, for example, the eye does not transmit to the brain a continuous series of duplicate impressions. The brain itself supplies much of the visual perception. Colors don't exist in objects, though we perceive that they do. In short, even if auras are perceived, that is not good evidence that there is an energy field in the physical or supernatural world corresponding to the perceptions.

That auras reflect health is a common notion among true believers. However, there is no consensus on what the colors mean. Edgar **Cayce** not only gave a meaning to each of seven colors and related the colors to possible health disorders, he also connected each color to a note on the musical scale and a planet in the solar system.

When aura readers have been tested under controlled conditions, they have failed to demonstrate that they can even *perceive* an aura, much less interpret it correctly.

See also **aura therapy**.

Further reading: Nickell 2000a; Randi 1995; Rosa et al. 1998; Sacks 1995.

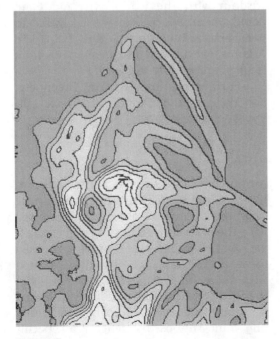

"Kirlian."

aura therapy

A type of New Age healing that detects and treats disease by reading and manipulating a person's **aura**. According to aura therapists, the aura is an **energy** field surrounding the body that exhibits signs of physical disease before the body itself exhibits either signs or disease.

The most popular form of aura therapy is **therapeutic touch**. It is taught in many nursing programs and practiced in many hospitals. Less well known is *aurasomatherapy* or *Aura-Soma*, described by Vicky Wall—who rediscovered this ancient healing practice in the mid-1980s—as "an holistic soul therapy in which the vibrational powers of color, crystals and natural aromas combine with light in order to harmonise body, mind and spirit of mankind" (home.snafu.de/guschi/e_einf.htm).

Beverli Rhodes uses laser **crystal** wand energy in her aura therapy. She says that crystals help in finding "disturbances in the auric field" and that

> by using your laser crystal wand energies and your own energies, which will fuse with that of the wand, you can bring about relief and in time a cure. As crystals have their very own special electromagnetic field [the aura] this can be used to balance our own aura's. As disease appears firstly in the auric field, it would seem logical that one would begin to heal and clear the problem at the source.

It also seems logical to Rhodes that the proper way to assist stressed clients is by concentrating the laser-wand on the third eye (sixth **chakra**) area for 1 minute. She notes that it is "necessary to re-programme your crystal first so that it may ready itself to clear the disharmony that exists in the auric field of the client in order to heal the specific illness."

Dr. J. M. Shah uses **Kirlian photography** and gem therapy to treat heart disease. Like other aura therapists, Dr. Shah believes that when disease enters one of our several bodies, there is reduction in **prana**. He takes Kirlian photographs of the fingers to discover disease. He assumes that changes in the Kirlian photos are due to changes in the aura rather than to changes in moisture or other natural phenomena. Once he has detected disease by photo mis-reading, he uses rubies to "open the heart" of those who have bad hearts. He advises, however, that the rubies have to be energized and their negativity removed before they can be effective in treatment. For supportive medical treatment, he puts photos of his patients in a "radiation cabinet" with rubies.

automatic writing (trance writing)

Writing allegedly directed by a spirit or by the **unconscious mind**. It is sometimes called "trance" writing because it is done quickly and without judgment, writing whatever comes to mind, "without consciousness," as if in a trance. It is believed that this allows one to tap into the subconscious mind, where "the true self" dwells. Uninhibited by the conscious mind, deep and mystical thoughts can be accessed. Trance writing is also used by some psychotherapists who think it is a quick way to release **repressed memories**. There is no compelling scientific evidence that trance writing has any therapeutic value.

Advocates of automatic writing claim that the process allows them to access other intelligences and entities for information and guidance. They further claim that it permits them to recall previously irretrievable data from the subconscious mind and to unleash spiritual energy for personal growth and revelation. According to **psychic** Ellie Crystal, entities from beyond are constantly trying to communicate with us. Apparently, we all have the potential to be as **clairaudient** as James Van Praagh and John Edward.

One 19th-century **medium**, Hélène Smith (Catherine Müller), specialized in automatic writing and even invented a Martian alphabet to convey messages from Mars to her clients in the Martian language. Martian has a strong resemblance to Ms. Smith's native language, French, according to Théodore Flournoy, a psychology professor who investigated her claim (Randi 1995: 22).

Skeptics consider automatic writing to be little more than a parlor game, although sometimes useful for self-discovery and for getting started on a writing project. While it is likely that many unconscious desires

Catherine Müller's Martian writing. From Théodore Flournoy, *From India to the Planet Mars: A Case of Multiple Personality with Imaginary Languages* (1901, English translation by Daniel B. Vermilye).

and ideas are expressed in automatic writing, it is unlikely that they are any more profound than one's conscious notions. There is no more evidence that the true self is in the unconscious than there is that the true self is revealed while drunk or in a psychotic break. Automatic writing may enhance personal growth if it is evaluated reflectively and with intelligence. By itself, automatic writing is no more likely to produce self-growth or worthwhile revelation than any other human activity. In fact, some people have even had such bad experiences doing automatic writing that they are convinced that **Satan** is behind it. For some minds, apparently it is better not to know what's lurking in the cellar. Others may be disappointed to find that the cellar is empty.

See also **channeling, medium, Ouija board, repressed memory therapy**, and **spiritualism**.

Further reading: Rawcliffe 1959.

avatar

A variant phase or version of a continuing basic entity, such as the incarnation in human form of a divine being. Avatar is also the name of a New Age self-help course based on changing a person's life by training the person to manage his or her beliefs.

Avatar awakens you to a natural ability you already have to create and discreate beliefs. With this skill, you can restructure your life according to the blueprint that you determine. One discovery many people on the Avatar course make is that what you are believing is less important than the fact that you are believing it. Avatar empowers you to realize that there aren't "good" beliefs and "bad" beliefs. There are only the beliefs that you wish to experience and the beliefs you prefer not to experience. Through the tools that the course presents you with, you create an experience of yourself as the source, or creator, of your beliefs. From that place, it's very natural and easy to create the beliefs that you prefer. [www.starsedge .com/store/]

These notions seem so obviously a mixture of the true, the trivial, and the false that one hesitates to comment on them. If there are no good or bad beliefs, then how did the people at Avatar come upon the belief that their course has any value? And what difference does it make whether anyone believes in Avatar belief management techniques?

Ayurvedic medicine

Ayurvedic medicine is an **alternative health practice** whose practitioners claim is the traditional medicine of India. Ayurveda is based on two Sanskrit terms: *ayu,* meaning *life,* and *veda,* meaning *knowledge* or *science.* Since the practice is said to be some 5,000 years old, what it considers to be knowledge or science may not coincide with the most updated information available to Western medicine. In any case, most ancient treatments are not recorded and what is called traditional Indian medicine is, for the most part, something developed in the 1980s by the Maharishi Mahesh Yogi (Barrett 1998), who brought **Transcendental Meditation** to the Western world.

Ayurvedic treatments are primarily dietary and herbal. According to Maharishi Ayurveda Products, Inc. (MAPI; www .mapi.com), patients are classified by body types, or *prakriti,* which are determined by proportions of the three *doshas,* which allegedly regulate mind-body harmony. Illness and disease are considered to be a matter of imbalance of *vata, pitta,* and *kapha* in the doshas. Treatment is aimed at restoring harmony or balance to the mind-body system.

Deepak Chopra is the foremost advocate of Ayurvedic medicine in America. He is a graduate of Harvard Medical School and a former leader of the Maharishi Mahesh Yogi's Transcendental Meditation program. Chopra claims that perfect health is a matter of choice. He says he can identify your *dosha* and its state of balance or imbalance simply by taking your pulse, and that allergies are usually caused by poor digestion. He claims you can prevent and reverse cataracts by brushing your teeth, scraping your tongue, spitting into a cup of water, and washing your eyes for a few minutes with this mixture. According to Chopra, "contrary to our traditional notions of aging, we can learn to direct the way our bodies metabolize time" (Wheeler: www .hcrc.org/contrib/wheeler/chopra.html).

Chopra also promotes **aromatherapy** based on the Ayurvedic metaphysical physiology. He sells oils and spices specifically aimed at appeasing *vata, pitta,* or *kapha,* but mostly he sells *hope:* hope to the dying that they will not die and hope to the living that they can live forever in perfect health. He says that it is unnecessary to do scientific tests of Ayurvedic claims since "the masters of Ayurvedic medicine can determine an herb's medicinal qualities by simply looking at it" (Wheeler). He is also fond of saying such things as

If you can wiggle your toes with the mere flicker of an intention, why can't you reset your biological clock?

If you could live in the moment you would see the flavor of eternity and when you metabolize the experience of eternity your body doesn't age.

Ayurveda is the science of life and it has a very basic, simple kind of approach, which is that we are part of the universe and the universe is intelligent and the human body is part of the cosmic body, and the human mind is part of the cosmic mind, and the atom and the universe are exactly the same thing but with different form, and the more we are in touch with this deeper reality, from where everything comes, the more we will be able to heal ourselves and at the same time heal our planet.

What any of this means is anyone's guess. His mind-body claims get even murkier as he tries to connect Ayurveda with quantum physics. He calls the connection "quantum healing":

> Quantum healing is healing the body-mind from a quantum level. That means from a level which is not manifest at a sensory level. Our bodies ultimately are fields of information, intelligence and energy. Quantum healing involves a shift in the fields of energy information, so as to bring about a correction in an idea that has gone wrong. So quantum healing involves healing one mode of consciousness, mind, to bring about changes in another mode of consciousness, body.

According to Chopra, "We are each a localized field of **energy** and information with cybernetic feedback loops interacting within a nonlocal field of energy and information." He claims we can use "quantum healing" to overcome aging. He believes that the mind heals by harmonizing or balancing the "quantum mechanical body" (his term for **chi** or **prana**). He says, "[S]imply by localizing your awareness on a source of pain, you can cause healing to begin, for the body naturally sends healing energy wherever attention is drawn." "If you have happy thoughts," says Chopra, "then you make happy molecules." This "quantum mysticism" has no basis in modern physics or medicine (Barrett 1998; Stenger 1997b).

The notion that ancient Hindu mysticism is just quantum physics wrapped in metaphysical garb seems to have originated with Fritjof Capra in his book *The Tao of Physics* (1975). The book has inspired numerous New Age energy medicine advocates to claim that quantum physics proves the reality of *prana* and **ESP**. The idea that there is such a connection is denied by most physicists, but books like Capra's and Gary Zukav's *The Dancing Wu Li Masters: An Overview of the New Physics* (1976) overshadow and are much more popular than more sensible books written by physicists such as Heinz R. Pagels's *The Cosmic Code: Quantum Physics as the Language of Nature* (1982). Pagels vehemently rejects the notion that there is any significant connection between the discoveries of modern physicists and the metaphysical claims of Ayurveda. "No qualified physicist that I know would claim to find such a connection without knowingly committing fraud," says Dr. Pagels.

The claim that the fields of modern physics have anything to do with the "field of consciousness" is false. The notion that what physicists call "the vacuum state" has anything to do with consciousness is nonsense.

Reading these materials authorized by the Maharishi causes me distress because I am a man who values the truth. To see the beautiful and profound ideas of modern physics, the labor of generations of scientists, so willfully perverted provokes a feeling of compassion for those who might be taken in by these distortions. I would like to be generous to the Maharishi and his movement because it supports world peace and other high ideals. But none of these ideals could possibly be realized within the framework of a philosophy that so will-

fully distorts scientific truth. [www
.trancenet.org/research/pagels.shtml]

In short, there isn't a quantum of truth in
quantum healing.

Meditation is also a significant therapy in Ayurveda. But except for the benefits of relaxation, there is no scientific evidence to support any of the many astounding claims made on behalf of Ayurvedic medicine. Even the claims made for the significant health benefits of **Transcendental Meditation** have been greatly exaggerated and distorted (Wheeler).

As would be expected of a guru spreading false hope, Chopra's trustworthiness has been compromised. In 1991, when president of the American Association of Ayurvedic Medicine, Chopra submitted a report to the *Journal of the American Medical Association,* along with Hari M. Sharma, M.D., professor of pathology at Ohio State University College of Medicine, and Brihaspati Dev Triguna, an Ayurvedic practitioner in New Delhi, India. Chopra, Sharma, and Triguna claimed they were disinterested authorities and were not affiliated with any organization that could profit by the publication of their article. But

they were intimately involved with the complex network of organizations that promote and sell the products and services about which they wrote. They misrepresented Maharishi Ayur-Veda as India's ancient system of healing, rather than what it is, a trademark line of "alternative health" products and services marketed since 1985 by the Maharishi Mahesh Yogi, the Hindu swami who founded the Transcendental Meditation (TM) movement. (Skolnick 1991)

Furthermore, Chopra has also admitted in so many words that his *Ageless Body, Timeless Mind: The Quantum Alternative to Grow-ing Old* plagiarized Professor Robert Sapolsky's contribution to *Behavioral Endocrinology.* Sapolosky is the author of chapter 10, "Neuroendocrinology of the Stress-Response." He sued Chopra in 1997 for lifting large chunks of his work without proper attribution.

Chopra spends much of his time writing and lecturing from his base in California. He charges $25,000 per lecture performance, giving spiritual advice while warning against the ill effects of materialism. Chopra is much richer and certainly more famous than he ever was as an endocrinologist or as chief of staff at New England Memorial Hospital. He now runs the Chopra Center for Well Being in La Jolla, where the mission is "to heal, to love, to transform and to serve." It is not a medical center, for Chopra has no license to practice medicine in California. It is a *spiritual* center, where you can come to "better understand the power of your body, mind and spirit connection to both your inner and outer universe." Because many of those who come to this center are sick, one might call it a *faith healing* center.

It is understandable that he would give up working in medicine in favor of working in religion. In medicine you are surrounded by sick people and constantly reminded of your own mortality. It is difficult work, often very stressful and unrewarding. As Chopra himself put it: "It's frustrating to see patients again and again, and to keep giving them sleeping pills, tranquilizers and antibiotics, for their hypertension or ulcers, when you know you're not getting rid of the problem or disease" (Redwood 1995). Also, while taking care of others, a physician might fail to take care of himself and come to require sleeping pills, tranquilizers, something to lower the blood pressure and relieve the stress in himself. In religion, on the other hand, you can surround yourself with sycophants. By turning to religion instead of biology, one avoids the risk of being proved wrong. It is much easier to dispense hope based on nothing to miserable people than it is to accept harsh and sometimes brutal reality while maintaining health, optimism, and happiness. It is much easier to find confirming evidence for a world-view than it is to do nuts-and-bolts research. It is certainly much more enjoyable to chat with Oprah Winfrey and rub elbows with the rich and famous than to watch another cancer patient die.

The road to La Jolla and international fame, however, was a long one that began on the east coast. Chopra left behind conventional medicine in 1981 when Triguna convinced him that if he didn't make a change he'd get heart disease. Shortly after that he got involved in Transcendental Meditation. In 1984 he met the Maharishi himself and in 1985 he became director of the Maharishi Ayurveda Health Center for Stress Management in Lancaster, Massachusetts. Soon he was an international purveyor of herbs and tablets through MAPI. When association with TM itself became too stressful and a hindrance to his success, he left. (Chopra had heard that Bill Moyers wouldn't include him in his PBS series "Healing and the Mind" because of Chopra's association with a "cult.")

Finally, perhaps the greatest deception of Ayurveda is its claim that it *cares for the person,* not just the *body* as conventional medicine does. As Chopra puts it, "The first question an Ayurvedic doctor asks is not, 'What disease does my patient have?' but, 'Who is my patient?' " That may be the question, but it is not a person that the doctor is healing. It is the *quantum body* or the *mind-body;* it is the *dosha* that needs balancing. Taking a person's pulse and telling them their *dosha* is unbalanced and they should eat more nuts or less spicy foods, for example, hardly shows concern for the patient as a person. Furthermore, even though his patients died while he was claiming he had given them perfect health, Chopra maintained his position (Barrett 1998). I suppose what this tells us is that it is the person, not the disease, that carries the purse or the wallet.

Further reading: Ankerberg and Weldon 1996; Butler 1992; Stalker and Glymour 1989.

Aztec (New Mexico) UFO hoax

This hoax was publicized by *Variety* columnist Frank Scully (1892–1964), who was himself hoaxed by two con men, Silas M. Newton and Leo A. Gebauer. Scully liked the hoax so much he wrote a book based on it: *Behind the Flying Saucers* (1950). Scully claimed that a UFO had landed in Hart Canyon, 12 miles northeast of Aztec, in March 1948. Sixteen humanoid bodies were discovered at the crash site inside a metal disk that was 99.99 (not 100) feet in diameter. A conspiring military secretly removed the craft and the bodies for their sinister research. Interestingly, no one in the area noticed the crash or the military activity. With no witnesses, Newton and Gebauer could play wildly with the truth.

Newton and Gebauer were involved in oil exploration finance schemes. Their hoax was perpetrated to get investors. They claimed they had built a machine that would find oil and natural gas deposits using alien technology. J. P. Cahn of the *San Francisco Chronicle* had some of the "alien" metal tested and determined it was aluminum. Cahn's account of the phony alien ship appeared in *True* magazine in 1952. Several people who had been swindled by Newton and Gebauer came forward. One of their victims, Herman Glader, a millionaire from Denver, pressed charges, and the pair was convicted of fraud and related charges in 1953. (They had charged

$18,500 for a "tuner" that could be bought at surplus stores for $3.50 at the time.)

William Steinman and Wendelle Stevens revived the Aztec story in 1986 in their privately published book *UFO Crash at Aztec*. It was revived again in 1998 when Linda Mouton Howe claimed she had government documents that proved the Aztec crash. What she had was a rumor eight times removed from the source. Silas Newton told George Koehler about 3-foot-tall aliens and their saucer; Koehler told Morley Davies, who told Jack Murphy and I. J. van Horn, who told Rudy Fick, who told the editor of the *Wyandotte Echo* in Kansas City where it was read by an Air Force agent in the Office of Special Investigations, who passed on the story to Guy Hottel of the FBI, who sent a memo to J. Edgar Hoover (Thomas 1998).

The citizens of Aztec have seen how **Roswell** has turned **UFO** mania into a profitable tourist attraction and have followed suit. In March 2002 they celebrated their 5th annual Aztec UFO Festival. The festival was started as a way to raise money for the town's library, to support, one hopes, the nonfiction section.

See also **alien abductions** and Area 51.

Further reading: Klass 1997b; Peebles 1994; Pflock 2000; Saler 1997.

B

Bach's flower therapy

A type of **aromatherapy** developed in the 1930s by British physician Edward Bach (1886–1936). Bach claimed to have psychically or intuitively discovered the healing effects of 38 wildflowers. Each of the 38 flowers of the Bach system is used to balance specific emotional pains or to abate physical symptoms. His discoveries were arrived at by inspirations. For example, while on a walk he had an inspiration that dewdrops on a plant heated by the sun would absorb healing properties from the plant. He claimed that all he needed to do was hold a flower or taste a petal and he could intuitively grasp its healing powers. From these intuitions he went on to prepare essences using pure water and plants.

Bach claimed that wildflowers have a **soul** or **energy** with an affinity to the human soul. The flower's spiritual energy is transferable to water. Devotees drink a **homeopathic** concoction of flower essence, mineral water, and brandy in order to get the flower soul to harmonize their own soul's energy. Bach thought that illness is the result of a contradiction between the soul and the personality. This internal war leads to negative moods and energy blocking, causing lack of harmony and physical diseases.

Backster effect

The alleged power of plants to understand human thought by reading *bioenergetic fields*. It is named after Cleve Backster for his work on **plant perception**.

backward satanic messages (backmasking)

Backward satanic messages are allegedly inserted in some musical recordings. For example, Led Zeppelin's "Stairway to Heaven" has a lyric that when played backward allegedly says: "Here's to my sweet Satan. The one whose little path would make me sad, whose power is Satan. He'll give you 666, there was a little tool shed where he made us suffer, sad Satan."

Madonna's "Like a Prayer" has the lyric "Life is a mystery," which allegedly is "O hear our savior Satan" when heard backward.

It is likely that many backward listeners are hearing what they want to hear or are hearing what others tell them they will hear.

The belief in the existence and efficacy of backward satanic messages may derive from the ancient practice of mocking Christianity by saying prayers backwards at the **witches'** Sabbath. The belief in backmasking is mainly popular among certain fundamentalist preachers who cannot look at anything without wondering how **Satan** is involved.

See also Aleister **Crowley**, **pareidolia**, and **reverse speech**.

ball lightning

A luminous sphere that seems to appear out of nowhere and vanish into thin air. It varies in size from 2 to 10 inches in diameter. It is usually seen shortly before, after, or during a thunderstorm. Its duration varies from a few seconds to a few minutes.

> The lifetime of ball lightning tends to increase with size and decrease with brightness. Balls that appear distinctly orange and blue seem to last longer than average. . . . Ball lightning usually moves parallel to the earth, but it takes vertical jumps. Sometimes it descends from the clouds, other times it suddenly materializes either indoors or outdoors or enters a room through a closed or open window, through thin nonmetallic walls or through the chimney. [Peter H. Handel: www.sciam.com/askexpert/physics/physics30.html]

Some have speculated that ball lightning is a plasma ball, but that theory has been dismissed because a "hot globe of plasma should rise like a hot-air balloon" and that is not what ball lightning does. Many physicists have speculated that ball lightning must be due to electrical discharges. For example, Russian physicist Pyotr Kapitsa thinks ball lightning is an electrodeless discharge caused by UHF waves of unknown origin present between the earth and a cloud. According to another theory, "Outdoor ball lightning is caused by an atmospheric maser—analogous to a laser, but operating at a much lower energy—having a volume of the order of many cubic kilometers" (ibid.).

Two New Zealand scientists, John Abrahamson and James Dinniss, believe ball lightning consists of "fluffy balls of burning silicon created by ordinary fork lightning striking the earth."

> According to their theory, when lightning strikes the ground, minerals are broken down into tiny particles of silicon and its compounds with oxygen and carbon. The tiny charged particles link up into chains, which go on to form filamentary networks. These cluster together in a light fluffy ball, which is borne aloft by air currents. There, it hovers as ball lightning or a burning orb of fluffy silicon emitting the energy absorbed from the lightning in form of heat and light, until the phenomenon burns itself out. [BBC News: news.bbc.co.uk/hi/english/sci/tech/newsid_628000/628709.stm]

Ball lightning has been observed since ancient times and by thousands of people in many different places. Most physicists seem to believe that there is little doubt that it is a real phenomenon. But there is still disagreement as to what it is and what causes it.

Further reading: Prenn, 1991; Stenhoff 2000.

Barnum effect

See **Forer effect**.

begging the question

A fallacious form of arguing due to assuming what one claims to be proving. The following argument begs the question: "We know God exists because we can see the perfect order of His Creation, an order that demonstrates supernatural intelligence in its design." The conclusion of this argument is that God exists. The premise assumes a creator and designer of the universe exists, that is, that God exists. In this argument, the arguer should not be granted the assumption that the universe exhibits **intelligent design**, but should be made to provide support for that claim.

The following is another example of begging the question: "Paranormal phenomena exist because I have had experiences that can only be described as paranormal." The conclusion of this argument is that paranormal phenomena exist. The premise assumes that the arguer has had paranormal experiences, and therefore assumes that paranormal experiences exist.

Here is a final example of begging the question: "Past-life memories of children prove that past lives exist because the children could have no other source for their memories besides having lived in the past." The conclusion of this argument is that past lives exist. The premise assumes that children have had past lives.

Bermuda (or Devil's) Triangle

A triangular area in the Atlantic Ocean bounded roughly at its points by Miami, Bermuda, and Puerto Rico. Vincent Gaddis was the first to call the area by this name in "The Deadly Bermuda Triangle," which appeared in the February 1964 issue of *Argosy,* a magazine devoted to fiction.

Legend has it that many people, ships, and planes have mysteriously vanished in this area. How many have mysteriously disappeared depends on who is doing the locating and the counting. The size of the triangle varies from 500,000 square miles to three times that size, depending on the imagination of the author. (Some include the Azores, the Gulf of Mexico, and the West Indies.) Some trace the mystery back to the time of Columbus. Even so, estimates range from about 200 to no more than 1,000 incidents in the past 500 years. Howard Rosenberg of the Naval Historical Center claims that in 1973 the U.S. Coast Guard answered more than 8,000 distress calls in the area and that more than 50 ships and 20 planes have gone down in the Bermuda Triangle within the last century.

Skeptics believe that there is no mystery to be solved and nothing that needs explaining. Given its size, location, and the amount of traffic it receives, the number of wrecks in this area is not extraordinary. Investigations to date have not produced convincing evidence of any unusual phenomena involved in the disappearances. Many of the ships and planes that have been identified as having disappeared mysteriously in the Bermuda Triangle were not in the Bermuda Triangle at all. The real mystery is how the Bermuda Triangle became a mystery.

The modern legend of the Bermuda Triangle began soon after five Navy planes (Flight 19) vanished on a training mission during a severe storm in 1945. The most logical theory as to why they vanished is that lead pilot Lt. Charles Taylor's compass failed. The trainees' planes were not equipped with working navigational instruments. The group was disoriented and simply, though tragically, ran out of fuel. No mysterious forces were likely to have been involved other than the mysterious force of gravity on planes with no fuel. One of the rescue planes blew up shortly after takeoff, but this was likely due to a faulty gas tank rather than to any mysterious forces.

Over the years there have been dozens of articles, books, and television programs promoting the mystery of the Bermuda Triangle. In his study of this material, Larry Kushe found that few did any investigation into the mystery. Rather, they passed on the speculations of their predecessors as if they were passing on the mantle of truth. There have been many uncritical accounts of the mystery of the Bermuda Triangle, but no one has done more to muddy the waters than Charles Berlitz, who had a bestseller on the subject in 1974. After examining the over 400-page-long official report of the Navy Board of Investigation on the disappearance of the navy planes in 1945, Kushe found that the board wasn't baffled at all by the incident. Furthermore, the board did not mention alleged radio transmissions cited by Berlitz in his book. According to Kushe, what Berlitz doesn't misinterpret, he fabricates. Kushe writes: "If Berlitz were to report that a boat were red, the chance of it being some other color is almost a certainty" (Kushe 1995).

In short, the mystery of the Bermuda Triangle became a mystery by a kind of **communal reinforcement** among uncritical authors and a mass media willing to pass on uncritically the speculation that something mysterious is going on in the Atlantic. The theories run the gamut from evil extraterrestrials to residue crystals from **Atlantis** to evil humans with anti-gravity devices or other weird technologies. Some blame vile vortices from the fourth dimension. Others blame strange magnetic fields and oceanic flatulence

(methane gas from the bottom of the ocean). There have been some accidents in the area, of course, but weather (thunderstorms, hurricanes, tsunamis, earthquakes, high waves, currents, etc.), bad luck, pirates, explosive cargoes, incompetent navigators, and other natural and human causes are preferred explanations among skeptical investigators.

Further reading: Randi 1982a; Rosenberg 1974.

Bible Code

A code allegedly embedded in the Bible by God. The code is revealed by searching for equidistant letter sequences (ELS). For example, start with any letter ("*L*") and read every *n*th letter ("*N*") thereafter in the book, not counting spaces. If an entire book, such as Genesis, is searched, the result is a long string of letters. Using different values for *L* and *N*, one can generate many strings of letters. Imagine wrapping the string of letters around a cylinder in such a way that all the letters can be displayed. Flatten the cylinder to reveal several rows with columns of equal length, except perhaps the last column, which might be shorter than all the rest. Now search for meaningful names in proximity to dates. Search horizontally, vertically, diagonally, any which way. A group of Israeli mathematicians did just this and claimed that when they searched for names in close proximity to birth or death dates (as published in the *Encyclopedia of Great Men in Israel*), they found many matches, for example, the date of the assassination of Yitzhak Rabin was in close proximity to letters spelling out his name. Doron Witztum, Eliyahu Rips, and Yoav Rosenberg (1994) published their findings under the title "On Equidistant Letter Sequences in the Book of Genesis." The editor of the journal commented:

When the authors used a randomization test to see how rarely the patterns they found might arise by chance alone they obtained a highly significant result, with the probability $p = 0.000016$. Our referees were baffled: their prior beliefs made them think the Book of Genesis could not possibly contain meaningful references to modern-day individuals, yet when the authors carried out additional analyses and checks the effect persisted.

The probability of getting the results they did was 16 out of one million, or one out of 62,500. The authors state: "Randomization analysis shows that the effect is significant at the level of 0.00002 [and] the proximity of ELS's with related meanings in the *Book of Genesis* is not due to chance." Harold Gans, a former cryptologist at the U.S. Defense Department, replicated the work of the Israeli team and agreed with their conclusion. Witzum later claimed that, according to one measure, the probability of getting these results by chance is one in 4 million. He has apparently changed his mind and now claims that the probability $p = 0.00000019$ (one out of 5.3 million).

As further evidence of the statistical significance of their results, the Israeli team analyzed the Hebrew version of the book of Isaiah and the first 78,064 characters of a Hebrew translation of Tolstoy's *War and Peace*. They found many names in close proximity to birth or death dates, but the results were statistically insignificant. (The book of Genesis used in their study, the Koren version, has 78,064 characters.)

What does this all mean? To some it means that the patterns in Genesis are intentional and that God is the ultimate author of the code. If so, should the book of Isaiah, and any other book in the Bible that fails the ELS test, be dumped? Should we conclude that these statistics verify the claim that the Jews are the chosen people

of God, or that no more names should be added to list of Great Men in Israel unless they pass the ELS test? Unless other religions can duplicate such statistically improbable results, the mathematically minded supernaturalist might well consider them to be imposters. Should we translate all the sacred books of all the religions of the world into Hebrew and see how many great men of Israel are encoded there?

Can a computer really read the mind of God? Apparently. For on this theory God dictated in His favorite language, Hebrew, a set of words that are more or less intelligible if taken at face value, containing stories of creation, floods, fratricide, wars, miracles, and so on, with many moral messages. But this Hebrew God chose his words carefully, encoding the Bible with prophecies and messages of absolutely no religious value.

Many, however, are not at a loss at all. Some Christian "creation scientists" claim the Bible Code provides scientific proof of God's existence. If they are right, they should convert to Judaism. Doron Witztum can't do that, since he is already a Jew. But he has taken the work done on Genesis a bit further than his colleagues. Witztum went on Israeli television and claimed that the names of the subcamps on a map of Auschwitz appeared remarkably close to the phrase "in Auschwitz." The odds of such occurring, he said, are "one in a million." Some of his students did the math and claim their mentor was off by "a factor of 289,149." Witztum's math may not be as good as his intentions, but it is difficult to see what those intentions might be. Was God revealing in an odd way that the subcamps of Auschwitz are in Auschwitz?

Michael Drosnin has written a book based on the ELS study. He claims in *The Bible Code* that decoding the Bible allegedly leads to the discovery of prophecies and profound truths of a secular nature, not all of which are related to the Jews. Drosnin claims that the Bible is the only text in which these encoded phrases are found in a statistically significant pattern, and that the chance of this being a random phenomenon is unlikely. Using the ELS method, Drosnin claims that the assassination of Yitzhak Rabin was foretold in the Bible. He also claims that the assassinations of Anwar Sadat and the Kennedy brothers are encoded in biblical ELS.

Not everybody agrees with the Drosnin hypothesis, including Harold Gans, a retired Defense Department cryptologist who corroborated the work of Witztum, Rips, and Rosenberg. Gans says that the

book states that the codes in the Torah can be used to predict future events. This is absolutely unfounded. There is no scientific or mathematical basis for such a statement, and the reasoning used to come to such a conclusion in the book is logically flawed. While it is true that some historical events have been shown to be encoded in the Book of Genesis in certain configurations, it is absolutely not true that every similar configuration of "encoded" words necessarily represents a potential historical event. In fact, quite the opposite is true: most such configurations will be quite random and are expected to occur in any text of sufficient length. Mr. Drosnin states that his "prediction" of the assassination of Prime Minister Rabin is "proof" that the "Bible Code" can be used to predict the future. A single success, regardless of how spectacular, or even several such "successful" predictions proves absolutely nothing unless the predictions are made and evaluated under carefully controlled conditions. Any respectable scientist knows that "anecdotal" evidence never proves anything.

Dr. Eliyahu Rips, one of the authors of the study that started the Bible Code craze, has also made a public statement regarding Drosnin's *Bible Code:*

I do not support Mr. Drosnin's work on the Codes, nor the conclusions he derives. . . . All attempts to extract messages from Torah codes, or to make predictions based on them, are futile and are of no value. This is not only my own opinion, but the opinion of every scientist who has been involved in serious Codes research.

Professor Menachem Cohen, a celebrated Bible scholar at Bar-Ilan University, has criticized Witzum et al. on two counts: (1) There are several other Hebrew versions of Genesis for which ELS does not produce statistically significant results; and (2) the appellations given to the Great Men in Israel was inconsistent and arbitrary. Other critics, such as Brendan McKay, have done their own analysis of *War and Peace* with remarkably different results than those reported by Witztum et al. Many critics, however, have done little more than use ELS to find names, dates, and so on in various books, a feat already known by even the weakest of statisticians to be unremarkable. Drosnin once said, "When my critics find a message about the assassination of a prime minister encrypted in *Moby–Dick,* I'll believe them." McKay promptly produced an ELS analysis of *Moby–Dick* predicting not only Indira Ghandi's assassination, but also the assassinations of Martin Luther King Jr., John F. Kennedy, Abraham Lincoln, and Yitzhak Rabin, as well as the death of Diana, Princess of Wales. Mathematician David Thomas did an ELS on Genesis and found the words "code" and "bogus" close together not once but 60 times. What are the odds of that happening? Does this mean that God put in a code to reveal that there is no code?

Further reading: Bar-Natan et al. 1997; McKay: cs.anu.edu.au/people/bdm/dilugim/index.html; Thomas 1997.

Bigfoot

An apelike creature reportedly sighted hundreds of times around the world since the mid-19th century. The creature is variously described as standing 7 to 10 feet (2 to 3 meters) tall and weighing over 500 pounds (225 kilograms), with footprints 17 inches (43 centimeters) long. The creature goes by many names, but in northern California it is known as "Bigfoot." (It is also known as the Abominable Snowman of the Himalayas, Mapinguari [the Amazon], Sasquatch, Yowie [Australia], and Yeti [Asia]). The creature is big business in the Pacific Northwest along a stretch of US 101 in southern Humboldt County known as the Redwood Highway. Numerous shops line the roadway, each with its own Bigfoot chainsaw-carved out of majestic redwood.

Most scientists discount the existence of Bigfoot because the evidence supporting belief in the survival of a prehistoric bipedal apelike creature of such dimensions is scant. The evidence consists mainly of **testimony** from Bigfoot enthusiasts, footprints of questionable origin, and pictures that could easily have been of apes or humans in ape suits. There are no bones, no scat, no artifacts, no dead bodies, no mothers with babies, no adolescents, no fur, no explanation for how a species likely to be communal has never been seen in a family or group activity. There is no evidence that any individual, much less a community of such creatures, dwells anywhere near any of the "sightings." In short, the evidence points more toward hoaxing and delusion than real discovery. Some believers dismiss all such criticism and claim that Bigfoot exists in another dimension and travels by **astral projection**. Such

claims reinforce the skeptic's view that the Bigfoot legend is a function of passionate fans of the **paranormal**, aided greatly by the mass media's eagerness to cater to such enthusiasm.

In addition to the eyewitness testimonials of enthusiastic fans, the bulk of the evidence provided by proponents of Bigfoot consists of footprints and film. Of the few footprints available for examination in plaster casts, there is such great disparity in shape and configuration that the evidence "suggests many independent pranksters" (M. Dennett, 1996).

Probably the most well-known evidence for belief in Bigfoot's existence is a film shot by Bigfoot hunters Roger Patterson and Bob Gimlin on October 20, 1967, at Bluff Creek in northern California. The film depicts a walking apelike creature with pendulous breasts. Its height is estimated at between 6'6" and 7'4" and its weight at nearly one ton. Over thirty years have passed, yet no **cryptozoologist** has found further evidence of the creature near the site except for one alleged footprint.

The North American Science Institute claims it has spent over $100,000 to prove

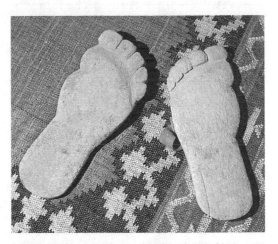

Ray Wallace's carved wooden feet for making Bigfoot tracks. Photo by Dave Rubert.

the film is of a genuine Bigfoot. However, according to veteran Hollywood director John Landis, "that famous piece of film of Bigfoot walking in the woods that was touted as the real thing was just a suit made by John Chambers," who helped create the ape suits in *Planet of the Apes* (1968). Howard Berger, of Hollywood's KNB Effects Group, also has claimed that it was common knowledge within the film industry that Chambers was responsible for a hoax that turned Bigfoot into a worldwide cult. According to Mark Chorvinsky, Chambers was also involved in another Bigfoot hoax (the so-called Burbank Bigfoot). According to Loren Coleman, however, Chambers denied the allegations about the Patterson hoax in an interview with Bobbie Short and claimed that Landis had in fact started the rumor about Chambers making the suit. Apparently, Short did not ask Chambers about the "Burbank Bigfoot" incident, nor did he interview Landis for his version of the story (Chorvinsky 1996). Short and Coleman remain convinced that the film is not of a human in an ape suit but is footage of a genuine Bigfoot.

According to David J. Daegling and Daniel O. Schmitt (1999), "it is not possible to evaluate the identity of the film subject with any confidence." Their argument centers on uncertainties in subject and camera positions, and the reproducibility of the compliant gait by humans matching the speed and stride of the film subject.

According to Michael Wallace, Bigfoot is a hoax that was launched in August 1958 by his father Ray L. Wallace (1918–2002), an inveterate prankster. Shortly after Ray's death, Michael revealed the details of the hoax, which were reported widely in the press. Ray had a friend carve him 16-inch-long feet that he could strap on and make prints with. Wallace owned a construction company that built logging roads at the time and he set the prints around one of his bulldozers in Humboldt County. Jerry Crew, a bulldozer operator, reported the prints and *The Humboldt Times* ran a front-page story about "Bigfoot." The legend was born. However, a former logger, 71-year-old John Auman, claims that Wallace left the giant footprints to scare away thieves and vandals who'd been targeting his vehicles. His hoaxes didn't begin until after he'd seen what a stir he'd created.

Over the years, Ray Wallace produced Bigfoot audio recordings, films, and photographs. At one time, he even put out a press release offering $1 million for a baby Bigfoot. He published one of his photos as a poster depicting Bigfoot having lunch with other animals. He also published photos and films of Bigfeet eating elk, frogs, and cereal. Michael Wallace claims that his mother told him that she participated in some of the pranks and had been photographed in a Bigfoot suit. Chorvinsky claims that Ray told him that the Patterson film was a hoax and that he had alerted Patterson of the sighting at Bluff Creek. According to Chorvinsky, Ray knew who was in the Patterson suit, but said he had nothing to do with it (Bob Young, "Lovable Trickster Created a Monster with Bigfoot Hoax," The *Seattle Times,* December 5, 2002).

The news of Wallace's 1958 hoax did not daunt Bigfoot enthusiasts such as Idaho State University anatomy professor Dr. Jeff Meldrum, who has casts of 40 to 50 big footprints. Meldrum believes such a large number of casts couldn't all be hoaxes (ibid.). The same has been said about the large number of **crop circles**, but it appears that hoaxers are not deterred from their activities by the belief that their numbers are small.

The interest in Bigfoot seems to have been succinctly captured in the saying of an old Sherpa: *There is a Yeti in the back of everyone's mind; only the blessed are not haunted by it.*

Further reading: Dennett 1982; Napier 1972; Randi 1982a, 1995.

bio-ching

A type of **New Age psychotherapy** that unites **biorhythms** with the *I Ching*. Bio-ching was created by Roderic Sorrell, D.D., and Amy Max Sorrell, D.D., who describe themselves as "therapists" on their web site. The Sorrells use a computer program that produces an electronic fortune cookie from the *I Ching* for each of 512 biorhythmic combos.

They prepared for their great innovation by living on a houseboat in San Francisco Bay for several years, "sampling the New Age Emporium that is California." There they learned of "the meridian energy of **acupuncture**, the power of the deep massage of **Rolfing**," and "the esoteric practices of Taoist meditation." They studied "herbal healing" and "were introduced to the newly emerging electronic approaches to the mind: sound and light stimulation of the mind's beta, **alpha**, theta, and delta waves, and biofeedback." On the side, they became **reiki** masters.

The Sorrells describe themselves as deliriously happy and at peace with the world in their home in Truth or Consequences, New Mexico. They want to make others happy, too, and help them achieve inner tranquility. They say they want to fix physical or spiritual inadequacies others might have. They offer private retreats (minimum of 3 days) for couples, partners, and friends. In addition to spiritual counseling, they offer room and board for a set fee per day per person. For a few hundred dollars more they offer a 3-day retreat on their houseboat at a nearby lake. While there, ask them about their union of water and reiki to form the new therapy of "aqua-reiki." Also ask about their "sound and light machine,"

their "bio-feedback machine," and their "**subliminal** tapes."

Bio-ching might well be called a *folie à deux deux*.

bioharmonics

Bioharmonics, according to its inventor, Linda Townsend, is "[t]he science that studies bioenergy motions and interactions with other energy sources." "If someone is eating a disharmonic diet," says Townsend on her web site www.bioharmonics.com, "then there is no harmony in the bioenergy." However, she does not understand *bioenergy* the way biochemists do. In conventional biochemistry, bioenergy refers to "the readily measurable exchanges of energy within organisms, and between them and their environment, which occur by normal physical and chemical processes" (Stenger 1999). Townsend writes (all quotes from Townsend are from her web site):

> In my personal research with bioenergy testing, I always find an irregularity that seems to be related to the physical condition whatever that condition may be. This has raised some questions about bioenergy being an expression of biochemistry. Therefore, if bioenergy and biochemistry have a mutual influence on each other, correcting bioenergy irregularities may also effect balancing the biochemistry.

These are not the claims of someone knowledgeable about biochemistry.

Townsend never quite defines "bioenergy," but her theory is that it needs to be "harmonized." She sells a Harmonizer for $1,295, which will help "to retune those weaken [*sic*] disharmonious areas of the body commonly found over sites of illnesses." She also recommends polarizers ($80 to $120) and **magnets**. The polarizers are "non-magnetic devices filled with kelp, other plant life and minerals specifically chosen for their ability to attract cosmic light **energy**, also called 'chi' or 'life force' energy." How she knows kelp attracts **chi** is not clear.

Ms. Townsend hasn't published any studies, but she implies that the Harmonizer can "help" with many ailments, including cancer, diabetes, heart conditions, Parkinson's disease, and paralysis. She says she has **testimonials** to back up these claims, though she is careful to disclaim any medical benefit for her products:

> We do not claim that any medical conditions have been improved by BioHarmonics; we only have seen that bioenergy imbalances can be improved. . . . This research is not medically related in any way and should not be considered in any matter to be beneficial for medical conditions. . . . There are no guarantees offered expressed or implied.

Presumably, she thinks such disclaimers protect her from lawsuits or from being criminally charged with practicing medicine without a license. She also claims that her Harmonizer is better than others because her harmonics are in twos and other devices are in threes and "the main harmonics of the bioenergy of the body are in twos." How she knows this, or what it even means, is not clear.

Townsend makes numerous unsubstantiated, meaningless, or inane claims such as the color blue "dominates the left side of a healthy body in the outer bioenergy layer and is found in the blood bioenergy. It is also found at the nerve branches of several vertebra in the spine." And "Red dominates the right side of the body in the outer bioenergy layer. It is opposite and attracting to Blue." She seems to have derived these notions from one of the great American quacks, Dinshah P. Ghadiali

(1873–1966), who invented Spectro-Chrome Therapy. She claims he influenced her early theories. According to the Bureau of Investigation of the American Medical Association, November 1935, Ghadiali claimed that the human body is composed of oxygen, hydrogen, nitrogen, and carbon. He called these *elements* and claimed that each "exhibits a preponderance of one or more of the seven prismatic colors," although four colors are dominant: blue, red, green, and yellow. In the healthy body, "the four colors are properly balanced; when they get out of balance we are diseased; ergo, to cure disease administer the lacking colors or reduce the colors that have become too brilliant" (www.mtn.org/~quack/amquacks/ghadiali.htm). He made quite a bit of money from his device until he was arrested for introducing a misbranded article into interstate commerce, found guilty on 12 counts, and sentenced to five years' probation. At his trial, he produced a witness whom he had allegedly cured of seizures. Unfortunately for Ghadiali, the witness had a seizure while on the witness stand (Schwarcz 2000).

Further reading: Beyerstein 1997.

biorhythms

Rhythmic cycles said to significantly affect our daily lives. However, scientists who study biological rhythms do not acknowledge these cycles.

At the moment of birth, the biorhythmic cycles are allegedly set to zero. According to classical biorhythm theory, there are three cycles: *intellectual* (33 days), *emotional* (28 days), and *physical* (23 days). Knowing only the date of your birth and the number of days you have lived, your present cycle status can be determined. A biorhythm chart for March 18, 2002, for someone born on February 15, 2002, would look like the one pictured. The line going through the middle is the zero line. A cycle is said to be in a positive phase when above the zero line, and in a negative phase when below the zero line. A cycle begins in an ascent for the first fourth of a cycle, then half of the cycle is in descent, then the last quarter of the cycle ascends back to the zero line. The cycles repeat until you die. Should you live something like 58 years and 66 days, you will reach the point at which the three cycles return to the same point on the zero line as at birth. For some, this is a moment of rebirth.

According to the theory, when certain points on the cycles are reached, a person may enjoy special strength or suffer special weakness. "Switch point" days, when cycles cross the zero line on the ascent or descent, are "critical" days. Performance on critical days is supposedly very poor. It has even been predicted that people are especially accident-prone on critical days. This empirical claim is easily testable and has been shown to be false (Hines 1998). However, any cycle with an odd number of days does not have an exact day in the middle, a fact that has led to some slippery math. For example, one study said to sup-

port biorhythm theory claims that about 60% of all accidents occur on critical days, but critical days make up only 22% of all days. If true, this statistic would probably not be due to chance, and biorhythm advocates might justifiably claim support for their theory from such data. However, biorhythmists include *the day before and the day after* a switch point day as critical days. Thus, critical days actually make up about *two-thirds of all days* and the data indicate what would be expected by chance (ibid.).

Despite the fact that critical days outnumber noncritical days two to one, advocates claim that critical days are days you want to know about in advance so you can prepare for them. For example, if you are scheduled to take a test that will measure your thinking ability, make sure you do not take the test on a day when your intellectual cycle is at a critical or low point. If you are a long distance runner, try to pick your next race date so that you are at a peak on your physical cycle.

The worst day of all, according to the classical theory, is the "triple critical," the day when all three cycles are at a switch point. Next worst is the "double critical,"

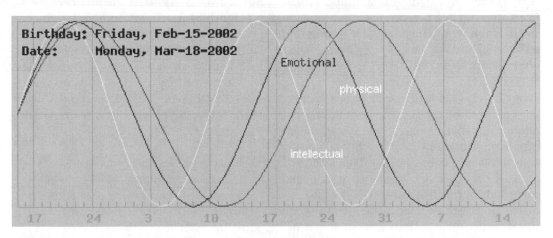

Biorhythm chart for March 18, 2002, for someone born on February 15, 2002. Biorhythm chart generated by a program provided on the Internet by Tay-Tec at www.tay-tec.de/biorhythm.

when two cycles meet at the switch point. It gets complicated tracing the cycles on their ascents, descents, switch points, and so on. But it does not take a mathematician to figure out that it is going to be easy to find cases that fit the theory. For example, the physical cycle is 23 days long. That means that every 11.5 days is a physical cycle switchover day. So the odds of, say, having a heart attack on a given physical switchover day are about one in 11. Most people would agree that having a heart attack is having a bad physical day. One valid empirical test of the theory would be to collect data on heart attack victims and see whether significantly more than 9% (one out of 11) had their heart attacks on physical switchover days. Instead, the usual evidence given by believers is an anecdote about Clark Gable or someone else who had a heart attack on a switchover day. There are thousands of heart attack victims each year, and one out of 11 of them would be predicted by chance to have the attack on a switchover day. So, finding several individual cases of people who have serious physical problems on a critical physical day is to be expected.

The ho-hum response that anecdotes such as the Clark Gable story should evoke from a reasonable person might put one to sleep when you consider that biorhythmists generally count the day before and after a critical day as being just as bad as critical days. This means that *six* out of every 23 days (26% of our days) are danger days for the body. Thus, the odds are about one in four that any given person who has a bad physical day is at a critical point. Anecdotes of people having bad physical days are particularly inconsequential given such odds. A meaningful test of the theory might be to study heart attack victims. If significantly greater than 25% of the sample have attacks on critical days, then you have a scoop.

Another typical but useless test of the theory is to keep track of how accurate the theory is by charting each day and keeping a diary of your days. Actress Susan St. James, a fervid believer in biorhythms, once described on a television talk show how she had done this. If her chart predicted a low emotional day, she was upset that day. If her chart predicted a physical high, she felt great that day. On a day when her intellectual cycle was at a low, she couldn't think straight about anything. In some circles this is known as the self-fulfilling prophecy, **confirmation bias**, or **subjective validation**. Whatever you call it, it isn't science.

In another experiment, the subject was given her own personal biorhythm chart. She was to keep a day-by-day diary for 2 months and to rate her chart for accuracy. She reported that the chart had been "at least ninety percent accurate," even though the chart was a false one (Randi 1982). When told a mistake had been made and asked to check her diary against her actual chart, the subject said that the new chart was even more accurate than the first. However, the second chart was also a false chart (ibid.) The subject had **shoehorned** the diary to fit the chart.

Biorhythm theory originated in the nineteenth century with Wilhelm Fliess, a Berlin physician, **numerologist**, and good friend and patient of Sigmund **Freud**. Fliess was fascinated by the fact that no matter what number he picked, he could figure out a way to express it in a formula with the number 23, 28, or both.

Fliess's basic formula can be written $23x + 28y$, where x and y are positive or negative integers. On almost every page Fliess fits this formula to natural phenomena, ranging from the cell to the solar system. . . . He did not realize that if any two positive integers that have no common divisor are substituted for 23 and 28 in his basic formula, it is possible to express any positive integer whatever. Little wonder that the formula could be so readily fitted to natural phenomena! (Gardner 1981: 134–135)

The latter number he associated with menstruation, and thus when he was convinced that all the world is governed by 23 and 28, he called the 28-day period "female" and the 23-day period "male." In 1904, several years after Fliess's discovery, Dr. Hermann Swoboda of the University of Vienna claimed he discovered these same periods on his own. In the 1920s, Alfred Teltscher, an Austrian engineering teacher, added the "mind" period of 33 days, based on his observation that his students' work followed a 33-day pattern.

Biorhythm theory was popularized in the 1970s by George Thommen (*Is This Your Day?: How Biorhythm Helps You Determine Your Life Cycles*) and Bernard Gittleson (*Biorhythm: A Personal Science*). Neither book provides scientific evidence for biorhythms. Both consist of little more than speculation and anecdotes. However, they replaced the static idea of *periods* with the dynamic notion of *cycles*, which they referred to as the *physical, emotional,* and *intellectual* cycles. Interestingly, not only did the female period become the emotional cycle, but both men and women are said to share the same physical and emotional cycles of 23 and 28 days, respectively. One might have expected that, given the different biological natures of males and females, the sexes might have at least some unique and distinct rhythmic cycles.

New cycles have been added in recent years. There is the 38-day *intuitional* cycle, the 43-day *aesthetic* cycle, and the 53-day *spiritual* cycle. Others claim there are cycles that are combinations of the three primary cycles. The *passion* cycle is the physical joined with the emotional cycle. The *wisdom* cycle is the emotional joined with the intellectual cycle. And the *mastery* cycle is the intellectual joined with the physical cycle. However many cycles there are, the function is the same: to predict what kind of day one is likely to have.

There have been several meaningful tests of biorhythm theory, all failing to support it (Hines 1998). Advocates of this theory have more **ad hoc hypotheses** to explain away disconfirming evidence than Galapagos has islands. One beauty is the hypothesis that some people are *arhythmic* some or all of the time. Any contrary case can be explained away by reference to the case being *arhythmic*. Another favorite ad hoc hypothesis concerns Thommen's claim that he could predict with 95% accuracy the sex of a child by the biorhythms of the mother. If during conception the mother's physical (masculine) cycle was at a high point, a boy was likely. If during conception the mother's emotional (female) cycle was at a high point, a girl was likely. A study done by W. S. Bainbridge, a professor of sociology at the University of Washington, concluded that using the biorhythm theory the chances of predicting the sex of the child were 50:50, the same as flipping a coin. A defender of the theory suggested to Bainbridge that the cases where the theory was wrong probably included many homosexuals, who have indeterminate sex identities!

Most of the support for biorhythms is in the form of anecdotes. One favorite concerns Mark Spitz (b. February 10, 1950), who won seven gold medals in the 1972 Olympics. Spitz's emotional and physical cycles peaked and converged on September 5, the day of the Munich massacre. His intellectual cycle was very low during this period. A logical person, if given the choice between attributing Spitz's performance to his biorhythms or to his swimming ability, would use **Occam's razor** to reject biorhythms in favor of the simpler explanation. On the other hand, Hall of Fame baseball player Reggie Jackson (b. May 18, 1946) had the greatest day in his brilliant career on October 18, 1977. On that day he hit three consecutive home runs on three consecutive pitches off three different pitchers to help the New York Yankees win the game and the World Series against the Los Angeles Dodgers. Jackson's cycles were all in the low end of their negative phases on that day. Rather than admit that Jackson's performance contradicts what would be predicted according to his biorhythm chart, one defender of the theory invented the ad hoc hypothesis that even though Jackson was at the low end of negative phases, his cycles were changing direction and beginning to come up. This brought about "increasing discharge" and Jackson was actually recharged and synchronized, which explains his great performance (www.ziplink.net/~rstriker/newsltr5.htm).

In this dynamic and energetic view, even days in the negative phase of a cycle can be good days, days in the positive cycle can be bad, and vice-versa, depending on whether they are ascending or descending, charging or discharging, available or unavailable. Such constructions may make it impossible to *refute* the theory, but they render it untestable and so slippery as to be of little use for predicting the future. Everything can be made to fit the theory, even contrary readings such as those of Mark Spitz and Reggie Jackson, who deserve more credit for their accomplishments than biorhythm theory can provide.

The cycles of biorhythm theory did not originate in the study of biological organisms, nor have they been supported by anything resembling a scientific study. The theory has been around for over a hundred years and there has yet to be a scientific journal that has published a single article supporting the theory. There have been some three dozen studies supporting biorhythm theory, but all of them suffer from methodological and statistical errors (Hines 1998). An examination of some 134 biorhythm studies found that the theory is not valid (ibid.). When advocates of an empirically testable theory refuse to give up the theory in the face of overwhelming evidence against it, it seems reasonable to call the theory **pseudoscientific**.

Further reading: Gardner 1981; Hines 1990, 1991; Masson 1985; Randi 1982.

Blavatsky, Helena Petrovna (1831–1891)

See **theosophy**.

Blondlot and N-rays

René Prosper Blondlot (1849–1930) was a French physicist who claimed to have discovered a new type of radiation shortly after Roentgen discovered x-rays. Blondlot called it the N-ray, after Nancy, the name of the town and the university where he lived and worked. He was trying to polarize x-rays when he discovered his new form of radiation. Dozens of other scientists confirmed the existence of N-rays in their own laboratories. However, N-rays don't exist. How could so many scientists be wrong? How had they deceived themselves into thinking they were seeing something that wasn't there?

The story of Blondlot is a story of **self-deception**. Because many people have the misguided notion that science should be infallible and a fount of absolutely certain truths, they look at the Blondlot episode as a vindication of their excessive skepticism toward **science**. They relish accounts such as the one regarding Blondlot and the phantom N-rays because it is a story of a famous scientist making a great error. However, if one properly understands science and scientists, the Blondlot episode indicates little more than the fallibility of scientists and the self-correcting nature of science.

Blondlot claimed that N-rays exhibit impossible properties and yet are emitted by all substances except green wood and certain treated metals. In 1903, he claimed that he had generated N-rays using a hot wire inside an iron tube. The rays were detected by a calcium sulfide thread that glowed slightly in the dark when the rays were refracted through a 60-degree-angle prism of aluminum. According to Blondlot, a narrow stream of N-rays was refracted through the prism and produced a spectrum on a field. The N-rays were reported to be invisible, except when viewed as they hit the treated thread. Blondlot moved the thread across the gap where N-rays were thought to come through and the thread was illuminated.

Nature magazine was skeptical of Blondlot's claims because laboratories in England and Germany had not been able to replicate the Frenchman's results. *Nature* sent American physicist Robert W. Wood of Johns Hopkins University to investigate. Wood suspected that N-rays were a delusion. To test his hypothesis, Wood removed the prism from the N-ray detection device, unbeknownst to Blondlot or his assistant. Without the prism, the machine couldn't work. Yet when Blondlot's assistant conducted the next experiment, he found N-rays. Wood then surreptitiously tried to replace the prism but the assistant saw him and thought he was removing the prism. The next time he tried the experiment, the assistant swore he could not see any N-rays. But he should have, since the equipment was in full working order.

According to Martin Gardner, Wood's exposure of Blondlot led to the French scientist's madness and death (Gardner 1957: 345). But were those who verified Blondlot's N-ray experiments stupid or incompetent? Not necessarily, since the issue isn't one of intelligence or competence but of the psychology of perception. Blondlot and his followers suffered "from self-induced visual hallucinations" (ibid.).

The Blondlot episode is a reminder that although scientists often make errors, even big ones, other scientists will uncover the errors and get science back on the right path to understanding nature. Those who think that science should be infallible do not understand its nature.

Recent examples of "Blondlot's Folly" include the discovery of cold fusion (1989) by Pons and Fleischmann and the discovery of ununoctium, or element 118, (1999) by the Lawrence Berkeley National Laboratory in California.

See also Charles **Fort**, **pathological science**, and the **Piltdown hoax**.

Further reading: Ashmore 1993; Asimov 1988; Klotz 1980.

blue sense

A cop's intuition about such things as impending danger, about whether a suspect is guilty, about whether someone's lying, or about hunches regarding cases or people. The term is used to refer to something akin to **psychic** power possessed by good cops. "It is that unknown quantity in the policeman's decision-making process that goes beyond what he can see and hear and smell" (Lyons and Truzzi 1991: 11).

Studies have not validated the "blue sense," but there is good evidence that some people, including some cops, reliably infer others' emotions, intentions, and thoughts by their demeanor and facial expressions (Ekman and Friesen 1975; Ekman and Rosenberg 1997).

See also **psychic detectives**.

brainwashing

See **mind control**.

breatharianism

See **inedia**.

bunyips

Legendary spirits or creatures of the Australian aboriginal tradition. Bunyips haunt rivers, swamps, creeks, and billabongs. Their main goal in life is to cause nocturnal terror by eating people or animals in their vicinity. They are renowned for their terrifying, bellowing cries in the night, and have been known to frighten aborigines to the point where they would not approach any water source where a bunyip might be waiting to devour them.

Some say the bunyip looks like a huge snake with a beard and a mane; others say it looks like a huge furry half-human beast with a long neck and a head like a bird. However, most Australians now consider the existence of the bunyip to be mythical. Some scientists believe the bunyip was a real animal, the diprotodon, extinct for some 20,000 years, which terrified the earliest settlers of Australia. According to Oodgeroo Noonuccal (Kath Walker) in *Stradbroke Dreamtime* (1972), the bunyip is an evil or punishing spirit from the Aboriginal Dreamtime. Today, the bunyip mainly appears in Australian literature for children and makes an occasional appearance in television commercials.

C

Cardiff Giant

A fake fossil of an antediluvian giant some 10 feet high with 21-inch feet. The "fossil" is actually a carved slab of gypsum, sculpted a year or two before its "discovery" in 1869. The fake was the idea of George Hull, a cigar manufacturer and **atheist**, and Stubb Newell, a distant relation who owned the farm in Cardiff, New York, where the hoax was perpetrated. Experts almost immediately suspected the alleged fossil was a hoax, but their warnings went unheeded. Rumor had it that the fossil was proof of the Bible's accuracy about giants such as Goliath. The curious came in the hundreds per day to the remote upstate New York farm for a view of Biblical history. Scientists declaring the giant a fake did not deter visitors, who shelled out 50 cents each to see the "Goliath."

Within a week of its "discovery," Newell sold three-fourths of his interest in the Giant to a syndicate in Syracuse, New York, for $30,000. Business was so good that P. T. Barnum wanted to get in on the action. He offered to rent the giant for just three months to take on the road with his circus, but Newell and the syndicate wouldn't deal. So Barnum had a duplicate made and charged people to see a fake of the fake. It is said that when both were displayed in New York City at the same time, Barnum's fake of the fake outdrew the real fake (Feder 2002).

Kenneth Feder sees the Cardiff Giant episode as a familiar one:

Trained observers such as professional scientists had viewed the Giant and pronounced it be an impossibility, a statue, a

The Cardiff Giant.

York, where it is labeled "America's Greatest Hoax."

Carlos hoax

"Carlos" is the name of a 2,000-year-old spirit allegedly channeled by José Alvarez when he toured Australia in 1988. Channeling was all the rage in Australia, and an Australian television program contacted James Randi about finding someone who might show Australians that channeling was something suspect. Randi approached Alvarez, a performance artist, who created "Carlos." He studied videotapes of alleged channelers who speak in strange voices, claim to be in touch with other worlds, and wear crystals and beads. He went to Australia and ultimately performed at the Sydney Opera House and charmed a rapt audience with the Spirit of Carlos. Randi (personal correspondence) notes:

> All of the material that he produced was spurious. In the press releases, he invented magazines and newspapers, he invented towns and cities and radio stations and TV channels and whatnot, that didn't even exist. He prepared videos of radio interviews and theater appearances that never happened. And just one phone call by the media back to the United States would have revealed the whole thing as a hoax. Even after it was all revealed on the Australian Sixty Minutes TV show, a week after the Opera House appearance, many continued to believe in "Carlos" and his uninspired messages.

clumsy fraud, and just plain silly. Such objective, rational, logical, and scientific conclusions, however, had little impact. A chord had been struck in the hearts and minds of many otherwise levelheaded people, and little could dissuade them from believing in the truth of the Giant. Their acceptance of the validity of the giant was based on their desire . . . to believe it. (ibid.: 37)

In short, skepticism toward scientific experts is often rooted in **wishful thinking**, a desire to believe what one wants to believe regardless of the evidence.

The "fossil" is now on display at the Farmer's Museum in Cooperstown, New

For Alvarez, the creation of the character "Carlos" was a performance/experiment to see how far he could take his creation, but his purpose was not to make people look foolish. He hoped to liberate them from a false belief. However, the result of the performance seemed to demonstrate how easy it is to create a cult from scratch and how, even when the truth is revealed to people, many still refuse to accept it. The "Carlos" hoax also demonstrates how gullible and uncritical the mass media are when covering paranormal or supernatural topics. Rather than having an interest in exposing the truth, the members of the media were obsessed with "Carlos" the phenomenon, and transformed his character from a hoax to a myth. The media should have known that "Carlos" was not genuine. He performed for free, and the first sign of an authentic fake guru is greed.

Alvarez continues to travel the world performing and expanding on his "Carlos" creation. He appears on global network TV and performs before large live audiences. Why? To bring people real enlightenment about gurus, belief, charisma, power, and how they intersect. His exploration of these ideas was featured at the 2002 Biennial Exhibition at the Whitney Museum of American Art in New York City.

See also **Ramtha**.

Further reading: Sagan 1994.

cartomancy

Literally, "**divination** from cards." Cartomancy is a type of fortune telling by reading cards such as the **tarot**.

Castaneda, Carlos (1925?–1998)

Carlos Castaneda was a best-selling author of books centering on a Mexican Yaqui shaman's pharmacologically induced visions. He called the shaman Don Juan Matus. Castaneda claimed he was doing anthropology, that his books were not fiction. He was granted a Ph.D. by the UCLA Anthropology Department in 1973 for his third book, *Journey to Ixtlan*. Critics say the work is not ethnographically accurate and is a work of fiction.

Castaneda's books are full of stories of magic, **sorcery**, and **out-of-body experiences**. His first books hit the market during the late 1960s when LSD guru Timothy Leary was advising the world to "turn on, tune in, drop out." The LSD gurus believe that the chemical changes in their brains, which cause them to perceive the world differently and to perceive different worlds, bring them into a divine realm. Getting high means opening the doors of perception to a higher reality. Castaneda could not have had better timing for his books.

Castaneda claims that he met Don Juan in 1960 at a bus station in Nogales, Arizona. At the time, Castaneda was a graduate student in anthropology doing research on medicinal plants used by Indians of the Southwest. He claims that Don Juan made him a sorcerer's apprentice and introduced him to the world of peyote and visions. It is unlikely that a great shaman would pick up a stranger at a bus stop and make him a disciple, but we'll never know since no one but Castaneda ever met Don Juan. Was Don Juan a hoax? Probably. Yet Castaneda's books have sold over eight million copies. How?

Castaneda obviously filled a need. He told good stories and gave enigmatic advice. He gave people hope, especially those who believe that the more modern civilization has become, the further it has driven human beings from their spiritual or true nature. The old shamans know the way. They know truths the modern scientist has not even dreamed of. They do hallucinogenic drugs, too. Maybe that is why they thought they could fly and transmogrify into birds and other animals.

In his later years, Castaneda introduced a new way to get high: *Tensegrity*. It involves meditation, exercises, a luminous egg, an assemblage point, depersonalization, **dreaming**, and other New Age magic. Tensegrity allegedly leads to the perception of "pure **energy**," breaking down the barriers to higher consciousness. It is supposed to be based on ancient magic, known to Indian shamans centuries ago. Sounds familiar.

Further reading: Fikes 1996; Lindskoog 1993.

cattle mutilations

The term "mutilations" is used by **UFO** devotees to describe animal corpses with "unusual" or "inexplicable" features. What counts as unusual or inexplicable is just about any cut, mark, wound, excision, incision, swelling, distention, organ or blood absence, abrasion, scrape, or bruise. These mutilations, we are told, are being done by bad aliens. No one has shown either that there are thousands of inexplicable animal deaths around the globe or that, if there are, they are related, much less that they are the result of alien experimentation. These facts, however, are no deterrent to those who are sure we are not alone. To them, these visitors from other worlds not only are responsible for the deaths and mutilations of thousands of cattle, horses, cats, and other domestic animals around the globe, they are also responsible for numerous human abductions for the purpose of experimental and reproductive surgery. Furthermore, some of these aliens are destroying crops around the globe in an effort to impress us with their artistic abilities or to communicate to us in strange symbols just how much they like our planet's cattle.

Believers in alien mutilators reject the notion that there could be an earthly and naturalistic explanation for unusual animal wounds, such as lactic acidosis. They are convinced that aliens need cow blood and organs for their experiments. What seems most convincing to the alien theorists is that wounds and missing organs such as the tongue and the genitalia seem completely inexplicable to them in any but mysterious terms, that is, alien surgeons. Naturalistic explanations in terms of predators (skunks, buzzards, weasels, etc.), insects (such as blowflies), and birds are to no avail, even though the most thorough examination of cattle mutilations concluded there was nothing mysterious that needed explaining (Rommel 1980). It is useless to note that insects and animals often devour the vulnerable mucous membranes and the softer parts of dead animals such as the genitalia rather than try to burrow through cowhide. It is pointless to note that incisions to a carcass by the teeth of predators or scavengers often resemble knife cuts. It is of no use to point out that there is little or no blood oozing from the wounds because blood settles, the heart does not pump when an animal is dead, and insects devour the blood that does spill out. Explanations in terms of lactic acidosis are to no avail, even though there are many common features between the effects of this disease and the list of strange signs cited by those who favor the alien mutilation hypothesis: death, missing body parts, missing tissue, no blood near the carcass, shape and texture of wound edges, and no signs of decomposition.

If there were earthly explanations, one could not plausibly invoke the favored government conspiracy theory to juice up the story. Cattle mutilations are not only associated with **UFO** sightings, **alien abductions**, and **crop circles**; they are linked to black helicopter sightings as well. This suggests to the well-trained conspiratorial mind that our military forces are testing new weapons on the livestock of unsuspecting citizen ranchers. It shouldn't surprise anyone to find that the military thinks it is their right to experiment on cattle surreptitiously, since they've done it on people in the past (Scheflin and Opton 1978). Another possibility, according to UFOlogists, is that the helicopters are UFOs disguised to appear as terrestrial craft. Some even think that the aliens have been in collusion with the Air Force for the past 30 years and that the animal mutilations are being performed by aliens with the full knowledge of our military. We allow it because of some sort of treaty the U.S. government supposedly signed with the aliens. You would think that such intelligent beings might have worked out a better arrangement.

Why would beings with the intelligence and power to travel billions of miles to our planet spend their time mutilating cows, experimenting on otherwise unremarkable people, and carving up wheat fields? Perhaps the aliens need cow blood and glands for food and experiments. Maybe they carve up wheat fields with ever more elaborate designs to impress us with their intelligence.

See also **flying saucers**, **Men in Black**, and **UFOs**.

Further reading: Frazier 1997; Hines 1996; Hitt 1997; Huston 1997; Kagan and Summers 1984; Perkins 1982; Stewart 1977.

Cayce, Edgar (1877–1945)

Edgar Cayce is known as one of America's greatest **psychics**. However, he wrongly predicted that California would slide into the ocean and that New York City would be destroyed in a cataclysm. He made many other erroneous predictions concerning such things as the Great Depression (1933 would be a good year!), the

Lindbergh kidnapping (most of it wrong, all of it useless), and the conversion of China to Christianity by 1968.

There are many myths and legends surrounding Cayce: that an **angel** appeared to him when he was 13 years old and asked him what his greatest desire was (Cayce allegedly told the angel that his greatest desire was to help people); that he could absorb the contents of a book by putting it under his pillow while he slept; that he passed spelling tests by using **clairvoyance**; that he was illiterate and uneducated. (The *New York Times* helped spread the illiteracy myth in an article entitled "Illiterate Man Becomes a Doctor When Hypnotized," October 9, 1910). He also claimed to be able see and read **auras**, but this power was never tested under controlled conditions.

Many of the myths were passed on unchecked by Thomas Sugrue, who believed Cayce had cured him of a disabling illness. In *There Is a River: The Story of Edgar Cayce* (1945), Sugrue asserts that Cayce cured his own son of blindness and his wife of tuberculosis, though both were also treated by medical doctors. It is as a psychic medical diagnostician and psychic reader of past lives that Cayce is best known. Because he would close his eyes and appear to go into a trance when he did his readings, he was called "the sleeping prophet" (Stern 1967). A stenographer took notes during his sessions. And at his death, he left some 30,000 accounts of past life and medical readings, which are under the protection of the Association for Research and Enlightenment. The documents are the only evidence for Cayce's psychic medical powers, but they are "worthless by themselves" because there is no way to distinguish what Cayce discerned by psychic ability from what information was provided to him by his assistants, by letters from patients, or by simple observation (Beyerstein 1996).

Cayce usually worked with an assistant, either an M.D. named John Blackburn, **homeopath** Wesley Ketchum, or **hypnotist** and mail-order **osteopath** Al Layne. **Osteopathy** was akin to **naturopathy** and folk medicine in Cayce's day.

Many of his readings would probably make sense only to an osteopath of his day. For example, here is his reading for his own wife, who was suffering from tuberculosis:

> from the head, pains along through the body from the second, fifth and sixth dorsals, and from the first and second lumbar . . . tie-ups here, floating lesions, or lateral lesions, in the muscular and nerve fibers which supply the lower end of the lung and the diaphragm . . . in conjunction with the sympathetic nerve of the solar plexus, coming in conjunction with the solar plexus at the end of the stomach. . . . (Gardner 1957: 217)

The fact that Cayce mentions the lung is taken by his followers as evidence of a correct diagnosis; it counts as a psychic "hit." But what about the incorrect diagnoses: dorsals, lumbar, floating lesions, solar plexus, and stomach? Why aren't those counted as diagnostic misses?

Cayce was the first to recommend laetrile as a cancer cure. (Laetrile contains cyanide and is known to be ineffective for cancer.) He also recommended "oil of smoke" for a leg sore, "peach-tree poultice" for convulsions, "bedbug juice" for dropsy, and "fumes of apple brandy from a charred keg" for tuberculosis. It is uncertain what he prescribed for terminal gullibility.

The *volume* and alleged accuracy of his "cures" provide the main basis for belief in Cayce as a psychic. Unfortunately, there is no way to demonstrate that Cayce used psychic powers even on those cases where there is no dispute that he was instrumental in the cure. Nevertheless, many people considered themselves cured by Cayce. Not every customer went away satisfied, however. Martin Gardner notes that Dr. J. B. Rhine, famous for his ESP experiments at Duke University, was not impressed with Cayce. Rhine felt that a psychic reading done for his daughter didn't fit the facts. Defenders of Cayce respond to such criticism by claiming that if a patient has any doubts about Cayce, the diagnosis won't be a good one.

Even though Cayce didn't have a formal education much beyond grammar school, he was a voracious reader, worked in bookstores, and was especially fond of **occult** and osteopathic literature. As noted above, he was in contact with and assisted by people with various medical backgrounds. He was not the illiterate ignoramus he is sometimes made out to be in order to make his diagnoses seem all the more miraculous.

Finally, Cayce is one of the main people responsible for some of the sillier notions about **Atlantis**. He predicted that in 1958 the United States would discover a death ray that had been used on Atlantis. He claimed that the Atlantaeans had a Great Crystal called the Tuaoi Stone. He said it was a huge cylindrical prism that was used to gather and focus "energy," allowing the Atlanteans to do many fantastic things. They got greedy and stupid, however, and tuned their crystal to too high a frequency, setting off volcanic disturbances that led to the destruction of that ancient world.

Further reading: Randi 1982a; Shermer and Benjamin 1992.

Celestine Prophecy, The

James Redfield wrote a novel called *The Celestine Prophecy,* now seen as a spiritual guide for the New Age. Even Redfield treats his novel as a spiritual guide and basis for a spiritual and material industry. He's started a newsletter for his followers: *The Celestine Journal: Exploring Spiritual Transformation.* He has a sequel, *The Tenth Insight,* said to be "a trip that will take you through portals into other dimensions." And he has a sequel to the sequel: *The Secret of Shambhala: In Search of the Eleventh Insight.* He also has audiotapes and CDs for sale.

Redfield starts with a notion shared by many New Age gurus: The world is emerging into a new spiritual awareness. He puts it this way:

> For half a century now, a new consciousness has been entering the human world, a new awareness that can only be called transcendent, spiritual. If you find yourself reading this book, then perhaps you already sense what is happening, already feel it inside.

What is the evidence for this New Age? There are vague references to vibrations and **energy**. For those who don't get it yet, there is vague advice to avoid the negative (you can tell good people by their eyes), stop doubting, follow your intuitions and premonitions, flow with coincidences, believe in the purposiveness of everything, join thousands of others on the quest, tune into your feelings, and evolve to a higher plane.

In the novel, the meaning of life is revealed in an ancient Peruvian manuscript written in Aramaic. It predicts a massive spiritual transformation of society in the late 20th century. We will finally grasp the secrets of the universe, the mysteries of existence, the meaning of life. The real meaning and purpose of life won't be found in religion and won't be found in material wealth, but rather in things such as **auras**. The manuscript is full of insights like this, and these insights are the way to transformation. How do we know this? Just look at the dissatisfaction and restlessness all around you. That's the key. We're like caterpillars ready for metamorphosis

into butterflies, to burst forth together into the New Age. Do you think it is a coincidence that coincidences are happening more and more frequently?

> [T]he Manuscript says the number of people who are conscious of such coincidences would begin to grow dramatically in the sixth decade of the twentieth century. He said that this growth would continue until sometime near the beginning of the following century, when we would reach a specific level of such individuals— a level I think of as a critical mass.

I'm not sure, but I think Redfield meant to say the *seventh* decade, not the sixth. The sixth decade of the 20th century would be the 1950s. Nobody seems to think that the '50s were a time of restlessness. The '60s, however, has entered historical consciousness as a very restless period: the Vietnam War and the antiwar movement, marijuana and LSD, the civil rights movement, assassinations of the Kennedy brothers and Martin Luther King, Jr., the Beatles, and so on. In any case, the novel has some good advice. Make love, not war. Be neither intimidator nor interrogator, aloof nor pitiable. We don't need fear, humiliation, guilt, or shame. Contemplate, meditate, and follow your intuitions and dreams as you go through your spiritual evolution. Fact or fiction, it doesn't matter.

This New Age subjectivism and relativism encourages people to believe that reality is whatever you want it to be. The line between fact and fiction gets blurry and obscured. Of course fiction has its place in a satisfying life, but so should fact.

cellular memory

The notion that human body cells contain clues to our personalities, tastes, and histories, independently of either genetic codes or brain cells. According to Dr. John

Schroeder of the Stanford Medical Center, "The idea that transplanting organs transfers the coding of life experiences is unimaginable." The idea may be absurd, but it has been imagined several times in films such as *Brian's Song*. In that film, the 26-year-old Brian Piccolo (played by James Caan) is dying of cancer when Gayle Sayers (played by Billy Dee Williams), his friend and Chicago Bears teammate, visits him in the hospital. Piccolo had been given a transfusion and he asks Sayers whether he had donated any blood. When Sayers says yes, Piccolo remarks that that explains his craving for chitlins.

In *Dianetics*, L. Ron Hubbard speculated that cellular memory might explain how engrams work.

Maybe the idea originated with Maurice Renard (1875–1939), whose *Les Mains d'Orlac* tells the story of a concert pianist who loses his hands in an accident and is given the hands of a murderer in a transplant operation. Suddenly, the pianist has an urge to kill. Several variations of Renard's story have made it into film, including *Orlacs Hände*, a 1935 silent Austrian film, *Mad Love* (1935), *Les Mains D'Orlac* (1960), and *Hands of a Stranger* (1962). A similar story is told by Pierre Boileau and Thomas Narcejac (authors of *Vertigo*) in *Et mon tout est un homme* (1965), which was made into the film *Body Parts* in 1991. A prison psychiatrist loses an arm in an accident and is given the arm of an executed psycho-killer. The arm then develops a mind of its own.

More recently, Claire Sylvia, a heart-lung transplant recipient, explained her sudden craving for beer and fried chicken by noting that her donor was an 18-year-old male who died in a motorcycle accident. She's even written a book about it, cleverly titled *A Change of Heart,* which the Lifetime Channel plans to turn into a movie produced by the actress Sally Field. Paul Pearsall, a psychologist and author of *The Pleasure Principle,* also casts his vote for the theory of cellular memory and transfer. Pearsall goes much further in his speculations, however, claiming that "the heart has a coded subtle knowledge connecting us to everything and everyone around us. That aggregate knowledge is our spirit and soul. . . . The heart is a sentient, thinking, feeling, communicating organ." How he knows this is anybody's guess. Perhaps he has been reading the fiction of Edna Buchanan *(Pulse)*, who asks: "What if the soul is contained in DNA? What if DNA is contained in the soul?"

Sylvia Browne, legendary **psychic** in her own mind, taught a course for an alternative education program in Sacramento entitled "Healing Your Body, Mind and Soul." In one two-hour session Browne said she would teach anyone "how to directly access the genetic code within each cell, manipulate that code and reprogram the body to a state of normalcy." Despite the preposterous nature of her claims, the course sold out.

See also **sympathetic magic.**

chain letters

See **pyramid schemes.**

chakras

According to **Tantric** philosophy and yoga, chakras are points of energy in the astral body. There are seven primary chakras, which are associated with various parts of the body, emotions, desires, thoughts, powers, and health. New Age gurus think chakras have colors and give rise to **auras,** which reveal one's spiritual and physical health, as well as one's **karma.** The alleged **energy** of the chakras is not scientifically measurable, however, and is at best a **metaphysical** chimera and at worst an anatomical falsehood.

channeling

A process whereby an individual (the "channeler") claims to have been invaded by a spirit entity for the purpose of communication. The channeling craze began in earnest in 1972 with the publication of *Seth Speaks* by Jane Roberts and her husband Robert Butts. Seth, a very wise "unseen entity," allegedly communicated his wisdom to the entranced Jane, who dictated messages to Butts.

Though Roberts, a somewhat accomplished poet, was obviously very literate and widely read in many religious and **occult** traditions (including **Jung**), her advocates portray her as communicating ideas beyond her ability. They take this as proof she was inspired. If she and Butts were inspired, it was probably by the depth of human credulity.

A big boost for belief in channeling came from actress Shirley MacLaine and the ABC television network. They gave this modern version of **spirits** speaking through a **medium** a modicum of credibility in 1987 when ABC showed a miniseries based on MacLaine's book *Out on a Limb.* MacLaine allegedly converses with spirits through channeler Kevin Ryerson. One of the spirits who speaks through Ryerson is a contemporary of Jesus called John, who doesn't speak Aramaic—the language of Jesus—but a kind of Elizabethan English. John tells MacLaine that she is cocreator of the world with God. MacLaine, a consummate egoist, becomes ecstatic to find out that she is right about a belief she'd expressed earlier, namely, that she *is* God (Gardner 1987).

One of MacLaine's favorite channelers is J. Z. Knight, who claims to channel a 35,000-year-old Cro-Magnon warrior from **Atlantis** called **Ramtha.**

See also **Carlos**, Edgar **Cayce**, Bridey **Murphy**, and **reincarnation.**

Further reading: Alcock 1996a; Gardner 1988; Gordon 1988; Schultz 1989.

Charcot, Dr. Jean-Martin (1825–1893)

See **hystero-epilepsy.**

charms

Things or words believed to possess magic power.

See also **amulets**, **fetishes**, and **talismans.**

chelation therapy

Consists of slow-drip IV injections of EDTA (ethylenediamine tetraacetic acid), a synthetic amino acid, combined with aerobic exercise, special diet, and no smoking. EDTA treatment has been around since the 1940s, when it was developed to treat lead poisoning. The word "chelate" is derived from the Greek word for "claw" and apparently refers to the alleged clawing away of plaque and calcium deposits from arteries and veins by EDTA.

Advocates claim that there is ample evidence to support the claim that chelation can prevent and cure heart disease, stroke, senility, diabetic gangrene, and many other vascular diseases. For example, the unpublished Cypher report by Philip Hoekstra, Sr., John M. Baron, D.O., et al. collected data from several physicians who used chelation to treat patients with vascular diseases. Over 19,000 cases were studied and about 86% showed "a significant enhancement in the arterial perfusion of the upper and lower extremities," according to James P. Carter, M.D., in *Racketeering in Medicine: Hippocrates Forsaken for Profit.* However, different physicians carried out the treatments independently and there were no control groups. Lack of adequate controls in studies purporting to demonstrate the effectiveness of chelation has been a consistent criticism of skeptics. The evidence in favor of chelation as a cure for heart disease consists mainly of **testimonials**, subjective patient/physician reports, and alleged conspiracies by the medical and pharmaceutical establishments. Advocates claim that it is too expensive to do scientifically **controlled studies** and that there is a conspiracy by the medical establishment to prevent such studies from being undertaken.

Critics of chelation in the American Medical Association, the American Heart Association, and the Federal Drug Administration claim that there is no good scientific evidence supporting the extravagant claims of advocates. Medicare does not cover chelation therapy nor will most insurance companies pay for it. Defenders of the therapy claim that the medical establishment has engaged in half a century of deceit and conspiracy to suppress chelation because of fear it would cut into the profits made by drug therapy and surgery. They claim that EDTA is 10 times cheaper than a coronary bypass with equal or better results, but since EDTA can't be patented, there is no big money to be made by pharmaceutical firms. Advocates claim that scientific medicine bases treatment decisions on politics and economics, not on evidence from controlled studies. They say the medical establishment prefers surgery because it is expensive. The evidence backing up these claims is lacking.

Finally, advocates of "oral chelation" claim it is much cheaper than traditional chelation therapy, but so far there is no charge of conspiracy by traditional chelation advocates, though they seem to consider "oral chelation" misleading and ineffective.

Further reading: Barrett and Butler 1992; Barrett and Jarvis 1993; Raso 1993.

chi (ch'i, qi)

Ch'i or qi (pronounced *chee* and henceforth spelled "chi") is the Chinese word used to describe "the **natural** energy of the Universe." This **energy**, though called "natural," is spiritual or supernatural, and is not part of an empirical belief system. Chi is thought to permeate all things, including the human body. Such metaphysical systems are generally referred to as types of **vitalism**. One of the key concepts related to chi is the concept of harmony. Trouble, whether in the universe or in the body, is a function of disharmony, of things being out of balance and in need of restoration to equilibrium.

Proponents claim to prove the existence and power of chi by healing people with **acupuncture** or **chi kung** (qi gong), by doing magic tricks such as breaking a chopstick with the edge of a piece of paper or resuscitating a "dead" fly, or by martial arts stunts such as breaking a brick with a bare hand or foot. When examined under controlled conditions, however, the seemingly paranormal or supernatural feats of masters of chi turn out to be quite ordinary feats of magic, deception, or natural powers, or natural feats requiring extraordinary physical training and discipline.

Vitalism is a popular philosophy in many cultures. Thus, chi has many counterparts: **prana** (India; see **therapeutic touch**), ki (Japan); Wilhelm Reich's **orgone energy**, **Mesmer's** animal magnetism, and Bergson's élan vital (vital force), to name just a few. The concept is very popular among New Age "alternative" healers, where it generally goes by the name of energy, though the concept bears no resemblance to the concept as used by physicists or chemists.

See also *I Ching* and **yin and yang**.

Further reading: Huston 1995; Livingston 1997; Randi 1995.

chi kung (ch'i kung, qi gong)

Ch'i kung or qi gong (pronounced "chee gung" and henceforth spelled "chi kung") is claimed to be "the science and practice" of **chi**. Chi kung literally means *energy cultivation*. Physical and mental health are allegedly improved by learning how to manipulate chi through controlled breathing, movement, and acts of will. Chi kung masters claim to be able to heal at a distance by manipulating chi. It is even said that one can strengthen the immune system by mastering one's chi, though this is doubtful (see **naturopathy**).

Most Westerners are vaguely familiar with kung fu and tai chi, both of which are related to chi kung. The former is a martial art and the latter is a type of exercise, or internal martial art. The former is sometimes known for demonstrations of breaking bricks with bare hands. The latter is known for the graceful poses of its practitioners. These demonstrations, and stories of even more powerful demonstrations, are offered as evidence of the **paranormal** or supernatural power that comes to those who master chi.

Asian martial arts schools have become very popular in the West. There is certainly a good side to these training centers for children and adults. They encourage attention to diet and physical exercise. They cultivate physical strength and mental self-discipline. Many focus on self-defense, and they boost self-confidence and self-esteem, even if they don't really make one invincible. However, they also often encourage students to believe they can achieve supernatural or paranormal powers, or heal just

about any illness by an act of will, by training and discipline under a "master."

What evidence is there for chi or its harnessing? **Testimonials** and self-validating statements, as well as demonstrations of strength, agility, and endurance are offered as proof. The acupuncturist is convinced he or she is unblocking chi. The **reiki** therapist and **therapeutic touch** nurse think they are directing or unblocking ki or **prana**. The Reichians think they can heal the body by harnessing and directing **orgone energy**. As a philosophy, chi kung and its relatives may provide one with a sense of harmony, power, and meaning. There is no way to disprove the existence of chi. However, explanations of events in terms of controlling and harnessing chi seem superfluous.

Further reading: Huston 1995; Lin 2000; Randi 1995.

chiromancy

See **palmistry**.

chiropractic

An **alternative health practice** based on the theory that *subluxations* are the cause of most medical problems. A subluxation is a misalignment of the spine that allegedly interferes with nerve signals from the brain. Chiropractors believe that by adjusting the misalignments they can thereby restore the nerve signals and allow the body to heal itself.

D. D. Palmer, a grocer in Davenport, Iowa, first proposed this theory in 1895. There is little scientific evidence to support the theory. Most support for chiropractic comes from **testimonials** of people whose back pain lessened after going to a chiropractor.

The theory of subluxations maintains that all health problems are due to "blockage" of nerves. It is true that nerves from the spine connect to the organs and tissues of the body and it is true that damage to those nerves affects whatever they connect to; for example, sever the spinal cord and your brain can't communicate with your limbs, though your other organs can still continue to function.

Chiropractic is based on the belief that the body is basically self-healing. Hence, drugs and surgery are not recommended except in extreme cases. Conventional medicine has opposed chiropractic for the most part. Chiropractors rarely are in joint practice with medical doctors, and they are almost never on staff at hospitals. Is there a conspiracy on the part of the American Medical Association (AMA) to prevent chiropractors from cutting into their profits, as many chiropractors maintain?

The AMA is partly responsible for chiropractic's reputation for **quackery**. For years, the AMA made no bones about their disapproval of chiropractic, which was featured in the report of its *Committee on Quackery*. But the chiropractors fought back and won a significant lawsuit against the AMA in 1976 for restraint of trade. Years later the American College of Surgeons issued a position paper on chiropractic advocating that the two professions work together. Publicly, the AMA no longer attacks chiropractic. Today, numerous so-called complementary medical techniques are being allowed to flourish in hospitals and medical clinics around the country without a word of protest from the AMA. The National Institutes of Health has a flourishing division for testing even the most unpromising of alternative health practices. Chiropractors and other "alternative" practitioners have learned one thing from the AMA: It pays to organize and to lobby Congress and state legislatures. The AMA is still the most powerful lobby among health care professionals,

but it is no longer flying solo. Even so, the AMA's lobbying is not the only reason that chiropractic's public image has suffered.

For years chiropractors relied more on faith than on empirical evidence in the form of **controlled studies** to back up their claims about the wonders of nerve manipulation. This is changing. There is some scientific evidence that chiropractic is effective in the treatment of many lower back ailments and neck injuries. There is some scientific evidence that chiropractic is effective for the treatment of certain kinds of headaches and other pains. The chiropractor is one of the few "alternative" health practitioners that medical insurance will generally cover. However, the likelihood that diseases such as cancer, for example, will ever be attributed to nerve blockage seems extremely remote. Making extravagant claims about the wonders of chiropractic, references to the flow of **energy** or "life forces" that heal the body, or to such notions as "bio-energetic synchronization" are not likely to contribute to the advancement of the discipline into mainstream medicine. Likewise, making claims that germ theory is wrong, a common chiropractic claim, does little to make chiropractors seem like advanced medical practitioners. To ignore bacteria and viruses or to underestimate the role of microbes in infections, as chiropractors are wont to do, is not likely to advance their cause. Every misdiagnosis or mistreatment by a chiropractor undermines the whole profession, rather than only the individual practitioner, because of the contentious nature of the theory of subluxations.

Chiropractic's defenders often rely on smokescreens such as trying to defame all medical doctors by noting how one physician has amputated the wrong leg. There are, of course, horror stories featuring medical doctors. However, very few people take such stories as indictments of the entire profession. They are seen as aberrations, not typical. This is probably not due to the better lobbying efforts of the AMA or to a conspiracy to control the press. It is most likely due to the experiences most people have had with medical doctors and the generally positive effects of modern medicine. In many cases, medical doctors take much greater risks than any chiropractor ever will. Hence, failures by an M.D. can be disastrous or even fatal; rarely will that be the case for a chiropractor or other "alternative" practitioner. This may well change, however, if the current push by chiropractic to become primary care practitioners for infants and children is successful. Pediatrics is much riskier than manipulating the spine of a middle-aged man who is there because he doesn't want surgery and wants to play golf that afternoon.

In short, chiropractic remains controversial, though not in all areas of its practice. It has established itself as being an effective treatment for lower back pain. It is attractive because there is no danger from side effects of drugs, since chiropractors don't generally recommend drugs to their patients. It is also attractive because it is seen as an alternative to surgery. And it is attractive because it is generally less expensive than treatment by a physician with drugs or surgery. Although, since chiropractic often requires repeated visits over many years, the overall cost of treatment can be substantial. Also, it should not be assumed that all medical doctors are quick to prescribe drugs or surgery. Many, like their chiropractic brothers and sisters, will recommend selected exercises or rest for specific back problems, both of which have been shown to be just as effective as chiropractic in relieving back pain.

Further reading: Jarvis 1991; Magner 1995; Ralph Smith 1984.

Chopra, Deepak

See **Ayurvedic medicine**.

chupacabra

Literally, "goat sucker." The chupacabra is an animal unknown to science, yet it is allegedly killing animals in places such as Puerto Rico, Mexico, and Chile. The creature's name originated with the discovery of some dead goats in Puerto Rico that had puncture wounds in their necks and their blood drained. According to *UFO Magazine* (March/April 1996), the chupacabra is responsible for more than 2,000 animal mutilations in Puerto Rico.

Puerto Rican authorities maintain that the deaths are due to attacks from groups of stray dogs or other exotic animals, such as the illegally introduced panther. The director of Puerto Rico's Department of Agriculture Veterinary Services Division, Hector Garcia, has stated that there is nothing unusual or extraordinary about the cases they've observed. One veterinarian said, "It could be a human being who belongs to a religious sect, even another animal. It could also be someone who wants to make fun of the Puerto Rican people."

Like other creatures in the **cryptozoologist's** barnyard, the chupacabra has been variously described. Some witnesses have seen a small half-alien, half-dinosaur, tailless vampire with quills running down its back; others have seen a pantherlike creature with a long snakelike tongue; still others have seen a hopping animal that leaves a trail of sulfuric stench. Some think the chupacabra is an alien animal, a pet, or an experiment gone awry. Such creatures are known as Anomalous Biological Entities (ABEs) in UFO circles.

Those who think the chupacabra is an ABE also believe that there is a massive government and mass media conspiracy to keep the truth hidden. This view is maintained despite the fact that the president of the Puerto Rico House of Representatives Agricultural Commission, Juan E. Lopez, has introduced a resolution asking for an official investigation to clarify the situation.

See also **cattle mutilations**.

clairaudience

An alleged **psychic** ability to hear things that are beyond the range of the ordinary power of hearing, such as voices or messages from the dead. Alleged clairaudients such as James Van Praagh and John Edward consider themselves *grief counselors*. They believe they give comfort to the living by providing "messages from the dead," even if the messages are sometimes absurd, for example, "Someone's nickname is Miss Piggy" or "Your dog wishes you hadn't given away his favorite dish."

See also **medium**.

Further reading: Frazier 1986, 1991; Kurtz 1985; Randi 1982a; Stein 1996b.

clairvoyance (second sight)

An alleged **psychic** ability to see things beyond the range of the power of vision. Clairvoyance is usually associated with **precognition** or **retrocognition**.

The faculty of seeing into the future is called "second sight" if it is not induced by **scrying**, drugs, trance, or other artificial means.

Further reading: Frazier 1991; Frazier and Randi 1981; Gardner 1989; Randi 1982a; Sagan 1979, 1995; Stein 1996b; Steiner 1996.

clustering illusion

The intuition that random events that occur in clusters are not really random events. The illusion is due to **selective**

thinking and is based on a false assumption. For example, it strikes most people as unexpected if heads comes up four times in a row during a series of coin flips. However, in a series of 20 flips, there is a 50% chance of getting four heads in a row (Gilovich 1993). It may seem unexpected, but the chances are better than even that any given neighborhood in California will have a statistically significant cluster of cancer cases (Gawande 1999).

What would be rare, unexpected, and unlikely due to chance would be to flip a coin 20 times and have each result be the alternate of the previous flip. In any series of such random flips, it is more *unlikely* than likely that short runs of 2, 4, 6, 8, and so on will yield what we know logically is predicted by chance. In the long run, a coin flip will yield 50% heads and 50% tails (assuming a fair flip and a fair coin). But in any short run, a wide variety of probabilities are expected, including some runs that seem contrary to intuition.

Finding a statistically unusual number of cancers in a given neighborhood—such as six or seven times greater than the average—is not rare or unexpected. Much depends on where you draw the boundaries of the neighborhood. Clusters of cancers that are *7,000 times* higher than expected, such as the incidence of mesothelioma in Karian, Turkey, are very rare and unexpected. The incidence of thyroid cancer in children near Chernobyl was *100 times* higher after the disaster (Gawande 1995).

Sometimes a subject in an **ESP** experiment or a **dowser** might be correct at a higher than chance rate. However, such results do not indicate that an event is not a chance event. In fact, such results are predictable by the laws of chance. Rather than being signs of nonrandomness, they are actually signs of randomness. ESP researchers are especially prone to take streaks of "hits" by their subjects as evidence that psychic power varies from time to time. Their use of **optional starting and stopping** is based on the presumption of **psychic** variation and an apparent ignorance or ignoring of the probabilities of random events. Combining the clustering illusion with **confirmation bias** is a formula for **self-deception** and delusion.

In epidemiology, the clustering illusion is known as the **Texas sharpshooter fallacy**. Khaneman and Tversky (1971) called it "belief in the Law of Small Numbers" because they identified the clustering illusion with the fallacy of assuming that the pattern of a large population will be replicated in all of its subsets. In logic, this fallacy is known as the fallacy of division, assuming that the parts must be exactly like the whole.

coincidences

See **law of truly large numbers**.

cold reading

A set of techniques used by **psychics**, **mentalists**, and professional manipulators to get a subject to behave in a certain way or to think that the cold reader has some sort of special ability that allows him to "mysteriously" know intimate things about the subject, such as personality traits, career choices, or plans for the future. The cold reader names dead relatives or living friends and identifies events from the subject's past and present. Subjects are often taken aback at the ability of a total stranger to know one's innermost secrets. Cold reading goes beyond the usual tools of manipulation, such as suggestion and flattery. In cold reading, one banks on the subject's inclination to find more meaning in a situation than there actually is. The desire to make sense out of our experience has led us to many

wonderful discoveries, but it has also led to many follies. The manipulator knows that his mark will be inclined to try to make sense out of whatever he is told, no matter how farfetched or improbable. He knows, too, that people are generally self-centered, that we tend to have unrealistic views of ourselves, and that we will generally accept claims about us that reflect not how we are or even how we really think we are but how *we wish we were* or *think we should be*. He also knows that for every several claims he makes that will be rejected as inaccurate, he will make one that is approved. He knows that the mark will remember the hits and forget the misses.

Thus, a good manipulator can provide a reading of a total stranger that will make the stranger feel that the manipulator possesses **paranormal** powers. For example, Bertram **Forer** has never met you, the reader, yet he offers the following cold reading of you:

Some of your aspirations tend to be pretty unrealistic. At times you are extroverted, affable, sociable, while at other times you are introverted, wary and reserved. You have found it unwise to be too frank in revealing yourself to others. You pride yourself on being an independent thinker and do not accept others' opinions without satisfactory proof. You prefer a certain amount of change and variety, and become dissatisfied when hemmed in by restrictions and limitations. At times you have serious doubts as to whether you have made the right decision or done the right thing. Disciplined and controlled on the outside, you tend to be worrisome and insecure on the inside.

Your sexual adjustment has presented some problems for you. While you have some personality weaknesses, you are generally able to compensate for them. You have a great deal of unused capacity which you have not turned to your advantage. You have a tendency to be critical of yourself. You have a strong need for other people to like you and for them to admire you.

This assessment satisfies many customers, but it was taken from a newsstand astrology book.

The selectivity of the human mind is always at work. It is not simply because we are gullible or suggestible that we are easily manipulated. Nor is it simply because the signs and symbols used by the manipulator are vague or ambiguous. Even when the signs are clear and we are skeptical, we can still be manipulated. In fact, it may even be the case that particularly bright persons are more likely to be manipulated when the language is clear and they are thinking logically. To make the connections that the manipulator wants you to make, you must be thinking logically.

Not all cold readings are done by malicious manipulators. Some readings are done by people who genuinely believe they have paranormal powers. They are just as impressed by their correct predictions or "insights" as their clients are. We should remember, however, that just as scientists may sometimes be wrong in their predictions, quacks may sometimes be right in theirs.

There seem to be three common factors in cold readings. One factor involves fishing for details. The psychic says something at once vague and suggestive, for example, "I'm getting a strong feeling about January here." If the subject responds, positively or negatively, the cold reader's next move is to play off the response. For example, if the subject says, "I was born in January," or "My mother died in January," then the reader says something like "Yes, I can see that," anything to reinforce the idea that the reader was more precise than he or she really was. If the subject responds negatively—"I can't think of anything particularly special about January"—the reader might reply, "Yes, I see that you've suppressed a memory about it. You don't want to be reminded of it. Something painful in January. Yes, I feel it. It's in the lower back [fishing] . . . oh, now it's in the heart [fishing] . . . umm, there seems to be a sharp pain in the head [fishing] . . . or the neck [fishing]." If the subject gives no response, the reader can leave the area, having firmly implanted in everybody's mind that the reader really did "see" something but the subject's suppression of the event hinders both the reader and the subject from realizing the specifics of it. If the subject gives a positive response to any of the fishing expeditions, the reader follows up with more of "I see that very clearly, now. Yes, the feeling in the heart is getting stronger."

Fishing is a real art and a good mentalist carries a variety of bait in his memory. For example, professional mentalist and author of one of the best books on cold reading, Ian Rowland (2002), says that he has committed to memory such things as the most common male and female names and a list of items likely to be laying about the house such as an old calendar, a photo album, newspaper clippings, and so on. Rowland also works on certain themes that are likely to resonate with most people who consult psychics: love, money, career, health, and travel. Since cold reading can occur in many contexts, there are several tactics Rowland covers. But whether one is working with **astrology**, **graphology**, **palmistry**, **psychometry**, or **tarot cards**, or whether one is **channeling** messages from the dead à la James Van Praagh, there are specific techniques one can use to impress clients with one's ability to know things that seem to require paranormal powers.

Finally, the subject will forget those occasions where the reader has guessed wrongly about the subject. What will be remembered are the seeming hits, giving the overall impression of "Wow, how else could she have known all this stuff unless she is psychic?" This same phenomenon of suppression of contrary evidence and **selective thinking** is so predominant in every form of psychic demonstration that it seems to be related to the old psychological principle: *We see what we want to see and disregard the rest.*

(The **psychic** entry has a detailed description of how to do a cold reading, from psychologist Ray Hyman.)

See also **confirmation bias** and **wishful thinking**.

Further reading: Dickson and Kelly 1985; Hyman 1977, 1989; Randi 1982a.

collective hallucinations

Sensory hallucinations in groups of people, induced by the power of suggestion. They generally occur in heightened emotional situations, especially among the religiously devoted. The expectancy and hope of bearing witness to a **miracle**, combined with long hours of staring at an object or place, make certain religious persons susceptible to seeing such things as weeping statues, moving icons and holy portraits, or the Virgin Mary in the bark of a tree or in the clouds.

The similarity of hallucinatory accounts of a miraculous vision is due to the witnesses having the same preconceptions and expectations, as well as having heard the same accounts from others. Furthermore, dissimilar accounts converge toward harmony as time passes and the accounts get retold. Those who say they see nothing extraordinary are dismissed as not having **faith**. Some, no doubt, see nothing but "rather than admit they failed . . . would imitate the lead given by those who did,

and subsequently believe that they had in fact observed what they had originally only pretended to observe" (Rawcliffe 1988: 114).

Not all collective hallucinations are religious. In 1897, Edmund Parish reported shipmates who had shared a ghostly vision of their cook who had died a few days earlier. The sailors not only saw the ghost, but also distinctly saw him walking on the water with his familiar and recognizable limp. Their ghost turned out to be a "piece of wreck, rocked up and down by the waves" (ibid.: 115).

See also Our Lady of **Watsonville** and **pareidolia**.

Further reading: Nickell 1993; Slade and Bentall 1988.

collective unconscious

See Carl **Jung**.

communal reinforcement

The process by which a claim becomes a strong belief through repeated assertion by members of a community. The process is independent of whether the claim has been properly researched or is supported by empirical data significant enough to warrant belief by reasonable people. Often, the mass media contribute to the process by uncritically supporting the claims. More often, however, the mass media provide tacit support for untested and unsupported claims by saying nothing skeptical about even the most outlandish of claims.

Communal reinforcement explains how entire nations can pass on ineffable gibberish from generation to generation. It also explains how **testimonials** reinforced by other testimonials within the community of therapists, sociologists, psychologists, theologians, politicians, philosophers, talk show hosts, and others can supplant and be more powerful than scientific studies or accurate gathering of data by disinterested parties.

confabulation

A fantasy that has unconsciously replaced fact in memory. A confabulation may be based partly on fact or be a complete construction of the imagination.

The term is often used to describe the "memories" of mentally ill persons, memories of **alien abduction**, and **false memories** induced by therapists or interviewers.

confirmation bias

A type of **selective thinking** whereby one tends to notice and to look for what confirms one's beliefs, and to ignore, not look for, or undervalue the relevance of what contradicts one's beliefs. For example, if you believe that during a full moon there is an increase in admissions to the emergency room where you work, you will take notice of admissions during a full moon, but be inattentive to the moon when admissions occur during other nights of the month. A tendency to do this over time unjustifiably strengthens your belief in the relationship between the full moon and accidents or other **lunar effects**.

This tendency to give more attention and weight to data that support our beliefs than we do to contrary data is especially pernicious when our beliefs are little more than prejudices. If our beliefs are firmly established upon solid evidence and valid confirmatory experiments, the tendency to give more attention and weight to data that fit with our beliefs should not lead us astray as a rule. Of course, if we become blinded to evidence truly refuting a favored hypothesis, we have crossed the line from reasonableness to closed-mindedness.

Numerous studies have demonstrated that people generally give an excessive amount of value to confirmatory information, that is, to positive or supportive data. The "most likely reason for the excessive influence of confirmatory information is that it is easier to deal with cognitively" (Gilovich 1993). It is much easier to see how a piece of data supports a position than it is to see how it might count against the position. Consider a typical **ESP** experiment or a seemingly **clairvoyant** dream: Successes are often unambiguous or data are easily massaged to count as successes, while negative instances require intellectual effort to even see them as negative or to consider them as significant. The tendency to give more attention and weight to the positive and the confirmatory has been shown to influence **memory**. When digging into our memories for data relevant to a position, we are more likely to recall data that confirm the position (ibid.).

Researchers are sometimes guilty of confirmation bias by setting up experiments or framing their data in ways that will tend to confirm their hypotheses. They compound the problem by proceeding in ways that avoid dealing with data that would contradict their hypotheses. For example, **parapsychologists** are notorious for using **optional starting and stopping** in their ESP research. Experimenters might avoid or reduce confirmation bias by collaborating in experimental design with colleagues who hold contrary hypotheses. Individuals have to constantly remind themselves of this tendency and actively seek out data contrary to their beliefs. Since this is unnatural, it appears that the ordinary person is doomed to bias.

See also **control group study** and **self-deception**.

Further reading: Evans 1990; Gould 1987; Martin 1998.

coning

See **ear candling**.

conjuring

The art of legerdemain, of magical tricks, of performance of feats seemingly requiring the assistance of supernatural or **paranormal** powers or forces.

To summon a demon or spirit by **invocation** or **incantation** is also called conjuring.

conspiracy theorists

See **Illuminati**.

control group study

A control group study compares an experimental group with a parallel, nonexperimental group (the control) in a test of a causal hypothesis. The control and experimental groups must be identical in all relevant ways except for the introduction of a suspected causal agent into the experimental group. If the suspected causal agent is actually a causal factor of some event, then logic dictates that that event should manifest itself more significantly in the experimental than in the control group. Significance is measured by relation to chance: If an event is not likely due to chance, then its occurrence is significant.

A double-blind test is a control group test where neither the evaluator nor the subject knows which group is the control group. A random test is one that randomly assigns subjects to the control or experimental groups.

The purpose of double-blind and random controlled studies is to reduce error due to such things as **communal reinforcement**, **confirmation bias**, **self-deception**, **subjective validation**, and **wishful thinking**.

The lack of testing under controlled conditions explains why many **astrologers**, **dowsers**, **graphologists**, New Age **psychotherapists**, and **psychics** believe in their abilities. To test a dowser, for example, it is not enough to have the dowser and his friends tell you that it works by pointing out all the wells that have been dug on the dowser's advice. One should perform a random, double-blind test, such as one done by Ray Hyman with an experienced dowser on the PBS program "Frontiers of Science" (first aired on November 19, 1997). The dowser claimed he could find buried metal objects, as well as water. He agreed to a test that involved randomly selecting numbers that corresponded to buckets placed upside-down in a field. The numbers determined which buckets a metal object would be placed under. The one doing the placing of the objects was not the same person who went around with the dowser as he tried to find the objects. The exact odds of finding a metal object by chance could be calculated. For example, if there are 100 buckets and 10 of them have a metal object, then getting 10% correct would be predicted by chance. That is, over a large number of attempts, getting about 10% correct would be expected of anyone, with or without a dowsing rod. On the other hand, if someone consistently got 80% or 90% correct, and we were sure he or she was not cheating, that would usually be considered a confirmation of the dowser's powers.

The dowser walked up and down the lines of buckets with his rod but said he couldn't get any strong readings. When he selected a bucket he qualified his selection with something to the effect that he didn't think he'd be right. He was right about never being right! He didn't find a single metal object despite several attempts. His performance is typical of dowsers tested under controlled conditions. His response was also typical: he was genuinely surprised. Others might explain his failure by the **ad hoc hypothesis** that **paranormal** abilities come and go, and they usually go when being tested by a skeptic because of the lack of faith or abundance of skepticism in the air.

Many control group studies use a **placebo** in control groups to keep the subjects in the dark as to whether they are being given the causal agent that is being tested. For example, both the control and experimental groups will be given identical looking pills in a study testing the effectiveness of a pharmaceutical drug or a natural substance. Only one pill will contain the agent being tested; the other pill will be a placebo. In a double-blind study, the evaluator of the results would not know which subjects got the placebo until his or her evaluation of observed results was completed. This is to avoid evaluator bias from influencing observations and measurements.

Further reading: Giere 1998; Kourany 1997; Sagan 1995.

cosmobiology

A method of **astrology** developed by Reinhold Ebertin in the 1920s. Cosmobiology shuns the use of traditional house systems and uses a complicated charting method that places special importance on *midpoints* to develop a *cosmogram*. A midpoint is a point halfway between two planets or other notable celestial objects. For example, "The distance between 0 degrees Aries and 0 degrees Cancer is 90 degrees. Half of 90 is 45, so the midpoint would be located at 15 degrees Taurus" (members.tripod.com/~junojuno2/urast.htm). Cosmobiologists consider "indirect midpoints" to be important, too. "The point opposite 15 degrees Taurus is 15 Scorpio: this is an indirect midpoint. In fact, it is common to use all indirect midpoints at 45 or even 22.5-degree intervals. Indirect midpoints carry nearly the same energy as a direct midpoint" (ibid.).

Ebertin's influence increased astronomically after the publication of *The Combination of Stellar Influences* in 1940, which gives interpretations for all possible planetary combinations and midpoints. Cosmobiologists, like other astrologers, are consulted for advice in personal and business matters, assisting in medical diagnoses, and in matters regarding fertility.

Further reading: Eysenck and Nias 1982; Kelly 1981, 1998.

Course in Miracles, A (ACIM)

The name of a book allegedly dictated by Jesus to Helen Schucman (1909–1981), a research psychologist. *ACIM* is Christianity improved: Jesus wants less suffering, sacrifice, separation, and sacrament. He also wants more love and forgiveness.

ACIM is a minor industry. To find out what Jesus really had in mind when he came to save the world, you can buy *ACIM* or one of a dozen similar books from the Foundation for Inner Peace (FIP). About 1.5 million copies of *ACIM* were sold worldwide between 1976 and 2002. FIP also sells audio and videotapes, and conducts workshops, seminars, and discussion groups. Its sister organization, Foundation for a Course in Miracles (FACIM), has an academy where you can get the Holy Spirit to help you understand the real message of

Jesus. You may then return to your every-day life situation with a deeper appreciation for the difference between appearance and reality, illusion and truth. The academy is known as "The Institute for Teaching Inner Peace Through a Course in Miracles" (TIP).

Why should anyone believe that the words of Helen Schucman are the words of Jesus? She was a clinical psychologist by training (she received a Ph.D. in 1957 from New York University). She claims that from 1965 to 1972 an inner voice dictated to her the three books that make up *ACIM*. She was assisted by a colleague, William Thetford (1923–1988). In 1972, another psychologist and his wife, Kenneth and Gloria Wapnick, assisted Schucman with her work. The Wapnicks are the ones who started FACIM.

Why is it called *A Course in Miracles*? According to Schucman, the voice said to her: "This is a course in miracles, please take notes."

ACIM got a big boost when Marianne Williamson, one of America's most popular New Age spirituality writers, began promoting her version of it in the 1990s. The teachings of *ACIM* are not new. They have been culled from various sources, east and west. That does not make them false or worthless. But to claim that they were dictated by someone who has been dead for 2,000 years is a bit much to swallow. Did she really hear voices or didn't she want to take responsibility for what she thought?

craniometry

The measurement of cranial features in order to classify people according to race, criminal temperament, intelligence, and so on. The underlying assumption of craniometry is that skull size and shape determine brain size, which determines such things as intelligence and capacity for moral behavior. Empirical evidence for this assumption is lacking, but that has not hindered some people from claiming they are members of a superior race or gender because the head size of their racial or gender group is larger on average than the head size of some other racial or gender group. These people also reason fallaciously that *any* member of a superior race or gender must be superior to *all* members of inferior races and to *all* members of the other gender. What is true of the whole or group, however, is not necessarily true of the parts or members of the group. Logicians call this kind of reasoning the *fallacy of division*.

In the 19th century, the British used craniometry to justify racist policies toward the Irish and black Africans. Irish skulls were said to have the shape of Cro-Magnon men and were akin to that of apes, proof of their inferiority along with black Africans. In France, Paul Broca demonstrated that women are inferior to men because of their smaller crania. He argued against higher education for

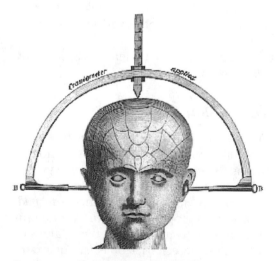

Craniometer from George Combe, *Elements of Phrenology*, 2nd ed. (1834).

women because their small brains couldn't handle the demands (Bem 1993). In the 20th century, the Nazis used craniometry and **anthropometry** to distinguish Aryans from non-Aryans.

Because of its association with racism, many anthropologists avoid craniometry, but others believe that by making some 90 measurements of a skull they can identify the continent of origin with a high degree of accuracy. By using such measurements, Dr. Corey S. Sparks of Pennsylvania State University claims that Kennewick Man—which Native Americans want returned for burial—most closely resembles the Ainu, the original inhabitants of Japan, and is not a distant relative of today's Native Americans.

See also **phrenology**.

Further reading: Bolt 1971; Gould 1993a, 1996; Singer 1993.

craniosacral therapy

A physical therapy that involves the manipulation of the skull bones (the cranium) and the sacrum to relieve pain and a variety of other ailments, including cancer. It is also known as *craniopathy* and *cranial osteopathy*. (The sacrum is part of the vertebral column, composed of five fused vertebrae that form the posterior pelvic wall.) Osteopath William G. Sutherland invented the therapy in the 1930s. Another osteopath, John Upledger, is the leading proponent of craniosacral therapy today. Like many quack therapies, this one emphasizes subjective concepts such as **energy**, *harmony, balance, rhythm,* and *flow*.

Craniosacral therapists claim to be able to detect a craniosacral "rhythm" in the cranium, sacrum, cerebrospinal fluid, and the membranes that envelop the craniosacral system. The balance and flow of this rhythm is considered essential to good health. The rhythm is measured by the therapist's hands. Any needed or effected changes in rhythm are also detected only by the therapist's hands. No instrument is used to measure the rhythm or its changes; hence no systematic objective measurement of healthy versus unhealthy rhythms exists. The measurement, the therapy, and the declared cure are all subjectively based. As one therapist puts it:

> During the treatment, the client is usually supine on a table. The therapist assesses the patterns of energy in the body through touch at several "listening stations" and then decides where to start that day and how to focus the treatment. [www.vicpain.com/therapy.htm#CRANIO]

The same therapist maintains that the therapy is "a waste of time and money" for people who do not have **faith** in the therapy.

Skeptics note that the skull does not consist of movable parts (unlike the jaw), and the only rhythm detectable in the cranium and cerebrospinal fluid is related to the cardiovascular system. When tested, several therapists were unable to consistently come up with the same measurements of the alleged craniosacral rhythm (Wirth-Pattullo and Hayes 1994).

See also **alternative health practices, chiropractic,** and **therapeutic touch.**

Further reading: Barrett: www.quackwatch.com/04ConsumerEducation/QA/osteo.html; Jarvis: www.ncahf.org/articles/c-d/cranial.html.

cranioscopy

See **phrenology**.

creationism, creation science

Creationism is the **metaphysical** theory that a supernatural being created the universe. Creation science is a **pseudoscientific** theory that claims that the stories in

Genesis are accurate scientific accounts of the origin of the universe and life on Earth, and that the Big Bang theory and the theory of evolution must be incorrect since they are incompatible with the stories in Genesis. "Creation science" is an oxymoron, since **science** is concerned only with **naturalistic** explanations of empirical phenomena and does not concern itself with *supernatural* explanations of anything.

Creationism is not necessarily connected to any particular religion. Millions of Christians and non-Christians believe there is a creator of the universe and that scientific theories such as the theory of evolution do not conflict with belief in a creator. However, those Christians calling themselves *creation scientists* have coopted the term "creationism," making it difficult to refer to creationism without being understood as referring to Scientific Creationism. Thus, it is commonly assumed that creationists are Christians who believe that the account of the creation of the universe as presented in Genesis is literally true in its basic claims about Adam and Eve, the six days of creation, making day and night on the first day even though He didn't make the sun and moon until the fourth day, making whales and other animals that live in the water or have feathers and fly on the fifth day, and making cattle and things that creep on the earth on the sixth day.

Creation scientists claim that the Big Bang theory and the theory of evolution are false and that scientists who advocate such theories are ignorant of the truth about the origins of the universe and life on Earth. They also claim that creationism is a *scientific* theory and should be taught in our science curriculum as a competitor to the theory of evolution.

One of the main leaders of creation science is Duane T. Gish of the Institute for Creation Research, who puts forth his views mainly in the form of attacks on evo-lution. Gish is the author of *Evolution: The Fossils Say No!* (1978), and admits:

> We do not know how the Creator created, what processes He used, for He used processes which are not now operating anywhere in the natural universe. This is why we refer to creation as special creation. We cannot discover by scientific investigations anything about the creative processes used by the Creator.

Gish also authored *Evolution: The Challenge of the Fossil Record* (1985) and *Evolution: The Fossils Still Say No!* (1985).

Another leader of this movement is Walt Brown of the Center for Scientific Creationism. Despite the fact that most of the scientific community considers evolution of species from other species to be a fact, creation scientists proclaim that evolution is not a fact but *just* a theory, and that it is false. The vast majority of scientists who disagree about evolution disagree as to *how* species evolved, not as to *whether* they evolved. As the eminent geneticist Theodosius Dobzhansky put it: "Nothing in biology makes sense except in the light of evolution."

Scientific creationists are not impressed that they are in the minority. After all, they note, the entire scientific community has been wrong before. That is true. For example, at one time the geologists were all wrong about the origin of continents. They thought the earth was a solid object. Now they believe that the earth consists of plates. The theory of plate tectonics has replaced the old theory. However, when the entire scientific community has been proved to be wrong in the past, it has been proved to be wrong by other scientists, not by religionists. They have been proved wrong by others doing empirical investigation, not by others who begin with **faith** in a religious dogma and who see no need to do any empirical investigation to support their theory. Erroneous scientific theories have been replaced by better theories, theories that *explain more* empirical phenomena and that increase our understanding of the natural world. Plate tectonics not only explained how continents can move, it also opened the door for a greater understanding of how mountain ranges form, how earthquakes are produced, and how volcanoes are related to earthquakes. Creationism is no more a scientific alternative to natural selection than the stork theory is a scientific alternative to sexual reproduction (Hayes 1996). The creationist theory has not led to a better understanding of biological or physical phenomena, and it is unlikely that it ever will.

Creation science bears little resemblance to what Charles Darwin did as a scientist. Darwin's theory of how evolution happened is called natural selection. That theory is quite distinct from the *fact* of evolution. Other scientists have different theories of evolution, but only a negligible few deny the fact of evolution. In the *Origin of Species,* Darwin provided vast amounts of data about the natural world that he and others had collected or observed. Only after providing the data did he argue that his theory accounted for the data much better than the theory of special creation. Gish, on the other hand, assumes that whatever data there are must be explained by special creation because, he thinks, God said so in the Bible. Furthermore, Gish claims that it is impossible for us to understand special creation, since the Creator "used processes which are not now operating anywhere in the natural universe." Thus, Gish, rather than gather data and demonstrate how special creation explains the data better than natural selection, must take another approach, the approach of apologetics. His approach, and that of all the other creation scientists, is to attack at

every opportunity what they take to be the theory of evolution. Rather than show the strengths of their own theory, they rely on trying to find and expose weaknesses in evolutionary theory. Gish and the other creation scientists actually have no interest in *scientific* facts *or* theories. Their interest is in defending the faith against what they see as attacks on God's Word.

Creation science is not science but pseudoscience. It is religious dogma masquerading as scientific theory. Creation science is put forth as being absolutely certain and unchangeable. It assumes that the world must conform to its understanding of the Bible. Where creation science differs from creationism in general is in its notion that once it has interpreted the Bible to mean something, no evidence can be allowed to change that interpretation. Instead, the evidence must be refuted.

Compare this attitude with that of the leading European creationists of the 17th century who had to admit eventually that Earth is not the center of the universe and that the sun does not revolve around our planet. They did not have to admit that the Bible was wrong, but they did have to admit that human interpretations of the Bible were in error. Today's creationists seem incapable of admitting that their interpretation of the Bible could be wrong. What should be an affront to many Christians and non-Christian creationists is the insinuation that if one does not adhere to the creation scientist's interpretation of the Bible, one is offending God. Many creationists believe that God is behind the beautiful unfolding of evolution (Haught 1999). There is no contradiction in believing that what appears to be a mechanical, purposeless process from the human perspective can be teleological and divinely controlled. Natural selection does not require that one "get rid of God as the creator of life" any more than heliocentrism requires one to get rid of God as the creator of the heavens.

What is most revealing about the creation scientists' lack of any true scientific interest is the way they willingly and uncritically accept even the most preposterous claims, if those claims seem to contradict traditional scientific beliefs about evolution. For example, the creationists welcome any evidence that seems to support the notion that dinosaurs and humans lived together. Also, the way creation scientists treat the second law of thermodynamics indicates either gross scientific incompetence or deliberate dishonesty. They claim that evolution of life forms violates the second law of thermodynamics, which "specifies that, on the macroscopic scale of many-body processes, the entropy of a closed system cannot decrease" (Stenger 2000):

> Consider simply a black bucket of water initially at the same temperature as the air around it. If the bucket is placed in bright sunlight, it will absorb heat from the sun, as black things do. Now the water becomes warmer than the air around it, and the available energy has increased. Has entropy *decreased*? Has energy that was previously *unavailable* become available, in a closed system? No, this example is only an apparent violation of the second law. Because sunlight was admitted, the local system was not closed; the energy of sunlight was supplied from outside the local system. If we consider the larger system, including the sun, entropy has *increased* as required. [Klyce: www .panspermia.org/seconlaw.htm]

Creation scientists treat the evolution of species as if it were like the bucket of water in the example above, which they incorrectly claim occurs in a closed system. If we consider the entire system of nature, there is no evidence that the second law of thermodynamics is violated by evolution.

Finally, although philosopher Karl Popper's (1959) notion that falsifiability distinguishes scientific from metaphysical theories has been much attacked by philosophers of science (Kitcher 1983), it seems undeniable that there is something profoundly different about such theories as creationism and natural selection. It also seems undeniable that one profound difference is that the metaphysical theory is consistent with every conceivable empirical state of affairs, while the scientific one is not. "I can envision observations and experiments that would disprove any evolutionary theory I know," writes Stephen Jay Gould, "but I cannot imagine what potential data could lead creationists to abandon their beliefs. Unbeatable systems are dogma, not science" (Gould 1983).

Creationism can't be refuted, even in principle, because everything is consistent with it, even apparent contradictions and contraries. Scientific theories allow definite predictions to be made from them; they can, in principle, be refuted. Theories such as the Big Bang theory, the steady state theory, and natural selection can be tested by experience and observation. Metaphysical theories such as creationism are "airtight" if they are self-consistent, that is, contain no self-contradictory elements. No scientific theory is ever airtight.

What makes scientific creationism a pseudoscience is that it attempts to pass itself off as science even though it shares none of the essential characteristics of scientific theorizing. Creation science will remain forever unchanged as a theory. It will engender no debate among scientists about fundamental mechanisms of the universe. It generates no empirical predictions that can be used to test the theory. It is taken to be irrefutable. And it assumes a priori that there can be no evidence that will ever falsify it.

The fact that Gish is unable to convert even a small segment of the scientific community to his way of thinking is a strong indication that his arguments have little merit. This is not because the majority must be right. Of course, the entire scientific community could be deluded. However, since the opposition issues from a religious dogmatist who is doing not scientific investigation but theological apologetics, it seems more probable that it is the creation scientists who are deluded rather than the evolutionary scientists.

Further reading: Dawkins 1995, 1996; Dobzhansky 1982; Gardner 1957; Gould 1979, 1993b; Pennock 1999; Plimer 1994; Schadewald 1986; Shermer 1997.

crop circles

Geometric patterns, some very intricate and complex, appearing in fields, usually wheat fields and usually in England. Most, if not all, are probably due to pranksters,

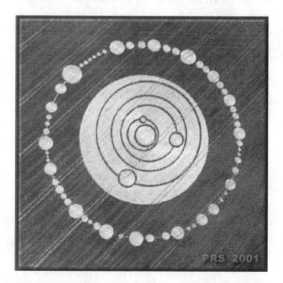

Crop circle art entitled "Earth Is Missing." © 2001 Peter Sorensen.

such as Doug Bower and David Chorley, who in 1991 admitted to creating approximately 250 circles over many years.

There is a segment of the population that believes the circles are messages from aliens. Some maintain that the aliens are trying to communicate with us using ancient Sumerian symbols or symbolic representations of alien DNA. Those who engage in such serious study and theorizing about crop circles are known as "cerealogists" or "croppies."

Even scientifically minded people have been brought into this fray. They have wisely avoided the thesis that aliens have been carving out messages in fields. But they have stretched their imaginations to come up with theories of vortexes, **ball lightning**, plasma, and other less **occult** explanations involving natural forces such as wind, heat, or animals. However, when looking for a **naturalistic** explanation of weird things, we should never omit from our checklist the possibility that the phenomenon we are studying is a hoax.

Had crop circles existed in the 13th century, they would have been attributed to **Satan**, who was said to have been responsible for many things, including the construction of Stonehenge and Hadrian's Wall. It was believed by many that the ancients could not possibly have accomplished such feats on their own. Today, Satan's power to produce weird or wondrous things has been usurped by aliens.

See also **alien abductions**, **ancient astronauts**, **cattle mutilations**, **flying saucers**, and **UFOs**.

Further reading: Huston 1997; Nickell 1992; Schnabel 1994.

Crowley, Aleister (1875–1947)

Aleister Crowley was a self-proclaimed drug and sex fiend, a mostly self-published author of books on the **occult** and **magick**, a poet and mountaineer, and a leader of a **cult** called Ordo Templi Orientis (OTO), whose tenets he detailed in one of his many writings, *The Book of the Law.* The latter contains his version of the *Law of Thelema,* which Crowley claims he **channeled** for a "praeterhuman intelligence" called Aiwass.

Do what thou wilt shall be the whole of the Law is his motto for OTO. In practice, for Crowley this meant rejecting traditional morality in favor of the life of a drug addict and womanizer. ("I rave; and I rape and I rip and I rend" is a line from one of his poems. *Diary of a Drug Fiend* is the title of one of his books.) He claimed to identify himself with the Great Beast 666 (from the *Book of Revelation*) and enjoyed the self-appellation of "wickedest man in the world." He had two wives; both went insane. Five mistresses committed suicide. According to Martin Gardner, "scores of his concubines ended in the gutter as alcoholics, drug addicts, or in mental institutions" (Gardner 1992: 198). However, Crowley should not be blamed for destroying the virtue of saintly young girls. His allure was such that the women who were attracted to him tended already to be alcoholics, drug addicts, or emotionally disturbed. His allure seems to have consisted of two main qualities: He inherited a fortune and he worked hard at being strange.

Crowley's *Magick in Theory and Practice* is a very popular book among occultists. When Dover was about to release a reprint of the book in 1990, an editor asked Martin Gardner to write a foreword for the reprinting. The 1976 Dover edition had been one of their best sellers. Gardner was an unlikely choice to write the foreword for Crowley's book, since he had already written that Crowley was a no-good fraud in his classic *Fads and Fallacies in the Name of Science.* Gardner wrote the foreword and

painted a picture of such a cruel, despicable, egotistical charlatan that Dover decided not to reprint the book. "His reputation had been that of a man who worshipped Satan," Gardner wrote of Crowley, "but it was more accurately said that he worshipped no one except himself." The foreword has been published in Gardner's *On the Wild Side* (1992).

Given Crowley's reputation, it is inevitable that he would have a certain appeal to certain rock musicians of the late 20th century. Jimmy Page, Led Zeppelin guitarist and occultist, bought Crowley's mansion, Boleskine House, near Foyers, Scotland, and owns a large collection of Crowley memorabilia. And Crowley's face is one of many on the album cover of the Beatles' *Sergeant Pepper's Lonely Hearts Club Band.*

Further reading: Crowley 1989.

cryptomnesia

Literally, "hidden memory." The term is used to explain the origin of experiences believed to be original but which are actually based on forgotten memories of events. **Past life regressions** induced through **hypnosis** may be **confabulations** fed by cryptomnesia.

Cryptomnesia may also explain how the apparent plagiarism of such people as Helen Keller or George Harrison of the Beatles might actually be cases of hidden memory. Harrison didn't intend to plagiarize the Chiffons' song "He's So Fine" in "My Sweet Lord." Nor did Keller intend to plagiarize Margaret Canby's fairy tale "The Frost Fairies" when she wrote "The Frost King." Both may simply be cases of not having a conscious memory of their experiences with the works in question.

See also Bridey **Murphy** and **memory**.

Further reading: Baker 1996a; Schick and Vaughn 1998; Schultz 1989.

cryptozoology

Literally, "the study of hidden animals." Cryptozoology is the study of such creatures as **Bigfoot**, the Australian **bunyips**, the **chupacabra**, the **Loch Ness Monster**, **Mokele-Mbembe**, and **unicorns**. It is not a recognized branch of the science of zoology.

Cryptozoology relies heavily on **testimonials** and circumstantial evidence in the form of legends and folklore, and the stories and alleged sightings of mysterious beasts by indigenous peoples, explorers, and travelers. Since cryptozoologists spend most of their energy trying to establish the existence of creatures, rather than examining actual animals, they are more akin to **psi** researchers than to zoologists. Expertise in zoology, however, is asserted to be a necessity for work in cryptozoology, according to Dr. Bernard Heuvelmans, who coined the term to describe his investigations of animals unknown to science.

Further reading: Bauer 1996a; Heuvelmans 1955; Shuker 1997.

crystal power

For centuries, crystals and other gems have been desired for their alleged magical healing and mystical **paranormal** powers such as **scrying**. These beliefs continue, even though there is no reliable evidence for them.

We can dismiss the prescientific belief in the magical powers of crystals and gemstones as due to the lack of scientific knowledge. Modern occultists, however, distort and falsify scientific knowledge in order to promote belief in their crystal products. Crystals are said to channel good **energy** and ward off bad energy. They carry "vibrations" that resonate with healing "frequencies," work with the **chakras**, and help balance **yin and yang**. Crystals allegedly help physical and emotional healing and can improve one's self-expression, creativity, meditation, and immune system.

Crystal wands are used to heal **auras** in **aura therapy** and can provide protection against harmful **electromagnetic forces** emitted from computer monitors, cellular phones, microwave ovens, hair dryers, power lines, and even other people. The latter claim is made by Charles Brown, a **chiropractor** from Montana and the inventor of the Bioelectric Shield. Brown claims he heard voices in his head and had visions in his bed as to how to arrange crystals in the shape of a flying saucer in order to provide this protection. Marketed as "jewelry with a purpose," his bioelectric shields are sold for anywhere from $139 to over a $1,000. Cherie Blair, the wife of England's prime minister, wears one of these magical pendants. They are said to be "medically proven" and "based on Nobel Prize winning physics." Even if the claims about the protective power of these pendants were true, it would be necessary to envelop your entire body in one to achieve the desired result. By hanging a little piece of jewelry around the neck, you might be able to protect a small part of the throat, however, but don't count on even that small amount of protection.

The New Age idea that crystals can harness and direct energy seems to be based on a misunderstanding of one of the more curious characteristics of certain crystals, namely, that they produce an electrical charge when compressed. This is known as the piezoelectric effect and was discovered in 1880 by Pierre and Jacques Curie. The piezoelectric effect, however, does not give crystals healing or protective power, despite the claims of those who use and sell crystals in New Age and neo-**pagan** shops. However, wearing crystals seems to give some people a *feeling* of protection. This feeling and their aesthetic qualities seem to be the only virtues of crystal jewelry.

There is no credible evidence that crystals work any better than animal organs for divining the future, although grinding crystals for **divination** is clearly more humane and sanitary than disembowelment of poor creatures who don't know yesterday from tomorrow.

See also **crystal skulls** and **divination**.

Further reading: Chittenden 1998; Jerome 1996; Randi 1995.

crystal skulls

Crystal skulls are stone carvings in the shape of human skulls. The sculptures vary in size from a few inches to life-size. Some are made of pure quartz **crystal**, but many are made of other types of stone found in abundance on Earth. Some stone skulls are genuine artifacts from Mesoamerican cultures such as the Aztec and are known as skull masks or death heads. But the crystal skulls that interest New Agers are extraterrestrial in origin or come from **Atlantis**. They allegedly are endowed with magical powers such as the spontaneous production of holographic images and the emission of weird sounds. Today, millions of skulls, made of various types of stones and metals, are manufactured in a variety of sizes for the New Age **paratrinket** market, as well as for the museum replica market. Despite the fact that replicas are easily made and are available from a variety of sources, advocates of the paranormal nature of crystal skulls such as Nick Nocerini claim that no one knows how these skulls were made and that they are impossible to duplicate. Nocerino is the founder of the Society of Crystal Skulls, International. His society uses **psychometry**, **remote viewing**, and **scrying** as part of its research methodology.

The myth of crystal skulls as extraterrestrial and extrapowerful seems to have begun with F. A. "Mike" Mitchell-Hedges (1882–1959), his adopted daughter Anna, and their "skull of doom," a clear quartz skull, with a detachable jaw, about 5 inches high and weighing about 11 pounds. Their creative fictions have been uncritically promoted by Frank Dorland, photographer for Phyllis Galde's *Crystal Healing: The Next Step* (1988), and Richard Garvin, author of *The Crystal Skull: The Story of the Mystery, Myth and Magic of the Mitchell-Hedges Crystal Skull Discovered in a Lost Mayan City During a Search for Atlantis* (1973). The myth has been carried on by Ellie Crystal, who likens the quest for crystal skulls to the quest for the Holy Grail, and Josh Shapiro, coauthor (with Nocerino and Sandra Bowen) of *Mysteries of the Crystal Skulls Revealed*.

The "skull of doom" was allegedly discovered by 17-year-old Anna Mitchell-Hedges in 1924 or 1927 while accompanying her adoptive father on an excavation of the ancient Mayan city of Lubaantun in Belize, where the elder Mitchell-Hedges believed he would find the ruins of Atlantis. The evidence proves beyond a reasonable doubt that Mitchell-Hedges bought the skull at a Sotheby's sale in 1943 for £400. The man who owned the piece, Sidney Burney, and those who were on the Lubannatun expedition deny that Mitchell-Hedges found the skull. Mitchell-Hedges himself never mentioned the skull until just after he bought it in 1943 (Nickell and Fischer 1991).

Much of the occult and sinister legend surrounding the so-called skull of doom originated with fabrications by Mitchell-Hedges, who claimed the skull is 3,600 years old and that "according to scientists it must have taken over 150 years, generation after generation working all the days of their lives, patiently rubbing down with sand an immense block of rock crystal until finally the perfect Skull emerged." He also claims that it "was used by the High

Priest of the Maya when performing eso-teric rites. . . . when he willed death with the help of the skull, death invariably fol-lowed" (crystalinks.com/skull.html).

Anna has continued the hoax. Even though there is no evidence that she was even at Lubaantun when the discovery was supposedly made, she has maintained that Burney had the piece on loan from her father only until he could pay off a debt he owed Burney. Anna has received some attention and made a few dollars over the years by putting her skull on display, claiming it came from outer space and was kept in Atlantis before it was brought to Belize. She is still in possession of the skull, but seems to have tired of the publicity and has retired it from public viewing.

In 1970, Anna let crystal carver Frank Dorland examine her skull. Dorland de-clared that it is excellent for scrying and emits sounds and light, depending on the position of the planets. He claimed that the skull originated in Atlantis and was car-ried around by the Knights Templar during the Crusades. He claims they had the skull examined at a Hewlett-Packard lab. D. Trull uncritically reports that the lab found that the skull

had been carved against the natural axis of the crystal. Modern crystal sculptors always take into account the axis, or ori-entation of the crystal's molecular sym-metry, because if they carve "against the grain," the piece is bound to shatter—even with the use of lasers and other high-tech cutting methods.

To compound the strangeness, HP could find no microscopic scratches on the crystal which would indicate it had been carved with metal instruments. Dor-land's best hypothesis for the skull's con-struction is that it was roughly hewn out with diamonds, and then the detail work was meticulously done with a gentle solu-tion of silicon sand and water. The exhausting job—assuming it could possi-bly be done in this way—would have required man-hours adding up to 300 years to complete. [D. Trull: www.parascope.com/en/1096/skull1.htm]

Dorland also claimed "crystal stimulates an unknown part of the brain, opening a psychic door to the absolute." Dorland's claims formed the basis of Garvin's book on crystal skulls.

The questionable origin of the Mitchell-Hedges skull has not deterred belief in the skull's mysterious properties. Rather, at least 13 other skulls have mys-teriously appeared over the years. Some of these skulls are claimed to have magical origins and healing powers. However, a study of several crystal skulls by the British Museum in 1996 indicates that the only magic involved in the creation of these skulls was in keeping their fraudu-lent origin a secret. The study concluded that the skulls were made in Germany within the past 150 years. The recent ori-gin explains how they were made with tools unavailable to the ancient Mayans or Aztecs.

A similar result occurred in 1992, when the Smithsonian received a crystal skull from an anonymous source who claimed it was an Aztec skull that had been bought in Mexico City in 1960. Research by the Smithsonian concluded that several crystal skulls popular with the New Agers originated with Eugene Boban, a French-man of dubious character. Boban dealt in antiques in Mexico City between 1860 and 1880, and seems to have acquired his skulls from a source in Germany. Jane MacLaren Walsh of the Smithsonian con-cluded that several crystal skulls held in museums were manufactured between 1867 and 1886.

Further reading: Lewallen 1997.

cults

The term "cult" expresses disparagement and is usually used to refer to unconventional religious groups, though the term is sometimes used to refer to nonreligious groups that appear to share significant features with religious cults. For example, there are some who refer to Amway and **Landmark Forum** as cults, but the term is best reserved for groups such as Scientology, the Order of the Solar Temple (74 suicides in 1984), Heaven's Gate (39 suicides in 1997), the **Raëlians**, the **Urantians**, the Maharishi Mahesh Yogi's **Transcendental Meditation** program, and the group that followed the Rev. Jim Jones to Guyana where more than 900 joined in a mass murder/suicide ritual in 1978.

Three ideas seem essential to the concept of a cult. One is thinking in terms of us versus them with total alienation from "them." The second is the intense, though often subtle, indoctrination techniques used to recruit and hold members. The third is the charismatic cult leader. Cultism usually involves some sort of belief that outside the cult all is evil and threatening; inside the cult is the special path to salvation through the cult leader and his teachings. The indoctrination techniques include:

1. Subjection to stress and fatigue
2. Social disruption, isolation, and pressure
3. Self criticism and humiliation
4. Fear, anxiety, and paranoia
5. Control of information
6. Escalating commitment
7. Use of auto-hypnosis to induce "peak" experiences. (Singer and Lalich 1995; Kevin Crawley: www.ex-cult.org/General/cult.definition)

Of course, there is a positive side to cults. One gets love, a sense of belonging, of ful-filling a special purpose, of being protected, of being free from the evils of the world, of being on the path to eternal salvation, of having power. If the cult did not satisfy needs that life outside the cult failed to satisfy, cults would probably not exist.

One common misconception about cults is that their members are either insane or brainwashed (see **mind control**). The evidence for this is insubstantial. It consists mainly of the subjective feeling that no one in their right mind could possibly choose to believe the things that cult members believe. For example, the 39 members of the Heaven's Gate cult believed a space ship was coming to get them to take them to a "higher level." They believed that their leader, Marshall Applewhite (a.k.a. Do), was Christ coming to take the chosen few to a better life somewhere in outer space, perhaps to work on a starship like the *Enterprise* one sees in *Star Trek* movies and on television. They believed they would be given new, asexual bodies in the new world, with no hair or teeth and vestigial eyes and ears (not those gross bug eyes one sees in so many alien pictures). To many people, these beliefs sound like the delusions of lunatics and it seems inconceivable that anyone in his or her right mind would accept such beliefs.

Examined closely, however, the beliefs of Heaven's Gate or Scientology are no stranger than the beliefs that billions of "normal" people hold dear in their sacred religions: heaven and hell, **angels**, **Satan**, crucified gods, resurrections, **reincarnation**, messiahs, trinities, transubstantiation, and so on. As has been noted by others: The delusions of one person is insanity, delusions by a few a cult, and by many a religion.

It is true that the cult leader or religious founder often shows signs of brain disease, such as hearing voices or having delusions of grandeur. But the *followers* need not be mad. Some are undoubtedly deranged, but most probably are not. The cult leader must be extremely attractive to those who convert. He or she must satisfy a fundamental need, most likely the need to have someone you can totally trust, depend on, and believe in: someone who can give sense and direction to your life and provide you with purpose and meaning. But above all, life with the messiah and the other cult members must satisfy some fundamental need. Some studies have found that a significant number of cult members are depressed before joining, and the cult lifts their spirits. Even if they aren't depressed, cult membership must be more satisfying than life in the real world with one's real family.

Cult members are certainly deluded and manipulated. Severe control tactics may be used to keep them in the flock, such as cutting them off from the rest of the world, especially from their family and friends, **communally reinforcing** the cult's dogmas, and inculcating paranoia. Isolation, communal reinforcement, and the inculcation of paranoia as a control tactic are used by some parents over their children, some political leaders over their citizens, and even some therapists over their patients. So, cults are not unique in attempting to control people using these tactics. And one should not overlook the possibility that one's loved one truly believes in what he or she is doing, and that the ones left behind simply do not see the truth.

Cult members may gradually become paranoid and be led to believe that the government, their family, and former friends can't be trusted. They may gradually become more isolated and militant. They may even begin to stockpile weapons for the coming Armageddon. They may turn themselves over completely to their savior and be willing to kill or die for him or her. This is not to say that they are leading meaningful lives, but they are not necessarily lunatics, morons, or **zombies**.

Further reading: Conway and Siegelman 1997; Langone 1995; Lifton 1989; Martin 1985, 1989.

curse

A curse is an invocation expressing a wish that harm, misfortune, injury, or great evil be brought upon another person, place, thing, clan, or nation. People are also said to be cursed if harm comes to them regularly or in seeming disproportion to the rest of us.

Curses seem to have been a regular part of ancient cultures and may have been a way to frighten enemies and explain the apparent injustices of the world. There is no evidence that anyone has successfully invoked **occult** powers to do harm to others, but there is evidence that those who *believe* they have been cursed can be made miserable if others exploit that belief. Fear and the human tendency to **confirmation bias** can sometimes lead the believer to fulfill the curse.

Belief in curses may make it easier to explain why bad things often happen to good people: They are cursed because of some bad thing an ancestor did. A little bit of reflection, however, should reveal that this is not a very satisfactory explanation. Whether it is God or Nature doing the cursing, neither seems very just in punishing the children for the sins of their mothers or fathers.

The curse is a favorite literary theme in Greek mythology. Modern writers such as William Faulkner use the family curse theme to great effect. The Old Testament is a litany of curses. In the New Testament, even a fig tree gets cursed.

The curse is also a favorite theme of the mass media whenever something bad happens to one of the Fitzgerald-Kennedy (FK) clan. The so-called Kennedy curse is a media creation. The FK clan is cursed no more than any African family destroyed by slavery or any Jewish family destroyed by Nazism. The media would have us believe that the FK clan has suffered a disproportionate amount of harm. Their harm is certainly disproportionately public, but that is because the clan is rich and famous, not because they are cursed. Their harm has been disproportionately influential because some members of the clan have been extremely influential.

In their attempt to bolster the myth of the Kennedy curse, the media have included self-caused harms as "tragedies." Getting drunk and leaving a girl to drown is a tragedy for the girl's family, not the FK clan. Dying in a plane crash because of inexperience, date rape, reckless behavior on a ski slope, having an affair with your babysitter, being arrested for possession of heroin, beating someone to death with a golf club, and dying of a drug overdose are not tragedies. The womanizing, the Bay of Pigs fiasco, working for Joe McCarthy, and involving the United States militarily in Vietnam were chosen behaviors. If there is a curse here, it is the curse of too much money, power, and leisure time combined with a disposition for risk taking.

If one considers the size of the FK clan, their wealth, their extraordinary achievements, and their propensity for taking risks, then their misfortunes do not seem disproportionate. The media would have us believe, however, that if a member of this clan dies in war, gets cancer, or has a mental disorder, it's because they're cursed. If they are cursed, then so are the millions of others who suffer the same fate.

See also **evil eye**.

D

Däniken, Erich von

See **ancient astronauts**.

déjà vu

French for "already seen." Déjà vu is an uncanny feeling or illusion of having already seen or experienced something that is being experienced for the first time. If we assume that the experience is actually of a remembered event, then déjà vu probably occurs because an original experience was neither fully attended to nor elaborately encoded in **memory**. If so, then it would seem most likely that the present situation triggers the recollection of a *fragment* from one's past. The experience may seem uncanny if the memory is so fragmented that no strong connections can be made between the fragment and other memories.

Thus, the feeling that one has been there before is often due to the fact that one *has* been there before. One has simply forgotten most of the original experience because one was not paying close attention the first time. The original experience may even have occurred only seconds or minutes earlier.

On the other hand, the déjà vu experience may be due to having seen pictures or having heard vivid stories many years earlier. The experience may be part of the dim recollections of childhood.

However, it is possible that the déjà vu feeling is triggered by a neurochemical action in the brain that is not connected to any actual experience in the past. One feels strange and identifies the *feeling* with a memory, even though the experience is completely new.

The term was applied by Emile Boirac (1851–1917), who had strong interests in **psychic** phenomena. Boirac's term directs our attention to the past. However, a little reflection reveals that what is unique about déjà vu is not something from the past but something in the *present,* namely, the *strange feeling* one has. We often have experiences the novelty of which is unclear. In such cases we may have been led to ask such questions as, "Have I read this book before?" "Is this an episode of Inspector Morse I've seen before?" "This place looks familiar; have I been here before?" Yet these experiences are not accompanied by an uncanny feeling. We may feel a bit confused, but the feeling associated with the déjà vu experience is not one of confusion; it is one of *strangeness*. There is nothing strange about not remembering whether you've read a book before, especially if you are 50 years old and have read thousands of books over your lifetime. In the déjà vu experience, however, we feel strange because we don't think we should feel familiar with the present perception. That sense of inappropriateness is not present when one is simply unclear whether one has read a book or seen a film before.

Thus, it is possible that the attempt to explain the déjà vu experience in terms of lost memory, **past lives**, **clairvoyance**, and so on may be completely misguided. We should be talking about the déjà vu *feeling*. That feeling may be caused by a brain state, by neurochemical factors during perception that have nothing to do with memory. It is worth noting that the déjà vu feeling is common among psychiatric patients. The déjà vu feeling also frequently precedes temporal lobe epilepsy attacks. When Wilder Penfield did his famous experiment in 1955 in which he electrically stimulated the temporal lobes, he found about 8% of his subjects experienced "memories." He

assumed he elicited actual memories. They could well have been hallucinations and the first examples of artificially stimulated déjà vu.

Further reading: Alcock 1990, 1996a; Reed 1988; Schacter 1996.

deport

The disappearance of an object during a **séance**. Believers attribute a deport to **paranormal** forces. Skeptics attribute it to **conjuring**.

See also **apport** and **teleportation**.
Further reading: Keene 1997.

dermo-optical perception (DOP)

The alleged ability to "see" without using the eyes. DOP is a **conjurer's** trick, often involving elaborate blindfolding rituals, but always leaving a pathway (usually down the side of the nose), which allows for unobstructed vision.

See also **extraordinary human function**.
Further reading: Benski 1998; Gardner 1981; Keene 1997.

devadasi

A religious practice in parts of southern India, including Andhra Pradesh, whereby parents marry a daughter to a deity or a temple. The marriage usually occurs before the girl reaches puberty and requires the girl to become a prostitute for upper-caste community members. Such girls are known as **jogini**. They are forbidden to enter into a real marriage. The practice was legal in India until 1988, yet it still continues, as evidenced by the testimony of a 35-year-old former jogini named Ashama. She ran away from her village and returned to lead the fight to abolish the illegal practice.

The local police do not enforce the law and the villagers themselves make no effort to abolish the heinous practice.

> Since the day of the initiation, I have not lived with dignity. I became available for all the men who inhabited Karni. They would ask me for sexual favours and I, as a jogini, was expected to please them. My trauma began even when I had not attained puberty. [www.tribuneindia .com/20010505/windows/main1.htm]

Ashama was seven when her parents married her to the local god. She was recently awarded the Neerja Bhanot award for courage. (Neerja Bhanot was a Pan Am attendant killed by Palestinian hijackers in 1986, but not before saving hundreds of lives by her heroic actions.)

The practice of religious prostitution is known as *basivi* in Karnataka and *matangi* in Maharastra. It is also known as *venkatasani, nailis, muralis,* and *theradiyan.*

The Atheist Centre of Vijayawada, India, has been most instrumental in the movement to eradicate this pernicious religious custom.

Further reading: Singh 1997.

devil

See **Satan**.

DHEA

Dehydroepiandrosterone (DHEA) is a natural steroid hormone produced from cholesterol by the adrenal glands. DHEA is chemically similar to testosterone and estrogen and is easily converted into those hormones. DHEA production peaks in early adulthood and declines in production with age. Thus, many diseases that correlate with age also correlate with low levels of DHEA production. Advocates of DHEA recommend it to prevent the effects of aging. There has been no scientific evidence, however, that low levels of DHEA is a significant causal factor in the development of diseases associated with aging. Nor is there any evidence that increasing DHEA slows down, stops, or reverses the aging process.

For years DHEA was promoted as a miracle weight loss drug, based upon some studies that indicated DHEA was effective in controlling obesity in rats and mice. Other rodent studies found similar promising results for DHEA in preventing cancer, arteriosclerosis, and diabetes. Studies on humans have not yet duplicated these results. Despite the lack of sufficient scientific evidence, DHEA supplements are being promoted as having therapeutic effects on many chronic conditions including Alzheimer's disease, cardiovascular disease, depression, diabetes, disorders of the immune system, hypercholesterolemia, multiple sclerosis, obesity, osteoporosis, and Parkinson's disease. The healthy truth is that very little is known about DHEA. Long-term effects of self-medicating by using DHEA supplements may be beneficial, neutral, or harmful, but it is unlikely that DHEA supplements will affect each individual in the same way. Increasing DHEA may well increase testosterone, which in men may lead to prostate enlargement and in women to facial hair. Increasing estrogen may help prevent osteoporosis or heart disease but may increase the risk of breast cancer. In short, taking DHEA is a high-risk gamble based on insubstantial evidence.

The research of Dr. Elizabeth Barrett-Connor, professor and chair of the department of family and preventive medicine at the University of California, San Diego, is cited by promoters of DHEA as evidence that DHEA is effective in fighting cardiovascular disease. However, Dr. Barrett-

Conner says, "DHEA is the snake oil of the '90s. It makes me very nervous that people are using a drug we don't know anything about. I won't recommend it" (Skerret 1996).

The main voices in favor of DHEA as a miracle drug are those who are selling it or who make a good living selling books or programs advocating "**natural** cures."

Further reading: Barrett and Butler 1992; Barrett and Jarvis 1993; Raso 1994.

Dianetics

In 1950, Lafayette Ronald Hubbard published *Dianetics: The Modern Science of Mental Health* (Los Angeles: American Saint Hill Organization; all page references are to this edition.) The book is the "bible" for Scientology, which calls itself a science, a church, and a religion. Hubbard tells the reader that Dianetics "contains a therapeutic technique with which can be treated all inorganic mental ills and all organic psycho-somatic ills, with assurance of complete cure." He claims that he has discovered the "single source of mental derangement" (6). However, in a disclaimer on the frontispiece of the book, we are told that "Scientology and its substudy, Dianetics, as practiced by the Church . . . does not wish to accept individuals who desire treatment of physical illness or insanity but refers these to qualified specialists of other organizations who deal in these matters." The disclaimer seems clearly to have been a protective mechanism against lawsuits and for practicing medicine without a license, for the author repeatedly insists that Dianetics can cure just about anything that ails you. He also repeatedly insists that Dianetics is a **science**. Yet anyone familiar with scientific texts will be able to tell from the first few pages of *Dianetics* that the text is no scientific work and the author no scientist. Dia-

netics is a classic example of a **pseudoscience**.

On page 5 of *Dianetics,* Hubbard asserts that a science of mind must find "a single source of all insanities, psychoses, neuroses, compulsions, repressions and social derangements." Such a science, he claims, must provide "invariant scientific evidence as to the basic nature and functional background of the human mind." This science, he says, must understand the "cause and cure of all psycho-somatic ills." Yet he also claims that it would be unreasonable to expect a science of mind to be able to find a single source of all insanities, since some are caused by "malformed, deleted or pathologically injured brains or nervous systems" and some are caused by doctors. Undaunted by this contradiction, he goes on to say that this science of mind "would have to rank, in experimental precision, with physics and chemistry." He then tells us that Dianetics is "an organized science of thought built on definite axioms: statements of natural laws on the order of those of the physical sciences" (6)

There are broad hints that this so-called science of the mind isn't a science at all in the claim that Dianetics is built on "definite axioms" and in his a priori notion that a science of mind must find a single source of mental and psychosomatic ills. Sciences aren't built on axioms and they don't claim a priori knowledge of the number of causal mechanisms that must exist for any phenomena. A real science is built on tentative proposals to account for observed phenomena. Scientific knowledge of causes, including how many kinds there are, is a matter of discovery, not stipulation. Also, scientists generally respect logic and would have difficulty saying with a straight face that this new science must show that there is a single source of all insanities except for those insanities that are caused by other sources.

There is other evidence that Dianetics is not a science. For example, Hubbard's theory of mind shares little in common with modern neurophysiology and what is known about the brain and how it works. According to Hubbard, the mind has three parts:

> The analytical mind is that portion of the mind which perceives and retains experience data to compose and resolve problems and direct the organism along the four dynamics. It thinks in differences and similarities. The reactive mind is that portion of the mind which files and retains physical pain and painful emotion and seeks to direct the organism solely on a stimulus-response basis. It thinks only in identities. The somatic mind is that mind, which, directed by the analytical or reactive mind, places solutions into effect on the physical level. (39)

According to Hubbard, the single source of insanity and psychosomatic ills is the engram. Engrams are to be found in one's "engram bank," that is, in the "reactive mind." The "reactive mind," he says, "can give a man arthritis, bursitis, asthma, allergies, sinusitis, coronary trouble, high blood pressure, and so on down the whole catalogue of psycho-somatic ills, adding a few more which were never specifically classified as psycho-somatic, such as the common cold" (51). One searches in vain for evidence of these claims. We are simply told: "These are scientific facts. They compare invariably with observed experience" (52).

An engram is defined as "a definite and permanent trace left by a stimulus on the protoplasm of a tissue. It is considered as a unit group of stimuli impinged solely on the cellular being" (60 note). We are told that engrams are recorded only during periods of physical or emotional suffering. During those periods the "analytical mind"

shuts off and the reactive mind is turned on. The analytical mind has all kinds of wonderful features, including being incapable of error. It has, we are told, standard memory banks, in contrast to the reactive bank. These standard memory banks are recording all possible perceptions, and, he says, they are perfect, recording exactly what is seen or heard, and so on.

What is the evidence that engrams exist and that they are "hard-wired" into cells during physically or emotionally painful experiences? Hubbard doesn't say that he's done any laboratory studies, but he says that

> in dianetics, on the level of laboratory observation, we discover much to our astonishment that cells are evidently sentient in some currently inexplicable way. Unless we postulate a human soul entering the sperm and ovum at conception, there are things which no other postulate will embrace than that these cells are in some way sentient. (71)

This explanation is not on the "level of laboratory observation" but is a false dilemma and **begs the question**. Furthermore, the theory of souls entering zygotes has at least one advantage over Hubbard's own theory: It is not deceptive and is clearly **metaphysical**. Hubbard tries to clothe his metaphysical claims in scientific garb.

> The cells as thought units evidently have an influence, as cells, upon the body as a thought unit and an organism. We do not have to untangle this structural problem to resolve our functional postulates. The cells evidently retain engrams of painful events. After all, they are the things which get injured. . . .
>
> The reactive mind may very well be the combined cellular intelligence. One need not assume that it is, but it is a handy structural theory in the lack of any

real work done in this field of structure. The reactive engram bank may be material stored in the cells themselves. It does not matter whether this is credible or incredible just now. . . .

The scientific fact, observed and tested, is that the organism, in the presence of physical pain, lets the analyzer get knocked out of circuit so that there is a limited quantity or no quantity at all of personal awareness as a unit organism. (71)

Hubbard asserts that these are scientific facts based on observations and tests, but the fact is there hasn't been any real work done in this field. The following illustration is typical of the kind of "evidence" provided by Hubbard for his theory of engrams.

A woman is knocked down by a blow. She is rendered "unconscious." She is kicked and told she is a faker, that she is no good, that she is always changing her mind. A chair is overturned in the process. A faucet is running in the kitchen. A car is passing in the street outside. The engram contains a running record of all these perceptions: sight, sound, tactile, taste, smell, organic sensation, kinetic sense, joint position, thirst record, etc. The engram would consist of the whole statement made to her when she was "unconscious": the voice tones and emotion in the voice, the sound and feel of the original and later blows, the tactile of the floor, the feel and sound of the chair overturning, the organic sensation of the blow, perhaps the taste of blood in her mouth or any other taste present there, the smell of the person attacking her and the smells in the room, the sound of the passing car's motor and tires, etc. (60)

How this example relates to insanity or psychosomatic ills is explained by Hubbard this way:

The engram this woman has received contains a neurotic positive suggestion. . . . She has been told that she is a faker, that she is no good, and that she is always changing her mind. When the engram is restimulated in one of the great many ways possible [such as hearing a car passing by while the faucet is running and a chair falls over], she has a feeling that she is no good, a faker, and she will change her mind. (66)

There is no possible way to empirically test such claims.

Hubbard claims that an extraordinary amount of data have been collected and not a single exception to his theory has been found (68). We are to take his word on this, apparently, for all the "data" he presents are in the form of anecdotes or made-up examples such as the one presented above.

Another indication that Dianetics is not a science, and that its founder hasn't a clue as to how science functions, is given in claims such as the following: "Several theories could be postulated as to why the human mind evolved as it did, but these are theories, and dianetics is not concerned with structure" (69). This is his way of saying that it doesn't concern him that engrams can't be observed, that even though they are defined as permanent changes in cells, they can't be detected as physical structures. It also doesn't bother him that the cure of all illnesses requires that these "permanent" engrams be "erased" from the reactive bank. He claims that they aren't really erased but simply transferred to the standard bank. How this physically or structurally occurs is apparently irrelevant. He simply asserts that it happens this way, without argument and without proof. He simply repeats that this is a scientific fact, as if saying it makes it so.

Another "scientific fact," according to Hubbard, is that the most harmful engrams occur in the womb. The womb turns out to be a terrible place. It is "wet, uncomfortable and unprotected" (130):

> Mama sneezes, baby gets knocked "unconscious." Mama runs lightly and blithely into a table and baby gets its head stoved in. Mama has constipation and baby, in the anxious effort, gets squashed. Papa becomes passionate and baby has the sensation of being put into a running washing machine. Mama gets hysterical, baby gets an engram. Papa hits Mama, baby gets an engram. Junior bounces on Mama's lap, baby gets an engram. And so it goes. (130)

We are told that people can have "more than two hundred" prenatal engrams and that engrams "received as a zygote are potentially the most aberrative, being wholly reactive. Those received as an embryo are intensely aberrative. Those received as the foetus are enough to send people to institutions all by themselves" (130–131). What is the evidence for these claims? How could one test a zygote to see whether it records engrams? "All these things are scientific facts, tested and rechecked and tested again," he says (133). We must take Hubbard's word for it. However, scientists generally do not expect others to take their word for such dramatic claims.

Furthermore, to get cured of an illness you need a Dianetic therapist, called an *auditor*. Who is qualified to be an auditor? "Any person who is intelligent and possessed of average persistency and who is willing to read this book [*Dianetics*] thoroughly should be able to become a dianetic auditor" (173). The auditor must use "Dianetic reverie" to effect a cure. The goal of Dianetic therapy is to bring about a "release" or a "clear." The "release" has had major stress and anxiety removed by Dianetics; the "clear" has neither active nor potential psychosomatic illness or aberration (170). The "purpose of therapy and its sole target is the removal of the content of the reactive engram bank. In a release, the majority of emotional stress is deleted from this bank. In a clear, the entire content is removed" (174). The "reverie" used to achieve these wonders is described as an intensified use of some special faculty of the brain that everyone possesses but which "by some strange oversight, Man has never before discovered" (167). Hubbard has discovered what none before him has seen, and yet his description of this "reverie" is of a man sitting down and telling another man his troubles (168). In a glorious non sequitur, he announces that auditing "falls utterly outside all existing legislation," unlike psychoanalysis, psychology, and hypnotism, which "may in some way injure individuals or society" (168–169). It is not clear, however, why telling others one's troubles is a monumental discovery. Nor it is clear why auditors couldn't injure individuals or society, especially since Hubbard advises them: "Don't evaluate data. . . . don't question the validity of data. Keep your reservations to yourself" (300). This does not sound like a scientist giving advice to his followers. This sounds like a guru giving advice to his disciples.

Further reading: Atack 1990; Gardner 1957.

divination (fortune telling)

The attempt to foretell the future or discover **occult** knowledge by interpreting omens or by using **paranormal** or supernatural powers. The list of items that have been used in divination is extraordinary. Some are listed below. Many end in "mancy," from the ancient Greek *manteia* (divination), or "scopy," from the Greek *skopein* (to look into, to behold).

- aeluromancy (dropping wheatcakes in water and interpreting the result)
- alectoromancy or alectryomancy (divination by a cock: grains of wheat are placed on letters and the cock "spells" the message by selecting grains)
- anthropomancy (divination by interpreting the organs of newly sacrificed humans)
- astragalomancy or astragyromancy (using knucklebones marked with letters of the alphabet)
- **astrology**
- axinomancy (divination by the hatchet: interpreting the quiver when whacked into a table)
- bronchiomancy (divination by studying the lungs of sacrificed white llamas)
- **cartomancy**
- cephalomancy (divination by a donkey's head)
- chiromancy (**palmistry**)
- cleidomancy (divination by interpreting the movements of a key suspended by a thread from the nail of the third finger on a young virgin's hand while one of the Psalms is recited)
- cromniomancy (divination by onions)
- **dowsing**
- **geomancy**
- gyromancy (divination by walking around a circle of letters until dizzy and one falls down on the letters or in the direction to take)
- haruspicy (inspecting the entrails of sacrificed animals)
- hepatoscopy or hepatomancy (divination by examining the liver of sacrificed animals)
- kephalonomancy (burning carbon on the head of an ass while reciting the names of suspected criminals; if one is guilty, a crackling sound will be heard when one's name is spoken)
- margaritomancy (divination by the pearl: if it jumps in the pot when a person is named, then he is the criminal)
- **metoposcopy** (interpreting frontal wrinkles)
- myrmomancy (divination by watching ants eating)
- necromancy (communicating with spirits of the dead to predict the future)
- oinomancy (divination by wine)
- omphalomancy (interpretation of the belly button)
- oneiromancy (interpretation of **dreams**)
- ornithomancy or orniscopy (interpreting the flights of birds)
- rhapsodmancy (divination by a line in a sacred book that strikes the eye when the book is opened after the diviner prays, meditates, or invokes the help of spirits)
- **scrying**
- skatharomancy (interpreting the tracks of a beetle crawling over the grave of a murder victim)
- splanchnomancy (reading cut sections of a goat liver)
- **stichomancy**
- tasseography (reading tea leaves)
- tiromancy (interpreting the holes or mold in cheese)
- uromancy (divination by reading bubbles made by urinating in a pot)

Further reading: de Givry 1971; Pickover 2001; Steiner 1996.

Dixon, Jeane, and the Jeane Dixon effect

Jeane Dixon (1917–1997) was an **astrologer** and alleged **psychic** who did *not* predict the assassination of President Kennedy. She was often featured in publications that engage in the entertaining pursuit of making predictions for the New Year. Ms. Dixon was never correct in any prediction of any consequence. She predicted that the Soviets would beat the United States to the moon, for example. But most of her predictions were equivocal, vague, or mere possibility claims. She achieved a reputation as a very good psychic, however, when the mass media perpetuated the myth that she had predicted President Kennedy's assassination. In 1956 she predicted in *Parade* magazine that the 1960 election would be won by a Democrat and that he would be assassinated or would die in office, "although not necessarily in his first term." However, in 1960, apparently forgetting or overriding her earlier prediction, she predicted unequivocally that "John F. Kennedy would fail to win the presidency."

Dixon was an FBI stooge who agreed to make claims about Russia being behind the civil rights movement and left-wing agitation on college campuses. She was chummy enough with J. Edgar Hoover that he agreed to serve as an honorary director to Children to Children, Inc., a foundation established by Dixon to help sick children.

The "Jeane Dixon effect" refers to the tendency of the mass media to hype or exaggerate a few correct predictions by a psychic, guaranteeing that they will be remembered, while forgetting or ignoring their many incorrect predictions.

Further reading: Hines 1990.

Dogon and Sirius

The Dogon are a people who number about 100,000 who dwell in western Africa. According to Robert Temple in *The Sirius Mystery* (1976), the Dogon had contact with some ugly amphibious extraterrestrials, the Nommos, some 5,000 years ago. The aliens came here for unknown reasons from a planet orbiting Sirius (8.6 light years from Earth). The alleged visitors from outer space seem to have done little else than give the earthlings some useless astronomical information.

One of Temple's main pieces of evidence for his preposterous claim is the tribe's alleged knowledge of Sirius B, a companion to the star Sirius. The Dogon are supposed to have known that Sirius B orbits Sirius and that a complete orbit takes 50 years. One of the pieces of evidence Temple cites is a sand picture made by the Dogon to explain their beliefs. The diagram that Temple presents, however, is not the complete diagram that the Dogon showed to the French anthropologists Marcel Griaule and Germaine Dieterlen, who were the original sources for Temple's story. Temple has either misinterpreted Dogon beliefs or distorted Griaule and Dieterlen's claims to fit his fantastic story.

> Griaule and Dieterlen describe a world renovation ceremony associated with the bright star Sirius (sigu tolo, "star of Sigui"), called sigui, held by the Dogon every sixty years. According to Griaule and Dieterlen the Dogon also name a companion star, po tolo "Digitaria star" (Sirius B) and describe its density and rotational characteristics. Griaule did not attempt to explain how the Dogon could know this about a star that cannot be seen without telescopes, and he made no claims about the antiquity of this information or of a connection with ancient Egypt. [Ortiz de Montellano: www.ramtops.demon.co.uk/dogon.html]

Temple lists a number of astronomical beliefs held by the Dogon that seem curious. They have a traditional belief in a heliocentric system and in elliptical orbits of astronomical phenomena. They seem to have knowledge of the satellites of Jupiter and rings of Saturn, among other things. Where did they get this knowledge, he asks, if not from extraterrestrial visitors? They don't have telescopes or other scientific equipment, so how could they get this knowledge? Temple's answer is that they got this information from amphibious aliens from outer space.

Afrocentrists such as F. C. Welsing, on the other hand, claim that the Dogon could see Sirius B without the need of a telescope because of their special eyesight due to quantities of melanin ("Lecture," 1st Melanin Conference, San Francisco, September 16–17, 1987). There is, of course, no evidence for this special eyesight, or for other equally implausible notions such as the claim that the Dogon got their knowledge from black Egyptians who had telescopes.

Carl Sagan agreed with Temple that the Dogon could not have acquired their knowledge without contact with an advanced technological civilization. Sagan suggests, however, that that civilization was terrestrial rather than extraterrestrial. Perhaps the source was Temple himself and his loose speculations on what he learned from Griaule, who based his account on an interview with one person, Ambara, and an interpreter.

According to Sagan, western Africa has had many visitors from technological societies located on planet Earth. The Dogon have a traditional interest in the sky and astronomical phenomena. If a European had visited the Dogon in the 1920s and 1930s, conversation would likely have turned to astronomical matters, including Sirius, the brightest star in the sky and the center of Dogon mythology. Furthermore, there had been a good amount of discussion of Sirius in the scientific press in the 1920s, so that by the time Griaule arrived, the Dogon may have learned of 20th-century technological matters from earthly visitors (Sagan 1979).

Or Griaule's account may reflect his own interests more than that of the Dogon. He made no secret of the fact that his intention was to redeem African thought. When the Belgian Walter van Beek studied the Dogon, he found no evidence they knew Sirius was a double star or that Sirius B is extremely dense and has a 50-year orbit.

> Knowledge of the stars is not important either in daily life or in ritual [to the Dogon]. The position of the sun and the phases of the moon are more pertinent for Dogon reckoning. No Dogon outside of the circle of Griaule's informants had ever heard of sigu tolo or po tolo. . . . Most important, no one, even within the circle of Griaule informants, had ever heard or understood that Sirius was a double star. [Ortiz de Montellano: www.ramtops.demon.co.uk/dogon.html]

According to Thomas Bullard, van Beek speculates that Griaule "wished to affirm the complexity of African religions and questioned his informants in such a forceful leading manner that they created new myths by **confabulation**." Griaule either informed the Dogon of Sirius B or "he misinterpreted their references to other visible stars near Sirius as recognition of the invisible companion" (Bullard 1996).

The only mystery is how anyone could take seriously either the notion of amphibious aliens or telescopic vision due to melanin.

See also **ancient astronauts** and **UFOs**.

Further reading: Griaule 1997; James and Thorpe 1999; Ortiz de Montellano 1991, 1993; Randi 1982a; Ridpath 1978; Sagan 1995; van Beek 1992.

dowsing (water witching)

Dowsing is the action of a person (the dowser) using a rod, stick, or other device—called a dowsing rod, dowsing stick, doodlebug (when used to locate oil), or divining rod—to locate such things as underground water, hidden metal, buried treasure, oil, lost persons, or golf balls. Since dowsing is not based on any known scientific or empirical laws or forces of nature, it should be considered a type of **divination**. The dowser tries to locate objects by **occult** means.

Map dowsers use a dowsing device, usually a pendulum, over maps to locate oil, minerals, persons or water. However, a typical dowser is the field dowser, who walks around an area using a forked stick to locate underground water. When above water, the rod points downward. (Some dowsers use two rods. The rods cross when above water.) Various theories have been given as to what causes the rods to move: electromagnetic or other subtle geological forces, suggestion from others or from geophysical observations, **ESP**, or other **paranormal** explanations. Most skeptics accept the explanation of William Carpenter (1852). The rods move due to involuntary motor behavior, which Carpenter dubbed **ideomotor** action.

Of more interest than why the rods move, however, is the issue of whether dowsing works. Obviously, many people believe it does. Dowsing and other forms of divination have been around for thousands of years. There are large societies of dowsers in America and Europe, and dowsers practice their art every day in all parts of the world. There are also scientists who have offered proof that dowsing works. There must be something to it, then, or so it seems.

The **testimonials** of dowsers and those who observe them provide the main evidence for dowsing. The evidence is simple: Dowsers find what they are dowsing for and they do this many times. This type of fallacious reasoning is known as **post hoc reasoning** and is a very common basis for belief in paranormal powers. It is essentially unscientific and invalid. Scientific thinking includes being constantly vigilant against **self-deception** and being careful not to rely on insight or intuition in place of rigorous and precise empirical testing of theoretical and causal claims. Every **controlled study** of dowsers, including the Scheunen, or barn, study (see below), has shown that dowsers do no better than chance in finding what they are looking for.

Most dowsers do not consider it important to doubt their dowsing powers or to wonder whether they are self-deceived. They never consider doing a controlled scientific test of their powers. They think that the fact that they have been successful over the years at dowsing is proof enough. When dowsers are scientifically tested and fail, they generally react with genuine surprise. Typical is what happened when James Randi tested some dowsers using a protocol they all agreed on. All the dowsers failed the test, though each claimed to be highly successful in finding water using a variety of nonscientific instruments, including a pendulum. Says Randi, "The sad fact is that dowsers are no better at finding water than anyone else. Drill a well almost anywhere in an area where water is geologically possible, and you will find it" (Randi 1982a).

Some of the strongest evidence for dowsing comes from Germany and the Scheunen, or barn, experiment. In 1987 and 1988, more than 500 dowsers participated in more than 10,000 double-blind tests set up by physicists in a barn near Munich. (*Scheune* is German for "barn.")

The researchers claim they proved "a real dowsing phenomenon." Jim Enright of the Scripps Institute of Oceanography evaluated the data and concluded that the "real dowsing phenomenon" can reasonably be attributed to chance. His argument is lengthy, but here is a taste:

> The long and the short of it is that dowsing performance in the Scheunen experiments was not reproducible. It was not reproducible inter-individually: from a pool of some 500 self-proclaimed dowsers, the researchers selected for their critical experiments 43 candidates whom they considered most promising on the basis of preliminary testing; but the investigators themselves ended up being impressed with only a few of the performances of only a small handful from that select group. And, even more troublesome for the hypothesis, dowsing performance was not reproducible intra-individually: those few dowsers, who on one occasion or another seemed to do relatively well, were in their other comparable test series usually no more successful than the rest of the "unskilled" dowsers. (Enright 1995)

The barn study itself is curious. It seems clearly to have been repudiated by another German study done in 1992 by a group of German scientists and skeptics. The Gesellschaft zur wissenschaftlichen Untersuchung von Parawissenschaften (GWUP; Society for the Scientific Investigation of the Parasciences) set up a three-day controlled test of some 30 dowsers, mostly from Germany. The test was done at Kassel, north of Frankfurt, and televised by a local television station. The test involved plastic pipe buried 50 centimeters in a level field through which a large flow of water could be controlled and directed. On the surface, the position of the pipe was marked with a colored stripe, so all the dowsers had to do was tell whether there was water running through the pipe. All the dowsers signed a statement that they agreed the test was a fair test of their abilities and that they expected a 100% success rate. "At the end of three days of testing, GWUP announced the results of almost a thousand bits of data to the assembled dowsers. A summary of their results produced just what would be expected according to chance" (Randi 1995). Defenders of dowsing do not care for these results, and continue to claim that the barn study provides scientific proof of dowsing.

The precursor to dowsing? From Dennis Diderot, *Essay on Blindness* (1749).

Further evidence for dowsing has been presented by the Deutsche Gesellschaft für Technische Zusammenarbeit (the German Society for Technical Co-operation), sponsored by the German government. They claim, for example, that in some of their water dowsing efforts they had success rates above 80%, "results which, according to responsible experts, could not be reached by means of classical methods, except with disproportionate input." Of particular interest is a report by University of Munich physicist Hans-Dieter Betz, "Unconventional Water Detection: Field Test of the Dowsing Technique in Dry Zones," published in the *Journal of Scientific Exploration* in 1995. (This is the same Betz who, with J. L. König, authored a book in 1989 on German government tests proving the ability of dowsers to detect **E-rays**.) The report covers a 10-year period and over 2,000 drillings in Sri Lanka, Zaire, Kenya, Namibia, Yemen, and other countries. Especially impressive was an overall success rate of 96% achieved in 691 drillings in Sri Lanka. "Based on geological experience in that area, a success rate of 30–50 percent would be expected from conventional techniques alone," according to Betz. "What is both puzzling yet enormously useful is that in hundreds of cases the dowsers were able to predict the depth of the water source and the yield of the well to within 10 or 20 percent. We carefully considered the statistics of these correlations, and they far exceeded lucky guesses."

Betz ruled out chance and the use of landscape and geological features by dowsers as explanations for their success. He also ruled out "some unknown biological sensitivity to water." Betz thinks that there may be "subtle electromagnetic gradients" resulting from fissures and water flows that create changes in the electrical properties of rock and soil. Dowsers, he thinks, somehow sense these gradients in a hypersensitive state. "I'm a scientist," says Betz, "and those are my best plausible scientific hypotheses at this point. . . . we have established that dowsing works, but have no idea how or why."

There are some puzzling elements to Betz's conclusions, however. Most of his claims concern a single dowser named Schröter. Who observed this dowser or what conditions he worked under remain unknown. Betz is a physicist and what knowledge he has of hydrogeology is unknown. Furthermore, Betz's speculation that dowsers are hypersensitive to subtle electromagnetic gradients does not seem to be based on scientific data. In any case, the hypothesis was not tested and I am not sure how one would go about testing such a claim. At the very least, one would expect that geological instruments would be able to detect such "electromagnetic gradients."

When others have done controlled tests of dowsers, dowsers' results were no better than chance and no better than those of nondowsers (Enright 1995, 1996; Hyman 1996a; Randi 1995; Vogt and Hyman 2000). Some of Betz's data are certainly not scientific, for example, the subjective evaluations of Schröter regarding his own dowsing activities. Much of the data is little more than a report that Schröter used dowsing and was successful in locating water. Betz assumes that chance or scientific hydrogeological procedures would not have produced the same or better results. It may be true that in one area they had a 96% success rate using dowsing techniques and that "no prospecting area with comparable sub-soil conditions is known where such outstanding results have ever been attained." However, this means nothing for establishing that dowsing had anything to do with the success. Analogous subsoil conditions seems to be an insufficient similarity to justify concluding that dowsing, rather than chance or use of landscape or geological features, must account for the success rate.

Betz seems to have realized that without some sort of testing, reasonable people would not accept that it had been established that dowsing is a real phenomenon based on the above types of data. He then presents what he calls "tests" to establish that dowsing is real. The first test involves Schröter again. A Norwegian drilling team dug two wells and each failed to hit water. The dowser came in and allegedly not only hit water but predicted the depth and flow. Apparently, we have the dowser's own word on this. In any case, this is not a test of dowsing, however impressive it might seem.

In the second test, Betz asserts that dowsers can tell how deep water is because "the relevant biological sensations during dowsing are sufficiently different to allow for the required process of distinction and elimination." He has no evidence for this claim. In any case, in this test Schröter again is asked to pick a place to dig a well and again he is successful. This time his well is near a well already dug and known to be a good site. Betz claims that there were some geological formations that would have made the dowser's predictions difficult, but again this was not a scientific test of dowsing.

The third test was a kind of contest between the dowser and a team of hydrogeologists. The scientific team, about whom we are told nothing significant, studied an area and picked 14 places to drill. The dowser then went over the same area after the scientific team had made their choices and he picked seven sites to drill. A site yielding 100 liters per minute was considered good. The hydrogeologists hit three good sources; the dowser hit six. Clearly, the dowser won the contest. This test does not prove anything about dows-

ing, however. Nevertheless, I think Herr Schröter should knock on James Randi's door and be allowed to prove his paranormal powers under controlled conditions. If he is as good as he and Betz say he is, he should walk away a very rich man (see **Randi Psychic Challenge**).

Further reading: Enright 1999; Feder 2002; Gardner 1957; Hyman 1996a.

dreams

Images and feelings occurring during sleep. Most dreams occur in conjunction with rapid eye movements; hence, they are said to occur during REM-sleep, a period typically taking up 20–25% of sleep time. Infants are believed to dream during about 50% of their sleep time. Dreams occurring during non-REM periods are said to occur during NREM-sleep.

Sleep researchers divide up sleep time into stages, mainly defined by the electrical activity of cortical neurons represented as brain waves by an electroencephalograph (EEG). The EEG records electrical activity in the brain through electrodes connected to the scalp. The stages of sleep occur in sequence and then go backward to stage 1 and REM-sleep about 90 minutes later. This cycle recurs throughout the night, with the REM period usually getting longer at each recurrence. Typically, a person will have four or five REM periods a night, ranging from 5 to 45 minutes each in duration. However, there is some evidence that REM-sleep evolved *before* dreaming and that the two are independent of one another (Siegel et al. 1999).

The REM-dream state is a neurologically and physiologically active state. When a person is in deep sleep there is no dreaming, and the waves (called delta waves) come at high amplitude, about three per second. In REM-sleep, the waves come at a rate of about 60 to 70 per second, and the brain generates about five times as much electricity as when awake. Blood pressure, heart rate, and breathing rate can change dramatically during REM-sleep. Since there is generally no external physical cause of these states, the stimuli must be internal, that is, in the brain, or external and nonphysical. The latter explanation—that dreams are a gateway to the **paranormal** or supernatural—seems to be largely without merit, although it is very ancient. Each of the following may have contributed to this misconception: dreams of dead persons, dreams of being in distant places or of traveling back or forward in time, dreams that seem prophetic, and dreams that are so strange, curious, or bizarre that they call out for a paranormal interpretation. The fact that the part of the brain that controls REM is the pons, a primitive section of the brain stem that controls reflexes such as breathing, supports the notion that the stimuli for the physiological changes that take place during REM originate internally.

Nowadays, hardly anyone believes that dreams are messages from the gods. But some parapsychologists, such as Charles **Tart**, believe that dreams offer entry into a paranormal universe of **out-of-body experiences** (OBEs). His main evidence seems to be his personal faith and an anecdote about his baby-sitter, "Miss Z." Tart claims she had the power to leave her body during sleep. He says he tested her in his sleep lab at the University of California, Davis, after she told him that she "thought everyone went to sleep, woke up in the night, floated up near the ceiling for a while, then went back to sleep." Tart hooked up "Miss Z" to an EEG machine, put a number on a shelf, and put her to bed with instructions to read the number while having an OBE. She claims that even though she didn't read the number on the shelf, she flew around the room the first few nights. She didn't get the number right until the fourth night. Skeptics think either Tart is making up the story or it took the girl four nights to figure out how to trick the scientist. (See **Tart**'s "A Psychophysiological Study of Out-of-the-Body Experiences in a Selected Subject," *Journal of the American Society for Psychical Research* 62 [1968]: 3–27.) Others have investigated the question of whether the mind is open to **telepathic** input during sleep and have failed to find evidence of **psychic** ability while dreaming. Scientific research by psychiatrist Montague Ullman and parapsychologist Charles Honorton in the early 1970s at Maimonides Hospital in Brooklyn, New York, obtained chance results after an initial testing that looked positive for **psi** (Baker 1996b).

It is possible that dreaming may be related to the OBE. In some dreams, the dreamer is an observer, even an observer of himself. Perhaps the brain mechanism that controls spectator dreams versus first-person dreams is the same mechanism that controls the illusion of leaving one's body in the OBE.

Tart and other parapsychologists who think that the dream state is a gateway to another world cite the distinct brain waves of the various stages of sleep as their evidence. They seem to think that brain waves represent states of consciousness and that sleep is an **altered state of consciousness**. Brain waves, however, represent electrical activity in the brain, not states of consciousness. Brain activity during dream-sleep is indeed curious. While dreaming, not only do we experience the equivalent of hallucinations, some of which would qualify as psychotic if we had them while awake, but most of us feel as if we are physically moving, acting, and being acted on, without the body actually moving. Brain stem mechanisms protect us during sleep from motor activities that could lead to self-injury or injury to others. That is, most of us are paralyzed during sleep. However, some people suffer a weakness or disruption of the brain stem that causes a sleep disorder where motor activities are not prevented. People who suffer from this disorder flail or sleepwalk and can be a danger to themselves or others. Such people do not leave their bodies, but they often leave their beds during sleep.

Another curious quality of brain activity during dreaming is that dream amnesia is the norm. This is not due to anything paranormal or supernatural, but to weak encoding. **Memory** depends on encoding the data of experience. Encoding depends on connections in parts of the brain, which in turn depend on connections in experience. A visual event with a strong emotional component is more likely to be remembered than one with no emotional component because emotional memories are recorded in different parts of the brain than visual components. Neural connections link them. We are likely to remember dreams if we wake shortly after they occur. Even so, if we do not encode the dream by making some effort to remember it, we are likely to forget it. Some people assist memory by getting up and writing down the dream. Others find that an easier method is to stay in bed and create some associations. One simple method of association is to give the dream a title with a purposive description. For example, a dream of being chased by a polar bear across the snow into a library might be labeled "Research the Polar Bear." Go back to sleep and you are likely to remember the dream by recalling the title.

Perhaps the most curious quality of dreams is that most of us most of the time are not aware that we are dreaming while we are dreaming. PET scans during dreaming have shown that there is reduced activity in the prefrontal cortex during REM-sleep, and this might account for several features of the dream state. The prefrontal cortex is where planning and self-awareness reside. Because of the reduced activity in this area, a person might not realize that unfeasible dream events are unreal. Reduced activity in the prefrontal cortex may also account for distortions in the dreamer's perception of time and for dream amnesia.

Some researchers cite the lack of prefrontal activity as a sign that the function of sleep is restorative. Sleep gives a rest to the frontal lobes, the most active part of the brain while awake. It may well be that **lucid dreaming**—being aware of dreaming while dreaming—is possible for some people because their frontal lobes don't completely shut down during dreaming. Most parapsychologists, however, are not interested in the physiology of dreaming. They focus instead on the *content* of dreams, which they believe reveals a passage to the paranormal.

The prophetic or **clairvoyant** dream is perhaps the strongest reason for believing that dreaming is a gateway to another world. Some dreams seem uncanny. They seem to foretell events. If a significant number of dreams of just a single person corresponded to future events, this would be a great benefit to humankind and we should try to find out what mechanism is at work here. However, no such person has yet been found. Individual dreams that occasionally seem clairvoyant provide very weak evidence for clairvoyant dreams.

While it is admitted by most parapsychologists that some amount of **coincidence** is to be expected between what a person dreams and what actually happens, it is argued that there are too many cases of seemingly prophetic dreams to reasonably explain them all away as coincidence. It is true that not all prophetic dreams are due to coincidence. Most of them probably should be so understood, but many of them may be explained as filling in memories of dreams *after* the fact, and many others should be explained as cases of lying. Prophetic dreams are impressive to those who lack understanding of the **law of truly large numbers**, **confirmation bias**, and how memory works. If the odds are a million to one that any given dream is truly prophetic, then, given the number of people on earth and the average number of dreams people have during each sleep period—250 dream themes a night (Hines 1990: 50)—we should expect that every day there will be more than 1.5 million dreams that seem clairvoyant. That is not including all the dreams had by cats, dogs, and other animals, who may well be having apparently psychic experiences while they sleep, though to what purpose we can only guess. Furthermore, one would think that if dreaming were a gateway to the paranormal, blind persons would not have their dream time restricted by their physical limitations any more than those with sight. Yet people blind from birth do not have visual dreams (Manfred Davidmann: www.solbaram.org/articles/humind.html).

There are also those who think that the dream state is a gateway to past lives. Others think the dreams we have today are due to fears our ancestors had. Universal dream themes, such as being chased or falling, are said to hearken back to our hunter-gatherer days. We have these dreams because our ancestors were chased by saber-toothed tigers and slept in trees. The evidence for such beliefs is negligible; although a strong case can be made that the *form* rather than the *content* of such dreams might well be due to an evolutionary development linked to exercising instinctive behavior necessary for survival.

If the dream state is a gateway to anything, it is probably a gateway to current fears and desires, rather than to ancient fears of ancestors. We assume dreaming has a purpose, but that purpose is more likely to be rooted in this life than in some other one. Any decent theory of dreams must try to explain why the brain stimulates the memories and **confabulations** that it does. It is most likely that dreams are a result of electrical energy that stimulates memories located in various regions of the brain. Why the brain stimulates and confabulates just the memories it does remains a mystery, though there are several plausible explanations.

One such hypothesis for sleep-related rhythms is that they are the brain's way of disconnecting the cortex from sensory input. When we are asleep, thalamic neurons prevent penetration of sensory information upward to the cortex (Sylvia Helena Cardoso: www.epub.org.br/cm/n02/mente/neurobiologia_i.htm). This gives the cortex a bit of a rest and explains why people who suffer sleep deprivation suffer a loss of critical thinking abilities and are prone to poor judgment. Another hypothesis is that dreaming plays a role in memory processing, especially with emotional memories. During REM-sleep, the amygdalae, which play a role in the formation and consolidation of memories of emotional experiences, are quite active. A related theory is that dreams are "watchdogs of the psyche" (Baker 1996b). Dreams are mechanisms that inform and guide our feelings and emotions. They are a way for us to express our desires and fears that need to be expressed but are not expressed when awake. If this were true, it would seem to follow that only one very intimate with the dreamer should attempt to interpret a particular dream. Dreams are very personal and speak to the specific emotional life of the dreamer. The "surest guide to the meaning of a dream is the feeling and judgment of the dreamer himself or herself, who, deep down inside, knows its real meaning" (ibid.). This theory seems to be based on the fact that most dreams are about things that have occurred within the past day or two and reflect the dreamer's present life and concerns, including unresolved feelings. This theory also implies that the interpretation of dreams can play a significant role in self-discovery; for dreams often reflect feelings and desires of which we are not conscious when awake. We may have anxieties or desires that only our dreams can reveal.

Most of us would have little difficulty in finding examples of "anxiety dreams" or "wish-fulfillment dreams" from our own experience. We may not have been aware of our desires or fears until they were awakened by the dream. Sometimes our symbolic dreams are so clear that we do not need outside assistance to help us interpret their meaning. Yet many dreams are so strange, irrational, or bizarre that we are at a loss to find meaning in them. We seek others who claim expertise in dream interpretation to help us ferret out the hidden meanings of our dreams. Those who engage in the interpretation of dreams should be especially careful not to impose their own pet theories onto the dreams of others. For example, the dream mentioned above of being chased by a polar bear into a library might be interpreted in many different ways, but only I, my wife, and one or two other persons familiar with the experience that the dream is rooted in are in a position to interpret it "correctly." I don't doubt that there are many possible interpretations and that some of them might seem quite plausible. But the "correct" one is one that has meaning for the dreamer. It was a frightening dream, just as the experience of dealing with a close relative with bipolar disorder (manic depression) was frightening. The experience led me to the library and to bookstores to get as much information about this brain disorder as I could. I have no doubt that a Freudian or Jungian could find some latent or symbolic meaning here that I do not note, but I have no interest in their interpretations because I have no way to check them against reality and do not share their assumptions regarding the psyche. I have no idea why my brain confabulated this dream, arousing fear and disturbing sleep.

There are some people who have experienced much more horrible things than I have, and who dream about them every single night of their lives (Barrett 1996; Restak and Grubin 2001; Sacks 1995) or who, while conscious, cannot get thoughts of the traumatic experience out of their minds (Ramachandran 1998). Why the brain should terrify its owner by repeating horrifying memories during sleep seems beyond comprehension. Such obsessive dreaming is of no more value than obsessive-compulsive behavior. Such people don't just have nightmares; they are too terrified to go to sleep. They need the help of a good therapist, but they are not in need of a dream interpreter. If such dreamers are to be helped, they must learn to control their dreams. There are various methods used to control dreaming, most involving visual or auditory preparations prior to sleep. Some therapists claim success with victims of recurrent nightmares by treating what is loosely called "post-traumatic stress disorder." Some patients claim that they have been helped to overcome the experience of repetitious nightmares by lucid dreaming. Few are likely to be helped by treating dreams as a gateway to some higher realm of consciousness or reality.

Further reading: Alcock 1990; Asserinsky and Kleitman 1953; Coren 1997; Farady 1985; Hobson 1988; Schacter 1996.

druids

The "wise men" of the Celts. Although dozens of books have been written about them, almost nothing is known about the druids. Their beliefs were esoteric and passed on orally. Their practices, for the most part, were not public. With no written tradition and no major temples where art might provide a key to some of the druids' activities, we must rely on the words and speculations of foreign observers.

The druids are portrayed by the ancient Roman authors Strabo, Diodorus, Posidonius, and Julius Caesar as overseeing bloody religious rituals. Hence, the druids are often thought of as having primarily a religious function and are often called "priests." Diodorus calls them "philosophers." Strabo calls them bards and soothsayers with a reputation for mediation. Whatever they were, the druids enjoyed a position of high status in Celtic society very unlike the position of modern druids, who find solace communing with grass or the wind while parading around stone circles.

Modern druids treat Stonehenge and other megalithic monuments of the British Isles as places of worship. All of the stone circles, menhirs, and dolmens of the British Isles were constructed by peoples who antedated the Celts by 1,000 to 3,000 years. Stonehenge, for example, was built over a period of centuries, from 2800 B.C.E. to 1550 B.C.E. The Celts did not arrive in the British Isles until long after the great

Drombeg stone circle, Country Cork, Ireland. Though favored by modern druids, such circles antedate the Celts.

megaliths had been erected. Hence, if the Celts used these ancient megaliths for religious services, they must have fit their rituals to the megaliths, for they certainly did not construct the megaliths to fit their rituals.

It seems likely that the Celtic druids were a class apart from the warriors in Celtic society. "They served the tribes and clans as judges, prophets, soothsayers, wise men and as keepers of the collective memory" (Herm 1977:61). They were the intellectuals in a warrior society and have been compared to the Brahmins of the Indian caste system.

The word "druid" is thought to derive from the Greek *drus* (oak) and the Indo-European *wid* (wisdom), "which produces the apparent absurdity of 'oak-knower'" (ibid.: 57). In any case, druids are typically associated with oak trees. Some say they held assemblies in sacred groves, that they prized the mistletoe growing on the oaks, or that they worshipped the trees themselves. However, the modern druids' nature worship is a fanciful connection to the ancient Celts. Finally, despite the attempt by some Wiccans to establish a historical link between the Celtic druids and modern **Wicca**, there is none.

Further reading: Delaney 1986; P. Ellis 1998; Lonigan 1996; 'O Hogain 1999.

dualism

The **metaphysical** doctrine that there are two substances, that is, distinct and independent types of being, one material and the other spiritual. Material substance is defined as physical and is asserted to be the underlying reality of the empirical world, that is, the world we see, hear, and measure with our senses and technical instruments that extend the range of the senses, such as electron microscopes, telescopes, and radar.

The spiritual world is usually described negatively as the nonphysical, nonmaterial reality underlying the nonempirical world, variously called the psychological or the mental world.

Dualists are fond of a belief in immortality. If there is another type of reality besides the body, this nonbody can survive death. The nonbody can conceivably exist eternally in a nonphysical world, enjoying nonphysical pleasures or pains distributed by a nonphysical God. This notion seems to be nonsense, but it apparently gives many people great comfort and hope.

See also **soul**.

Further reading: Churchland 1986; D. Dennett 1978, 1991, 1996; Ryle 1949; Ryle and Myer 1993.

E

ear candling (coning)

A method of cleaning the ears and the mind. A hollow candle is stuck into the ear and lit, allegedly sucking out ear wax and negative **energy**, alleviating a host of physical, emotional, and spiritual ailments. The process is ineffective for earwax removal (Seely et al. 1996) and may result in harm from burning, infection, obstruction of the ear canal, or perforation of the eardrum.

What wax appears in the cone is from the melted candle, not from the ears. The suction created by the coning flame is insufficient to remove wax, which, by the way, is good for you. It traps dust and dirt and helps fight infections.

The origin of this unnatural practice has been given variously as ancient Tibet, China, India, Egypt, pre-Columbian

America, and **Atlantis**, depending on which purveyor of coning products one asks.

Further reading: Dryer 2001.

ectoplasm

The stuff oozing from **ghosts** or **spirits** that makes it possible for them to materialize and perform feats of **telekinesis**. For some strange reason, ectoplasm is often not visible to the naked eye but appears in photographs. This may be due to a number of physical factors having to do with reflection, refraction, film processing, or other natural phenomena.

In the heyday of **séances**—the 19th and early 20th centuries—ectoplasm was often produced by the **medium**. James Randi claims that in such cases what was produced was painted cheesecloth and other common physical substances. In short, he thinks the **psychics** cheated. Of course, Randi cannot prove that every psychic cheated every time; therefore, some psychics may not have cheated. This *possibility* is enough for the **true believer** to warrant belief in the reality of ectoplasm. A true skeptic, however, would have to conclude that the probability of some ectoplasm being real material from the spirit world is near zero.

See also **spiritualism**.

Further reading: Keene 1997; Randi 1995.

electromagnetic field (EMF)

A region through which a force produced by electric current is exerted. Many people fear that EMFs cause cancer; however, a causal connection between EMFs and cancer has not been established. The National Research Council (NRC) spent more than three years reviewing more than 500 scientific studies that had been conducted over a 20-year period and found "no conclusive and consistent evidence" that electromagnetic fields harm humans. According to the chairman of the NRC panel, neurobiologist Dr. Charles F. Stevens, "Research has not shown in any convincing way that electromagnetic fields common in homes can cause health problems, and extensive laboratory tests have not shown that EMFs can damage the cell in a way that is harmful to human health."

In 1997, the *New England Journal of Medicine* published the results of the largest, most detailed study ever done of the relationship between EMFs and cancer. Dr. Martha S. Linet, director of the study, said: "We found no evidence that magnetic field levels in the home increased the risk for childhood leukemia." The study took eight years and involved measuring the exposure to magnetic fields generated by nearby power lines. A group of 638 children under age 15 with acute lymphoblastic leukemia were compared with a group of 620 healthy children. "The researchers measured magnetic fields in all the houses where the children had lived for five years before the discovery of their cancer, as well as in the homes where their mothers lived while pregnant." The study was criticized because it is impossible to know exactly what the EMFs were at the times the mothers or their children were exposed. All measurements were made only after the exposure had taken place and assumptions were made that the levels of EMFs were not substantially different during exposure. It is unlikely, however, that anyone except the intellectual descendants of Nazi doctor Joseph Mengele will ever do a **control study** on humans that systematically controls exposure to EMFs from the moment of conception through early childhood.

A report published in the *Journal of the American Medical Association* on a study of

891 adults who used their cell phones between 1994 and 1998 stated that there was no increased risk of brain cancer associated with cell phone use (Muscat 2000). Yet many people believe that living near power lines or using cellular phones cause cancer. Why? Some lawyers, the mass media, and a scientifically illiterate public can take the credit here.

Robert Pool claims popular opinion has been aroused against EMFs by unscientific sources such as *The New Yorker* magazine (Pool 1990). The fear that cell phones might be causing brain tumors was also aroused by ABC's *20/20* (October 1999) in a story that focused on the claims of Dr. George Carlo, who, for the previous 6 years, ran the cell phone industry's research program on the effects of radiation from cell phones. Gordon Bass also relied heavily on Carlo for his alarmist piece in *PC Computing*, "Is Your Cell Phone Killing You?" (November 30, 1999). Carlo contradicts the conclusions of most other researchers in the field and maintains that "we now have some direct evidence of *possible* harm from cellular phones" (emphasis added). Contrast Carlo's view with the following:

> The epidemiological evidence for an association between RF [radio frequency] radiation and cancer is found to be weak and inconsistent, the laboratory studies generally do not suggest that cell phone RF radiation has genotoxic or epigenetic activity, and a cell phone RF radiation–cancer connection is found to be physically implausible. Overall, the existing evidence for a causal relationship between RF radiation from cell phones and cancer is found to be weak to nonexistent. (Moulder et al. 1999)

In a press release on October 20, 1999, the Federal Communications Commission (FCC) responded to *20/20* and claimed that the "values of exposure reported by ABC were well within that safety margin, and, therefore, there is no indication of any immediate threat to human health from these phones." Furthermore, the *20/20* story claimed that *cell phone antennae* emit microwave radiation into the brain, which is misleading. Microwaves, which are in the 1–300 GHz range, are emitted by the antennae of cellular *towers* that transmit the messages, but the phone's antennae do not emit microwaves. Cellular phones emit in the 800–900 MHz range, according to the FCC. (For comparison, microwave ovens have a frequency of about 2450 MHz.)

Similar arousal of fear has been evoked by talk show hosts such as Larry King, who introduced the nation to a widower who claims that his wife's fatal brain tumor was caused by the EMF emitted from her cellular phone. There is a lawsuit, of course. The evidence? The tumor was located near where she held the phone to her ear. The major networks reported the story about the lawsuit and the brain tumor and the cellular phone. Scientists were interviewed to give the story more "depth" and credibility. However, no scientist has yet found a causal connection between EMF and cancer, much less between cellular phones and brain tumors. So, a scientist who has exposed *existing* tumors to EMF was interviewed. He reported that his research indicates that tumors grow faster when exposed to EMF. Sales of cellular phones dropped and stock in companies that manufacture them dropped. Because tumors exposed to EMF grow more rapidly than tumors not so exposed does *not* indicate that EMF *causes* tumors, cancerous or otherwise.

It is *possible* that cellular phones are causing brain tumors, but the likelihood is small. The phones emit very low EMF levels and exposure to them is intermittent. It is *possible* that a person with a brain tumor who uses a cellular phone is running a significant risk that the tumor will grow faster than it otherwise would. As yet, however, there is no evidence to support the view that there is a reasonable probability of either. This is not surprising since, as physicist Robert Park has noted: "All known cancer-inducing agents, including radiation, certain chemicals and a few viruses, act by breaking chemical bonds to produce mutant strands of DNA. Photons with wavelengths longer than the near ultraviolet do not have enough energy to break a chemical bond in DNA" (Brody 2002).

It is true that "there have been numerous scientific reports of elevated levels of leukemia in people who are exposed to high EMF levels on the job, such as power-line repairmen and workers in aluminum smelters" (Pool 1991). While the scientific jury is still out on the causal connection, if any, between living near power lines and cancer, the lawsuits are starting to come in. Over 201 challenges to utility projects were made in 1992 in which EMF was an issue. At least three suits have been filed in federal courts claiming exposure to utility lines caused cancer (ibid.). Utility companies are running scared. They are pouring billions of dollars into efforts to cut EMF exposure from their power lines. Lawyers intimidate power companies with a Swedish study that found leukemia rates were 400% higher among children living near power lines. Dr. Robert Adair, a physicist at Yale University, calls the reaction "electrophobia" and says that it would take EMF levels 150 times higher than those measured by the Swedish researchers to pose a hazard.

There is currently a great push to bury all power lines. Better safe than sorry? The cost goes up twentyfold to bury the lines.

We would have to bury our electrical wires even deeper than our power poles are high if we are to make a significant difference in shielding us from the magnetic fields of power lines.

It is not very likely that the average person has anything to worry about from power lines. Most of us do not get that close to them to be significantly affected by their EMFs. Our exposure to them, even if they are nearby, is not direct, up close, and constant. We're probably in more danger of EMF pollution from the wiring in our homes and the electrical appliances we use than from the wires overhead. No one can avoid electromagnetic radiation. It is everywhere. We are constantly exposed to it from light, radio and television transmissions, police two-way transmissions, walkie-talkies, and so on. Furthermore, "while electrical fields are easily screened, magnetic fields make their way unimpeded through most substances" (Pool 1990).

It is curious that while fear of EMFs is on the rise, so is **magnet therapy**.

Further reading: Edwards 1988; Livingston 1997; Richards 1993; Sagan 1992.

electronic voice phenomenon (EVP)

The alleged communication by **spirits** through tape recorders and other electronic devices. The belief in EVP in the United States seems to have mushroomed thanks to Sarah Estep, president of the American Association of Electronic Voice Phenomena. Estep claims that in the 1970s she started picking up voices on her husband's Teac reel-to-reel recorder. She is sure the voices are spirits, proving there is life after death. Estep also claims to hear voices of aliens on some of her tapes. Aliens don't speak English, however, so she is not sure what they are saying.

Interest in EVP apparently began in the 1920s. An interviewer from *Scientific American* asked Thomas Edison about contacting the dead. Edison said that nobody knows whether "our personalities pass on to another existence," but "it is possible to construct an apparatus which will be so delicate that if there are personalities in another existence . . . who wish to get in touch with us . . . this apparatus will at least give them a better opportunity to express themselves than the tilting tables and raps and ouija boards of mediums and other crude methods now purported to be the only means of communication" (Clark 1997: 235). There is no evidence, however, that Edison ever designed or tried to construct such a device.

While it is impossible to prove that all EVPs are due to natural phenomena, skeptics maintain that they are probably due to such things as interference or cross modulation.

Further reading: Marion: www .ghostshop.com.

energy

In physics, the basic idea of energy is the capacity of a physical system to do "work." In physics, "work" is defined as the product of a force times the distance through which that force acts. "Energy" is a term to express the power to move things, either potential or actual. New Age spirituality is all about empowerment. The New Age is about enhancing your energy, tapping into the energy of the universe, manipulating energy so that you can be healthy, happy, fulfilled, successful, and lovable; so life can be meaningful, significant, and endless.

Of course, New Age energy has nothing to do with mechanics, electricity, or the nuclei of atoms. New Age energy has more to do with things like **chi** or **prana**.

New Age energy isn't measurable by any known scientific instrument and is believed to be the source not only of life, but of health as well. There are no ergs, joules, electron-volts, calories, or foot-pounds of New Age energy. This energy is outside the bounds of scientific control or study. Only healers with special powers at "unblocking," "harmonizing," "unifying," "tuning," "aligning," "balancing," "channeling," or otherwise manipulating New Age energy can measure this energy. How? They measure it by *feeling* it. Energy medicines are based on variants of the metaphysical theory known as **vitalism**, a theory that has been dead in the West for over a century. But New Age **quackery** often maintains that the older a theory is, the more one should have faith in it.

Few things are more intimidating to the nonscientist than modern physics. Even an educated person has difficulty comprehending the most basic claims made about the entities and possible entities of the subatomic world, not to mention the exotic claims about entities and possible entities at the edges of the universe. Even the concepts of "subatomic" and "edge of the universe" boggle the mind. Perhaps it is because of the obscurity and inaccessibility of modern physics that many uneducated people scoff at science and find solace in fundamentalist religious interpretations of the origin and nature of the universe. Another response to the seemingly transcendental nature of concepts in modern physics has been to interpret those concepts in terms of ancient metaphysical doctrines popular for thousands of years in exotic places (to the Western mind) such as India and China. This notion of a "harmony" between ancient metaphysics and modern physics is attractive to those who accept science and reject the Christian sects they were raised in, but still have spiritual longings. Believing in this notion of harmony between the ancient East and the modern West has the virtue of allowing one to avoid appearing to be an imbecile who rejects science in order to accept religion.

Acting much like nuclear accelerators on atoms, New Age theorists smash concepts into bits, only the bits are interfered with in ways the physicist Werner Heisenberg never foresaw. We may as well talk about "alternative" physics; for what they have done to the concepts of modern physics is to refashion them into a **metaphysics** with its own technology and product line. Nothing demonstrates this more clearly than the New Age conceptions of energy.

Fittingly, an experiment by a 9-year-old neatly measures the strength of New Age energy. Emily Rosa tested 21 **therapeutic touch** practitioners to see whether they could *feel* her life energy when they could not see its source. The test was very simple and seems to clearly indicate that the subjects could not detect the life energy of the little girl's hands when placed near theirs. They had a 50% chance of being right in each test, yet they correctly located Emily's hand only 44% of the time in 280 trials. If they can't *detect* the energy, how can they manipulate or transfer it?

See also **pathological science**.

enneagram

A drawing with nine lines. Figuratively, however, the enneagram is a New Age mandala, a mystical gateway to personality typing. The drawing is based on a belief in the mystical properties of the numbers 7 and 3. It consists of a circle with nine equidistant points on the circumference. The points are connected by two figures: one connects the number 1 to 4 to 2 to 8 to

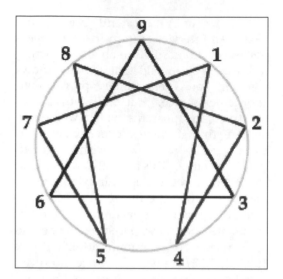

The enneagram.

5 to 7 and back to 1; the other connects 3, 6, and 9. The 1-4-2-8-5-7 sequence is based on the fact that dividing 7 into 1 yields an infinite repetition of the sequence 142857. In fact, dividing 7 into any whole number not a multiple of 7 will yield the infinite repetition of the sequence 142857. Also, $142857 \times 7 = 999999$. And, 1 divided by 3 yields an infinite sequence of threes. The triangle joining points 3, 6, and 9 links all the numbers on the circle divisible by 3. To ascribe **metaphysical** or mystical significance to the properties of numbers is mere superstition and a throwback to an earlier time in human history when ignorance was considered a point of view (apologies to Scott Adams and "Dilbert").

The enneagram represents nine personality types. How the types are defined depends on whom you ask. Some define them by a fundamental weakness or sin. Others define them by a fundamental **energy** that drives one's entire being. Some follow classical **biorhythm** theory and classify the nine types according to three types of types: mental, emotional, and physical. Others classify the nine types according to three types of instinctual drives: the Self-Preserving drives, the Social drives, and the Sexual drives. Some follow G. I. **Gurdjieff**, who claims to have followed Sufism, and identify the types as mental, emotional, and instinctual. Gurdjieff's disciple **Ouspensky** claimed:

> All knowledge can be included in the enneagram and with the help of the enneagram it can be interpreted. And in this connection only what a man is able to put into the enneagram does he actually know, that is, understand. What he cannot put into the enneagram makes books and libraries entirely unnecessary. Everything can be included and read in the enneagram. (*In Search of the Miraculous,* www.natcath.com/NCR_Online/documents/ennea2.htm)

The father of the enneagram seems to be Oscar Ichazo (b. 1931), who spoke of enneagons (nine-pointed figures, enclosed in a circle, with straight lines connecting each point to two others) and *ego fixations* corresponding to each of the nine points. (Ichazo learned of the enneagram through Ouspensky's writings of Gurdjieff.) He called his system Arica, after the coastal city in northern Chile, near the Peruvian border, where he opened his first school. In the early 1990s, there were 40 or so Arica training centers located in South America, the United States, Europe, and Australia.

> The Arica system constitutes a body of practical and theoretical knowledge in the form of a nine-level hierarchy of training programs aimed at the total development of the human being. . . . The Arica system observes that the human body and psyche is composed of nine independent yet interconnected systems. Particular imbalances within these systems are called "fixations". . . . These nine separate components are represented by enneagons—nine pointed figures that map the human psyche. . . . [T]here are seven fundamental enneagons associated with the nine ego fixations. Thus, the enneagons constitute the structural maps of a human psyche . . . [and] provide a guide through which a person may better understand oneself and one's interactions with others. . . . An ego fixation is an accumulation of life experience organized during one's childhood and which shapes one's personality. Arica training seeks to overcome the control and influence of the ego fixations so that the individual may return to the inner balance with which he or she was born. [Arica Institute, Inc., plaintiff-appellant, v. Helen Palmer and Harper & Row Publishers, Incorporated, defendants-appellees, no. 771, docket 91-7859, United States Court of Appeals, Second Circuit, argued Jan. 30, 1992, decided July 22, 1992; florida-lawfirm.com/arica.html]

Ichazo makes claims such as "the dominant passion of the Indolent fixation is Sloth; the dominant passion of the Resentment fixation is Anger; and the dominant passion of the Flattery fixation is Pride." In short, he developed a typology of ego fixations based on the classical Christian notion of the seven capital sins plus fear and deceit.

Ichazo claims to have been trained in the mystical arts of Sufism, the cabala, and Zen and to have studied martial arts, yoga, Buddhism, Confucianism, the *I Ching*, and **alchemy**. He was called the "continuation of Gurdjieff" by filmmaker Alexandro Jodorowsky (*El Topo, The Holy Mountain*), who claims to have spent a weekend expanding his consciousness with Ichazo by using LSD. Ichazo claims he began teaching the enneagram after spending a week in a "divine coma" (Keen 1973).

Ichazo never claimed to have a scientific basis for his theory of personality types, ego fixations, and so on. His notions were based on visions and insights taken from numerous eclectic sources and freely mixed into an amalgam of mystical psychobabble.

> Ichazo claimed to have discovered the personality type meaning of the enneagram while in some kind of ecstatic state or trance under the influence of some spirit or angelic being: the Archangel Gabriel, the "Green Qu'Tub" [a Sufi spiritual master] or Metatron, the prince of the archangels. [*National Catholic Reporter,* www.natcath.com/NCR_Online/documents/ennea2.htm]

Like Gurdjieff, Ichazo claimed we are born with an essence (nature) that conflicts with our personality (nurture), and we must struggle to harmonize the two and return to our true essence. He founded his Arica Institute in the late 1960s. The institute continues to exist, though it has contracted somewhat from its heyday in the early 1990s, and now offers training in "Nine Hypergnostic Systems" and tai chi chuan in centers in New York and Europe.

Several former disciples have modified Ichazo's teachings during the past twenty years. Claudio Naranjo attended Ichazo's lectures on ennead personality types in Santiago, Chile, in the 1970s and published a book called *Enneatypes in Psychotherapy* in 1995. A Jesuit priest named Bob Ochs got the enneagrams from Naranjo and taught courses on enneagrams at Loyola University in Chicago in 1971. Naranjo also taught Helen Palmer, who claims to be carrying on the esoteric oral tradition in her writings. By the time the enneagram got to Palmer, it was imbedded with Western psychological notions. Nevertheless, it remained a set of teachings without any scientific foundation.

Helen Palmer is the author of *The Enneagram: Understanding Yourself and the Others in Your Life* (1988). Arica sued Palmer for copyright violations but lost. Nevertheless, she seems to have based her work on Ichazo's, and simply changing the terminology seems to be her main contribution. *Enneagram* replaced *enneagon,* and *personality type* replaced *ego fixation,* for example.

Palmer says that the "enneagram is a psychological and spiritual system with roots in ancient traditions." She types people by fundamental weakness or sin: anger, pride, envy, avarice, gluttony, lust, sloth, fear, and deceit. She calls these weaknesses "capital tendencies." Each of us has a personality that is dominated by one of the nine capital tendencies. Knowing what type you are, and what type others are, will put you on the road to "self-understanding and empathy, giving rise to improved relationships," says Palmer.

Each personality type is numbered and labeled.

The Nine Personality Types and the Nine Capital Tendencies

The One	The Perfectionist	anger
The Two	The Giver	pride
The Three	The Performer	deceit
The Four	The Romantic	envy
The Five	The Observer	avarice
The Six	The Trooper	fear
The Seven	The Epicure	gluttony
The Eight	The Boss	lust
The Nine	The Mediator	sloth

Personality typing, of course, is somewhat arbitrary. The classification systems used by Ichazo, and modified by Palmer and others according to their own idiosyncratic beliefs, are not without merit. For example, one could certainly learn much of importance about oneself by focusing on one's central fault or faults, but those who advocate using the enneagram seem to be interested in much more than a bit of self-knowledge. Entire metaphysical systems, psychologies, religions, cosmologies, and New Age springboards to higher consciousness and fuller being are said to be found by looking into the enneagram. There is seemingly no end to what one can find in these nine lines.

Some, for example, have developed personality profiles for different "styles" of personalities. For example:

Style Five

The life of the style Five centers on their thinking. Healthy Fives are both highly intellectual and involved in activity. They can be, if not geniuses, then extraordinarily accomplished. As the most intellectual of the nine types, they are often superb teachers and/or researchers. Many healthy Fives are fine writers because of their acute observational skills and a developed idealism. They are highly objective and able to see all sides of a question and understand them.

When Fives become less healthy, they tend to withdraw. Instead of dealing with their sensitivity by being emotionally detached from results, they split off from reality, living in worlds of their own creating and not answering the demands of active living. Their natural independence as a thinker degenerates into arrogance. They can become quite arrogant or eccentric. In the movies, Fives are the "mad professors."

Fives you may know: Bill Gates, Scrooge, Buddha, T. S. Eliot, John Paul Sartre, Rene Descartes, Timothy McVeigh,

Joe DiMaggio, Albert Einstein, H. R. Haldeman, Ted Kaczynski, Jacqueline Onassis and Vladimir Lenin. [Enneagram Central, www.enneagramcentral.com/testc_bl.htm]

What this typology is based on is anybody's guess. But the style profile is probably best evaluated in terms of the **Forer effect**. Nothing in the typology resembles anything approaching a scientific interest in personality, and there does not seem to be any way to validate this typology.

The above style was said to be mine as a result of a test I took. However, the test came with the following advisory:

> Does this fit you? If it does not, go back over the test, rethink some of your answers and see if you come up with your style. This is not easy. Your enneagram style is an energy you have been using without knowing all your life. You have a vested interest in not knowing this energy because it may slightly alter what you have considered your motivation for many things. Besides, this energy has a down side you may not like to acknowledge

If the style doesn't fit, go back and change some answers until it fits, but be careful because you may be deceiving yourself when you answered the questions the first time or you may be deceiving yourself with your revisions! Note also how the profile contains several weasel words: "can be," "are often," "tend to," "can become." The central feature of the Five is *thinking*. Nobody needs a personality test to determine whether his or her dominant energy, drive, fixation, or passion is the intellectual. Thinkers are observers and intellectuals are often arrogant. This is not a scoop. Nor is it very useful, as is evident by the listing of people who are allegedly Fives.

The limits of the enneagram are the

limits of the imagination of those who work with them. One master claims that the Fives' "primary passion is avarice in terms of their time and possessions, and their chief feature is withdrawal from experience." Another expert describes the Five as The Thinker and identifies this type by its dominant fear: fear of being overwhelmed by the world. We are told that if we want to get along with a Five,

> Be independent, not clingy. Speak in a straightforward and brief manner. I need time alone to process my feelings and thoughts. Remember that if I seem aloof, distant, or arrogant, it may be that I am feeling uncomfortable. Make me feel welcome, but not too intensely, or I might doubt your sincerity. If I become irritated when I have to repeat things, it may be because it was such an effort to get my thoughts out in the first place. Don't come on like a bulldozer. Help me to avoid my pet peeves: big parties, other people's loud music, overdone emotions, and intrusions on my privacy. [www .9types.com]

This is good advice for getting along with just about anybody, except for those who would rather be at a big party in the evening after spending the afternoon alone with a book.

We are also told that for a Five to reach his potential he must go against the grain and strive to be like an Eight, whose main vice is lust. The scientific studies supporting this claim seem to have been lost, however.

Further reading: McGuire and Hull 1977.

E-rays (Erdstrahlen, earth rays)

Evil rays emitted from below ground and allegedly detectable only by **dowsers** with **paranormal** powers. These evil rays are invisible and undetectable by ordinary people using ordinary scientific equipment. E-rays are blamed for everything from cancer in humans to wilting in plants. E-rays are especially bad for one's **aura**.

The belief in E-rays is especially popular in Germany, where some people sleep with protective sheets of black plastic under their beds for protection. Specialists in E-ray detection practice a German variant of **feng shui**, advising individuals and government employees on safe furniture arrangement.

Further reading: Randi 1995.

ESP

ESP is an acronym for extrasensory perception and refers to perception occurring independently of sight, hearing, or other sensory processes. People who have extrasensory perception are said to be **psychic**. J. B. Rhine, who began investigating the phenomenon at Duke University in 1927, coined the term "ESP" to refer to **clairvoyance**, **precognition**, and **telepathy**. In recent years, the term has been extended to refer to **clairaudience** and **remote viewing**. Collectively, ESP and other **paranormal** powers such as **telekinesis** are known as **psi**. The existence of psi is disputed, though systematic experimental research on these subjects has been ongoing for over a century.

Skeptics dismiss most of the evidence for ESP. Apparent successes of ESP research are most likely due to one or several of the following:

- incompetence or fraud by **parapsychologists** or believers in psi
- trickery by **mentalists**
- **cold reading**
- **subjective validation**
- **selective thinking** and **confirmation bias**
- poor grasp of probabilities and of the **law of truly large numbers**

- **shoehorning**, **retroactive clairvoyance**, and **retrospective falsification**
- gullibility, **self-deception**, and **wishful thinking**

The following case is typical of those cited as proof of ESP. It is unusual only in that it involves belief in a psychic dog, rather than a psychic human. The dog in question is a terrier, Jaytee, who has achieved fame as having ESP because of his alleged ability to know when his owner, Pam Smart, is deciding to come home when she is away shopping or on some other business. Jaytee has been featured on several television programs in Australia, the United States, and England. He resides in England with Pam and her parents, who were the first to perceive the dog's psychic abilities. They observed that the dog would run to the window facing the street at precisely the moment Pam was deciding to come home from several miles away. (How the parents knew the precise moment Pam was deciding to come home is unclear.) Parapsychologist Rupert Sheldrake (see **morphic resonance**) investigated and declared the dog is truly psychic. Two scientists, Dr. Richard Wiseman and Matthew Smith of the University of Hertfordshire, tested the dog under controlled conditions. The scientists synchronized their watches and set video cameras on both the dog and its owner. Alas, several experimental tries later, they had to conclude that the dog wasn't doing what had been alleged. He went to the window and did so quite frequently, but only once did he do so near the exact time his master was preparing to come home, and that case was dismissed because the dog was clearly going to the window after hearing a car pull up. Four experiments were conducted and the results were published in the *British Journal of Psychology* (89 [1998]: 453).

Much of the belief in ESP is based on apparently unusual events that seem inexplicable. However, we should not assume that every event in the universe can be explained. Nor should we assume that what is inexplicable requires a paranormal (or supernatural) explanation. Maybe an event can't be explained because there is nothing to explain.

Parapsychologists have attempted to verify the existence of ESP under controlled conditions. Some, like Charles **Tart** and Raymond **Moody**, claim success; others, such as Susan J. Blackmore, claim that years of trying to find experimental proof of ESP have failed to turn up any proof of indisputable, repeatable psychic powers. Defenders of psi claim that the **ganzfeld experiments**, the CIA's remote viewing experiments, and attempts to influence randomizers at **Princeton Engineering Anomalies Research** have produced evidence of ESP. Psychologists who have thoroughly investigated parapsychological studies, such as Ray Hyman and Blackmore, have concluded that where positive results have been found, the work was fraught with fraud, error, incompetence, or statistical legerdemain.

See also **optional starting and stopping** and **psi-missing**.

Further reading: Alcock 1990; Frazier 1986; Gardner 1957, 1981; Gordon 1987; Hansel 1989; Hines 1990; Hyman 1989; Keene 1997; Randi 1982a; Stein 1996a; Wiseman and Smith 1998.

est

Werner Erhard's est (Erhard Seminar Training and Latin for "it is") was one of the more successful entrants in the human potential movement. est is an example of what psychologists call a **Large Group Awareness Training** program.

The first est seminar was held in October 1971 at the Jack Tar Hotel in San Francisco, with nearly 1,000 in attendance. Erhard and est were known for training people to get "It," a concept taken from author, teacher, and expert communicator Alan Watts. When Erhard arrived in the San Francisco Bay area, Watts was teaching his version of Zen to small groups on his houseboat in Sausalito. Erhard, like Watts, would teach people to "get it." Watts, however, did most of his teaching through books. His seminars were small. Erhard would not teach through books, but in large hotel ballrooms to hundreds at a time.

est adopted, in part, the Zen master approach, which was often abusive, profane, demeaning, and authoritarian. While many participants did not perceive the training as particularly abusive, some were not used to the discipline requested of them. Some have claimed that one typically abusive approach was the requirement of extraordinary bladder control in est training. Participants were advised not to leave the room, even to go to the toilet, during training. According to one est participant, however, "bathroom breaks were scheduled at regular and reasonable intervals. . . . Two or three rows at the back of the room were reserved for those who required more frequent bathroom breaks (and I think either some sort of documentation or personal insistence were required to qualify). No one was ever physically required to stay in the room at any time" (personal correspondence). In any case, one should expect some sort of discipline and order for this kind of training. Having people come and go as they please is distracting and not conducive to the concentration necessary for any training program.

In the late 1960s, Erhard studied **Dianetics** and Scientology. L. Ron Hubbard became a significant influence. Scientologists accuse Erhard of stealing his main ideas for est from Hubbard. When Erhard set up est he considered making it a

church, as Hubbard had done with Dianetics and the Church of Scientology. But Erhard decided to incorporate as an educational firm for profit in a broad market.

Erhard and his supporters accuse Scientology of being behind attempts to discredit Erhard, including hounding by the IRS and accusations of incest by his children. Erhard won a lawsuit against the IRS, and the incest accusations may have been based on **false memories** induced in therapy. Erhard has even claimed that Scientologists have hired hit men to kill him, though the most logical explanation for his continued survival is probably that no one is really trying to kill him.

est bears little resemblance to Dianetics or Scientology, however. est is a hodgepodge of philosophical bits and pieces culled from the carcasses of existential philosophy, motivational psychology, Maxwell Maltz's psycho-cybernetics, Zen Buddhism, Alan Watts, **Freud**, Abraham Maslow, Hinduism, Dale Carnegie, Norman Vincent Peale, P. T. Barnum, and anything else that Erhard's intuition told him would work in the burgeoning human potential market. What did Erhard promise those who would pay hundreds or thousands of dollars for his programs? He promised he would "blow their minds" and raise them to a new level of consciousness. In short, he would make them special. He would first tell them that their problem was that they needed to have their consciousness "rewired" and that his program would do the rewiring. Once they got their consciousness on straight, life would be good or at least different. They would be powerful, confident, and successful because they would be independent and in control. They would learn to see things in radically different ways. Nothing would change and yet everything would change. (The same promise was made by Watts for the disciples of Zen.) Nothing could stand in their way and deprive them of all those opportunities in life they had heretofore been denied because of bad programming or wiring. Through est they would be set free and born again. All problems and limitations are in the mind. Just rewire the mind, that is, deconstruct personality, exorcise all negativity, quit blaming others, and learn to accept things. Happiness will follow.

Where did Erhard get his training? Mostly, he is self-taught. His study was undirected and accidental. In 1960 he was John Rosenberg, 25 years old, married with children. Apparently dissatisfied with his life but with no Large Group Awareness Training available to him, he did what many unhappy men have done: he abandoned his family. He left Philadelphia and went to St. Louis, changed his name, and sold cars. Some might find it interesting that a Christianized Jew (his parents had him baptized in the Episcopal Church) would come to identify himself with a German name. Of more interest to his transformation, however, are the books that influenced him.

Erhard was "profoundly dissatisfied with the competitive and meaningless status quo" and was deeply affected by Napoleon Hill's *Think and Grow Rich* (Bartley 1978). Hill's three basic principles are: Every achievement begins with an idea; plans call for their implementation; and what you think is what you do. Think positive, you will do positive deeds. Hill also advised visualizing objectives and selecting similar-minded friends. Hill gives good advice, but it is vague and not systematic. It doesn't offer much to people who haven't got a clue what their objectives are or should be. Some of his ideas can even be harmful, if not properly applied. For example, some people are taught that they should always talk positive, even if this means lying. Even if you haven't made a sale in two years, you must put on a positive front and tell everyone that business couldn't be better. Even if you know nothing about the product you are selling, you must praise it beyond belief. Even if you are experiencing one failure after another, you must lie to yourself and tell yourself that you are doing great. You must never blame the product for not selling. You must try harder, have more faith, and be more positive. Maybe you need to take advanced courses to help you succeed. By the time you wake up, you are bankrupt and those who were cheering you on (your sponsors) are nowhere to be found.

Another significant influence on Erhard was Maxwell Maltz's *Psycho-Cybernetics* (1960). As a young man, Erhard apparently had a lot of negatives in his self-image and was deeply affected by Maltz, who emphasized, among other things, self-hypnosis. Erhard put his new ideas and new self to work as a traveling salesman for a correspondence school. Maltz stimulated his interest in hypnotism, but Erhard's focus would be on "programming" and "reprogramming." The idea is not without merit, though the language is unnecessarily cumbersome. The basic idea Erhardt came to espouse is that bad habits are programmed into us: We have been "hypnotized" during normal consciousness and that's where our problems arise. Unconsciously, we've developed debilitating habits and beliefs. The point is to get rid of them by replacing them with positive and life-enhancing beliefs and habits. Again, however, the language is very vague, probably too vague to do any meaningful scientific appraisal of them.

By the time Erhard arrived in San Francisco, he'd had jobs selling and managing salespersons for *Great Books* and *Parents* magazine. He became part of the self-help movement after hiring Robert Hardgrove, who introduced Erhard to the work of

Abraham Maslow and Carl Rogers. Maslow and Rogers were unique in psychology at the time, for they emphasized not the disturbed or ill person but the healthy, happy, satisfied, accomplishing person. The human potential movement was just getting started and Erhard would be in on the ground floor.

It is estimated that some 700,000 people did the training before the seminars were halted in 1991, when Erhard packed up and left the country (Faltermayer 1988). He sold the est "technology" to some followers, who established **Landmark Forum**. Erhard's brother, Harry Rosenberg, heads Landmark Education Corp. (LEC), which at the turn of the century was doing some $50 million a year in business and had attracted some 300,000 participants. LEC is headquartered in San Francisco, as was est, and has 42 offices in 11 countries. Apparently, however, Erhard is not involved in the operation of LEC.

See also **firewalking** and **neurolinguistic programming**.

Further reading: Ankerberg and Weldon 1996; Barry 1997; Pressman 1993.

evil eye

One who possesses the evil eye is able to put a **curse** on a child, livestock, crops, and so on. There does not seem to be any particular reason why some people are born with and others without the evil eye. The curse is usually unintentional and caused by praising and looking enviously at the victim. In Sicily and southern Italy, however, it is believed that some people—*jettatore*—are malevolent and deliberately cast the evil eye on their victims. Belief in the evil eye is not necessarily associated with witchcraft or **sorcery**, although "Evil Eye" was on the list of things to look for during the Spanish Inquisition.

The superstitious belief in the evil eye

Apotropaic paratrinket from Greece.

is ancient and widespread, although not universal. It is thought to have originated in Sumeria. Its origins are obscure but the belief may have its roots in fear of strangers or other social concerns, and simple **post hoc** reasoning, for example, praise is given or a stranger passes and later a child is sick or the crops fail. Some folklorists believe that the evil eye belief is rooted in primate biology (dominance and submission are shown by gazing and averting the gaze) and relates to our dislike of staring.

Various rituals have developed to counteract the effects of the evil eye, such as defusing the praise, putting spit or dirt on a child who is praised, averting the gaze of strangers, or reciting some verses from the Bible or the Koran.

Belief in the evil eye is especially prevalent today in the Mediterranean and Aegean, where apotropaic **amulets** and **talismans** are commonly sold as protection against the evil eye.

Further reading: Dundes 1992; Stevens 1996.

exorcism

An exorcism is a religious rite for driving **Satan** or evil spirits out of a possessed person, place, or thing. In ancient times, many cultures had such rites. Today, the Roman Catholic Church still believes in diabolic possession and its priests still practice what is called "real exorcism," following a 27-page ritual and using holy water, **incantations**, **prayers**, incense, relics, or Christian symbols such as the cross to drive out evil spirits. The Catholic Church has at least ten official exorcists in America today (Cuneo 2001). The Archbishop of Calcutta, Henry Sebastian D'Souza, says he ordered a priest to perform an exorcism on Mother Teresa, winner of the Nobel Peace prize, shortly before she died in 1997 because he thought she was being attacked by the devil.

Most Protestant sects also believe in Satanic possession and exorcism. Michael Cuneo, a sociologist at Fordham University, claims, "By conservative estimates, there are at least five or six hundred evangelical exorcism ministries in operation today, and quite possibly two or three times this many." I have witnessed videotaped exorcisms by three evangelicals, Brian Connor, Tom Brown, and Bob Larson. I must say that the participants being exorcized seem to have seen the movie *The Exorcist* or one of the sequels. They all fell into the role of smirking and snarling and

speaking in a husky voice, as Linda Blair did in the film. The similarities in speech and behavior among the possessed have led some psychologists such as Nicholas Spanos to conclude that both exorcist and possessed are engaged in learned role-playing. They certainly seem to play off one another.

The behavior of the exorcists is at least as interesting as that of the possessed. Connor works with several assistants on one subject, whereas Brown and Larson work alone but with groups of troubled people, some of whom they think may be possessed. All use the Bible and the cross as their only props. They look their subjects in the eye while making physical contact with their hands and commanding the demons to leave. After several hours, in some cases, a catharsis is reached. There is much hugging and praising of the Lord. Those doing the exorcism declare the subject free of the demon.

Michael Cuneo watched a film of the exorcism performed by Brian Connor and concluded that the group was suggesting to the possessed how he should respond, and that he saw no evidence of either demonic possession or demons being exorcized. A psychiatrist was shown the same film and he announced that he couldn't evaluate what he observed as a psychiatrist but as a "believer" he thought that there might be something real going on involving demonic possession. When asked what he based his belief on, he replied tersely: **faith**. This man was a member of the American Psychiatric Association's Committee on Religion and Psychiatry.

Believing in demons is one thing; believing you have the ability to call up a supernatural being with infinite power and perfection who will cause demons to move on at your behest seems certifiable. The whole coven of exorcists and exorcized seems deluded. The former clearly felt great pride at their achievement and shared in a glorious victory over Satan. The latter was coddled and cuddled, hugged and loved, and eventually praised and rewarded with the good feelings of caring people when he released Satan and said, "Jesus is Lord." There doesn't seem to be anything deeply complicated about what happened. The exorcists convinced the subject he was possessed. They cued him as to how to behave and rewarded him and themselves when he let the demon go. **Communal reinforcement** and **self-deception** will go a long way toward explaining how the exorcists came to believe they could exorcize demons. The exorcists clearly enjoy their work and get great satisfaction out of "helping" people in this powerful way. I am sure that many evangelicals who saw the program are wondering where they can sign up to be an exorcist's helper.

Exorcisms can be done on inanimate objects or places as well as on people. These need not be "real exorcisms" but can be "simple exorcisms" (usually thought of as baptizing the infant or "blessing" the house or place). Satan is everywhere, it seems, but the specialist in real exorcism is needed only when the Evil One starts acting up.

Most, if not all, cases of alleged demonic possession of humans probably involve either people with brain disorders ranging from epilepsy and depression to schizophrenia and Tourette's syndrome, or people whose brains are more or less healthy but who are unfortunate enough to be sucked into playing a social role with very unpleasant consequences. In any case, the behaviors of the possessed resemble very closely the behaviors of those with electrochemical, neurochemical, or other physical or emotional disorders.

Some therapists practice a secularized version of exorcism. They specialize in unveiling and ridding their patients of "entities" that, the therapists believe, are the cause of the patient's troubles. Entity release therapists engage in this work even though there is about as much evidence for the entities as there is for the devils exorcized by Catholic priests or Protestant evangelicals. Many people however, are very resistant to the idea that demonic possession is a myth, especially since they have seen or read fictional works such as *The Exorcist* or *The Amityville Horror.* They can't imagine how anyone could make such stuff up; yet it would seem to take much more imagination to give credence to such tales.

Finally, many people fear possession by demons, but seem oblivious to the fact that exorcists can cause great harm. In San Francisco, Pentecostal ministers who were trying to drive out demons pummeled a woman to death in 1995. In 1997, exorcists stomped to death a Korean Christian woman in California, and in New York City a 5-year-old girl died after being forced to swallow a mixture containing ammonia and vinegar and having her mouth taped shut during an exorcism. In 1998, a woman suffocated her 17-year-old daughter with a plastic bag in Sayville, New York, while trying to destroy a demon inside her. In 2001, a woman in New York City drowned her 4-year-old daughter while performing an exorcism. Such exorcists are nearly sufficient to make an atheist believe in Satan.

Further reading: Cuneo 2001; Singer and Lalich 1996; Spanos 1996.

extraordinary human function (EHF)

Various activities would count as extraordinary human functions, such as the ability to read messages with one's ears, forehead, fingers, or some other part of the anatomy besides the eyes. There have even been accounts of reading by *sitting* on the message. The latter was popular in China in the late 1970s, when the study of EHFs became a major research topic at Beijing University and the Chinese Academy of Sciences. The scientists seemed particularly interested in finding a link between EHF and **chi**, believed by many Chinese to be the fundamental life force. Their research, like similar research in the Soviet Union and United States, covered everything from using **paranormal** powers to catch criminals to training astronauts to use such powers for spying or guiding missiles. Occasionally, someone will claim to be able to read while blindfolded. Some claim that techniques have been developed in **dermo-optical perception** to teach blind people to read through their forehead or fingers using paranormal powers. Braille is a much better bet, however.

Further reading: Benski 1998; Gardner 1981; Keene 1997; Lyons and Truzzi 1991.

extraterrestrials

See **alien abductions**

Eye Movement Desensitization and Reprocessing (EMDR)

A psychotherapeutic technique in which the patient moves his or her eyes back and forth while concentrating on whatever problem he or she might be having. The therapist waves a stick or light in front of the patient and the patient is supposed to follow the moving stick or light with his or her eyes. Dr. Francine Shapiro discovered the therapy while on a walk in the park. (Her doctorate was earned at the now defunct and never accredited Professional School of Psychological Studies. Her undergraduate degree is in English literature.) EMDR is a scientifically controversial tech-

nique, but this has not prevented thousands of practitioners from being certificated to practice EMDR by Shapiro and her disciples.

It is claimed that EMDR can "help" with "phobias, generalized anxiety, paranoid schizophrenia, learning disabilities, eating disorders, substance abuse, and even pathological jealousy" (Lilienfeld 1996), but its foremost application has been in the treatment of post-traumatic stress disorder (PTSD). So far, there has been no adequate explanation as to how EMDR works. Some think the rapid eye movements unblock "the information-processing system." Some think it works by a sort of ping-pong effect between the right and left sides of the brain, which somehow restructures memory. One therapist suggests that it works by sending signals that tame and control the part of the brain causing the psychological problems. Another thinks EMDR "activates the healing process of the brain, much as what occurs in sleep. As a result, the painful memories are re-processed and the original beliefs that sprang up from them are eliminated. New, healthy beliefs replace these" (Viviano: www.wowpages.com/viviano). In short, the healing occurs by activating the healing process.

Evidence for the *effectiveness* of EMDR is not much stronger than the theoretical explanations for how EMDR allegedly "works." The evidence has the virtue of being consistent, unlike the theoretical explanations, but it is mainly anecdotal and very vague. It has not been established beyond a reasonable doubt by any controlled studies that positive effects achieved by an EMDR therapist are not likely due to chance, the **placebo effect**, patient expectancy, spontaneous healing, posthypnotic suggestion, or other aspects of the treatments besides the eye movement aspect. Some claim that "what is new in EMDR does not appear to be helpful, and what is helpful is what we already know about relaxation, education, and psychotherapy" (Norwood, et al.: www .psych.org/pract_of_psych/principles_and _practice3201.cfm). This is not to say that there have not been controlled studies of EMDR. Dr. Shapiro cites quite a few on her web site, including her own (www.emdr .com/studies.htm).

A study by Wilson, Becker, and Tinker published in the *Journal of Consulting and Clinical Psychology* (65, no. 6 (1997): 1047–1056) reports a "significant improvement" in PTSD subjects treated with EMDR. The study also provides significant evidence that spontaneous healing cannot account for this improvement. Nevertheless, the study is unlikely to convince critics that *eye movement* is the main causal agent in measured improvement of PTSD subjects. Until a study is done that isolates the eye movement part from other aspects of the treatment, critics will not be satisfied. It may well be that those using EMDR are effecting the cures they claim and benefiting many victims of horrible experiences such as rape, war, terrorism, or murder or suicide of a loved one. It may well be that those using EMDR are directing their patients to restructure their memories so that the horrible emotive aspect of an experience is no longer associated with the memory of the experience. But for now the question still remains whether the rapid eye movement part of the treatment is essential. In fact, one of the control studies cited by Shapiro seems counterindicative:

In a controlled component analysis study of 17 chronic outpatient veterans, using a crossover design, subjects were randomly divided into two EMDR groups, one using eye movement and a control group that used a combination of forced eye fixation, hand taps, and hand waving. Six sessions

were administered for a single memory in each condition. Both groups showed significant decreases in self-reported distress, intrusion, and avoidance symptoms. [Pitman et al. 1996: www.emdr .com/studies.htm]

Maybe hand taps will work just as well as eye movements. Maybe both are unnecessary. According to one EMDR practitioner, "taps to hands, right and left, sounds alternating ear-to-ear, and even alternating movements by the patient can work instead. The key seems to be the alternating stimulation of the two sides of the brain" (Hume: www.pshrink.com/ emdrfile.html). When evidence came in that therapists were getting similar results to standard EMDR with *blind* patients whose therapists used tones and hand snapping instead of finger wagging, Shapiro softened her stance a bit. She admits that eye movement is not essential to eye movement desensitization processing, but claims that attacks on her are ad hominem and without merit. According to Dr. Hume, Shapiro now calls the treatment reprocessing therapy and says that eye movements aren't necessary for the treatment! Maybe none of these movements are needed to restructure memory.

EMDR is not accepted practice by the American Psychological Association. Advocates claim that EMDR is "a widely validated treatment for post-traumatic stress disorder" and other ailments such as "traumatic memories of war, natural disaster, industrial accidents, highway carnage, crime, terrorism, sexual abuse, rape and domestic violence" (David Drehmer, Ph.D., Licensed Clinical Psychologist and Associate Professor of Management, DePaul University, personal correspondence). So far, the validation referred to by Dr. Drehmer is mainly in the form of unconvincing research studies and **testi-**

monials by practitioners relating anecdotes and their interpretations of them.

Finally, I don't know whether Shapiro will be gratified to learn that Ranae Johnson has founded the Rapid Eye Institute on a blueberry farm in Oregon where she teaches Rapid Eye Technology (www .rapideyetechnology.com). This "amazing new therapy" is used "to facilitate releasing and clearing of old programming, opening the way to awareness of our joy and happiness." It helps us "find light and spirituality within us that has always been there." Apparently, people are paying $2,000 for the training and all the blueberries they can eat.

See also **thought field therapy**.

Further reading: Foreman 1998; Gaudiano and Herbert 2000; Lilienfeld 1996; Lohr et al. 1998; McNally 1999; Rosen and Lohr: www.hcrc.org/ncahf/newslett/ nl20-1.html#emdr; Singer and Lalich 1996.

F

face on Mars

An image seen in some photographs of the Cydonia region of Mars taken in 1976 by

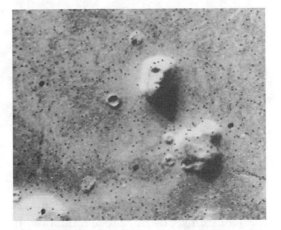

Enhanced images from Viking I, 1976.

the *Viking 1 Orbiter*. The enhanced image is of a natural formation, but looks like a face or a building.

Richard C. Hoagland, author of *The Monuments of Mars: A City on the Edge of Forever* (1987), is the one most responsible for the view that the face on Mars is an alien construction (Posner 2000b). NASA claims that the photos are just a play of light and shadow. Some took this explanation as a sure sign of a cover-up. Some NASA engineers and computer specialists digitally enhanced the images. This soon gave birth to the claim that the face was a sculpture of a human being located next to a city whose temples and fortifications could also be seen. Some began to wonder:

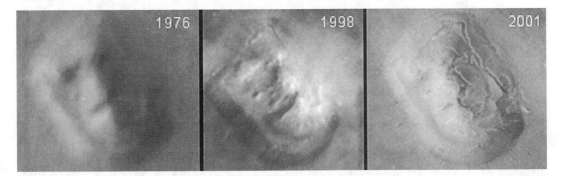

Images from *Viking I* (*left*), and *Mars Global Surveyor* (*center and right*).

Were these built by the same beings who built the ancient airports in **Nazca**, Peru, and who were now communicating to us through elaborate symbols carved in **crop circles**? Not according to Carl Sagan, who claimed that the "face" on Mars is the result of erosion, winds, and other natural forces (Sagan 1995: 52–55). Others see belief in the face on Mars as an example of **pareidolia**. The people at NASA know it started with a caption in a press release on July 31, 1976, which described one image thusly: "The picture shows eroded mesa-like landforms. The huge rock formation in the center, which resembles a human head, is formed by shadows giving the illusion of eyes, nose and mouth." Later and better images support the mesa thesis, but the legend considers them part of a conspiracy to hide the fact that there once was a civilization on Mars.

facilitated communication (FC)

A technique that allegedly allows communication by those who were previously unable to communicate by speech or signs due to autism, mental retardation, brain damage, or diseases such as cerebral palsy. FC involves a facilitator who places her hand over that of the patient's hand, arm, or wrist, which is placed on a board or keyboard with letters, words, or pictures. The patient is allegedly able to guide the facilitator's hand to a letter, word, or picture, spelling out words or expressing complete thoughts. Through their facilitators, previously mute patients recite poems, carry on high-level intellectual conversations, or simply express their feelings or beliefs. Parents are grateful to discover that their child is not hopelessly retarded but is either normal or above normal in intelligence. FC allows their children to demonstrate their intelligence; it provides them with a vehicle heretofore denied them.

But is it really the patient who is communicating? Most skeptics believe that the only one doing the communicating is the facilitator. The American Psychological Association (APA) has issued a position paper on FC, stating "Studies have repeatedly demonstrated that facilitated communication is not a scientifically valid technique for individuals with autism or mental retardation." According to the APA, FC is "a controversial and unproved communicative procedure with no sci-entifically demonstrated support for its efficacy."

FC therapy began in Australia with Rosemary Crossley, a nurse. The center for FC in the United States is Syracuse University, which since 1992 has housed the Facilitated Communication Institute (FCI) in their School of Education. The FCI conducts research, provides training to teach people to become facilitators, hosts seminars and conferences, publishes a quarterly newsletter, and produces and sells materials promoting FC.

While several studies have indicated that FC taps into the mind of a person who heretofore had been incommunicado, most studies have shown that FC taps only into the beliefs and expectations of the facilitator. Defenders of FC routinely criticize as insignificant or malicious those studies that fail to validate FC. Yet it is unlikely that there is a massive conspiracy on the part of all those who have done research on this topic and have failed to arrive at findings agreeable to the FCI. A very damaging, detailed criticism was presented on PBS's *Frontline* (first aired on October 19, 1993). The program was repeated December 17, 1996, and added that since the first showing, Syracuse University did three studies that verify the reality and effectiveness of FC, while 30 studies done elsewhere have concluded just the opposite.

The *Frontline* program showed facilitators allegedly describing what their clients were viewing, when it was clear their clients' heads were tilted so far back they couldn't have been viewing anything but the ceiling. When facilitators could not see an object that their client could see (a solid screen blocked each from seeing what the other was seeing) they routinely typed out the wrong answer. Furthermore, FC clients routinely use a flat board or keyboard, over which the facilitator holds the client's pointing finger. Even the most expert typist could not routinely hit correct letters without some reference as a starting point. Facilitators routinely look at the keyboard; clients do not. The messages' basic coherence indicates that someone looking at the keyboard produced them.

Nevertheless, there are many **testimonials** supporting FC: letters from clients who are grateful to FC for allowing them to show to the world that they are not retarded or stupid. Some of them may be from people who have been genuinely helped by FC. It seems that the FCI treats the retarded, autistic, and those with cerebral palsy. Anyone familiar with Helen Keller, Stephen Hawking, or Christy Brown knows that blindness, deafness, cerebral palsy, multiple sclerosis, amyotrophic lateral sclerosis (ALS), or physical or neurological disorders do not necessarily affect the intellect. There is no necessary connection between a physical handicap and a mental handicap. We also know that such people often require an assistant to facilitate their communication. But what facilitators do to help the likes of a Hawking or a Brown is a far cry from what those in the FC business are doing.

It may well be that some of those helped by FC suffer from cerebral palsy and are mentally normal or gifted. Their facilitators help them communicate their thoughts. But the vast majority of FC clients apparently are mentally retarded or autistic. Their facilitators appear to be reporting their own thoughts, not their patient's thoughts. Interestingly, the facilitators are genuinely shocked when they discover that they are not really communicating their patient's thoughts. Their reaction is similar to that of **dowsers** and others with "special powers" who, when tested under controlled conditions, find they don't have any special powers at all.

If FC worked, one would think that it would be easy to test by letting several different facilitators be tested with the same client under a variety of controlled conditions. If different "personalities" emerged, depending on the facilitator, that would indicate that the facilitator is controlling the communication. But believers in FC claim that it works only when a special bond has been established between facilitator and patient. It is interesting that the parents and other loved ones who have been bonding with the patient for years are unable to be facilitators with their own children. FC needs a kind stranger to work. But when the kind strangers and their patients are put to the test, they generally fail. We are told that is because the conditions made them nervous.

Despite much criticism and many experiments demonstrating that the messages, poems, and brilliant discourses being transmitted by the facilitators originate in the facilitators themselves, the FCI is going strong. With support groups all over the world and a respectable place at a respectable university, there is little chance that FC will soon fade away. Those within the FC movement are convinced FC "works." Skeptics think the evidence is in and that FC is a dangerous delusion.

Critics have noted a similarity between FC therapy and **repressed memory therapy**: Patients are accusing their parents and others of having sexually abused them. Facilitators are taught that some 13% of their clients have been sexually abused. This information may unconsciously influence their work. The facilitator cannot imagine that he or she is the source of the horrible charges being expressed; neither can the many school administrators or law enforcement authorities who believe FC is a magical way to tap into the thoughts of the autistic or the severely retarded. With repressed memory therapy, the evidence emerges when a repressed memory is brought to light or when therapists trained to treat sexually abused children interrogate a child. There is overwhelming evidence that many repressed memories of sexual abuse, as well as many "memories" of interrogated children, originate in the minds and words of therapists who suggest and otherwise plant them in their patients' minds. Similar findings have been made with FC: Facilitators report sexual abuse and their messages have been used to falsely charge parents and others with sexual abuse of mentally and physically handicapped persons.

The criticisms of FC as another therapy leading to a witch-hunt, turning decent parents into accused molesters of their handicapped children, are not without justification. How is one to defend oneself against an allegation made by someone who can never be interrogated directly? Missy Morton, an expert from the FC Institute, suggests the following:

> One facilitator can in any given case be mistaken, or can be influencing the person, and as a precaution it is helpful to have the message repeated to a second facilitator. If this is not immediately feasible a decision has to be taken as to whether the situation will allow any decision to wait until a second facilitator can be introduced. If with a second facilitator the message is confirmed in detail then it may be taken as confirmed that an allegation has been made. ("Disclosures of Abuse Through Facilitated Communication: Getting and Giving Support," Missy Morton, Facilitated Communication Institute, Syracuse University Division of Special Education and Rehabilitation, May 1992)

If there were evidence that facilitators were reporting the thoughts of their clients, there would still be concern for ensuring that the rights of the accused were not abused. But as the evidence is overwhelming that in most cases of FC the facilitator is reporting his or her own thoughts, the effort to ensure against false accusations should be enormous in order to prevent facilitators from unjustly accusing parents of heinous acts against their children.

Further reading: Gorman 1998; Green 1994; Jacobson et al. 1995; Shane 1994; Singer and Lalich 1996.

fairies

Mythical beings of folklore and romance, often depicted as diminutive winged humans with magical powers. The tooth fairy exchanges presents, usually coins, for teeth left out or under one's pillow at night. Fairy godmothers are protective beings, like guardian **angels**.

Fairies should not be confused with gnomes, which are also mythical diminutive humans but are deformed and live underground. Pixies, on the other hand, might be considered a type of fairy known for their cheerful nature and playful mischievousness. An elf might be thought of as a big pixie, often depicted as a mischievous dwarf, such as the Irish leprechaun known for his pranks but who is also believed to know where treasure is hidden. Elves are sometimes depicted as helpers of magicians, for example, Santa's helpers.

Belief in such mythical beings seems common in rural peoples around the

Frances Griffiths with fairy cutouts. Photo by 16-year-old Elsie Wright of her 10-year-old cousin taken in 1917.

world. Occasionally, an urbanite who should know better is duped into believing in fairies. An infamous example of such a dupe is Sir Arthur Conan Doyle, who was conned by a couple of schoolgirls and their amateur photographs of paper fairies (known as the "Cottingley Fairies") taken in their Yorkshire garden. Doyle even published a book on the fairies, *The Coming of the Fairies,* with Edward Gardner, a **Theosophist**. The photos were taken by 16-year-old Elsie Wright of her 10-year-old cousin, Frances Griffiths, who posed with Elsie's cutouts of fairies. Doyle and Gardner proclaimed that the photos were not fakes. The real howler, though, was the debate that ensued over whether these were photos of real fairies or **psychic photographs**

that recorded the thoughts of the girls projected onto the film! Doyle, like many who have come before and after him, longed for any proof of a world beyond the material world. His desire to find support for **spiritualism** led him to a number of delusions. Even so, he wrote great detective stories and in Sherlock Holmes created a mythical being much more interesting than any fairy, even if he didn't know the difference between induction and deduction.

Further reading: Bourke 1999; Cooper 1982; Gardner 1981; Randi 1982a, 1995.

faith

A nonrational belief in a proposition that is contrary to the sum of evidence for that belief. A belief is contrary to the sum of evidence for a belief if there is overwhelming evidence against the belief or if one commits to one of two or more equally supported propositions.

A common misconception regarding faith—or perhaps it is an intentional attempt at disinformation and obscurantism—is made by Christian apologists who make claims such as the following:

> A statement like "There is no God, and there can't be a god; everything evolved from purely natural processes" cannot be supported by the scientific method and is a statement of faith, not science. (Richard Spencer, Ph.D., quoted in *The Davis Enterprise,* January 22, 1999)

The error or deception here is to imply that anything that is not a scientific statement—that is, supported by evidence—is a matter of faith. To use "faith" in such a broad way is to strip it of any theological significance the term might otherwise have.

Such a conception of faith treats belief in all nonempirical statements as acts of faith. Thus, belief in the external world, belief in the law of causality, or even fundamental principles of logic such as the principle of contradiction or the law of the excluded middle would be acts of faith on this view. There seems to be something profoundly deceptive and misleading about lumping together as acts of faith such things as belief in the virgin birth and belief in the existence of an external world or in the principle of contradiction.

Dr. Spencer claims that the statement "there is no God and there can't be a god; everything evolved from purely natural processes" is a statement of faith. First, we must note that there are three distinct statements here. One, "There is no God." Two, "There can't be a god." And three, "Everything evolved from purely natural processes." Dr. Spencer implies that each of these claims is on par with such statements as "There is a God," "Jesus Christ is the Lord and Savior," "Jesus's mother was a virgin," "A piece of bread may have the substance of Jesus Christ's physical body and blood," "God is one Being comprised of three persons," and so on.

The statement "There cannot be a god" is clearly not an empirical statement but a conceptual one. Anyone who would make such a claim would make it by arguing that a particular concept of god contains contradictions, and so is meaningless. For example, to believe that "Some squares are circular" is a logical contradiction. Circles and squares are defined so as to imply that circles can't be square and squares can't be circles. James Rachels, for one, has argued that god is impossible, but at best his argument shows that the concepts of "an all-powerful God" and "one who demands worship from His creations" are contradictory. The concept of requiring worship, Rachels argues, is inconsistent with being all-powerful (Rachels 1971).

Rachels makes an argument. Some find it convincing; others don't. But it seems that his belief is not an act of faith in the same sense that it is an act of faith to believe in the Incarnation, the Trinity, transubstantiation, or the virgin birth. The first three articles of faith seem to be on par with believing in round squares, for they require belief in logical contradictions. Virgin births we now know are possible, but the technology for the implantation of fertilized eggs did not exist 2,000 years ago. The belief in the virgin birth involves belief that God miraculously impregnated Mary with Himself. Such a belief also defies logic. All arguments regarding these articles of faith are quite distinct from Rachels's argument. To defend these articles of faith, the best one can hope for is to show that they cannot be shown to be impossible. However, the consequence of arguing that logical contradictions may nevertheless be true seems undesirable. Such a defense requires the abandonment of the very logical principles required to make any argument, and is therefore self-annihilating. The fact that neither arguments such as Rachels's nor arguments defending articles of religious faith are empirical or resolvable by scientific methods hardly makes them equally matters of faith.

The statement "There is no God" is quite different from the claim that there can't be a god. The latter makes a claim regarding possibility; the former is an actuality claim. I doubt that there are many theologians or Christian apologists who would claim that all their faith amounts to is a belief in the *possibility* of this or that. One can believe there is no God on the grounds that there *can't be* a god, but one might also disbelieve in God while admitting the *possibility* of the Judeo-Christian God or any other god for that matter. Disbelief in God is analogous to disbelief in **Bigfoot**, the **Loch Ness Monster**, **Santa Claus**, and the Easter Bunny. Disbelievers argue that the evidence is not strong at all and does not deserve assent to the proposition that Bigfoot, Nessie, or others exist. Disbelievers in Bigfoot or God disbelieve not as an act of faith, but because the evidence for belief is not persuasive.

Finally, the claim "Everything evolved from natural processes" is not necessarily an act of faith. If the only alternatives are that everything evolved from either supernatural or natural forces, and one is unconvinced by the arguments and evidence presented by those who believe in supernatural forces, then, logically, the only reasonable belief is that everything evolved from natural forces. Only if the evidence supporting a supernatural being were superior or equal to the evidence and arguments against such a belief, would belief that everything evolved from natural forces be a matter of faith.

Those of us who are **atheists**, and believe that everything evolved from natural forces, nearly universally maintain that **theists** and supernaturalists have a very weak case for their belief, weaker even than the case for Bigfoot, Nessie, or Santa Claus. Thus, our disbelief is not an act of faith and, therefore, not nonrational, as is the belief in the God of Christian apologists. However, if Christian apologists insist on claiming that their version of Christianity and the rejections of their views are equally acts of faith, I will insist that the apologists have an *irrational* faith, while their opponents have a *rational* faith. Though I think it would be less dishonest and less misleading to admit that atheists and naturalists do not base their beliefs on faith in any sense close to that of religious faith.

fakir

An initiate in a mendicant Sufi order. The word derives from the Arab word for poverty. By extension, the word is used to refer to ascetic Indian *sadhus* (holy men). The term is also used to refer to itinerant Indian **conjurers** and alleged god-men who travel from village to village, performing alleged **miracles**, such as materializing *vibhuti* (holy ash) or jewelry. These fake fakirs do conjuring stunts such as walking on hot coals (**firewalking**), laying on a bed of nails, eating fire, sticking their hands in boiling "oil," piercing their faces with long needles, and putting large hooks through the flesh of their backs attached by ropes to wooden carts that they pull. Some pretend to **levitate**; others are said to have been buried alive for months and lived to tell about it. Some pretend to cut off their tongues and restore them. Others seem to materialize fire out of nothing. These conjurers sometimes have accomplices and they pretend to do **exorcisms** or other strange feats. After each performance, they pass the hat, collect what they can, and move on to the next village. Some become very famous and are considered to be god-men, such as **Sai Baba**.

B. Premanand of the Indian Skeptics has spent over 50 years exposing the tricks of the itinerant tricksters. His method is simple. He demonstrates how the "miracles" are done by performing them himself. Abraham Kovoor and Prabir Ghosh of the Indian Rationalist Association (IRA) have carried on the work of Premanand in exposing the deceptions of Sai Baba and others of like ilk, **astrologers, clairvoyants**, and **psychics**. The IRA was featured in the British documentary "Guru Busters" (Equinox) and in the Discovery Channel's *Science Mysteries* episode entitled "Physical Feats" (October 20, 2001). The cameras followed members of the IRA as they went from village to village and pretended to be god-men. The IRA walk over fiery coals and explain that anyone can do it without a need for supernatural intervention. They walk on glass, lay on nails, pull automobiles with hooks poked through the flesh on their backs, and jab long needles through their cheeks and tongues, just like the god-men. The goal of the IRA is to debunk the fakirs and reduce superstition among the villagers. They obviously have a long way to go, as is evidenced by the monkey-man hysteria that gripped New Delhi in the spring of 2001. Witnesses reported to the mass media that they had seen a giant ape that could jump 40 feet into the air and fly through windows. Others claimed they saw a 4-foot monkey that turned into a cat. Mass hysteria led to deserted streets and panic. One pregnant woman fell down a staircase and died as a result of trying to escape from the monkey-man. It was all a hoax that played on the religious superstitions of the people. One commentator put it bluntly:

> [H]ad we not been a nation nurtured on Hindu epics to become Hanuman worshippers, most people would have laughed at the very idea of a monkey-man and would have considered the so-called "witnesses" liars or demented maniacs from the outset, instead of waiting for the scientific community to debunk this hoax in its own, soft way, and lay it to rest. For, when it comes to religion the first "principle" that is taught by preachers to believers is that of blind acceptance. One may not question any religious dogma if he/she is a believer, and, he/she must necessarily accept the dogma/doctrine in toto. . . . (Mehul Kamdar: www .themronline.com/200107m11.html)

Amen.

false memory

A memory that is a distortion of an actual experience, or a **confabulation** of an imagined experience. Many false memories involve confusing or mixing fragments of memory events, some of which may have happened at different times but which are remembered as occurring together. Many false memories involve an error in source **memory**. Some involve treating **dreams** as if they were playbacks of real experiences. Still other false memories are believed to be the result of the prodding, leading, and suggestions of therapists and counselors. Dr. Elizabeth Loftus has shown not only that it is possible to implant false memories, but also that it is relatively easy to do so (Loftus 1994).

A memory of your mother throwing a glass of milk on your father when in fact it was your father who threw the milk is a false memory based on an actual experience. You may remember the event vividly and be able to "see" the action clearly, but only corroboration by those present can determine whether your memory of the event is accurate. Distortions such as switching the roles of people in one's memory are quite common. Some distortions are quite dramatic. For example, a woman accused memory expert Dr. Donald Thompson of being the man who raped her. Thompson was in another city doing a live interview for a television program just before the rape occurred. The woman had seen the program and "apparently confused her memory of him from the television screen with her memory of the rapist" (Schacter 1996: 114).

Jean Piaget, the child psychologist, claimed that his earliest memory was of nearly being kidnapped at the age of two. He remembered details such as sitting in his baby carriage, watching the nurse defend herself against the kidnapper, scratches on the nurse's face, and a police officer with a short cloak and a white baton chasing the kidnapper away. The nurse, the family, and others who had heard it told reinforced the story. Piaget was convinced that he remembered the event. However, it never happened. Thirteen years after the alleged kidnapping attempt, Piaget's former nurse wrote to his parents to confess that she had made up the entire story. Piaget later wrote: "I therefore must have heard, as a child, the account of this story . . . and projected it into the past in the form of a visual memory, which was a memory of a memory, but false" (Tavris 1993).

Remembering being kidnapped when you were an infant (under the age of three) is a false memory, almost by definition. The left inferior prefrontal lobe is undeveloped in infants, but is required for long-term memory. The elaborate encoding required for classifying and remembering such an event cannot occur in the infant's brain.

The brains of infants and very young children are capable of storing fragmented memories, however. Fragmented memories can be disturbing in adults. Schacter notes the case of a rape victim who could not remember the rape, which took place on a brick pathway. The words "brick" and "path" kept popping into her mind, but she did not connect them to the rape. She became very upset when taken back to the scene of the rape, though she didn't remember what had happened there (Schacter 1996: 232). Whether a fragmented memory of infant abuse can cause significant psychological damage in an adult has not been scientifically established, though it seems to be widely believed by many psychotherapists.

What is also widely believed by many psychotherapists is that many psychological disorders and problems are due to repressed memories of childhood sexual abuse. On the other hand, many psychologists maintain that their colleagues doing **repressed memory therapy** (RMT) are encouraging, prodding, and suggesting false memories of abuse to their patients. Many of the recovered memories are of being sexually abused by parents, grandparents, and ministers. Many of those accused claim the memories are false and have sued therapists for their alleged role in creating false memories.

It is unlikely that recovered memories of childhood sexual abuse are *all* false or *all* true. What is known about memory makes it especially difficult to sort out true from distorted or false recollections. However, some consideration should be given to the fact that certain brain processes are necessary for any memories to occur. Thus, memories of infant abuse or of abuse that took place while one was *unconscious* are unlikely to be accurate. Memories that have been directed by dreams or **hypnosis** are notoriously unreliable. Dreams are not usually direct playbacks of experience. Furthermore, the data of dreams are generally ambiguous. Hypnosis and other techniques that take advantage of a person's suggestibility must be used with great caution lest one create memories by suggestion rather than pry them loose by careful questioning.

Furthermore, memories are often mixed; some parts are accurate and some are not. Separating the two can be a chore under ordinary circumstances. A woman might have consciously repressed childhood sexual abuse by a neighbor or relative. Some experience in adulthood may serve as a retrieval cue, leading her to remember the abuse. This disturbs her and disturbs her dreams. She has nightmares, but now it is her father or grandfather or a priest who is abusing her. She enters RMT and within a few months she recalls vividly how her father, mother, grandfather, grandmother, or priest not only sexually abused her but engaged in horrific **satanic rituals** involving human sacrifices and cannibalism. Where does the truth lie? The patient's memories are real and horrible, even if false. The patient's suffering is real whether the memories are true or false. And families are destroyed whether the memories are true or false.

Obviously, it would be unconscionable to ignore accusations of sexual abuse based on recovered memories. But should such memories be taken at face value and accepted as true without any attempt to prove otherwise? It is equally unconscionable to be willing to see lives and families destroyed without at least trying to find out whether any part of the memories of sexual abuse is false. It also seems inhumane to encourage patients to recall memories of sexual abuse (or of being abducted by aliens, another common theme of RMT) unless one has a very good reason for doing so. Assuming all or most emotional problems are due to repressed memories of childhood sexual abuse is not a good enough reason to risk harming a patient by encouraging delusional beliefs and damaging familial relationships. A responsible therapist has a duty to help a patient sort out delusion from reality, dreams and confabulations from truth, and real abuse from imagined abuse. If good therapy means the encouragement of delusion as standard procedure, then good therapy is not worth it.

Finally, those who find that it is their duty to determine whether a person has been sexually abused or whether a memory of such abuse is a false memory should be well versed in the current scientific literature regarding **memory**. They should know that all of us are pliable and suggestible to some degree, but that children are especially vulnerable to suggestive and leading questioning. They should also remember that children are highly imaginative and that just because a child says he or she remembers something does not mean that he or she does. However, when children say they do not remember something, to keep questioning them until they do remember it is not good interrogation.

Investigators, counselors, and therapists should also remind themselves that many charges and memories are heavily influenced by media coverage. People charged with or convicted of crimes have noticed that their chances of gaining sympathy increase if others believe they were abused as children. People with grudges have also noticed that nothing can destroy another person so quickly as being charged with sexual abuse, while at the same time providing the accuser with sympathy and comfort. Emotionally disturbed people are also influenced by what they read, see, or hear in the mass media, including stories of repressed abuse as the cause of emotional problems. An emotionally disturbed adult may accuse another adult of abusing a child, not because there is good evidence of abuse but because the disturbed person imagines or fears abuse. In short, investigators should not rush to judgment.

See also **multiple personality disorder** and the **unconscious mind**.

Further reading: Baker 1996a; Cooper 1993; de Rivera 1993; Johnston 1999; Loftus 1980; Ofshe and Watters 1994; Sacks 1995; Schacter 1997, 2001.

falun gong (falun dafa)

Li Hongzhi's version of **ch'i kung** (qi gong), an ancient Chinese practice of **energy** cultivation. *Falun* means "wheel of law"; *falun dafa* is Buddha law (www .faluncanada.net/faq_eng.htm#Q9). According to Li, falun gong "is a cultivation system aiming at cultivating both human life and nature. The practitioner is required to attain enlightenment (open his cultivation energy) and achieve physical immortality in this mortal world when his energy potency and *Xinxing* [mind-nature] have reached a certain level" (www.falundafa .org/book/eng/flg_1.htm).

Li claims to have taken energy cultivation to a new level. He also claims to have some 100 million followers worldwide, though he also claims that he keeps no records. He claims that falun gong is not a **cult**, religion, or sect. His popularity is great enough in China to have led to the arrest of tens of thousands of practitioners. There is a formal ban on falun gong, apparently for little more than being popular and thereby posing a threat to the stability of the repressive Communist regime.

Li left China in the early 1990s and lives in New York City. He promotes his beliefs in books. His teachings are also available on the Internet, which has significantly affected his international status and popularity. While much of falun gong is a rehash of traditional Chinese notions regarding meditation and exercise, Li has emphasized an *antiscientific* approach to disease and medicine. He says disease "is a black energy mass" that he can dissipate with his powers. Those who use medicine for their illnesses lack **faith** in falun gong. True believers don't need medicine. They understand that disease exists in some other space beyond physical space and that only those with "supernormal capabilities" can truly heal. True healing involves "cultivation energies . . . in the form of light with very tiny particles in great density." He claims that he does not tell people not to use medicine, but that he has cured thousands of terminally ill people. He also claims that he advises terminally ill and mentally ill people not to practice falun gong. The former are too focused on their illness and the latter are not clear-minded enough to practice properly.

Li claims that falun gong is one of 84,000 cultivation ways of the Buddha's school. He claims that it has only been used once before, in prehistory, but that he is making it available again "at this final period of the Last Havoc."

Falun is the miniature of the universe with all the abilities of the universe. It can automatically move in rotation. It will forever rotate in your lower abdomen area. Once it is installed in your body, it will no longer stop and will forever rotate like this year in and year out. During the time when it rotates clockwise, it can automatically absorb energy from the universe, and it can also transform energy from itself to supply the required energy for every part of your body transformation. At the same time, it can emit energy when it rotates counter-clock and releases the waste material which will disperse around your body. When it emits energy, the energy can be released to quite a distance and it brings in new energy again. The emitted energy can benefit the people around you. . . . When Falun rotates clockwise, it can collect the energy back because it rotates forever. . . .

Because Falun rotates forever, it cannot be stopped. If a phone call comes or someone knocks on the door, you may go ahead and take care of it immediately without having to finish the practice. When you stop to do your work, Falun will rotate at once clockwise and take back the emitted energy around your body. [falundafa.org/intro/0-falun.htm]

How Li knows about these rotations is a mystery, but he has many followers throughout the world who feel enlightened by these teachings.

In short, falun gong is antiscience, anti–medical establishment, and antimaterialism; thus, falun gong is attractive to many people who are fed up with the world as it is and their position in it.

It is difficult to understand why the Communist party in China fears falun gong. Their practices would relieve the demand for medical assistance, thereby saving the government millions of yuan. Falun gong encourages truthfulness, forbearance, and compassion. Of course, members may not be very useful to society, since they are not materialistic and would prefer to spend their days meditating, exercising in the park, and cultivating energies, rather than working in factories.

feng shui

"Feng shui" (pronounced "foong shway") means, literally, "wind, water." Feng shui is part of an ancient Chinese philosophy of nature. Feng shui is often identified as a form of **geomancy**, or **divination** by geographic features, but it is mainly concerned with understanding the relationships between nature and ourselves so that we might live in harmony within our environment.

Feng shui is related to the very sensible notion that living *with* rather than *against* nature benefits both humans and our environment. It is also related to the equally sensible notion that our lives are deeply affected by our physical and emotional environs. If we surround ourselves with symbols of death, contempt, and indifference toward life and nature, with noise and various forms of ugliness and disorder, we will corrupt ourselves in the process. If we surround ourselves with beauty, gentleness, kindness, sympathy, music, and various expressions of the sweetness of an ordered life, we ennoble ourselves as well as our environment.

Alleged masters of feng shui, those who understand the five elements and the two energies such as **chi** and *sha* (hard energy, the opposite of chi), are supposed to be able to detect **metaphysical** energies and give directions for their optimal flow. Feng shui has been exploited by some landscape architects and interior decorators. They use a metaphysical map (called a "bagua") and compass to determine what area of a house, room, garden, and so on, should be devoted to one of nine categories such as health, creativity, love, and wealth. They declare where bathrooms should go, which way doorways should face, where mirrors should hang, which room needs green plants and which one needs red flowers, which direction the head of the bed should face, and so on.

Feng shui has become another New Age **energy** scam with arrays of metaphysical products from paper cutouts of half moons and planets to octagonal mirrors to wooden flutes offered for sale to help you improve your health, maximize your potential, and guarantee fulfillment of some fortune cookie philosophy.

According to Sutrisno Murtiyoso of

Indonesia, in countries where belief in feng shui is still very strong, feng shui has become a hodgepodge of superstitions and unverified notions that are passed off in the university curriculum as scientific principles of architecture or city planning. Mr. Murtiyoso wrote me about a university lecturer who had written an article in Indonesia's biggest newspaper "advocating feng shui as a guiding principle to Indonesia's future architecture." This upset Mr. Murtiyoso: "if it is done by a so-called '**paranormal**,' I wouldn't be that mad. But a 'colleague,' an architect . . . I just can't imagine how my people can face the next millennium still under this ancient spell. How can we progress . . . through this techno-jungle." If I were Mr. Murtiyoso, I wouldn't worry until the architects start advocating ignoring the laws of physics in favor of **prayer**.

See also **vastu**.

fetishes

Objects such as stones, teeth, or carvings supposedly possessing magical powers that can protect one from harm, cure disease, and so on. Some fetishes are thought to be magical in themselves; others get their magic from some divinity. Some fetishes are believed to be so powerful that only special individuals are allowed to handle them. For all others, the fetish is taboo.

firewalking

The activity of walking on hot coals, rocks, or cinders. In some cultures (e.g., India), firewalking is part of a religious ritual and is associated with the mystical powers of **fakirs**. In America, firewalking is part of a New Age self-empowering motivational activity.

Tony Robbins popularized firewalking as an activity for demonstrating it is possi-

ble for people to do things that seem impossible to them. He sees the firewalk as a technique for turning fear into power. Robbins doesn't consider the power of the mind to overcome fear of getting burnt as **paranormal**, however. Overcoming this fear is presented as a step in restructuring one's mind, almost as if this trial by fire was some sort of initiation into an esoteric and very special group of risk-takers. To the timid and those who feel powerless among the dynamic firebrands around them, a feat such as walking on hot coals must seem a significant event.

Robbins may have popularized firewalking, but Tolly Burkan, founder of the Firewalking Institute for Research and Education, claims he was the first to introduce the practice to North America. According to Burkan, firewalking is "a method of overcoming limiting beliefs, phobias and fears."

Walking across hot coals without getting burned does seem impossible to many people, but in fact it is no more impossible than putting your hand in a hot oven without getting burned. As long as you keep your hand in the air and don't touch anything in the oven, you won't get burned even if the oven is extremely hot. Why? Because "the air has a low heat capacity and a poor thermal conductivity" while "our bodies have a relatively high heat capacity." (Leikind and McCarthy 1991: 188). Thus, even if the coals are very hot (around 1,000°F), a firewalker won't get burned as long as he or she doesn't take too long to walk across the coals and as long as the coals used do not have a very high heat capacity. Volcanic rock and certain wood embers will work just fine. Also, "both hardwood and charcoal are good thermal insulators. . . . Wood is just as good an insulator even when on fire, and charcoal is almost four times better as an insulator than is dry hardwood. Further, the ash that is left after the charcoal has burnt is just as poor a conductor as was the hardwood or charcoal" (Willey: www.pitt.edu/~dwilley/fire.html).

Nevertheless, some people do get burned walking across hot coals, not because they lack faith or willpower but because the coals are too hot or have a relatively high heat capacity, or because the firewalker's soles are thin, or he or she doesn't move quickly enough. But even very hot coals with a high heat capacity can be walked over without getting burned if one's feet are insulated, for example, with a liquid such as sweat or water. (Think of how you can wet your finger and touch a hot iron without getting burned.) Again, one must move with sufficient speed or one will get burned.

However, even armed with this knowledge, it still takes courage to firewalk. When Michael Shermer of *Skeptic* magazine did a firewalk for *The Unexplained* television program, he had the knowledge, but the fear was obviously still there. Our instincts are telling us: *don't do this, you idiot!* Firewalking requires some faith as well as knowledge: faith that the coals were properly prepared, that you can move fast enough to avoid getting burned, and that something will work in practice as you know it should in theory. Even so, whether the firewalker gets burned depends on how the coals were prepared and on how fast the firewalker moves, rather than on willpower, the power of the mind to create a protective shield, or on any paranormal or supernatural force.

flying saucers

On June 24, 1947, Kenneth Arnold claimed that he'd seen nine "crescent shaped" aircraft flying erratically at incredible speeds near Mount Rainier. He said they reminded him of saucers skimming over water. An editor of the *Eastern Oregonian* reported that Arnold saw "round" objects. Other reports noted "disc-shaped" objects. Within a few weeks, there were hundreds of reports nationwide of sightings of flying "saucers."

The fact that so many **UFO** and alien sightings conform to rather standard depictions is taken by some as evidence that the observers are not mistaken. They must be seeing the same things. It is more likely that they see what they see because of their *expectations,* which are based on stereotypes created largely by the mass media. In this respect, and maybe some others as well, UFO and alien sightings might be compared to **Santa Claus** sightings.

See also **alien abductions**.

Further reading: Frazier 1997; Gardner 1957.

Forer effect

The tendency to accept vague and general personality descriptions as uniquely applicable to oneself without realizing that the same description could be applied to just about anyone. Psychologist B. R. Forer gave a personality test to his students, ignored their answers, and returned to each participant the following evaluation.

> You have a need for other people to like and admire you, and yet you tend to be critical of yourself. While you have some personality weaknesses you are generally able to compensate for them. You have considerable unused capacity that you have not turned to your advantage. Disciplined and self-controlled on the outside, you tend to be worrisome and insecure on the inside. At times you have serious doubts as to whether you have made the right decision or done the right thing. You

prefer a certain amount of change and variety and become dissatisfied when hemmed in by restrictions and limitations. You also pride yourself as an independent thinker; and do not accept others' statements without satisfactory proof. But you have found it unwise to be too frank in revealing yourself to others. At times you are extroverted, affable, and sociable, while at other times you are introverted, wary, and reserved. Some of your aspirations tend to be rather unrealistic. (Forer 1949)

Forer then asked the students to evaluate the evaluation on a scale of 0 to 5, with 5 meaning the recipient felt the evaluation was an "excellent" assessment. The class average evaluation was 4.26. The test has been repeated hundreds of times with psychology students and the average is still around 4.2.

In short, Forer convinced people he could successfully read their character. His accuracy amazed his subjects, though his personality analysis was taken from a newsstand **astrology** column and was presented to people without regard to their sun sign. The Forer effect seems to explain why there are so many satisfied customers convinced of the accuracy of their astrologer, **graphologist**, **palm reader**, or **psychic**.

Forer thought that gullibility accounted for the customers' tendency to accept the same assessments of their personalities. The most common explanations given today to account for the Forer effect are **self-deception**, vanity, **wishful thinking**, and the tendency to try to make sense out of experience. People tend to accept claims about themselves in proportion to their desire that the claims be true rather than in proportion to the empirical accuracy of the claims as measured by some nonsubjective standard. We tend to accept questionable, even false statements about ourselves, if we deem them positive or flattering enough. We often give very liberal interpretations to vague or inconsistent claims about ourselves in order to make sense out of the claims. Subjects who seek counseling from psychics often ignore false or questionable claims and, in many cases, provide most of the information they erroneously attribute to the psychic counselor.

"Hope and uncertainty evoke powerful psychological processes that keep all **occult** and pseudoscientific character readers in business" (Beyerstein 1996b). We are constantly trying "to make sense out of the barrage of disconnected information we face daily" and "we become so good at filling in to make a reasonable scenario out of disjointed input that we sometimes make sense out of nonsense" (ibid.). We often connect the dots and provide a coherent picture of what we hear and see, even though a careful examination of the evidence would reveal that the data is vague, confusing, obscure, inconsistent, and even unintelligible. Psychic mediums, for example, often ask so many disconnected and ambiguous questions in rapid succession (**shotgunning**) that they give the impression of having access to personal knowledge about their subjects. Furthermore,

once a belief or expectation is found, especially one that resolves uncomfortable uncertainty, it biases the observer to notice new information that confirms the belief, and to discount evidence to the contrary. This self-perpetuating mechanism consolidates the original error and builds up an overconfidence in which the arguments of opponents are seen as too fragmentary to undo the adopted belief. (Marks and Kamman 1979)

Having a pseudoscientific counselor go over a character assessment with a client is wrought with snares that can easily lead the most well intentioned of persons into error and delusion.

Barry Beyerstein suggests the following test to counteract the tendency to self-deception about the accuracy of a character assessment or life profile:

> a proper test would first have readings done for a large number of clients and then remove the names from the profiles (coding them so they could later be matched to their rightful owners). After all clients had read all of the anonymous personality sketches, each would be asked to pick the one that described him or her best. If the reader has actually included enough uniquely pertinent material, members of the group, on average, should be able to exceed chance in choosing their own from the pile. (Beyerstein 1996b)

Beyerstein notes that "no occult or pseudoscientific character reading method . . . has successfully passed such a test."

The Forer effect, however, only partially explains why so many people accept as accurate occult and pseudoscientific character assessment procedures. **Cold reading**, **communal reinforcement**, and **selective thinking** also underlie these delusions. Also, it should be admitted that while many of the claims in such assessments are vague and general, some are specific. However, many specific claims apply to large numbers of people and some specific assessment claims should be expected to be accurate by chance.

There have been numerous studies done on the Forer effect. Dickson and Kelly (1985) have examined many of these studies and concluded that overall there is significant support for the general claim that Forer profiles are generally perceived to be accurate by subjects in the studies. Further-

more, there is an increased acceptance of the profile if it is labeled "for you." Favorable assessments are "more readily accepted as accurate descriptions of subjects' personalities than unfavorable" ones. But unfavorable claims are "more readily accepted when delivered by people with high perceived status than low perceived status." It has also been found that subjects can generally distinguish between statements that are accurate (but would be so for large numbers of people) and those that are unique (accurate for them but not applicable to most people). There is also some evidence that personality variables such as neuroticism, need for approval, and authoritarianism are positively related to belief in Forer-like profiles. Unfortunately, most Forer studies have been done only on college students.

The Forer effect is also known as the "Barnum effect." (Both are also referred to as *subjective validation* or *personal validation*.) The expression seems to have originated with psychologist Paul Meehl, in deference to circus man P. T. Barnum's reputation as a master psychological manipulator.

Further reading: Beyerstein and Beyerstein 1991; Dickson and Kelly 1985; Thiriart 1991.

Fort, Charles (1874–1932)

A man who fancied himself a true skeptic: one who opposes all forms of dogmatism, believes nothing, and does not take a position on anything. He claimed to be an "intermediatist," one who believes nothing is real and nothing is unreal, that "all phenomena are approximations one way or the other between realness and unrealness." In fact, Fort was an antidogmatist who collected weird and bizarre stories.

Fort spent a good part of his adult life in the New York City public library examining newspapers, magazines, and scientific journals. He was looking for accounts of anything weird or mysterious that didn't seem to fit with current scientific beliefs. He collected accounts of frogs and other strange objects raining from the sky, **ghosts**, **psychic** abilities, **spontaneous human combustion**, **stigmata**, **UFOs**, and so on. He published four collections of weird tales and **anomalies** during his lifetime: *Book of the Damned* (1919), *New Lands* (1923), *Lo!* (1931), and *Wild Talents* (1932). In these works, he does not seem interested in questioning the reliability of his sources, which is odd given that he had worked as a news reporter for a number of years before embarking on his quest to collect stories of the weird and bizarre. He does reject one story about a talking dog who disappeared into a puff of green smoke. He expresses his doubt that the dog really went up in green smoke, though he doesn't question its ability to speak.

Fort did not seem interested in making any sense out of his collection of weird stories. He seemed particularly uninterested in scientific testing, yet some of his devotees consider him to be the founding father of modern **paranormal** studies. His main interest in scientific hypotheses was to criticize and ridicule the very process of theorizing. His real purpose seems to have been to embarrass scientists by collecting stories on "the borderland between fact and fantasy" that **science** could not explain or explain away. Since he did not generally concern himself with the reliability or accuracy of his data, this borderland also blurs the distinction between open-mindedness and gullibility.

Fort was skeptical about scientific explanations because scientists sometimes argue "according to their own beliefs rather than the rules of evidence" and they suppress or ignore inconvenient data. He seems to have understood that scientific theories are models, not pictures, of reality,

but he considered them to be little more than superstitions and myths. He seems to have had a profound misunderstanding of the nature of scientific theories. He criticized them for not being able to accommodate anomalies and for requiring data to fit. He took particular delight when scientists made incorrect predictions and he attacked what he called the "priestcraft" of science. Fort seems to have been opposed to science as it really is: fallible, human, and tentative, searching for probabilities rather than absolute certainties. He seems to have thought that since science is not infallible, any theory is as good as any other.

Fort was a prolific writer. He is said to have written 10 novels, but only one was published: *The Outcast Manufacturers* (1906). At least twice in his life he is said to have burned thousands of pages of notes and writings while severely depressed. Two early works of fiction, entitled *X* and *Y*, both burned, dealt with Martians controlling life on Earth and an evil civilization existing at the South Pole. When he was about 25 years old, Fort wrote his autobiography, *Many Parts*. Fragments of it have been preserved, but Fort himself came to recognize that there is little to recommend it and described it as "the work of an immature metaphysician, psychologist, sociologist, etc."

One of Fort's amusements as an adult seems to have been to speculate about such things as frogs falling from the sky. He postulated that there is a Super-Sargasso Sea above Earth (which he called *Genesistrine*) where living things originate and are periodically dumped on Earth by intelligent beings who communicate with secret societies down below, perhaps using **teleportation**.

Fort had very few friends, but one of them, Tiffany Thayer, created the Fortean Society to promote and encourage Fort-like attacks on science and scientists. When

Fort died in 1937, he left over 30 boxes of notes, which the Fortean Society began publishing in the *Fortean Society Magazine* (later *Doubt* magazine). In 1959 Thayer died and the Fortean Society came to an end. Fort's influence, however, became even greater. *Fortean Times* magazine is advertised as exploring "the wild frontiers between the known and the unknown" and features articles on topics such as the government's alleged suppression of evidence regarding crashed UFOs, synaesthesia, a mysterious undersea structure, and other things the editors think are strange or weird. The International Fortean Organization publishes *INFO Journal* several times a year. It features stories on such topics as anomalous astronomical phenomena, anomalies in the physical sciences, scientific hoaxes, and **cryptozoology**. The Society for the Investigation of the Unexplained collects data on unexplained events and publishes a magazine called *Pursuit*. The *Anomalist* magazine publishes articles on mysteries in science and nature. *Strange* magazine has articles, features, and columns covering all aspects of the anomalous and unexplained. William R. Corliss founded the Sourcebook Project (a catalog of anomalies) and *Science Frontiers*, a newsletter that since 1976 has been providing digests of reports that describe scientific anomalies. There are many other Fortean groups as well, but it is worth noting that Fort opposed the idea of a Fortean Society. He thought that such a group would attract **spiritualists** and crackpots.

Further reading: Gardner 1957; Lippard 1996.

fortune telling

See **divination**.

Forum, the

See **Landmark Forum**.

Freemasons

An international secular fraternal order, Freemasons are organized into lodges. Groups of lodges belong to a Grand Lodge or Grand Orient, but there is no single governing body that directs all the Grand Lodges. The origins of the Freemasons are disputed, but the first organized lodges date from 1717 in England. Members consider others in their lodge as "brothers" or "brethren," but consider members of other lodges as brothers only if their lodges officially recognize each other. Freemasons are sometimes accused of being secretive societies because they have "signs of recognition such as handshakes, passwords, and references that only initiated members would understand" (Masonic FAQ: www.mcdanielsells.com/frequent.htm# Religion).

Freemasonry is not a secret society, **cult**, religion or anti-Christian sect, nor is it behind the **Illuminati**, although it is often accused of being such. Membership does require belief in a Supreme Being, and there is a Masonic Bible, usually the King James Version of the same book accepted by those Christians who accuse the Masons of being anti-Christian.

Much anti-Masonic sentiment has been aroused by various tracts and books. In 1827, for example, William Morgan, who had been denied membership, joined with printer David Miller to published a diatribe entitled "Freemasonry Exposed." The tract itself may not have caused as much anti-Masonic sentiment as did the ensuing stories in the press that Morgan had been kidnapped and murdered by Masons in retaliation for exposing their secret beliefs and rituals. The evidence strongly indicates, however, that Morgan escaped from jail, where he was being held for a bad debt, and left town unscathed (King 2002).

Typical of recent attacks on Freemasonry are the works of Jim Shaw (1988: *The Deadly Deception: Freemasonry Exposed by One of Its Top Leaders*) and Charles Madden (1995: *Freemasonry—Mankind's Hidden Enemy: With Current Official Catholic Statements*). These malicious writings seem primarily motivated by opposition to Masonic beliefs in the brotherhood of man and the belief that strong moral character has nothing to do with organized religion. The hatred of the group is kept alive by Christian evangelists such as Pat Robertson and talk show hosts like Art Bell (Goeringer 1998).

In 1868, the National Christian Association (NCA) was formed in Pittsburgh for the sole purpose of blaming secretive societies for most of the world's ills. The NCA still exists and still puts the Freemasons at the top of their list of secret societies behind political assassinations, promotion of sexual immorality, and other evils (n.d.: Wheaton College Archives & Special Collections, www.wheaton.edu/learnres/ARCSC/collects/sc29).

Despite this long history of attack and abuse, the Freemasons continue to flourish. There are over 4 million members worldwide. There are also several Masonic-affiliated organizations, including the Shriners, which extend the social and charitable work of the Freemasons. Notable Masons include George Washington, Harry Houdini, Benjamin Franklin, and Thurgood Marshall.

Freud, Sigmund

See **psychoanalysis**.

friggatriskaidekaphobia

See **paraskevidekatriaphobia**.

Fritz, Dr.

A German **ghost** that allegedly has invaded the bodies of several Brazilians and turned them into healers. Zé Arigó (1918–1971) informed the world in 1950 that he was **channeling** Dr. Adolf Fritz, a German doctor who had died in World War I. The search for Adolf Fritz has been even less successful than the search for Bridey **Murphy**. In short, no proof of his ever existing has been brought forth. No matter. Faith healing is very big in Brazil and Arigó made quite a name for himself as a witch doctor or shaman. Some thought he was possessed by the devil, not realizing that a dead German doctor had taken over Arigó's body and begun writing illegible prescriptions for sick people. Fortunately, there was one person on earth who could read Arigó's writing: his brother, who by chance was a pharmacist. Dr. Fritz's reputation soared after it was alleged that he removed a cancerous tumor from the lung of a well-known Brazilian senator by **psychic surgery**. For 20 years Arigó's fame spread as he "cured" thousands of people, including the daughter of Brazil's president. Despite his fame, he was twice convicted of illegally practicing medicine. Arigó died in an automobile crash in 1971.

Dr. Fritz was not done with his work, however, and soon slipped into the body of another Brazilian, and when this healer also died in a violent crash, Dr. Fritz picked yet another body to invade. He has done this several times. Two of his most famous invasions have been in the bodies of Edson Queiroz from Recife and Rubens Farias, Jr. (1954–), of São Paulo, the current channeler of Dr. Fritz. The latest version of Dr. Fritz is well educated and heals the **astral body**. Farias seems to have abandoned his Catholic training for the teachings of Rudolf Steiner's **anthroposophy** or Madame Blavatsky's **theosophy**. Like them, he favors a mysticism that maintains the astral body is a duplicate of the physical body but comprised of a finer substance, and is what needs to be treated when one is ill. The physical body can be cured only by treating the astral body with "**energy** healing." But only special mystics can do this. Unfortunately, Dr. Fritz predicts a violent death for Farias, so he won't be practicing his mystical magic for much longer.

Despite being accused of practicing medicine without a license, Farias has unending lines of people waiting to be cured. The strong belief in witch doctors in Brazil is traced to the African influence on Candomblés, but the latest Dr. Fritz has shown that New Age mystical notions can dazzle Brazilians as well.

Further reading: Randi 1995.

G

ganzfeld ("total field") experiment

A kind of sensory deprivation test for **psi**. Halved ping-pong balls cover the subject's eyes and headphones cover the ears. While white noise is played through the headset, a bright red light is shown through the ping-pong balls. Soon the subject begins to hallucinate. At this point, others in another room are shown a visual stimulus such as a short video clip. They try to transmit **telepathically** what they are seeing to the subject in the ganzfeld.

The video clips or other visual stimuli are selected randomly from a large set of items. While the sender concentrates on the target, the receiver provides a continuous verbal report of his or her images. Images seen by the viewers are compared with the images seen by the subjects. Finally, "at the completion of the ganzfeld period, the receiver is presented with several stimuli (usually four) and, without knowing which stimulus was the target, is asked to rate the degree to which each matches the imagery and mentation experienced during the ganzfeld period. If the receiver assigns the highest rating to the target stimulus, it is scored as a 'hit.' Thus, if the experiment uses judging sets containing four stimuli (the target and three decoys or control stimuli), the hit rate expected by chance is 0.25" (Bem and Honorton 1994).

The hypothesis of **parapsychologist** Charles Honorton, the creator of the ganzfeld experiments, is that if psi exists, there should be a greater than chance match between the images of the senders and receivers. Honorton has reported studies such as one with 240 subjects who were right 34% of the time. This is not likely to be due to chance. The question is, is it due to psi? It could be. It could also be due to something else, something that correlates strongly with the selection of the video clips or other selected visual stimuli, such as age or gender of the subjects, or content or theme of the visual images. In any case, before Honorton gets his Nobel Prize, others need to replicate the study and make very persuasive arguments that the only plausible explanation is psi. In fact, others have claimed to have replicated Honorton's work.

Rick E. Berger, Ph.D., the creator of the automated ganzfeld (and coauthor with Honorton on several ganzfeld papers) claims that according to Dean Radin,

From 1974 to 1997, some 2,549 ganzfeld sessions were reported in at least forty publications by researchers around the world. After a 1985 meta-analysis established an estimate of the expected hit rate, a six-year replication was conducted that satisfied skeptics' calls for improved procedures. That "autoganzfeld" experiment showed the same successful results.

"The overall hit rate for the 2,549 sessions was 33.2% (where 25% was expected) yielding odds against chance of a million billion to one," says Berger. This sounds impressive until you examine the claim ever so slightly. The ganzfeld requires an *interpretation* of a verbal report from the test subject to be matched against an *image* sent telepathically to the subject. Thus, even if an image bears little or no resemblance to the verbal description, if it is selected as the one most closely resembling the image verbally described, then it counts as a hit. For example, here is a verbal description taken from Dr. Berger's website on the ganzfeld:

> I see the Lincoln Memorial . . . And Abraham Lincoln sitting there . . . It's the 4th of July . . . All kinds of fireworks . . . Now I'm at Valley Forge . . . There are fireworks . . . And I think of bombs bursting in the air . . . And Francis Scott Key . . . And Charleston. [www.psicxplorcr.com/ganzint2.htm[

There are quite a few images that would "match" this description, since the description itself contains several distinct images (the Lincoln memorial, memorials, Lincoln, the Fourth of July, holidays, fireworks, explosions, Valley Forge, bombs, Francis Scott Key, Charleston, the South) to which one could easily add a couple more, such as the American flag, Washington, D.C., the Star-Spangled Banner, and, oh yes, *George Washington,* which was the image selected as most closely resembling the verbal description. We're not told what the other three choices were.

One wonders, if this 8.2% difference is evidence of telepathy, why aren't the verbal descriptions more precise? For example, why didn't the **psychic** "see" George Washington, since that was what the image was? Why did he see the Lincoln memorial and a bunch of other things? How can they be sure of what they are measuring? Why isn't the subject allowed to choose "none"? Shouldn't the experimenters have some cases where the sender doesn't really send anything? And shouldn't the receiver be able to say, "I'm not getting any message at all"? If Berger and Honorton would do a ganzfeld where the sender sends no messages at all throughout the entire experiment, my guess is that the receiver would still "receive" and give a verbal description of his vision. What would his vision be of? Would these scientists say that the vision is one of the imagination or would they say that someone, somewhere, sent some message and the subject picked it up? How can they be sure, in fact, that their subjects are not picking up messages from others besides the sender? Perhaps the reason the subjects fail 66.8% of the time is because they are picking up messages from the wrong senders! Maybe there is 100% telepathy. Or maybe something else is going on besides telepathy.

Other researchers are not as enthusiastic as Honorton, Berger, or Radin. For example, Julie Milton and Richard Wiseman recently published their own meta-analysis of ganzfeld studies and concluded that "the ganzfeld technique does not at present offer a replicable method for producing **ESP** in the laboratory" (Milton and Wiseman 1999).

Further reading: Hyman 1985, 1989, 1995, 1996b; Nisbett 2000; Schick and Vaughn 1998.

Geller, Uri

A Hungarian/Austrian self-proclaimed **psychic** who was born in Israel but lives in

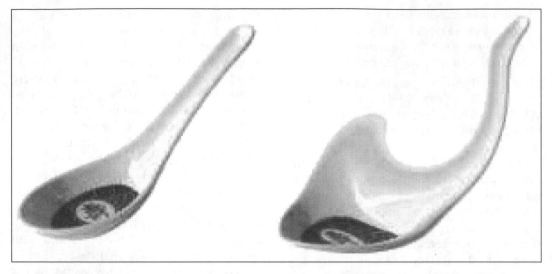

Bending a ceramic spoon.

England. Geller's reputation for **telepathically** bending spoons and keys and causing clocks and watches to stop running has brought him international fame. James Randi says, "If Uri Geller bends spoons with divine powers, then he's doing it the hard way." Geller claims he's had visions and may get his powers from extraterrestrials. He has sued several people for millions of dollars for saying he is not a psychic. What is a mystery is how he has been able to build a career out of breaking things.

Many magicians do what Geller does, but they call themselves **conjurers** or **mentalists**. Good magicians are good tricksters and good tricksters can fool the wisest of men. They can amaze people with their ability to seemingly move objects with an act of will, suspend objects in space, view objects that are remote, read minds, predict the future, identify the content of hidden messages or drawings, and so on. What is amazing is that those claiming to be psychics don't amaze people by winning the lottery or finding a cure for cancer. They don't bypass airports and paranormally transport themselves to gigs.

They take the car to a mechanic when it breaks down. Why do they waste their time moving a wire in a glass bottle instead of moving a waterfall over a forest fire? The answer is obvious. Such useful feats would require more than distraction and legerdemain.

Geller's use of deception was caught on videotape and revealed to the general public by Massimo Polidoro, one of the founders of the Italian Committee for the Investigation of Claims of the Paranormal. The video, *Alla ricerca dell'arca* (Search for the Ark), aired on Italian television on March 18, 1989.

Why do parlor tricks convince even intelligent people that they have witnessed a **paranormal** event rather than a bit of magic? Because most really intelligent people are too foolish to realize that they are not so knowledgeable as to be beyond being fooled. One really intelligent person who would not be fooled was Richard Feynman, who met Uri Geller and witnessed some key bending. Feynman said, "I'm smart enough to know that I'm dumb." He was intelligent enough to realize that a good magician can make it seem as if the laws of nature have been violated and that even a great physicist might not be able to figure out the trick.

Others are not so intelligent and have hired Geller as a **psychic detective** and as a psychic geologist to do pendulum **dowsing** for oil and minerals. He has also recently ventured into the lucrative New Age self-help/personal growth industry. For sale from his web site is his Mind-Power Kit. The kit includes an audiotape, a crystal, and a book to help you develop your psychic powers. We can guarantee that with these items you will become as psychic as Mr. Geller.

See also **telepathy**.

Further reading: Polidoro: www.fi .muni.cz/sisyfos/geller.htm; Randi 1982a, 1982b.

geomancy

A form of **divination** that is based on throwing dirt on the ground and interpreting the result. Geomancy also refers to divination that uses geographic features or lines from which to divine the future.

ghosts and poltergeists

Ghosts are alleged disembodied spirits of dead persons. Ghosts are often depicted as inhabiting **haunted houses**, especially houses where murders have occurred. Why some murder victims would stick around for eternity to haunt a place, while others seem to evaporate, is one of the great mysteries of existence.

A poltergeist is, literally, a "noisy spirit." Poltergeists make their presence known by rapping sounds and by throwing furniture or pots and pans around.

Further reading: Finucane 1996; Randi 1986; Sagan 1995; Schick and Vaughn 1998.

glossolalia

Glossolalia is speech that is semantically and syntactically unintelligible and meaningless. According to Dr. William T. Samarin, professor of anthropology and linguistics at the University of Toronto,

> glossolalia consists of strings of meaningless syllables made up of sounds taken from those familiar to the speaker and put together more or less haphazardly. . . . Glossolalia is language-like because the speaker unconsciously wants it to be language-like. Yet in spite of superficial similarities, glossolalia fundamentally is not language. (Nickell 1993: 108)

When spoken by schizophrenics, glossolalia is recognized as gibberish. In charismatic Christian communities glossolalia is sacred and referred to as "speaking in tongues" or having "the gift of tongues." In *Acts of the Apostles,* tongues of fire are described as alighting on the Apostles, filling them with the Holy Spirit. Allegedly, this allowed the Apostles to speak in their own language but to be understood by foreigners from several nations. Glossolalics, on the other hand, speak in a foreign language and are understood by nobody.

Glossolalics behave in various ways, depending on the social expectations of their community. Some go into convulsions or lose consciousness; others are less dramatic. Some seem to go into a trance; some claim to have amnesia of their speaking in tongues. All believe they are possessed by the Holy Spirit and that the gibberish they utter is meaningful. However, only one with **faith** and the gift of interpretation is capable of figuring out the meaning of the gabble. Of course, this belief gives the interpreter unchecked leeway in "translating" the meaningless utterances. Nicholas Spanos notes: "Typically, the interpretation supports the central tenets of the religious community" (Spanos 1996: 147).

Uttering gibberish that is interpreted as profound mystical insight by holy men is an ancient practice. In Greece, even the priestess of Apollo, god of light, engaged in prophetic babbling. The ancient Israelites did it. So did the Jansenists, the Quakers, the Methodists, and the Shakers.

See also **xenoglossy**.

Further reading: Baker 1996.

graphology

The study of handwriting as a means of analyzing character. Real handwriting experts are called "forensic document examiners," not graphologists. They examine handwriting to detect authenticity or forgery.

Graphologists examine loops, dotted i's and crossed t's, letter spacing, slants, heights, ending strokes, and so on. They believe such handwriting minutiae manifest **unconscious** mental functions. However, there is no evidence that the unconscious mind is a reservoir of truth about a person, much less that graphology provides a gateway to that reservoir.

Graphologists apply their craft to everything from understanding health issues to morality to past experiences to hidden talents and mental problems. However, "in properly controlled, blind studies, where the handwriting samples contain no content that could provide non-graphological information upon which to base a prediction (e.g., a piece copied from a magazine), graphologists do no better than chance at predicting . . . personality traits" (British Columbia Civil Liberties Association 1988). Even nonexperts are able to correctly identify the gender of a writer about 70% of the time (Furnham 1991: 204).

Graphologists use a variety of techniques. Even so, the techniques seem to be reducible to impressions from such things as the pressure exerted on the page, spacing of words and letters, crossed t's, dotted i's, size, slant, speed, and consistency of writing. Graphologists deny it, but the *content* of the writing is the most important factor in character assessment, even though content is independent of handwriting and can be easily analyzed without any special training.

Barry Beyerstein (1996a) considers many of the notions of graphologists to be based on **sympathetic magic**, for example, the notion that leaving wide spaces between letters indicates a proneness to isolation and loneliness because the wide spaces indicate someone who does not mix easily and is uncomfortable with closeness. One graphologist claims that a person betrays his sadistic nature if he crosses his t's with lines that look like whips.

Since there is no useful theory as to how graphology might work, it is not surprising that there is no empirical evidence that any graphological characteristics significantly correlate with any interesting personality traits. Adrian Furnham writes, "Readers familiar with the techniques of **cold reading** will be able to understand why graphology appears to work and why so many (otherwise intelligent) people believe in it" (Furnham 1991: 204). Add to cold reading **communal reinforcement**, **confirmation bias**, and the **Forer effect** and you have a fairly complete explanation for graphology's popularity.

Graphology's appeal as another quick and dirty decision-making process ranks it high in a long list of quack substitutes for hard work. It is appealing to the impatient who wish to avoid such troublesome matters as research, evidence analysis, reasoning, logic, hypothesis testing, and fairness.

Further reading: Basil 1991; Beyerstein and Beyerstein 1991; Gardner 1957.

Gurdjieff, G. I. (1872?–1949)

George S. Georgiades, a Greco-Armenian born in Russia, made a name for himself in Paris as the charismatic mystic George Ivanovitch Gurdjieff. In Russia, he established what he called the Institute for the Harmonious Development of Man (1919), which he reestablished in France in 1922. Gurdjieff promoted a litany of preposterous **occult** and mystical notions about the universe, which he said he learned from wise men while traveling and studying in Central Asia. His "insights" fill books with titles such as *Meetings with Remarkable Men, All and Everything,* and *Beelzebub's Tales to His Grandson: An Objectively Impartial Criticism of the Life of Man.*

His disciple Petyr Demianovich **Ouspensky** presented Gurdjieff's unintelligible musings in more accessible language, but they remain mostly tedious. To some devotees of Gurdjieff, Ouspensky was an incomplete mystic. Other disciples find Gurdjieff and Ouspensky to be co-gurus.

Gurdjieff attracted followers with notions such as "most human beings who are awake act as if they are asleep" and "most people are dead on the inside." Perhaps he meant that most people are trusting, gullible, easily led, very suggestible, not very reflective or suspicious of their fellow creatures, and need a guru to give their lives vitality and meaning. He seems to have taken advantage of the fact that many people are neither skeptics nor self-motivated and are easily duped by gurus because they want someone to show them the way to live a meaningful life. Gurdjieff offered a way to true wakefulness, a state of awareness and vitality that transcends ordinary consciousness. He was able to attract a coterie of writers, artists, wealthy widows, and other questing souls to work his farm for him in exchange for sharing his wisdom. He offered numerous claims and explanations for everything under the moon, rooted in little more than his own imagination and never tempered with concern for what science might have to say about his musings.

Gurdjieff obviously had a powerful personality, but his disdain for the mundane and for natural **science** must have added to his attractiveness. He allegedly exuded extreme self-confidence and exhibited no self-doubt, traits that must have been comforting to many people. His teachings and methods are ambiguous enough to be given depth by some disciples. For example, Fritz Peters tells the following story about Gurdjieff:

> To explain "the secret of life" to a wealthy English woman who had offered him £1,000 for such wisdom, Gurdjieff brought a prostitute to their table and told her he was from another planet. The food he was eating, he told her, was sent

G. I. Gurdjieff.

to him from his home planet at no small expense. He gave the prostitute some of the food and asked her what it tasted like. She told him it tasted like cherries. "That's the secret of life," Gurdjieff told the English lady. She called him a charlatan and left. Later that day, however, she gave him the money and became a devoted follower. (Peters 1976)

To those on a quest for spiritual evolution or transformation, guides such as Gurdjieff and Ouspensky promise entry into an esoteric world of ancient mystical wisdom. The Fellowship of Friends operates Gurdjieff Ouspensky Centers in over 30 countries around the world.

See also **enneagram**.

Further reading: Storr 1996.

H

haunted houses

When **Satan** or **ghosts** take up residence, a house is said to be "possessed" or "haunted." It is not clear why Satan or ghosts would confine themselves to quarters, since with all their alleged powers they could be anywhere or everywhere at any time. If they really wanted to terrorize the neighborhood, they could take turns haunting different houses.

Ideas about haunted houses often originate in movies such as *The Amityville Horror,* a fictional movie based on a true fraud. While it is quite common for a Catholic priest to bless a house or perform what is called a "routine exorcism," it is not common to perform what is called a "real exorcism" on houses, despite what was depicted in the movie. In the case of Amityville, the real devils were George and

Kathy Lutz, who concocted a preposterous story to help them out of a mortgage they couldn't afford and a marriage on the rocks (Schick and Vaughn 1998: 269–270).

Not all hauntings are obvious frauds. Some are hoaxes instigated by disturbed teenagers trying to get attention by scaring the devil out of their parents and siblings (Randi 1986, 1995).

Some cases involve otherwise normal people hearing strange noises or having visions of dead people or of objects moving with no visible means of locomotion. Hearing strange noises in the night and letting the imagination run wild are quite natural human traits and not very indicative of diabolical or **paranormal** activity. Likewise, visions and hallucinations are quite natural, even if unusual and infrequent, in people with normal as well as with very active imaginations (Sagan 1995).

Nevertheless, the market for "ghost-busters" flourishes. They go to allegedly haunted houses for television programs such as *Sightings*. They walk around with an electronic device that picks up electromagnetic fields. If the needle moves, they claim they have evidence of poltergeist activity, even though just about anything gives off a measurable level of electromagnetic radiation.

Many people report physical changes in haunted places, especially a feeling of a presence accompanied by a temperature drop and hearing unaccountable sounds. They are not imagining things. Most hauntings occur in old buildings, which tend to be drafty. Scientists who have investigated haunted places account for both the temperature changes and the sounds by finding sources of the drafts, such as empty spaces behind walls or currents set in motion by low-frequency sound waves produced by such mundane objects as extraction fans.

See also **exorcism**.

Further reading: Frazier 1979–80; Randi 1986, 1995.

herbal fuel

Ramar Pillai, from Tamil Nadu in India, claims he has an herb that can turn water into virtually pollution-free diesel fuel or kerosene for about 23 cents a gallon. Pillai has managed to convince a few zealous followers that he is the new Isaac Newton, but skeptics believe he has been exposed as a fake. In one demonstration of his magic herb, it was alleged that his stirring stick was hollowed out and filled with gasoline. When his mixture was heated up, a wax plug at the end of the stick melted, allowing gasoline to flow into the mixture.

Pillai, a high school dropout from a village near Rajapalayam, has intrigued scientists at the Indian Institute of Technology (IIT) by his claims and demonstrations. He says he doesn't want a lot of money, that he wants only a processing plant built near his home village and some protection for his family. He claims he was kidnapped and tortured for several days for refusing to tell his tormentors how he turns water and herbs into fuel. Pillai says his abductors suspended him from a ceiling fan and burned him with cigarettes.

To produce his fuel, Pillai boils water and cooks leaves and bark from his special plant for about 10 minutes. After adding a little salt, citric acid, and traces of a few unknown chemicals, the mixture is cooled and stirred. Once allowed to settle, the liquid fuel, which is lighter than water, floats to the top and is separated by filtering. The entire process takes less than 30 minutes.

According to the Department of Science and Technology at IIT, laboratory tests have conclusively shown that the herbal fuel is a pure hydrocarbon similar to kerosene and diesel fuel. Engineers at IIT in Madras conducted static tests in two-stroke engines and concluded that the herbal fuel offered better fuel economy than petrol. The fuel "will have good potential in a four-stroke petrol as well as diesel engines," according to the engineers in Madras.

If he is not using trickery, how is Pillai doing it? One theory, offered by Ratna Choudhury of IIT, is that atmospheric carbon dioxide is sucked in during the reaction. The carbon dioxide combines with hydrogen liberated from water and forms the hydrocarbon fuel. She admitted, however, that she was just guessing. *The Times of India* has a different theory. It published a report that claimed that the entire exercise of promoting Mr. Pillai was to legitimize the sale of stolen petrol and diesel from tankers of Indian oil companies in Rajapalayam. He has no magic powers and no magic herb, but plenty of followers and even a few scientists willing to try to explain what has never happened.

holistic medicine

"Holistic medicine" is another term for **alternative health practices** that claim to treat "the whole person." To holistic practitioners, a person is not just a body with physical parts and systems, but a spirit and mind as well. But many holistic practitioners seem to ignore the patient's body altogether or treat it as if it were an unnecessary appendage to the **astral body** or **soul**.

Holistic practitioners are truly alternative in the sense that they often reject science and avoid surgery or drugs as treatments, though they are quite fond of the manipulation of **energy** (**chi** or **prana**), meditation, **prayer**, herbs, vitamins, minerals, and exotic diets as treatments for a variety of ailments.

See also **Ayurvedic medicine**.

Further reading: Barrett and Jarvis 1993; Gardner 1957, 1991; Glymour and

Stalker 1985; Randi 1989a; Raso 1994, 1995; Trafford 1995.

hollow Earth

A theory that holds not only that the Earth is hollow, but that it has openings at the poles that allow the Agartha, an advanced race living within Earth, to venture into the atmosphere in their **UFOs**. The idea sounds like fiction, but the first to float the notion was the British astronomer Edmund Halley. In the late 17th century, he proposed that Earth consists of four concentric spheres and that the interior is populated. He even speculated that gas escaping from the openings at the poles causes the aurora borealis. The fiction came later, beginning with such works as Edgar Allan Poe's *The Narrative of Gordon Pym of Nantucket* (1838) and Jules Verne's *Journey to the Center of the Earth* in 1864. Edgar Rice Burroughs (1875–1950), the creator of Martian adventures and Tarzan of the Apes, also wrote novels set in the hollow Earth.

In the early 19th century, John Symmes (1780–1829) promoted the idea of interior concentric spheres so widely that the alleged opening to the inner world was named "Symmes Hole." Admiral Byrd's flights over the North Pole in 1926 and over the South Pole in 1929 should have put an end to that notion, but later hole advocates hail Byrd as having actually gone into the hollow Earth at both poles.

The idea of the hollow Earth has been promoted by a variety of characters. Cyrus Reed Teed, an herbalist and self-proclaimed **alchemist** and visionary, promoted the idea in pamphlets and speeches in the late 19th century. In 1906, William Reed published *The Phantom of the Poles,* in which he claimed that the North and South Poles don't exist because of the entrances to the hollow Earth. In 1913, Marshall B. Gardner

privately published *Journey to the Earth's Interior,* claiming that inside the hollow Earth is a sun 600 miles in diameter and that there are huge holes 1,000 miles wide at the poles. Oddly, no satellite photographs show holes at the poles.

In the 1940s, Ray Palmer, cofounder of *Fate, Flying Saucers from Other Worlds,* and many other pulp publications, teamed up with Richard Shaver to create a legend that included the story of Shaver living with the inner Earth people and writing of their advanced civilization. Shaver is the one who has given us the tale of the inner Earth creatures venturing into space in flying saucers.

In 1964, Rosicrucian leader Raymond W. Bernard summed up the entire hollow Earth mythos in the title of his book: *The Hollow Earth: The Greatest Geographical Discovery in History Made by Admiral Richard E. Byrd in the Mysterious Land Beyond the Poles— The True Origin of the Flying Saucers*. Bernard also authored *Flying Saucers from the Earth's Interior*. Dr. Bernard (his real name was Walter Seigmeister) died of pneumonia on September 10, 1965, while searching for the openings to the interior of the Earth in South America. Bernard accepted without question Shaver's claim that he learned the secret of relativity before Einstein from the hollow Earth people.

Further reading: Gardner 1957, 1988; Kossy 1994; Prytz 1970; Toronto: www.parascope.com/nb/articles/shaverMystery.htm.

Holocaust denial

The mass extermination of the Jews and other "undesirables" at the hands of the Nazis during World War II is referred to as the Holocaust. It has become a symbol of evil in our time. Like many symbols, the Holocaust has become sacrosanct. To many people, both Jews and non-Jews, the Holocaust symbolizes the horror of genocide against the Jews. Some modern anti-Semites have found that attacking the Holocaust causes as much suffering to some Jews as attacking Jews themselves. "Holocaust denial" refers to attacking the accuracy of any aspect of the symbology or history of the Holocaust.

Holocaust denial seems to be the main motivation of the Institute for Historical Review and its *Journal of Historical Review.* Since 1980 this journal has been publishing articles attacking the accuracy of various claims about the Holocaust. There is clearly an agenda when a journal is devoted almost exclusively to the single issue of making the Holocaust seem like an exaggeration of biased historians. If truth and historical accuracy were the only goals of this group, it would be praised rather than despised. However, it seems that its promoters are more concerned with hatred than with truth. Thus, even the inaccuracies that they correctly identify are met with scorn and derision. For they never once deal with the central question of the Holocaust. They deal with details and technical issues: Were there *six* million or *four* million Jews who died or were killed? Could this particular shower have been used as a gas chamber? Were these deaths due to natural causes or not? Did Hitler issue a Final Solution order or not? If so, where is it? These are legitimate historical issues. However, the Holocaust deniers do not deal with the questions of racial laws that led to the arrest and imprisonment of millions of Jews in several countries for the "crime" of race. They do not concern themselves with the policy of herding people like animals and transporting them to "camps" where millions died of disease or malnutrition, or were murdered. They don't address the moral issues of medical experimentation on humans or of persecution of homosexuals and the infirm. Why not?

Michael Shermer devotes two chapters of *Why People Believe Weird Things* (1997) to the arguments of the Holocaust deniers. (In *Denying History: Who Says the Holocaust Never Happened and Why Do They Say It?* [2000] Shermer and coauthor Alex Grobman devote nine chapters to the subject.) Shermer takes up many of the deniers' arguments and refutes them one by one. For example, one of the favorite appeals of the Holocaust deniers is to demand some proof that Hitler gave the order for the extermination of the Jews (or the mentally retarded, mentally ill, and physically handicapped). Holocaust deniers point to Himmler's telephone notes of November 30, 1941, as proof that there was to be no liquidation of the Jews. The actual note says: "Jewish transport from Berlin. No liquidation." Whatever the note meant, it did not mean that Hitler did not want the Jews liquidated. The transport in question, by the way, was liquidated that evening. In any case, if Hitler ordered no liquidation of the Berlin transport, then liquidation was going on and he knew about it. Hitler's intentions were made public in his earliest speeches. Even as his regime was being destroyed, Hitler proclaimed: "Against the Jews I fought open-eyed and in view of the whole world. . . . I made it plain that they, this parasitic vermin in Europe, will be finally exterminated." Hitler at one time compared the Jews to the tuberculosis bacilli that had infected Europe. It was not cruel to shoot them if they would not or could not work. He said: "This is not cruel if one remembers that even innocent creatures of nature, such as hares and deer when infected, have to be killed so that they cannot damage others. Why should the beasts who wanted to bring Bolshevism be spared more than these innocents?"

See also **Protocols of the Elders of Zion**.

Further reading: Lipstadt 1994; Shermer and Grobman 2000.

homeopathy

A system of medical treatment based on the use of minute quantities of remedies that in larger quantities produce effects similar to those of the disease being treated. Homeopathy seems to be based on little more than a belief in **sympathetic magic**. The term is derived from two Greek words: *homeo* (similar) and *pathos* (suffering). The 19th-century German physician Samuel Hahnemann (1755–1843) is considered the father of homeopathy, allegedly being inspired to the notion that "like cures like" from the treatment of malaria with cinchona bark. The bark contains quinine, which helps in the treatment of malaria but also causes fevers. Advocates of homeopathy think that concoctions with as little as one molecule per million can stimulate the body's healing mechanism. Critics such as *Consumer Reports* (January 1987) say: "Unless the laws of chemistry have gone awry, most homeopathic remedies are too diluted to have any physiological effect."

Homeopaths tend to believe in such things as "vital forces" being in harmony (resulting in health) or out of harmony (disease). They tend to advocate **holistic medicine**, treating the **soul** as well as the body. Homeopaths like to say that they treat "persons," not "bodies" or "diseases," and are fond of falsely accusing the majority of medical doctors as not treating "persons."

One criticism of homeopathy is that it takes a "cookie cutter" approach to treatment: one size fits all. No matter what ails you, treatment with a diluted like agent is the cure. Experience teaches otherwise. For example, the treatment for scurvy is not a tiny bit more scurvy, but vitamin C; the treatment for diabetes is not a granule or two of sugar, but insulin. There seem to be countless examples one could come up with that would contraindicate homeopathy as a reasonable approach to the treatment of disease. Thus, simply because it is sometimes reasonable to treat like with like (e.g., polio vaccines), it does not follow that it is *always* reasonable to treat like with like. It is misleading, however, to compare the use of vaccines in medicine to homeopathic remedies. Medical vaccines would be ineffective if they were as diluted as homeopathic remedies. Dr. Stephen Barrett says, "If the FDA required homeopathic remedies to be proved effective in order to remain on the market, homeopathy would face extinction in the United States."

One of the stranger tenets of homeopathy, proposed by Dr. Hahnemann himself, is that the potency of a remedy increases as the drug becomes more and more dilute. Some drugs are diluted so many times that they don't contain *any* molecules of the substance that was initially diluted, yet homeopaths claim that these are their most potent medications! It is not surprising to find that there is no explanation as to how this happens or is even possible, though some homeopaths have speculated that the water used to dilute a remedy has a "memory" of the initial substance.

> How do homeopaths explain this supposed potency of infinitesimal doses, even when the dilution removes all molecules of a drug? They invoke mysterious vibrations, resonance, force fields, or radiation totally unknown to science. (Gardner 1957)

Homeopathy's supporters point to clinical trials that indicate a homeopathic efficacy that cannot be explained by the **placebo effect**. Critics contend that such studies are poorly designed, methodologically biased, or statistically flawed.

Homeopathic advocates give ardent **testimonials** to the curative powers of their remedies. How can so many case histories be dismissed? Easily: The "cures" are probably the result of (1) misdiagnosis (the patient wasn't cured since the disease it "cured" wasn't present); (2) spontaneous remission (the body healed itself and we don't have a clue how it did it); (3) some other treatment used along with homeopathy; or (4) the placebo effect. The many testimonials given as proof that homeopathy "works" are of little value as empirical evidence for the effectiveness of homeopathic remedies. Even so, such "cures" are not meaningless. Left alone, the body often heals itself. Unlike conventional medicine with its powerful drugs and antibiotics, the likelihood of an adverse reaction to a homeopathic remedy is remote. (To prove this point, James Randi once gulped down a box of homeopathic sleeping pills before launching into a long and energetic lecture.) The main harm from homeopathy is not likely to come from its remedies, which are probably safe but ineffective. One potential danger is in the encouragement to self-diagnosis and treatment. Another is not getting proper treatment by a medical doctor in those cases where the patient could be helped by such treatment, such as for a bladder or yeast infection or for cancer.

In short, the main benefits of homeopathy seem to be that its remedies are not likely to cause harm in themselves, and they are generally inexpensive. The main drawbacks seem to be that its remedies are most likely inert and they require acceptance of metaphysical baggage incapable of scientific analysis. Homeopathy "works," just as **astrology, biorhythms, chiropractic**, or conventional medicine, for that matter, "work": It has its satisfied customers. Homeopathy does not work, how-

ever, in the sense of explaining pathologies or their cures in a way that not only conforms to known facts but that promises to lead us to a greater understanding of the nature of health and disease.

Homeopathy is said to be a $200 million a year industry in the United States. It is very popular in Europe, especially among the Royal Family of Britain. It is also very popular in India, where there are more than 100 schools of homeopathy.

See also **alternative health practices**.

Further reading: Barrett 1991; Barrett and Butler 1992; Barrett and Jarvis 1993; Linde et al. 1997; Park 2000; Philp 1996; Raso 1994; Reilly et al. 1994.

houris

Beautiful black-eyed virgins believed by some Muslims to be waiting in heaven for the enjoyment of the faithful, especially men who die as martyrs. According to the *Qu'ran* (55.56), the houris have never been touched by man or **jinni**.

Many Muslims and non-Muslims alike believe that terrorists who commit murder and suicide in the name of Islam commit their sins in order to gain instant admission to heaven, where they will enjoy many houris. According to Sheikh Abdul Hadi Palazzi, an Islamic scholar and secretary general of the Italian Muslim Association, the belief in a sensuous afterlife filled with houris is based upon a hadith collected by Imam at-Tirmidhi in "Sunan" (vol. 4, chapters on "The Features of Heaven as described by the Messenger of Allah," chapter 21, "About the Smallest Reward for the People of Heaven," hadith 2687) and also quoted by Ibn Kathir in his Tafsir (Koranic commentary) of Surah Rhman (55), ayah (verse) 72:

> It was mentioned by Daraj Ibn Abi Hatim, that Abu al-Haytham 'Adullah Ibn Wahb narrated from Abu Sa'id al-Khudhri, who

heard the Prophet Muhammad (Allah's blessings and peace be upon him) saying, "The smallest reward for the people of Heaven is an abode where there are eighty thousand servants and seventy two wives, over which stands a dome decorated with pearls, aquamarine and ruby, as wide as the distance from al-Jabiyyah to San'a." [www.naomiragen.com/Columns/MartyrsForSex.htm]

This sounds like hearsay three times removed, but the reward is obviously not restricted to those who die for the faith. Hence, it seems unlikely that anyone but the terminally perplexed would interpret this to mean that committing murder or suicide for Islam will automatically gain one many servants and a harem of houris.

Houston, Dr. Jean, and the Mystery School

Dr. Jean Houston's Mystery School is another in a long line of New Age self-help or personality transforming programs. According to Houston, "the purpose of the Mystery School is to engender the passion for the 'possible' in our human and global development while discovering ways of transcending and transforming the local self so that extraordinary life can arise!" She thinks we are all unhappy because we have suffered and have not achieved our full potential. Here is an excerpt from Houston's Mystery School Lecture One:

> Regardless of how difficult and estranged your life may have been, you've done that one. You've done estrangement. You've learned from it. You've done difficulty. You've even done derangement probably. You've done angst and anxiety and existential dread. You've done toxic mayhem. Yes? You've done breakdown. Now it's time to try the next level.
>
> You've had all this suffering. Great! It

has given you a wealth and depth of experience and compassion, if you frame it that way. If you don't frame it that way, then all you've got is galloping angst.

> Your energies, your powers, your stamina, your moral force seem limited only because you and your habituations and the habituations and expectations of your culture set limits. Therefore, what Mystery School tries to do here, is to go beyond the limits and create a consensual reality in which the horizon of the limits is greatly expandable and more becomes possible. [www.JeanHouston.org]

The lecture gives only a glimpse of what is in store for the disciple on the road to self-transformation. Her hook is baited with New Age angst: the pain, suffering, and dissatisfaction with life that needs to be relieved. Houston tells her listeners: "You've been wounded up the gazoo. I always say—'You're so full of holes from being so wounded, you're holy.' You're utterly available now." She speaks again and again to the pain and suffering of her audience, of their dissatisfaction with their lives. She tells them that this is necessary for transformation, that out of the evil she will bring good.

Here is Houston's own blurb on her Mystery School:

> It is my 20th Century version of an ancient and honorable tradition, the study of the world's spiritual mysteries. Once upon a time there were such schools in Egypt, Greece, Turkey, Afghanistan, Ireland, England, France, Hawaii, India, China, Japan and many other places on the globe.
>
> Mystery School is both experiential and experimental. I weave together the things I love most: sacred psychology, music, history, theatre, cultural wisdom, science (fact, fiction and fantasy), neurophysiology, philosophy, anthropology, theology,

poetry, laughter, cosmology, metaphysics and innovative ideas to provide a multi-faceted, multi-level Time out of time.

Exercises include . . . energy resonance . . . high play and mutual empowerment.

She claims that her school is part of a tradition that has probably existed ever "since humans have been humans." This claim seems to imply that the mystery schools have made very little progress. There is ample evidence she is correct about that. The reason for this is obvious: Mystery schools don't exist to discover the mysteries of life, but to encourage belief that life is a mystery.

Although Houston does not claim to be psychic, that is how *Newsweek,* the *Sacramento Bee,* and the CBS evening news referred to her in covering a story about Houston working with Hillary Clinton. An Associated Press story ambiguously referred to her as a "psychic researcher." (She and her husband, Robert Masters, once tested the **ESP** of subjects under the influence of LSD at their Foundation for Mind Research.)

Houston has a Ph.D. in philosophy of religion from Columbia University, according to *Newsweek* (July 1, 1996). According to the *Washington Post,* she has a Ph.D. in psychology. In an interview with Stone Phillips of NBC's *Dateline* she claimed to have several doctorates but was most proud of the one in psychology from Union Graduate School. Off camera she admitted that this is really the only doctorate she has. She said she made a mistake and blamed it on overwork or stress, but it seems probable that she lied. She might forget how many doctors she has, but not how many doctorates. How much of her biography is "mythos," as she might call it? Was she really chums with Einstein and Teilhard de Chardin? Was she really Margaret Mead's

adopted daughter? Did she really meet and get the inspiration for her primary teaching method from Edgar Bergen and Charlie McCarthy when she was 8 years old? It should not surprise anyone if it turns out that Jean Houston's autobiography is a myth spun by her imagination out of the fabric of her desires. She is one of the New Age philosophers for whom "deep" truth is something you create.

Houston likes metaphors that have proven successful in similar eclectic transformational endeavors by L. Ron Hubbard (*Dianetics*), Richard Bandler (**neuro-linguistic programming**), Werner Erhard (**est**), Frederick Lenz (**Rama**), and Tony Robbins (**firewalking**). For example, she says that her Mystery School "provides practices which have the effect of both rewiring your brain, body and nervous system, and eliciting the evolutionary latencies in your physical instrument. These latencies have been there like a fetal coding for perhaps tens of thousands of years, but could not be activated until various aspects of complexity emerged, joined to crisis. We find that emergence generally only occurs in emergencies. It's only when you really have to survive that you really turn on enough mindfulness and wakefulness to activate these different latencies."

Yes, and truth is whatever you want it to be.

Further reading: Gardner 2000.

Hubbard, L. Ron

See **Dianetics**.

hundredth monkey phenomenon

A sudden, spontaneous, and mysterious leap of consciousness achieved when an allegedly "critical mass" point is reached.

The expression "hundredth monkey" comes from an experiment on monkeys done in the 1950s. Lyall Watson alleged in his book *Lifetide* that one monkey taught another to wash potatoes, who taught another, who taught another, and soon all the monkeys on the island were washing potatoes where no monkey had ever washed potatoes before. When the hundredth monkey learned to wash potatoes, suddenly and spontaneously and mysteriously monkeys on other islands, with no physical contact with the potato-washing cult, started washing potatoes! Was this monkey **telepathy** at work or just monkey business on Watson's part?

It makes for a cute story, but it isn't true. At least, the part about spontaneous transmission of a cultural trait across space without contact is not true. There really were some monkeys who washed their potatoes. One monkey started it and soon others joined in. But even after 6 years not all the monkeys saw the benefit of washing the grit off of their potatoes by dipping them into the sea. Lyall made up the part about the mysterious transmission. The claim that monkeys on other islands had their consciousness raised to the high level of the potato-washing cult was a lie.

The notion of raising consciousness through reaching "critical mass" is being promoted by a number of New Age spiritualists. Ken Keyes Jr. has published a book on the Internet that calls for an end to the nuclear menace and the mass destruction that surely awaits us all if we do not make a global breakthrough soon. The title of his treatise is *The Hundredth Monkey*. In his book he writes that "there is a point at which if only one more person tunes-in to a new awareness, a field is strengthened so that this awareness is picked up by almost everyone!"

It seems to be working for spreading the word about the hundredth monkey phenomenon.

See also **morphic resonance**.

Further reading: Amundsom 1985, 1987; Possel and Amundson 1996.

hypnagogic state

The transition state of semiconsciousness between being awake and falling asleep. For some people, it is a time of visual and auditory hallucination.

See also **alien abductions**, **hypnopompic state**, and **sleep paralysis**.

hypnopompic state

The transition state of semiconsciousness between sleeping and waking up. For some people, it is a time of visual and auditory hallucination.

See also **alien abductions**, **hypnagogic state**, and **sleep paralysis**.

hypnosis

A process involving a hypnotist and a subject who agrees to be hypnotized. Being hypnotized is usually characterized by intense concentration, extreme relaxation, and high suggestibility.

The versatility of hypnosis is unparalleled. Hypnosis occurs under dramatically different social settings: the showroom, the clinic, the classroom, and the police station.

Showroom hypnotists usually work bars and clubs. The subjects of clinical hypnotists are usually people with problems who have heard that hypnotherapy works for relieving pain or overcoming an addiction or fear. Others use hypnosis to recover **repressed memories** of sexual abuse or of **past lives**. Some psychologists and hypnotherapists use hypnosis to discover truths hidden from ordinary consciousness by tapping into the **unconscious mind** where these truths allegedly reside. Finally, some hypnotic subjects are people who have been victims or witnesses of a crime. The police encourage them to undergo hypnosis to help them remember details from their experiences.

The common view of hypnosis is that it is a trance-like **altered state of consciousness**. Many who accept this view also believe that hypnosis is a way of accessing an unconscious mind full of repressed memories, **multiple personalities**, mystical insights, or memories of **past lives**. This view of hypnosis as an altered state and gateway to occult knowledge about the self is considered a myth by many psychologists. There are two distinct, though related, aspects to this mythical view of hypnosis: the myth of the altered state and the myth of the unconscious occult reservoir.

Those supporting the altered state theory often cite studies that show that, during hypnosis, the brain's electrical states change and brain waves differ from those that occur during waking consciousness. The critics of the mythical view point out that these facts are irrelevant to establishing hypnosis as an altered state of consciousness. One might as well call daydreaming, concentration, imagining the color red, and sneezing "altered states," since the experience of each will show electrical changes in the brain and changes in brain waves from ordinary waking consciousness.

Those supporting the unconscious occult reservoir theory support their belief with anecdotes of numerous people who, while hypnotized, recall events from their present or past life of which they have no conscious memory or relate being in distant places and/or future times while under hypnosis.

Most of what is *known* about hypnosis, as opposed to what is *believed,* has come from studies on the *subjects* of hypnosis. We know that there is a significant correla-

tion between being imaginative and being responsive to hypnosis. We know that those who are fantasy-prone are likely to make excellent hypnotic subjects. We know that vivid imagery enhances suggestibility. We know that those who think hypnosis is rubbish can't be hypnotized. We know that hypnotic subjects are not turned into **zombies** and are not controlled by their hypnotists. We know that hypnosis does not enhance the accuracy of **memory** in any special way. We know that a person under hypnosis is very suggestible and that memory is easily "filled in" by the imagination and by suggestions made under hypnosis. We know that **confabulation** is quite common while under hypnosis and that many courts do not allow testimony that has been induced by hypnosis because it is intrinsically unreliable. We know the greatest predictor of hypnotic responsiveness is what a person believes about hypnosis.

If hypnosis is not an altered state or gateway to a mystical and occult unconscious mind, then what is it? Why do so many people, including those who write psychology textbooks, or dictionary and encyclopedia entries, continue to perpetuate the mythical view of hypnosis as if it were established scientific fact? For one thing, the mass media perpetuates this myth in countless movies, books, and television shows, and there is an entrenched tradition of hypnotherapists who have faith in the myth, make a good living from it, and see many effects from their sessions that, from their point of view, can only be called successes. They even have a number of scientific studies to support their views. Psychologists such as Robert Baker think such studies are about as valid as the studies that supported the belief in phlogiston or the ether. Baker claims that what we call hypnosis is actually a form of learned social behavior (Baker 1990).

The hypnotist and subject learn what is expected of their roles and reinforce each other by their performances. The hypnotist provides the suggestions and the subject responds to the suggestions. The rest of the behavior—the hypnotist's repetition of sounds or gestures, the use of a soft, relaxing voice, and the trance-like pose or sleep-like repose of the subject—are just window dressing, part of the drama that makes hypnosis seem mysterious. When one strips away these dramatic dressings, what is left is something quite ordinary, even if extraordinarily useful: a self-induced, psyched-up state of suggestibility.

Psychologist Nicholas Spanos agrees with Baker: "hypnotic procedures influence behavior indirectly by altering subjects' motivations, expectations and interpretations" (Spanos 1996). This has nothing to do with putting the subject into a trance and exercising control over the subconscious mind. Hypnosis is a learned behavior, according to Spanos, issuing out of a specific context of social learning. We can accomplish the same things in a variety of ways: going to college or reading a book, taking training courses or teaching oneself a new skill, listening to pep talks or giving ourselves a pep talk, enrolling in motivation courses or simply making a willful determination to accomplish specific goals. In short, what is called hypnosis is an act of social conformity rather than a unique state of consciousness. The subject acts in accordance with expectations of the hypnotist and hypnotic situation and behaves as he or she thinks one is supposed to behave while hypnotized. The hypnotist acts in accordance with expectations of the subject (and/or audience) and the hypnotic situation, and behaves as he or she thinks one is supposed to behave while playing the role of hypnotist.

Further reading: Loftus 1979; Schacter 1996, 1997; Spanos 1987–88; Spanos and Chaves 1989; Thornton 1976.

hystero-epilepsy

Dr. Jean-Martin Charcot (1825–1893) was one of the founders of modern neurology. Students came from all over the world to study under him in Paris, including Sigmund Freud in 1885. Charcot used **hypnosis** as a diagnostic tool in his study of hysteria, which influenced Freud's views on the origin of neurosis. Charcot made a number of important medical discoveries and even has a disease named after him (neurogenic arthropathy is also known as "Charcot's joints").

At one point in his career Charcot believed that he had discovered a new disease, which he called "hystero-epilepsy." The symptoms included "convulsions, contortions, fainting, and transient impairment of consciousness" (McHugh: www .psycom.net/mchugh.html).

He showed his students several examples of this new disease during his rounds at Salpêtrière Hospital.

A skeptical student, Joseph Babinski, decided that Charcot had invented rather than discovered hystero-epilepsy. The patients had come to the hospital with vague complaints of distress and demoralization. Charcot had persuaded them that they were victims of hystero-epilepsy and should join the others under his care. Charcot's interest in their problems, the encouragement of attendants, and the example of others on the same ward prompted patients to accept Charcot's view of them and eventually to display the expected symptoms. These symptoms resembled epilepsy, Babinski believed, because of a municipal decision to house epileptic and hysterical patients together (both having "episodic" conditions). The hysterical patients, already vulnerable to suggestion and persuasion, were continually subjected to life on the ward and to Charcot's neuropsychiatric examinations. They began to imitate the epileptic attacks they repeatedly witnessed. [ibid.]

Babinski convinced Charcot that hystero-epilepsy was not a disorder and that doctors can induce symptoms in their patients. They separated the "hystero-epileptic" patients from each other and from staff members who had treated them. The patients were moved to the general ward of the hospital. The doctors then treated the patients by ignoring their hysterical behavior and encouraging the patients to work on their recovery. "The symptoms then gradually withered from lack of nourishing attention" [ibid.]

The lesson of Charcot seems lost on many therapists today, in particular the **repressed memory therapists** and those who treat **multiple personality disorder**.

See also **psychotherapy**.

Further reading: Spanos 1996.

I

Ica stones

A collection of andesite stones allegedly discovered in a cave near Ica, Peru. Engraved on the oxidized surfaces of the stones are depictions that call into question just about everything science has taught us about the origin of our planet, ourselves, and other species. For example, some of the stones depict men who look like ancient Incas or Aztecs riding on and attacking dinosaurs with axes. Other stones seem to depict surgeons and astronomers using modern instruments.

The stones were allegedly discovered by a local farmer in a cave. The farmer was arrested for selling the stones to tourists. He told the police that he didn't really find them in a cave, but that he made them himself. In 1975, Basilio Uchuya and Irma Gutierrez de Aparcana claimed that they sold Dr. Javier Cabrera stones they'd engraved themselves and that they'd chosen their subject matter by copying from "comic books, school books, and magazines" (Polidoro 2002). Other modern Ica artists, however, continue to carve stones and sell forgeries of the farmer's forgeries.

Skeptics consider the stones to be a pathetic hoax, created for a gullible tourist trade. Nevertheless, three groups in particular support the authenticity of the stones: those who believe that extraterrestrials are an intimate part of Earth's real history; fundamentalist **creationists** who drool at the thought of any possible error made by anthropologists, archaeologists, and evolutionary biologists; and mytho-historians who claim that ancient myths are accurate historical records.

The Ica stone craze began in 1996 with Dr. Javier Cabrera Darquea, a Peruvian physician who allegedly abandoned a career in medicine in Lima to open up the Museo de Piedras Grabadas (Engraved Stones Museum) in Ica. Several thousand stones are on display. The Peruvian National Chamber of Tourism lists Dr. Cabrera's museum as a tourist site, though the chamber leaves open the question of the authenticity of the stones. Dr. Cabrera's authority in the matter of the stones seems to have originated from his declaration that a particular stone depicts an extinct fish. The depiction is stylized, as

Ica stone depicting an extinct fish? Species unknown.

are most of the drawings of ancient Peruvian cultures. It must be admitted that knowledge of extinct fish is rare among physicians, even those who have studied biology. Those who are impressed with this knowledge of extinct fish don't seem to be interested in exactly what fish this is supposed to be, when it became extinct, or which telltale marks allow for this identification.

It is argued by extraterrestrialists and creationists that this depiction of an extinct fish proves either that the Indians who made these stones were given information by aliens about extinct fish (for they could not possibly have found any fossils and copied the fossils) or that the timeline that places extinctions of animals such as this fish millions of years in the past are clearly wrong. The Indians lived within the past millennium or two, and so the extinctions must be recent.

Creationists and mytho-historians argue that since the stones depict men riding and attacking dinosaurs, dinosaurs must have existed alongside humans. Thus, either humans existed during the Jurassic period or dinosaurs existed until very recently. Thus, evolutionists are wrong and either God created all species a few thousand years ago or we are all descendents of aliens.

Cabrera has his own theory about the creators of the stones. His theory is based on the premise that the stones are not a hoax. This is understandable, since, if the stones are a hoax, Cabrera is the key hoaxer. Cabrera says the stones depict the first Peruvian culture as an extremely advanced technological civilization. How advanced? The stones allegedly depict open-heart surgery, brain transplants, telescopes, and flying machines. When did they exist? They came from the **Pleiades** about 1 million years ago. How does he know this? That is anybody's guess, but you can read about it in Cabrera's book, *The Message of the Engraved Stones of Ica.*

Why don't scientists simply date the stones and settle the matter? Stones without organic material trapped in them can be dated only by dating the organic material in the strata in which they are found. Since Cabrera's stones come from a mystery cave that has never been identified, much less excavated, there is no way to date them.

That no one has ever found any other remnant of this great culture is troublesome, however. Such a great society must have left at least something, such as a bone or two, a grave here or there, a temple, a hospital, an observatory, an airport, a mummy. But this great civilization, unlike every other great civilization of the past (except **Atlantis**, of course) has vanished without a trace, except for Cabrera's stones. Of course, there are the **Nazca lines**. Unfortunately, the creators of the Nazca figures didn't depict any Indians attacking dinosaurs or doing brain transplants, something that might have tied the Ica stones to the Nazca lines.

There is, of course, an explanation for the cleanliness of this great people. They left our planet and took everything with them except the stones. Perhaps they left the stones as a kind of puzzle for later generations of humans to solve. Maybe they went on to Nazca or to Lubaantun (see **crystal skulls**) to create more puzzles. Maybe these aliens are giving us an I.Q. test. Or the stones may be another test of **faith** given to mankind by the God of the Bible. Or maybe they're just another hoax.

The proof that the stones are not a hoax, says Dr. Cabrera, is in their number. There are too many stones for a single farmer, or even a collective of hoaxers, to have scratched out. He claims that the locals have unearthed about 50,000 stones and that they showed him a "tunnel" where there are another 100,000. However, so far no scientific expedition has set out to explore this tunnel.

Furthermore, says Cabrera, who apparently fancies himself an expert on volcanic stone as well as on extinct fish, andesite is too hard to carve well by mere mortals using stone tools. True, but the stones aren't carved. They are engraved, that is, a surface layer of oxidation has been scratched away. Dr. Cabrera assumes that the creators of the stones only had stone tools available to them. The Inca, Maya, and Aztec cultures all had advanced metallurgy by the time the Spanish arrived. Cabrera and the Ica locals certainly have more than stone tools available to them. Basilio Uchuya explained how he and Irma achieved the ancient look on their stones: They laid out the engraved stones in a chicken pen and the "chickens did the rest" (Polidoro 2002).

Are the stones authentic? If by authentic one means that they were engraved by pre-Columbians, then the answer has to be an unqualified "certainly not all of them." Some engraved stones are said to have been brought back to Spain in the 16th century. It is possible that some of the stones are truly examples of pre-Columbian art. However, it is known that

Ica stone depicting dinosaur riding?

some such stones are forgeries. Tourists, not just in Peru, but everywhere on earth where there are antiquities, are prime targets for forgers. Local con artisans are aware of the market for "forbidden" antiquities. Pre-Columbians certainly were fascinated with monsters, as were ancient European cultures, but do the stones depict dinosaurs? That is open to interpretation. If they do depict dinosaurs and humans together, what is more likely, that they are accurate historical documents or that they are part of a clever hoax? In light of the lack of corroborating evidence, a reasonable person must conclude that the stones are a hoax. Furthermore, the Ica stone episode indicates that some creationists, mytho-historians, and extraterrestrialists seem to think it is their duty to make the preposterous seem plausible in their *jihad* against belief in evolution.

Further reading: Feder 2002; Stein 1993.

I Ching (Book of Changes)

An ancient Chinese text used as an **oracle** to find out the answers to troubling questions such as, "What does the future hold for me?" The book consists of 64 hexa-

Hexagram.

grams, which is the number of possible combinations of six broken or unbroken lines. The lines represent the two primal cosmic principles in the universe, **yin** (the broken lines) and **yang** (the unbroken lines). Why the *I Ching* has *six* lines, however, is curious, since the ancient Chinese believed there were *five* elements (wood, fire, earth, metal, and water), *five* planets, *five* seasons, *five* senses, as well as *five* basic colors, sounds, and tastes. In Chinese **numerology**, however, six is associated with longevity because "the words six and longevity are tone variations of the same basic word" (www.chcp.org/Vnumbers .html). Another source claims that the words for "six" and "wealth" sound similar (thegeomancer.netfirms.com/numerology .htm)

The meanings of the hexagrams, divined many years ago by Chinese philosopher-priests in tune with the *tao* (Chinese for "path" or "way"), consist of such bits of fortune-cookie wisdom as: "If you are sincere, you have light and success. Perseverance brings good fortune." Or "the superior man discriminates between high and low."

One may consult the *I Ching* by flipping numbered coins and adding up the numbers to determine the hexagram.

Another method involves dividing up bundles of yarrow stalks.

ideomotor effect

The ideomotor effect describes the influence of suggestion on involuntary and unconscious motor behavior. William Carpenter coined the term "ideomotor action" in 1852 in his explanation for the movements of rods and pendulums by **dowsers** and table turning by spirit **mediums**. The movement of pointers on **Ouija** boards is also due to the ideomotor effect.

Carpenter argued that muscular movement can be initiated by the mind independently of volition or emotions. We may not be aware of it, but suggestions can be made to the mind by others or by observations. Those suggestions can influence the mind and affect motor behavior.

Scientific tests by American psychologist William James, French chemist Michel Chevreul, English scientist Michael Faraday, and American psychologist Ray Hyman have demonstrated that many phenomena attributed to spiritual or paranormal forces, or to mysterious energies, are actually due to ideomotor action. The tests show that "honest, intelligent people can [and do] unconsciously engage in muscular activity that is consistent with their expectations" (Hyman 1999). They also show that suggestions that can guide behavior can be given by subtle clues (Hyman 1977).

The movement of pointers on **Ouija** boards, of a facilitator's hands in **facilitated communication**, of hands and arms in **applied kinesiology**, and of some behaviors attributed to **hypnotic suggestion** are due to ideomotor action. Ray Hyman (1999) has demonstrated the seductive influence of ideomotor action on medical quackery, where it has produced such appliances as the "Toftness Radiation

Detector" (used by **chiropractors**) and "black boxes" used in medical radiesthesia and radionics (popular with **naturopaths** to harness "**energy**" used in diagnosis and healing.) Hyman also argues that such things as **chi kung** and "pulse diagnosis," popular in **Ayurvedic medicine**, are best explained in terms of ideomotor action and require no supposition of mysterious energies such as **chi**.

See also **cold reading** and **spiritism**.

Further reading: Hyman 1999.

Illuminati

A secret society founded in Bavaria in the late 18th century. They had a political agenda that included republicanism and abolition of monarchies, which they tried to institute by means of "subterfuge, secrecy, and conspiracy," including the infiltration of other organizations (NESS: www.theness.com/encyc/illuminati-encyc .html). They fancied themselves enlightened, but had little success and were destroyed within fifteen years of their origin (Pipes 1997).

Paranoid conspiracy theorists (PCTs) believe the Illuminati cabal still exists, either in its original form or as a model for later cabals. Many PCTs believe "that large Jewish banking families have been orchestrating various political revolutions and machinations throughout Europe and America since the late 18th century, with the ultimate aim of bringing about a satanic New World Order" (Harrington 1996). There are two main factions of PCTs, the militant Christian fundamentalists and the UFO/alien cult, but they seem to agree on only one thing: We are nearing the end of civilization as we know it.

According to Jay Whitley, the Illuminati are hastening the coming of the anti-Christ and the end of the world.

For those of us who still accept the Bible as God's revealed will to man, it's a matter of great concern to see the increasing propaganda for, and emergence of, a New World Order. . . . [B]oth Old and New Testaments warned us that the culmination of history would be marked by the reunion of the nations of the old Roman Empire in Europe; the restoration of the state of Israel (and the increasing hostility of all nations toward her); the implementation of a one-world governmental system; the imposition of a world-wide cashless monetary system; the development of a synchretistic [sic] world religion, based upon man, and presided over by a false prophet; the rise to power of a benign world dictator, who (once firmly in control) would eliminate individual freedoms, demonstrate iron-willed ferocity and cruelty, and make himself the object of worship; and world-wide apostasy, coupled with active persecution and execution of believing Jews and Christians. [www.inforamp.net/~jwhitley/nwo .htm]

Mr. Whitley is prepared for Armageddon, however. He sells Emergency Dehydrated Food Kits from his web site.

Fritz Springmeier has reviewed his own book, an exposé of the Illuminati:

Who really controls world events from behind-the-scene? Years of extensive research and investigation have gone into this massively documented work [*Bloodlines of the Illuminati*]. In almost 600 pages, Fritz Springmeier discloses mind-boggling facts and never before revealed truths about the top Illuminati dynasties. Discover the amazing role these bloodlines have played—and are now wielding—in human history, with family names such as Astor, DuPont, Kennedy, Onassis, Rockefeller, Rothschild, Russell, Van Duyn, and

Krupp. You'll also learn of the secretive, Chinese Li family, which operates with impunity in the U.S.A. and around the world. Along the way you'll find out why President John F. Kennedy and actress Grace Kelly were killed; who created the United Nations; who controls the two major U.S. political parties; how the Rothschilds invented and control modern-day Israel; who secretly founded false religions such as the Jehovah Witnesses; and much, much more. A literal encyclopedia of rare, unbelievable information! [www .texemarrs.com/bloodln.htm]

David Icke, another expert on the Illuminati, provides even more unbelievable information. Icke says he gets messages from alien "Illuminati-reptilians" who explain to him such things as the Gregorian calendar.

The whole senario [sic] was planned centuries ago because the reptilians, operating from the lower fourth dimension, and indeed whatever force controls them, have a very different version of "time" than we have, hence they can see and plan down the three-dimensional "time"-line in a way that those in three-dimensional form cannot. [www.davidicke.com/icke/articles/ illrituals.html]

Icke calls himself the "most controversial author and speaker in the world." There was a time when a man who claimed to be in contact with alien reptiles would have been shunned by the world. In today's open society, such a man is as likely to become a **cult** hero, guest lecturer at universities, or be given his own talk show as he is to be committed to an asylum.

Another expert on the Illuminati, Jim Keith, rivals Icke for imaginative power. Keith, a former executive Scientologist and author of nine conspiracy books (including *Saucers of the Illuminati*), could see things

the rest of us don't. An ad for Coca-Cola became a pornographic display of perversion in his eyes (www.spunk.org/ library/altern/pub/keith/sp000437.txt). He died on September 7, 1999, during surgery to repair a leg he injured at the Burning Man Festival.

Ken Adachi, on the other hand, does not need alien lizards or UFOs to expose the plot to take over the world and hasten the Apocalypse. The Illuminati, however, is only one aspect of the **occult** cabal. According to Mr. Adachi,

An extremely powerful civilian dominated cabal, the New World Order, includes Majesty Twelve [MJ-12], The Illuminati, Order of the Quest, The Bilderberg Group, The Trilateral Commission, The Executive Committee of The Council on Foreign Relations, The PI-40 Committee, The Jason Group, The Club of Rome, The Group, The Royal Institute of International Affairs, The Open Friendly Secret Society, The Rosicrucians, The Brotherhood of the Dragon (or Snake), The Russell Trust, The Black Families (of Europe), Skull & Bones, the Scroll & Key, The Knights of Malta, the Illuminati arm of The **Freemasons**, and many, many other secretive groups. (www.educate-yourself .org/nwoindex.html)

Adachi is not fiddling while Rome burns, however. He publishes a newsletter in which he extols the virtues of natural health cures and the evils of **allopathy**. Apparently, he wants to be healthy when the end comes.

Adachi may have a fine conspiracy web page but he seems to have borrowed everything from Myron Fagan, who undertook to explain all of world history as a plot of the Illuminati to establish the New World Order. Waterloo, Diamond Jim Brady, the French Revolution, any war you care to name, homosexuals in the State

Department, JFK, the United Nations, the ACLU, Jewish bankers, the Communist conspiracy to control Hollywood and make films that would hasten the arrival of the New World Order—these are all explained by Fagan as part of a vast conspiracy.

Fagan, born around 1888, was a playwright, director, producer, editor and public relations director for Charles Hughes, and Republican candidate for president in 1916. In 1930, Fagan came to Hollywood and worked as a writer and director. In 1945, he says he saw some secret documents that led him to write the plays *Red Rainbow* (1946) and *Thieves Paradise* (1946). The former portrays Roosevelt, Stalin, and others at Malta plotting to deliver the Balkans, Eastern Europe, and Berlin to Stalin. The latter portrays the same group plotting to create the United Nations as a Communist front for one world government. Until his death, Fagan relentlessly uncovered plots for almost every historical event of any note.

Another luminary among the PCTs was Milton William "Bill" Cooper, leader of the Arizona militia movement until he was killed in a shootout with a sheriff's deputy in November 2001. Cooper was being served a warrant for assaulting a neighbor when he shot the deputy in the head. Cooper wrote "The Secret Government: A Covenant with Death—The Origin, Identity, and Purpose of MJ-12," a paper given in Las Vegas at a MUFON (Mutual UFO Network) meeting in 1989. The paper focuses on his belief of a cover-up of an alien crash at **Roswell**. He also wrote *Secret Societies/New World Order*. He claims that he got his information "directly from, or as a result of [his] own research into the top secret/Majic material which [he] saw and read between the years 1970 and 1973 as a member of the Intelligence Briefing Team of the Commander in Chief of the Pacific Fleet." Cooper's veracity about his career in the navy and his access to secret documents has been questioned publicly, as have other aspects of his personality. Cooper ran a web site (williamcooper.com) that promotes his many rants, including an autobiographical page that might be of interest to mental health professionals.

Cooper's "investigations" uncover the usual conspiracies, although he also includes some new ones such as the conspiracy to use AIDS to thin out the population of blacks, Hispanics, and homosexuals, a notion he put forth in a book called *Behold a Pale Horse*.

Many PCTs consider the Great Seal of the United States and the motto *Novus Ordo Seclorum* to be Masonic and to mean New World Order. These false claims are considered evidence in the argument to prove the vast conspiracy of the Illuminati. It is useless to argue against these claims with PCTs. They consider us dupes, foolish for believing that the Latin means "New World of the Ages" and that the symbol of the eye in the pyramid relates to a poem in the Egyptian *Book of the Dead*.

The Great Seal of the United States (back side).

To enter the world of the PCTs is to enter Bedlam. It would be pointless here to examine, much less attempt to refute, the delusions of people who think they have been turned into assassins by **mind control** techniques so that they can carry out the will of inbred dynasties, that reptilian aliens are controlling the world, that none of the laws of science are actual, that the imagination and the thought of what is possible are better guides than the "physically manifested world." A rational person might think many of the PCTs are joking. There are Internet sites that seem to be parody sites, but it is difficult to tell, since there seems to be no belief, however inane or absurd, that the PCTs can't **shoehorn** to their bizarre worldview. A rational person unaware of Pat Robertson might well read his *New World Order* (1994) and think it a joke. However, Robertson's rambling paranoia regarding Jewish bankers, **Freemasons**, Muslims, homosexuals, foreigners, and so on looks like pabulum compared with the rants of other PCTs.

See also **Protocols of the Elders of Zion**.

Further reading: Abanes 1998; Boston 1996b; Camp 1997; Coughlin 1999; Roberts 1972; Vankin and Whalen 1998.

incantation

An incantation is a **spell** or verbal **charm** used in **magick** rituals.

incorruptible bodies

Whole human bodies, or parts of human bodies, that allegedly do not decay after death. The Catholic Church claims there are many incorruptible bodies and that they are divine signs of the holiness of the persons in question. It is more likely that they are signs of careful or lucky burial, combined with ignorance regarding the factors that affect rate of decay.

On March 7, 1952, Paramahansa Yogananda, an Indian Hindu who founded the Self-Realization Fellowship (SRF) in California, entered **mahasamadhi**. According to SRF,

> His passing was marked by an extraordinary phenomenon. A notarized statement signed by the Director of Forest Lawn Memorial-Park testified: "No physical disintegration was visible in his body even twenty days after death. . . . This state of perfect preservation of a body is, so far as we know from mortuary annals, an unparalleled one. . . . Yogananda's body was apparently in a phenomenal state of immutability." [www.yogananda-srf.org/py-life/index.html]

The director of Forest Lawn may have given an accurate statement, but calling lack of physical disintegration "an extraordinary phenomenon" is misleading. One wonders how much digging into the mortuary annals SRF did. The state of the yogi's body is not unparalleled, but common. A typical embalmed body will show no notable desiccation for 1 to 5 months after burial without the use of refrigeration or creams to mask odors. According to Jesus Preciado, who has been in the mortuary business for 30 years, "in general, the less pronounced the pathology [at the time of death], the less notable are the symptoms of necrosis." Some bodies are well preserved for years after burial (personal correspondence, Mike Drake). Some, under extraordinary conditions, are well preserved for hundreds, even thousands, of years, as in mummies found in the mountains of Argentina and in a Chinese desert.

Immutable human bodies are ultimately cases of *apparent* immutability. All human bodies and body parts disintegrate with time unless they are preserved by special conditions such as absence of oxygen, bacteria, worms, heat, and light.

See also **Januarius**.

Further reading: Nickell 1993; Quigley 1998.

Indian rope trick

This trick, reportedly witnessed by thousands of people, involves an Indian **fakir** who throws a rope to the sky, but the rope does not fall back to the ground. Instead, it mysteriously rises until the top of it disappears into thin air, the darkness, the mist, whatever. That would be trick enough for most people, but this one continues. A young boy climbs the unsupported rope, which miraculously supports him until he disappears into thin air, the mist, the darkness, whatever. That, too, would be trick enough for most of us, but this one continues. The fakir then pulls out a knife, sword, scimitar, whatever, and climbs the rope until he, too, disappears into thin air, mist, darkness, whatever. Again, this would a great trick even if it stopped here. But, no, it continues.

Body parts fall from the sky into a basket next to the base of the rope. The fakir allegedly then slides down the rope and empties the basket, throws a cloth over the scattered body parts and the boy miraculously reappears with all his parts in the right places. That *would* be a great trick, especially since it must be done in the open without the use of engineers, technicians, electronics, satellite feeds, television cameras, mirrors, or lasers.

Explanations abound as to how the trick is done: mass hypnosis or **mesmerism**; **levitation**; a magic trick involving a hanging invisible rope, a thrown rope that hooks onto the invisible rope somehow, shaved monkey limbs for body parts, and twins (one to go up the rope and one to appear under the blanket).

However, the only thing needed for this trick is human gullibility. According to Peter Lamont, a researcher at the University of Edinburgh and a former president of the Magic Circle in Edinburgh, the Indian rope trick was a hoax played by the *Chicago Tribune* in 1890. Lamont claims the newspaper was trying to increase circulation by publishing this ridiculous story as if there were eyewitnesses to the event. The *Tribune* admitted the hoax some 4 months later, expressing some astonishment that so many people believed it was a true story. After all, they reasoned, the byline was "Fred S. Ellmore."

Further reading: Lamont and Wiseman 1999.

Indigo Children, The

A book by Lee Carroll and his wife Jan Tober. Carroll is a **channeler** for an entity he calls Kryon, who has revealed that "love is the most powerful force in the entire universe." Carroll and Tober travel the world putting on Kryon seminars. Kryon wants to empower us by teaching us how to calibrate and balance our **electromagnetic field**. He has a number of products for sale to help us understand the benefits of controlling our **energy**.

The main thesis of *The Indigo Children* is that many children diagnosed as having attention deficit disorder (ADD) or attention deficit hyperactivity disorder (ADHD) are actually space aliens. These children don't need drugs like Ritalin (methylphenidate), but special care and training. The book consists of dozens of articles by authors from many walks of life. It is, accordingly, inconsistent and uneven in quality of analysis and advice. One of the authors is Robert Gerard, Ph.D., whose piece is called "Emissaries from Heaven." He believes his daughter is an Indigo Child. He also thinks, "Most Indigos see **angels** and other beings in the etheric." He runs Oughten House Foundation, Inc., and sells angel cards. Another contributor is Doreen Virtue, an advocate of **angel therapy**. Still another is Nancy Ann Tappe, who wrote a channeled book called *Understanding Your Life Through Color.* Not all the contributors are on the fringe of New Age metaphysics, however. For example, Dr. Judith Spitler McKee is a former preschool and elementary teacher and retired Eastern Michigan University professor. She spends her time trying to interest children in reading.

The Indigo Child is recognizable by his or her **aura** and by certain other traits, according to *The Indigo Children* web site (owned by Kryon Writings):

- They come into the world with a feeling of royalty (and often act like it).
- Self-worth is not a big issue. They often tell the parents "who they are."
- They have difficulty with absolute authority (authority without explanation or choice).
- They simply will not do certain things; for example, waiting in line is difficult for them.
- They get frustrated with systems that are ritually oriented and don't require creative thought.
- They often see better ways of doing things, both at home and in school, which makes them seem like "system busters" (nonconforming to any system).
- They seem antisocial unless they are with their own kind. If there are no others of like consciousness around them, they often turn inward, feeling like no other human understands them. School is often extremely difficult for them socially.
- They will not respond to "guilt" discipline ("Wait till your father gets home and finds out what you did").
- They are not shy in letting you know what they need. (www.indigochild .com)

One can understand why many parents would not want their child to be labeled as ADD or ADHD. The label implies imperfection. Some may even take it to mean the child is "damaged." Specifically, it means your child's behavior is due to a neuro-biological condition. To some, this is the same as having a malfunctioning brain or a mental disorder. Understandably, emotions run high here. Treatment of children with problems is a hot button issue for the mass media, attack lawyers, talk show hosts, columnists, and others not known for their role in clarifying complicated scientific or medical matters. Many jump on the bandwagon and attack the drug industry and psychiatrists for overdrugging our children. Opposition is fruitless, because few will listen to those who would defend those who "abuse" children. Fewer still will bother to investigate to see whether the critics know what they are talking about.

The National Institutes of Mental Health says that ADHD is the most commonly diagnosed childhood disorder. It affects some 3–5% of all school-age children. With so many children affected, it should be easy to find cases of misdiagnosis, inappropriate treatment, adverse drug reaction, and so on. Anecdotes of abuse, however, should not substitute for scientific studies or clinical observations by the professionals who treat these children on a daily basis. But we all know that an anecdote told on *Oprah* or *Larry King Live* by Arianna Huffington or Hillary Clinton is much more powerful than a controlled scientific study. Yet those scientific studies must be done. Ritalin has been around since 1950, yet there are no long-term studies I am aware of that show it is safe, effective, or better than any alternative. The support for its prescription comes mostly from those in the trenches, the practitioners who treat the millions of children and adults with ADHD. Support also comes from Ritalin's manufacturer, New Jersey–based Novartis Pharmaceuticals Corp., which says the drug "has been used safely and effectively in the treatment of millions of ADHD patients for over 40 years," attested by the results of 170 studies (Donohue 2000). However, Novartis is hardly a disinterested party.

In any case, no matter how many long-term studies are done that find nothing spectacularly wrong with Ritalin, there will always be the possibility that the next one will find something horrible. For example, "researchers at the University of California, Berkeley, say their study, tracking ADHD youths into adulthood, has found a connection between Ritalin use and later abuse of tobacco, cocaine and other stimulants" (Donohue 2000). Is the connection strong enough to warrant worry? How can we be sure it wasn't the ADHD, rather than the Ritalin, that was the main basis for the connection?

The hype and near-hysteria surrounding prescribing Ritalin for children has contributed to an atmosphere that makes it possible for a book like *The Indigo Children* to be taken seriously. Given the choice, who wouldn't rather believe that their children are special and chosen for some high mission rather than that they have a brain disorder? On the bright side, at least Kryon doesn't prescribe blue-green algae, a popular "alternative" to Ritalin, even though there are no long-term studies on what effects algae might have on a developing child's brain ("The Algae AD/HD Connection: Can Blue Green Algae Be of Help with Attention Deficit/Hyperactivity Dis-

order?" by John Taylor, Ph.D., www .adhdoutreach.com/Tayloradhd.htm).

inedia (breatharianism)

Inedia is the alleged ability to live without food or drink. Some inediates become breatharians, like Therese Neumann (1898–1962) of Bavaria, who said that "one can live on the Holy Breath alone." She claims to have done this from 1926 to 1962, during which time she allegedly suffered the **stigmata** and consumed only her daily serving of transubstantiated bread.

Fasting has long been considered a way to purify one's body and mind. Fasting reminds us of our dependence and weakness, and links us to those who suffer hunger as part of their daily lives. Inediates strive to be spiritual beings and carry fasting to an inhuman level. If restraint, self-control, and reducing one's intake of food and water are good, then eliminating all physical nourishment must be better. Spiritual beings don't need food, water, or sleep. However, food, water, and sleep are not optional for human beings.

One inediate who has been attracting followers to breatharianism is Australian Ellen Greve, a.k.a. Jasmuheen. According to Greve, a former financial adviser, we can get all the nutrition we need from **prana**, the universal life force. She is the author of *Living on Light: A Source of Nutrition for the New Millennium,* a 21-day program that will allow the body to stop aging and attain immortality by living solely on light.

Greve claims she hasn't eaten since 1993; yet she admits that "she drinks herbal teas and confesses to the occasional 'taste orgasm' involving chocolate or ice cream." She also admits, "[I]f I feel a bit bored and I want some flavour, then I will have a mouthful of whatever it is I'm wanting the flavour of. So it might be a piece of chocolate or it might be a mouthful of a cheesecake or something like that" (Willis: www.abc.net.au/science/correx/archives/ jasmuheen.htm). Several interviewers have found her house full of food, but she claims the food is for her husband, who once went to prison for misappropriating a pension fund. Apparently he hasn't seen the light and is unable to live on prana yet (Walker and O'Reilly 1999).

Greve runs the Cosmic Internet Academy and claims to have 5,000 followers worldwide. People pay over $2,000 to attend her seminars. There are many, apparently, who are not bothered by the contradiction of saying one needs only prana (or is it light?) but admits to the odd sweet and cup of tea, and has a house full of food. This "diet" is changing her chromosomes, she says. Her "DNA is changing to take up more hydrogen and is developing from 2 to 12 strands" (Willis, op. cit.). Greve also claims that the starving of the world would be just fine if they could only be "re-programmed." They starve to death, she says, because the mass media has tricked them into thinking they need food (Trull: www.parascope.com/en/lightlch .htm). Such gibberish would get some people into treatment; instead, she makes world tours promoting her book. At least three of Greve's followers have starved to death while trying to purify themselves with total fasting. Despite the dangerousness of her insane teachings, in the fall of 1999, the Australian television program *60 Minutes* tested her ability to live on prana, the "light of God."

After four days of fasting, Dr. Berris Wink, president of the Queensland branch of the Australian Medical Association, urged her to stop the test. According to the doctor, Greve's pupils were dilated, her speech was slow, she was dehydrated, and her pulse had doubled. The doctor feared kidney damage if she continued with the

fast. The test was stopped. Greve claimed that she failed because on the first day of the test she had been confined in a hotel room near a busy road, which kept her from getting the nutrients she needs from the air. "I asked for fresh air. Seventy percent of my nutrients come from fresh air. I couldn't even breathe," she said. However, the last three days of the test took place at a mountainside retreat, where she could get plenty of fresh air and where she claimed she could now live happily (www.gospelcom.net/apologeticsindex/ an991028.html#19). Clearly, had the test continued, she would have died. Instead, she lived to lead others to their deaths.

Another inspiration for breatharianism is Wiley Brooks, who heads the Breatharian Institute of America. For the past 30 years or so, Brooks has been claiming that we don't need food, water, or sleep. He asks, "[I]f food is so good for you, how come the body keeps trying to get rid of it? . . . Man was not designed to be a garbage can." He claims that adepts and yogis have been living on air for millennia. Brooks offers weekend workshops at a Sierra Nevada mountain retreat for $425, meals included.

intelligent design (ID)

The theory that design is empirically detectable in nature and in living systems, and that parts of the universe can be explained only as the result of intelligent causes. Advocates claim ID should be taught in the **science** classroom because it is a scientific alternative to the theory of natural selection. The main purpose of ID, however, is **metaphysical**: to prove the existence of God.

ID advocates falsely present natural selection as implying the universe could not have been designed or created. Denying God can create using natural selection contradicts the notion of an omnipotent creator.

One of the first ID promoters was University of California at Berkeley law professor Philip E. Johnson, who seems to have completely misunderstood Darwin's theory of natural selection as falsely implying (1) God doesn't exist, (2) natural selection could have happened only randomly and by chance, and (3) whatever happens randomly and by chance cannot be designed by God. Natural selection requires none of these beliefs. Belief in God the creator of the universe does not logically preclude natural selection. God could have designed natural selection.

Scientists Michael Behe, author of *Darwin's Black Box* (1996), and William Dembski, author of *Intelligent Design: The Bridge between Science and Theology* (1998) also carry the ID torch. Dembski and Behe, fellows of the Discovery Institute, a Seattle research institute funded largely by Christian foundations, couch their arguments in scientific terms backed by scientific competence. Even so, their arguments are identical in function to that of the **creationists**: Rather than provide positive evidence for their own position, they mainly try to find weaknesses in natural selection. But even if their arguments are successful against natural selection, that would not increase the probability of ID. Oddly, one of the main weaknesses the ID people find with natural selection is that it can't explain the ID hypothesis!

Behe is an associate professor of biochemistry at Lehigh University. He claims that biochemistry reveals a cellular world of such precisely tailored molecules and such staggering complexity that it is not only inexplicable by gradual evolution, but that it can be plausibly explained only by assuming an intelligent designer, that is,

God. Some systems, he thinks, can't be produced by natural selection because "any precursor to an irreducibly complex system that is missing a part is by definition nonfunctional" (Behe 1996: 39). He says that a mousetrap is an example of an irreducibly complex system, that is, all the parts must be there in order for the mousetrap to function. This **argument from design** is no more scientific than any other variant of the argument from design and, like the others, this one **begs the question**. He must assume design in order to prove a designer.

Behe's argument hinges on the notion of "irreducibly complex systems," systems that could not function if they were missing just one of their many parts. "Irreducibly complex systems . . . cannot evolve in a Darwinian fashion," he says, because natural selection works on small mutations in just one component at a time. He then leaps to the conclusion that intelligent design must be responsible for these irreducibly complex systems. Evolutionary geneticist H. Allen Orr does not agree. He writes:

The multiple parts of complex, interlocking biological systems do not evolve as individual parts, despite Behe's claim that they must. They evolve together, as systems that are gradually expanded, enlarged, and adapted to new purposes. As Richard Dawkins successfully argued in *The Blind Watchmaker,* natural selection can act on these evolving systems at every step of their transformation (Miller 1996).

An irreducibly complex system can be built gradually by adding parts that, while initially just advantageous, become—because of later changes—essential. The logic is very simple. Some part (A) initially does some job (and not very well, perhaps). Another part (B) later gets added because it helps A. This new part isn't essential, it merely improves things. But later on, A (or something else) may change in such a way that B now becomes indispensable. This process continues as further parts get folded into the system. And at the end of the day, many parts may all be required. [1996/1997: bostonreview.mit.edu/br21.6/orr.html]

Finally, Behe's argument assumes that natural selection will never be able to account for anything it cannot account for now. This begs the question.

William Dembski, a professor at Baylor University, also claims he can prove that life and the universe could not have happened by chance and by natural processes. He, too, concludes that they must be the result of intelligent design by God. He also claims that "the conceptual soundness of a scientific theory cannot be maintained apart from Christ" (Dembski 1998: 209), a claim that belies his **metaphysical** interest in the issue.

According to physicist Victor Stenger, Dembski uses math and logic to derive what he calls the law of conservation of information:

> He argues that the information contained in living structures cannot be generated by any combination of chance and natural processes. . . . Dembski's law of conservation of information is nothing more than "conservation of entropy," a special case of the second law [of thermodynamics] that applies when no dissipative processes such as friction are present. However, the fact is that entropy is created naturally a thousand times a day by every person on Earth. Each time any friction is generated, information is lost. (Stenger 2000)

The science in Dembski's argument, like that of Behe's, is a smokescreen. ID isn't a scientific theory. It isn't a theory at all. It's a philosophical argument that begs the question. It isn't an alternative to natural selection or any other scientific theory. The universe would appear the same to us whether designed by God or not. Empirical theories are about *how* the world appears to us; metaphysical theories may speculate *why* the world appears this way.

Believing the universe or some part of it was designed does not help understand how it works. Science is open to both **theists** and **atheists** alike. Nevertheless, belief in design does spawn its own difficulties.

If we grant that the universe is possibly or even probably the result of intelligent design, what is the next step? For example, assume a particular ecosystem is the creation of an intelligent designer. Unless this intelligent designer is one of us, that is, human, and unless we have some experience with the creations of this and similar designers, how could we proceed to study this system? If all we know is that it is the result of ID, but that the designer is of a different order of being than we are, how would we proceed to study this system? Wouldn't we be limited in always responding in the same way to any question we asked about the system's relation to its designer? *It is this way because of ID.* Such a theory explains everything but illuminates nothing.

Scientific progress is possible in part because scientists attempt to describe the workings of natural phenomena without reference to their creation, design, or ultimate purpose. God may well have created the universe and the laws of nature, but Nature is still a machine, mechanically changing and comprehensible as such. God is an unnecessary hypothesis in the quest to explain *how* nature works.

Further reading: Arnhart 2001; Barlow 2001; Bartelt 1999; Dawkins 1988; Dennett

1995; Edis 2001; Gould 1979, 1983; Paulos 1990; Pennock 1999); Stenger, 1997a.

intuitives

Persons who claim to have **psychic** abilities. They are sometimes called "sensitives." *Intuitive healers* use "insight" to diagnose illness. They are sometimes called "medical intuitives" or "psychic healers."

Some intuitive healers claim that their abilities allow them to make accurate diagnoses over the telephone. Some prefer the radio or the Internet for their healings. Intuitive Judith Orloff, M.D., claims she can diagnose mental illness at a distance. She calls her ability "second sight." The sensitive who founded the Barbara Brennan Healing Science claims she can do "astral healing" and psychic surgery on **auras**. Intuitive healer Carolyn Myss, on the other hand, has abandoned healing for the less dangerous and more lucrative work of giving lectures and holding workshops. (Her web site says she's booked through 2004.)

Further reading: Beyerstein 1997; Stenger 1999.

invocation

A special set of words used to call on some spirit or **occult** power for protection or assistance.

iridology

The study of the iris to diagnose disease. Iridology is based on the questionable assumption that every organ in the human body has a corresponding location within the iris and that one can determine whether an organ is healthy or diseased by examining the iris rather than the organ itself. The Canadian Institute of Iridology says, "Iridology is one of the fastest growing fields in **alternative health care** in Canada today."

Medical doctors see the iris as the colored part of the eye that regulates the amount of light entering by a contractile opening in the center, the pupil. The lens brings the light rays to a focus, forming an image on the retina where the light falls on the rods and cones, causing them to stimulate the optic nerve and transmit visual impressions to the brain. Medical doctors also recognize that certain symptoms of nonocular disease can be detected by an eye exam. Ophthalmologists and optometrists can identify nonocular health problems by examining the eye. If a problem is suspected, these doctors then refer their patients to an appropriate specialist for further examination. However, recognizing symptoms of disease by looking in the eyes is not what iridology is about. In fact, when iridologists have been tested to see whether they could distinguish healthy from sick people by looking at slides of their eyes, they have failed. In a study published in the *Journal of the American Medical Association* (1979, vol. 242: 1385–1387), three iridologists incorrectly identified nearly all of the study slides of the irises of 143 healthy and diseased people. "In fact, they often read the irises of the sickest people as being healthy and vice versa. They did not even agree with each other" (Lisa Niebergall, M.D., altmed.creighton.edu/Iridology/scientific_review.htm). Similar results involving five Dutch iridologists were published in the *British Medical Journal* (1988, vol. 297: 1578–1581).

Iridology goes way beyond the claim that the eyes often provide signs of disease. Iridologists maintain that each organ has a counterpart in the eye and that you can determine the state of the organ's health by looking at a particular section of the eye. This belief did not originate with scientific investigation but with one man's intuition.

Ignatz von Péczely, a 19th-century Hungarian physician, invented iridology. He got the idea for this novel diagnostic tool when he saw a dark streak in the eyes of a man he was treating for a broken leg and it reminded him of a similar dark streak in the eyes of an owl whose leg he had broken years earlier. Von Péczely went on to document similarities in eye markings and illnesses in his patients. Others completed the map of the eye. A typical map divides the eye into sections, using the image of a clock face as a base. So, for example, if you want to know the condition of a patient's thyroid gland, you need not touch the patient to feel for any enlargement of the gland. Nor do you need to do any tests of the gland itself. All you need to do is look in the iris of the right eye at about 2:30 and the iris of the left eye at about 9:30. Discolorations, flecks, or streaks in those parts of the eyes are all you need concern yourself with if it is the condition of the thyroid you wish to know. For problems with the vagina or penis, look at 5 o'clock in the right eye. And so on. An iridologist can do an examination with nothing more than an iridology map, a magnifying glass and a flashlight.

If von Péczely's reasoning is typical, we can surmise that he and other iridologists deceived themselves by looking for and finding correlations between eye markings and illness (**confirmation bias**). They were working with vague notions of "markings" and "illness." Diseases may not have been precisely or accurately diagnosed in many cases. They were able to validate iridology by finding many correlations that in fact were not established as causal relationships by rigorously defined **controlled studies**. Some of their correlations may be accurate, but many are undoubtedly bogus, due to very broad interpretations of "markings" and "disease." They found patterns where in fact there are no patterns (**apophenia**). They misinterpreted data and gave extraordinary significance to confirmations, while ignoring or not seeking disconfirmations. Many of their confirmations may have been merely matters of self-validation. We do not know how much the power of suggestion played in their patients' illnesses. Many diagnoses were probably wrong, but no objective tests were done to check out the validity of the diagnoses. Some diagnoses may have been correct, but the iridologists may have been using other signs besides eye markings to make their diagnoses.

What is most peculiar about the iris is that each is unique and unchangeable, so much so that many claim that the iris is a better identifier of an individual than fingerprints.

See also **acupuncture** and **reflexology**.

Further reading: Barrett and Jarvis 1993; Gardner 1957; Gilovich 1993; Hines 1990; Raso 1994; Worrall 1986.

J

jamais vu

The contrary of **déjà vu**. In *jamais vu*, an experience feels like it's the first time, even though the experience is a familiar one. *Jamais vu* occurs in certain types of amnesia and epilepsy.

Januarius (St. Gennaro)

The patron saint of Naples, Italy. His dried blood is said to miraculously liquefy twice a year: on his feast day of September 19 and on the first Saturday in May. On those occasions, a vial allegedly containing the saint's dried blood is removed from the cathedral in Naples and taken on procession through the city streets. The ritual used to be performed on December 16, "but the liquefaction occurred relatively rarely on those occasions—apparently due to the colder temperature—and those observances have been discontinued" (Nickell 1993: 81).

This so-called **miracle** has been occurring for some 600 years without fail, according to the faithful. Believers and uncritical reporters repeatedly confirm that the powdery substance kept in the vial is blood and that scientists cannot explain why it liquefies. However, Italian scientists who examined the vial in 1902 and in recent years were not allowed to take a sample to the lab. They were allowed to shine a light through the vial, and on the basis of a spectroscopic analysis concluded the substance is blood (ibid.: 78). It is not true, however, to say that scientists can't explain why the stuff in the vial liquefies regularly. A professor of organic chemistry at the University of Pavia, Luigi Garlaschelli, and two colleagues from Milan offered as an explanation thixotropy, a property of certain gels to become fluid when shook or otherwise disturbed. They made their own "blood" that liquefied and congealed, using chalk, hydrated iron chloride, and salt water. Joe Nickell did the same with oil, wax and "dragon's blood" (a resinous dark red plant product).

The Neapolitans are a superstitious people. There are about 20 allegedly miraculous vials of blood of various saints and nearly all of them are in the Naples region, "indicative of some regional secret" (ibid.: 79). Neapolitans believe that if the blood fails to liquefy, disaster will soon follow. They claim that on at least five occasions after the blood failed to liquefy there were terrible events, such as a plague in 1527 and an earthquake in southern Italy that killed 3,000 people in 1980. The proponents of this alleged miracle do not mention how many times disaster *didn't* happen after the blood failed to liquefy, nor do they note how many times disasters happened after the blood did liquefy. Even though the vial is taken on parade only twice a year, apparently the powder liquefies more than a dozen times a year.

According to traditional Catholic hagiography, Januarius was a bishop beheaded during the reign of the emperor Diocletian (284–305). Yet there is no historical record of his alleged blood relic before 1389, more than 1,000 years after his martyrdom. Some doubt that Januarius even existed (ibid.: 79).

Most skeptics are convinced that whatever is in the vial is reacting to some natural phenomenon, such as temperature change or motion. Many religious thinkers consider such "miracles" frivolous and unworthy of God.

See also **incorruptible bodies**.

jinni (genie)

A jinni (pl. jinn) is a spirit in Arabic and Muslim demonology that is capable of assuming human or animal form and exercising supernatural influence over people for good or ill. The jinn were popular in Middle Eastern literature, as in the stories in *A Thousand and One Nights*.

The jinn make many appearances in the Koran.

jogini

Women forced into prostitution by a religious custom known as **devadasi** in India. Young girls are married to a local deity and afterward it becomes their religious duty to provide sexual favors to the local men, usually those of the higher castes. This religious practice was banned in 1988, but the law is not being enforced in all parts of India.

joy touch

A meditative technique developed by Pete Sanders. He says it can help people lose weight, feel good, relax, quit smoking, eliminate life-threatening diseases, overcome fear of paranoia, transcend the body, get off drugs, and face the dentist.

The technique is reminiscent of meditating on the third eye in the middle of the forehead while silently humming "Om." Sanders's twist is to have one imagine a line from the center of the forehead to the center of the brain (the site of the septum pellucidum). Then, one imagines gently brushing that region of the brain. He may have been influenced by reports that neurosurgeons had elicited intense pleasure in patients zapped in the septum (Ramachandran 1998: 175).

Sanders teaches his discovery of joy touch in Sedona, Arizona, a New Age mecca for those in search of higher forms of consciousness. He claims to have an undergraduate degree from MIT in biomedical chemistry and is the author of *You Are Psychic!*

According to Sanders, the septum pellucidum is used as a remote control for the hypothalamus, generally considered the brain's pleasure center. The septum pellucidum has nerve connections to the hypothalamus and stimulates it directly. Sanders promises exhilarating relief in 2 or 3 seconds. He claims the pleasant state may last from 5 to 30 minutes. However, he advises against trying to visually massage the hypothalamus directly since the hypothalamus is very close to the rage and anxiety centers of the limbic system within the brain. Instead of finding oneself in a state of stoic serenity one might find oneself catatonic or enraged beyond the point of recovery.

Jung, Carl (1875–1961)

A Swiss psychiatrist and colleague of Sigmund Freud. He broke away from Freudian **psychoanalysis** over the issue of the **unconscious mind** as a reservoir of repressed sexual trauma that causes all neuroses. Jung founded his own school of analytical psychology.

Jung believed in **astrology**, **ESP**, **spiritualism**, and **telekinesis**. In addition to believing in a number of **occult** and **paranormal** notions, Jung contributed two new ones: *synchronicity* and the *collective unconscious*.

Synchronicity is an explanatory principle: It explains meaningful coincidences such as a beetle flying into his room while a patient was describing a dream about a scarab. The scarab is an Egyptian symbol of rebirth, he noted. Therefore, the propitious moment of the flying beetle indicated that the transcendental meaning of both the scarab in the dream and the insect in the room was that the patient needed to be liberated from her excessive rationalism. His notion of synchronicity is that there is an acausal principle that links events having a similar meaning by their coincidence in time rather than their occurrence in sequence. He claimed that there is a synchrony between the mind and the phenomenal world of perception.

What evidence is there for synchronicity? None. Jung's defense is so inane I hesitate to repeat it. He argues that "acausal phenomena must exist . . . since statistics are only possible anyway if there are also exceptions" (1973: vol. 2, p. 426). He asserts that "improbable facts exist—otherwise there would be no statistical mean" (vol. 2, p. 374). Finally, he claims that "the premise of probability simultaneously postulates the existence of the improbable" (vol. 2, p. 540).

Even if there were a synchronicity between the mind and the world such that certain coincidences resonate with transcendental truth, there would still be the problem of figuring out those truths. What guide could one possibly use to determine the correctness of an interpretation? There is none except intuition and insight, the same guides that led Jung's teacher, Sigmund Freud, in his interpretation of **dreams**. The concept of synchronicity is an expression of **apophenia**.

According to psychiatrist and author Anthony Storr, Jung went through a period of mental illness during which he thought he was a prophet with "special insight." Jung referred to his "creative illness" between 1913 and 1917 as a voluntary confrontation with the unconscious. His great insight was that he thought all his patients over 35 years old suffered from loss of religion, and he had just the thing to fill up their empty, aimless, senseless lives: his own metaphysical system of *archetypes* and the *collective unconscious*.

Synchronicity provides access to the archetypes, which are located in the collective unconscious and are characterized as universal mental predispositions not grounded in experience. Like Plato's forms, the archetypes do not originate in the world of the senses, but exist independently of that world and are known directly by the mind. Unlike Plato, however, Jung believed that the archetypes arise spontaneously in the mind, especially in times of crisis. Just as there are meaningful coincidences, such as the beetle and the scarab dream, which open the door to transcendent truths, so too a crisis opens the door of the collective unconscious and lets

out an archetype to reveal some deep truth hidden from ordinary consciousness.

Mythology, Jung claimed, bases its stories on the archetypes. Mythology is the reservoir of deep, hidden wondrous truths. Dreams and psychological crises, fevers and derangement, chance encounters resonating with meaningful coincidences—all are gateways to the collective unconscious, which is ready to restore the individual psyche to health with its insights. Jung maintained that these **metaphysical** notions are scientifically grounded. But they are not empirically testable in any meaningful way, a claim revealing the **pseudoscientific** nature of his enterprise.

Further reading: Gallo 1994, 1996; McGowan 1994; Noll 1997a, 1997b; Storr 1996.

K

Kabalarian philosophy (KP)

The brainstorm of Alfred J. Parker, who claimed that a numerological analysis of one's name reveals one's destiny. Since the death of Mr. Parker in 1964, the Kabalarians, headquartered in Vancouver, B.C., have been led by Ivon Shearing, who was sentenced to five years in prison in 1997 for sexually abusing several teenage girls over a 25-year period.

According to KP,

Every alphabet has a consistent mathematical order, which allows it to be measured. An analysis of the letters in your name will determine the qualities of your personality.

Your name determines your every experience. It defines your personal strengths and weaknesses both mentally and physically. It interprets your whole nature. It shows your position in life and your measure of success or failure. When you are named your destiny is created.

This sounds similar to another tangible reality known as **numerology**. Here is KP's revelation about my name:

> Your name of Robert creates a refined, diplomatic nature that elicits the co-operation and respect of others. You appreciate a high standard of living and the attendant luxuries and comforts; however, the name does spoil ambition and confidence, and creates a self-depreciating quality that is a deterrent when you must force an issue or even carry out an important decision.
>
> Your name of Carroll creates extreme sensitivity to the thoughts of others and a total lack of expression of your own thoughts and feelings, except through the medium of writing. Under this name you seldom find anyone who, you feel, under-stands you. You are easily embarrassed and then you become quiet and secretive in order to avoid being laughed at or mis-understood. You keep your thoughts and feelings inside and only feel really relaxed and at peace at such times when out in nature or listening, alone, to beautiful music. This name also gives you a weak-ness in your chest which could lead to pleurisy or heart murmurs, or any trouble affecting the bronchial organs.

Of course, some of these descriptions are correct. They would be true of many people, regardless of their names. Some of the more specific claims are correct, but most are not. An analysis of this sort of person-ality profiling is taken up in the entry on the **Forer effect**.

What is unique about KP is its advo-cates believe that *you can change your des-tiny by changing your name*. You can call them and they will help you do this legally, should you find another name with a bet-ter destiny than the one you were given at birth.

karma

A law in Hinduism that maintains that every act done, no matter how insignifi-cant, will eventually return to the doer with equal impact. Good will be returned with good; evil with evil. Since Hindus believe in **reincarnation**, karma knows no simple birth or death boundaries. If good or evil befall you, it is because of some-thing you did in this or a previous lifetime. Christians may wonder whether Adam and Eve had navels; Hindus may wonder whether the law of karma applied to the first sentient creatures.

Karma is sometimes referred to as a "moral law of cause and effect." Karma is both an encouragement to do good and to avoid evil, as well as an explanation for whatever good or evil befalls a person.

On one level, karma serves to explain why good things happen to bad people and bad things happen to good people. The injustices of the world, the seeming random distribution of good and evil, are only apparent. In reality, everybody is get-ting what he or she deserves. Even the child brutalized by drugged adults deserves the horror. The mentally ill, the retarded, the homosexuals, and the millions of Jews killed by the Nazis deserved it because of evil they must have done in the past. The slave beaten to within a breath of death deserved it, if not for what he did today, then for what he did in some previous life-time. The rape victim is just getting what she deserves. All suffering is deserved, according to the law of karma.

The idea of karma is popular among many in Western cultures where it has become detached from its Hindu roots. The **Theosophists**, for example, believe in karma and reincarnation. So does James Van Praagh, who claims to be a **psychic** conduit for all the billions of people who have died over the centuries.

> Let's say someone kills someone . . . at a bank machine. . . . It could be two things. It could be, the person who committed the crime used their free will to do that. Or this might sound weird, but it could have been a karmic situation where that person who was murdered had to be paid back for murdering the other person in a previous incarnation. (Amazon.com interview with James Van Praagh)

Van Praagh makes it clear that he thinks it is karma, not free will, that leads people to kill one another. If Van Praagh is right, we may as well dismantle our ethical and criminal justice systems. Everybody is just playing out his or her karma. Nobody is really good or evil. Nobody is really responsible for anything they do. We're all just karmic pawns doing a dance with destiny.

Why would such an amoral principle such as karma be paraded forth as if it explained the ultimate justice of an indifferent universe? Because, says Van Praagh, "We are on this earth to learn lessons. This is our schoolroom here. . . . We must go through certain lessons in order to grow." According to Van Praagh, life on Earth is actually life in purgatory. We are here working out our sins, evolving our souls, burning off some karma. Van Praagh's version of karma is not likely to be accepted by Hindus or Buddhists. They would maintain that when a person does evil, they are acting freely. And when a person suffers evil, it is because of some evil freely done by that person in the past.

Karma as understood by Van Praagh seems to make life trivial, a mere working out of a metaphysical "law" that reduces all humans to dehumanized creatures, devoid of morality and responsibility, mere causes and effects in a pointless system.

Further reading: Edwards 1996a.

Kirlian photography

In 1939, Semyon Kirlian discovered by accident that if an object on a photographic plate is subjected to a high-voltage electric field, an image is created on the plate. The image looks like a colored halo or coronal discharge. This image is said to be a physical manifestation of the spiritual

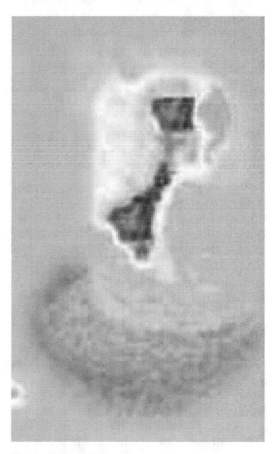

"Aura of a bunyip."

aura or life force that allegedly surrounds each living thing.

Actually, what is recorded is due to quite natural phenomena such as pressure, electrical grounding, humidity, and temperature. Changes in moisture (which may reflect changes in emotions), barometric pressure, and voltage, among other things, will produce different auras.

> Living things . . . are moist. When the electricity enters the living object, it produces an area of gas ionization around the photographed object, assuming moisture is present on the object. This moisture is transferred from the subject to the emulsion surface of the photographic film and causes an alternation of the electric charge pattern on the film. If a photograph is taken in a vacuum, where no ionized gas is present, no Kirlian image appears. If the Kirlian image were due to some paranormal fundamental living energy field, it should not disappear in a simple vacuum. (Hines 1990)

There have even been claims of Kirlian photography being able to capture phantom limbs, for example, when a leaf is placed on the plate and then torn in half and photographed, the whole leaf shows up in the picture. This is not due to paranormal forces, however, but to residues left from the initial impression made by the whole leaf.

Parapsychologist Thelma Moss popularized Kirlian photography as a diagnostic medical tool with her books *The Body Electric* (1979) and *The Probability of the Impossible* (1983). She was convinced that the Kirlian process was an open door to the bioenergy of the **astral body**. Moss came to UCLA in midlife and earned a doctorate in psychology. She experimented with and praised the effects of LSD and was in and out of therapy for a variety of psychological problems, but managed to overcome her personal travails and become a professor at UCLA's Neuropsychiatric Institute. Her studies focused on paranormal topics, such as auras, **ghosts**, and **levitation**. One of her favorite subjects at UCLA was Uri **Geller**, whom she photographed several times.

Moss paved the way for other parapsychologists to speculate that Kirlian photography was parapsychology's Rosetta stone. They would now be able to understand such things as **acupuncture**, **chi**, **orgone energy**, and **telepathy**, as well as diagnose and cure whatever ails us. For example, Bio-Electrography claims to be

> a method of investigation for biological objects, based on the interpretation of the corona-discharge image obtained during exposure to a high-frequency, high-voltage electromagnetic field which is recorded either on photopaper or by modern video recording equipment. Its main use is as a fast, inexpensive and relatively non-invasive means for the diagnostic evaluation of physiological and psychological states. [www.psy.aau.dk/bioelec]

The reliability of diagnosing illnesses by photographing auras is not very high, yet in addition to Bio-Electrography, there is Esogetic Colorpuncture, Peter Mandel's therapy, which unites acupuncture and Kirlian photography to detect "imbalances in **energy** flow."

Kirlian methods of diagnosis should not be confused with other types of medical photography, for example, roentgen-ray computed tomography, magnetic resonance imaging, single photon/positron emission computed tomography, and other useful types of medical imaging. These have nothing to do with auras.

Further reading: Abell 1981; Randi 1982a; Watkins and Bickel 1991.

koro (shook yang)

A psychological disorder characterized by delusions of penis shrinkage and retraction into the body, accompanied by panic and fear of dying. This delusion is rooted in Chinese **metaphysics** and cultural practices. The disorder is associated with the belief that unhealthy or abnormal sexual acts (such as sex with prostitutes, masturbation, or even nocturnal emissions) disturb the **yin and yang** equilibrium that allegedly exists when a husband has sex with his wife, that is, during "normal intercourse." This disturbance of metaphysical harmony (loss of yang) manifests itself in penis shrinkage. Yang is the vital essence of the male and when inappropriately expelled, it is believed, the result is a potentially fatal case of koro.

Koro is also thought to be transmitted through food. In 1967, there was a case of koro hysteria in Singapore after newspapers reported cases of koro due to eating pork that came from a pig that had been inoculated against swine fever.

Further reading: Bartholomew 1998; Rubin 1982.

L

Landmark Forum (the Forum)

A **Large Group Awareness Training** program in which up to 150 people take a seminar together aimed at helping them realize their true potential.

Landmark Forum began in 1985, introduced by those who had purchased the **est** "technology" from Werner Erhard. In 1991 the group changed its name to Landmark Education Corporation (LEC), which continues to offer the Landmark Forum training, along with several other programs emphasizing communication and productivity. Erhard's brother, Harry Rosenberg, heads LEC, which does some $50 million a year in business and has attracted some 300,000 participants. LEC is headquartered in San Francisco, as was est, and has 42 offices in 11 countries. According to LEC, Erhard is not involved in its operation.

LEC aims to help people transform their lives by teaching them specific communication and life skills along with some heavy philosophical training. The advertised goals of LEC seem very grand and very vague. The programs are hailed as "original, innovative and effective." They "allow participants to produce extraordinary and even miraculous results, and provide a useful, practical new freedom which brings a quality of effectiveness and plan to one's everyday life." Landmark is dedicated to "empowering people in generating unlimited possibilities and making a difference." Landmark claims to provide "limitless opportunities for growth and development for individuals, relationships, families, communities, businesses, institutions and society as a whole." Landmark is "successful" and "internationally recognized." It is "committed to generating extraordinary communication—powerful listening and committed speaking that results in self-expression and fulfillment." Landmark is "exciting, challenging and enjoyable." Landmark says it stands upon the tenets of "well-being, self-expression, accountability and integrity," which accounts for its "extraordinary customer, assistant and employee satisfaction." LEC wants to help you fulfill your entire human potential, your "capacity to create, generate, invent and design from nothing" (Landmark Education Charter).

Jill P. Capuzzo writes, "Other seminars may offer supportive hugs; this one hits you between the eyes." She also claims, "One of the most irritating aspects of The Forum is the hard sell to sign up future participants. Leaders encourage people to bring friends and family to [a] session to help celebrate their newfound love of life and invite them to enroll in the next available weekend." Capuzzo claims that 20 percent of the participants in her sessions brought visitors to the open session and nearly half the original participants signed up for an advanced course (*The Philadelphia Inquirer,* March 22, 1996).

Andy Testa, on the other hand, posted an account of his experience with Landmark Forum, in which he claims that he was hounded by recruiters who insisted that his resistance was proof he needed their help (www.religio.de/therapie/landmark/landmark.html#22).

Another participant describes his Forum experience as "the most powerful and dangerous experience in my life." He claims that he was so disoriented after the seminar that he couldn't work for three days. He claims that those "three days after my Forum were a living hell, unlike anything I had experienced in 21 years of formal training and six years of medical residency in New York City." However, this participant also said that he would do it again! (www.inlink.com/~dhchase/kirk.htm).

A Hare Krishna devotee advises fellow spiritual travelers to take the LEC training. He claims that the average person reports that Landmark seminars change lives for the better by bringing about improved effectiveness in relating to others, increased personal productivity, greater self-confidence, help in making good decisions by learning how to identify what's really important in life, and help in learning how to live a more satisfying life without making life more complicated. He thinks Landmark can help people achieve transcendental realization (vnn.org/world/9801/02-1450/index.html).

Some people claim to have had breakdowns after attending such programs as Landmark Forum (rickross.com/reference/landmark22.html). Yet some of those who seek out **cults** such as **Scientology** or self-help programs such as Landmark are troubled already. Some are deeply troubled and the training might send them over the edge. But whose fault is that? Such people might have gone to the movies and been pushed over the edge, like Heinrich Pommerenke, who started murdering women after seeing Cecil B. DeMille's film *The Ten Commandments* and becoming convinced that women are the root of all evil.

Y. Klar et al. reported in the *Journal of Consulting and Clinical Psychology* (58, no. 1 [1990]: 99–108):

> A study was conducted to assess the psychosocial characteristics of individuals who become involved in large group awareness training (LGAT) programs. Prospective participants in The Forum, which has been classified as an LGAT, were compared with nonparticipating peers and with available normative samples on measures of well-being, negative life events, social support, and philosophical orientation. Results revealed that prospective participants were significantly more distressed than peer and normative samples of community residents and had a higher level of impact of recent negative life events compared with peer (but not normative) samples.

People who are having problems, are dissatisfied, or feel unfulfilled often desire direction and are usually the kind of people who sign up for seminars that will help them. We would expect many people to have upswings and experience fewer problems, be more satisfied and fulfilled, and feel less lost after these periods of distress. It is predictable that many participants in self-growth programs will attribute their sense of improvement to the programs they've taken, but much of their reasoning may be **post hoc** (**regressive fallacy**). Furthermore, their *sense of improvement* might not be matched by improved behavior. Just because they *feel* they've benefited doesn't mean they have. Research has shown that the feelings of having benefited greatly from participation in an LGAT do not correspond to beneficial changes in behavior (Langone 1998).

While some Landmark participants may have had breakdowns after their training, it would be a mistake to infer that large numbers of emotionally unbalanced people are signing up for the program. "Based on psychic distress (symptoms) and impairment measures, those who sought out growth groups were not overly represented by those who were particularly disturbed or impaired in their lives" (Leibermann 1994).

In fairness, it is also post hoc reasoning to assume that very disturbed individuals who deteriorate rapidly after attending LGAT seminars do so because of their participation. Those in need of psychotherapy should not participate in LGAT programs, which may be too intense for the emotionally fragile. It is not without good reason that Landmark Forum requires prospective participants to sign a statement declaring that they are mentally and physically well. This gives notice that the program is not for the mentally or physically unstable. It also may protect Landmark from legal action should a client have a breakdown after attending the Forum, but there is no guarantee that such a signed statement would necessarily exculpate Landmark should it be charged with causing someone's mental or physical breakdown.

The training emphasizes not only how to communicate better but how to relate better to those around you, as it forces the participant to reflect on and examine his or her life. Rabbi Yisroel Persky claims, however, that the Forum just teaches "commonsense concepts cloaked in esoteric packaging" (Faltermayer 1998). Still, what is common sense to one person may seem like a golden insight to another. There must be something of substance to the content of the message (culled from the great minds, after all), but the importance of the messenger and the way the message is delivered cannot be overestimated. The messenger must be believable. He or she must appear sincere. He must exude confidence. She must know how to use her voice and body to get her message across. He must be a master of communication skills. She must have wit and humor. He must be a raconteur. She must not only talk the talk but appear to clearly walk the walk as well. He must do it with a large group and utilize the energy and enthusiasm of the group members to infect each other. If she is successful, the participants will leave charged up and ready to change their lives. Many will see positive effects immediately. In fact, many are so impressed that they want to share the experience with others. They become zealots and recruiters for the program. Part of their zealotry, however, derives from the intense pressure put on them to bring their friends and family into the program and to sign up for follow-up courses. The main marketing tool Landmark uses is high-pressure contact with participants, including phone calls that border on harassment, according to some participants. Some critics think that recruitment is the main goal of the program (Faltenmayer 1998).

See also **firewalking** and **neurolinguistic programming**.

Further reading: Ankerberg and Weldon 1996; Barry 1997; Lell 1997.

Large Group Awareness Training (LGAT)

Personal development programs in which dozens to hundreds of people are given several hours to several days of intense instruction aimed at helping participants to discover what is hindering them from achieving their full potential and living more satisfied lives. LGAT programs have also been developed for corporations and public agencies, where the focus is on improving management skills, conflict resolution, general institutional strengthening, and dealing with problem employees.

LGAT gurus claim to know how to help people become more creative, intelligent, healthy, and rich. They focus primarily on the role interpersonal communication plays in self-esteem and in defining our relationships with others.

LGAT gurus claim to know why their participants are not happy or why they are not living fulfilled lives. They assume the same things are hindering everyone and that one approach will suit all. Some LGAT gurus use public television and books as their vehicles. Others give seminars in hotel ballrooms. Some use infomercials and peddle books and tapes to the masses to get them on the path to self-realization and success.

The U.S. Army might think it takes a few years to become "all that you can be," but the gurus of self-help think it can be done in a few hours or days. These gurus might take a cookie cutter approach to self-help, but the founders of such programs as **est**, **Dianetics**, **Landmark Forum**, **neurolinguistic programming**, and Tony Robbins seminars, use their own distinct cookie cutters.

Though some advocate visualization, self-hypnosis, and other techniques for achieving self-realization, most LGAT programs focus on communication skills and the effect of language on thought and behavior. Those running the programs must excel in those skills. The trainers are motivators. They must use their powerful communication skills to persuade others to believe that they (the trainers) know something valuable about fulfilling one's potential; that the valuable knowledge can be transmitted to the participant in a short time; that the trainee can expect to reap tangible, even if subjective, benefits in a short time (such as improved relationships with others or feeling better about oneself); and that the trainee has experienced only a small taste of the wonderful pleasure and fulfillment that awaits those who sign up for advanced training. In short, the trainers are not just teachers; they are sellers. Their main job is to motivate participants to buy more services, that is, sign up for more courses. The fact that trainers are unlikely to do any follow-up on their trainees, except to try to persuade them to take more courses, indicates that their main interest is not in helping people lead more fulfilling lives. The trainers have a sales job to do. They are paid commissions for the number of people they recruit and train, not for the number of people they truly help. It is not in their interest to do follow-up studies on their trainees. It is in their interest to do follow-up recruiting calls.

A short amount of reflection should make it apparent that the gurus of personal development training are like those infomercial stars who promise to share with you their secrets on how to make millions of dollars by taking out classified ads or by buying repossessed properties. The real money is not in taking out classified ads or buying repossessions; otherwise, that is what the infomercial star would be doing instead of making infomercials. The real money is in selling the idea to others.

If the trainers who work for Tony Robbins or Landmark Forum could realize their true potential in a meaningful, lucrative way, would they take a sales commission job? Would they work for a guru for a relatively small sum of money, while investing a rather extensive amount of time in the hopes of some sort of breakthrough? No. If they want to reach their own true potential they must break away and start their own personal training program. Which is exactly what some of them do.

Personal training programs are likely to be successful, however, if only because the participants are strongly motivated toward self-improvement and the trainers force participants to reflect on themselves, their lives, and their relationships. Such motivation and reflection will result in either perceived insights or renewed effort to gain such insights. Being surrounded by many others in search of the Promised Land of self-fulfillment serves to energize participants and give them hope. Ultimately, the main product being sold by human potential gurus is *hope* itself. That in itself is not a bad thing. We all need hope. Without hope, there is no point in making plans for the future; without hope, there is no point in working on a relationship or setting goals. Insofar as participation in LGAT increases one's hope for finding one's way and for achieving one's goals, it is good. Sometimes, false hope may be better than no hope at all.

Since fear is a major obstacle to hope, the human potential trainers must help participants overcome those fears that hinder development. For example, there can be no hope of achieving a goal if the fear of failure is so strong that one avoids setting goals in order to avoid failing. Likewise, no troubled relationship can improve if one fears rejection by the other to such a degree that one will not even try to heal the wounds. One must overcome fear of failure, rejection, ridicule, humiliation, and so on if one is to have any hope of achieving a very meaningful existence as a human being. One is powerless to achieve anything if one is paralyzed by fear. Empowerment to achieve requires empowerment to overcome one's fears and thereby gives one hope. The most direct way to empower someone is to convince them that if what they most fear were to happen, not only would nothing be worse than it already is, but most likely things would be even better than they are. Another way is to convince people that their own beliefs are hindrances to success and that they can replace those beliefs at will.

No one knew this better than Leo Buscaglia, one of the more successful LGAT gurus of the 1970s and 1980s. He used books, lectures, and public television programs to promote the idea that the key to everything is *love*. He popularized notions that Nietzsche, Bertrand Russell, and B. F. Skinner had written about, for example, the psychological power of loving those you fear. "Love your enemies," he would say. "It'll kill them!" Your enemy doesn't have to be another person, however. Your own fears can be your enemies. Embrace your fears; it'll kill them. If your relationship fails, what is the worst that can happen? The relationship ends. You can dwell on it, crawl into yourself, withdraw, surrender. Or you can learn from it, grow, develop, be prepared for a better relationship in the future. It's up to you. As the Stoics said, know what's in your power and what is not. Don't try to change what is not in your power to change. You can't control what others say or do, but you can control your attitude, your emotional response, to what they say or do. In short, if you don't try, you can't succeed. If you try and fail, you can still succeed. It's up to you. It is up to the human potential guru or trainer to convince you of this.

LGAT programs have many satisfied customers, but there is little or no research done by the promoters of these programs to test causal claims that might establish some degree of effectiveness to their methods; to establish clear criteria for what counts as successful training; or to keep records of failures, those who feel taken advantage of or harmed by the programs.

Nevertheless, despite the lack of proof that these programs work the way their advocates claim, and despite the fact that many trainers are overly zealous in their recruitment of participants for seminars and advanced seminars, many participants feel they benefit greatly from such programs. However, research has shown that the feelings of having benefited greatly from participation in an LGAT do not correspond to beneficial changes in behavior (Langone 1998). Also, many of those who feel they have benefited do not understand that others may not benefit at all from such programs. To their healthy friends and family members, the zealot may appear to have been **brainwashed**. Their enthusiasm seems unnatural and disproportionate. Furthermore, if a person is unbalanced *before* taking the program, he or she may go beyond "breakthrough" into "breakdown."

Further reading: Barry 1997.

law of truly large numbers (coincidences)

With a large enough sample, many seemingly unlikely coincidences are *likely* to happen.

For example, the *New York Times* ran a story in 1990 about a woman who won the New Jersey lottery twice within four months calling her chances "1 in 17 trillion." However, statisticians Stephen Samuels and George McCabe of Purdue University calculated the odds of someone winning the lottery twice to be something like one in 30 for a 4-month period and better than even odds over a 7-year period. Why? Because players don't buy one ticket for each of two lotteries, they buy multiple tickets every week over an extended period of time (Diaconis and Mosteller 1996).

Some people find it surprising that there are more than 16 million others on the planet who share their birthday. At a typical football game with 50,000 fans, most fans are likely to share their birthday with about 135 others in attendance. (The notable exception will be those born on February 29. There will only be about 34 fans born on that day.) You may find it even more astounding that "in a random selection of twenty-three persons there is a 50 percent chance that at least two of them celebrate the same birthdate" (Martin 1998).

On the other hand, you might say that the odds of something happening are a million to one. Such odds might strike you as being so large as to rule out chance or coincidence. However, with over 6 billion people on earth, a million-to-one shot will occur frequently. Say the odds are a million to one that when a person has a dream of an airplane crash, there is an airplane crash the next day. With 6 billion people having an average of 250 dream themes each per night (Hines 1990: 50), there should be about 1.5 million people a day who have **dreams** that seem **clairvoyant**. The number is actually likely to be larger, since we tend to dream about things that legitimately concern or worry us, and the data of dreams are usually vague or ambiguous, allowing a wide range of events to count as fulfilling our dreams.

Finally, clusters of coincidences can seem designed, or the result of a preordained pattern, to someone who is very selective in his thinking, such as Uri **Geller**. After the anti-American terrorist attacks on September 11, 2001, Geller posted his thoughts on the number 11. He asked everyone to pray for 11 seconds for those in need. Why? He was convinced that there was a cryptic, **numerological** message in the events that occurred that day. In fact, he admits that he has had a long-term relationship with the number 11. He thinks that 11 "represents a positive connection and a gateway to the mysteries of the universe and beyond." But here is what he had to say about the terrorist attack and how 11 relates to this day of infamy:

- The date of the attack: 9/11: 9 + 1 + 1 = 11
- September 11 is the 254th day of the year: 2 + 5 + 4 = 11
- After September 11 there are 111 days left to the end of the year.
- 119 is the area code to Iraq/Iran. 1 + 1 + 9 = 11 (reverse the numbers and you have the date) [Note: this is not true. Iran's country code is 98 (9 + 8 = 17), Iraq's is 964 (9 + 6 + 4 = 19).]
- Twin Towers: standing side by side, looks like the number 11
- The first plane to hit the towers was Flight 11 by American Airlines or AA: A = 1st letter in alphabet so we have again 11:11
- State of New York: The 11th state added to the Union
- New York City: 11 letters
- The USS *Enterprise* is in the Gulf during the attack; its ship number is 65N: 6 + 5 = 11
- Afghanistan: 11 letters
- The Pentagon: 11 letters
- Ramzi Yousef: 11 letters (convicted of orchestrating the attack on the WTC in 1993)
- Flight 11 had 92 on board: 9 + 2 = 11

- Flight 77 had 65 on board: 6 + 5 = 11
- The number of stories is 11 (Remember that the zero is not a number)
- The house where the hijackers lived had the number 10001: again don't count the zeroes
- Names that have 11 letters: Air Force One, George W. Bush, Bill Clinton, Saudi Arabia, ww [worldwide?] terrorism, Colin Powell (U.S. Secretary of State), Mohamed Atta (the pilot who crashed into the World Trade Center)

Geller also claimed that "there will be more information coming in" and urged readers to e-mail him with more findings of this kind. Geller also wrote: "I would encourage everyone to send out this message to all your family, friends or business acquaintances and try to put this in the right perspective."

It is not too difficult to put this in the right perspective.

Accumulating more findings like these coincidences between the number 11 and other things should be easy, since there are countless items that can be made to relate in some fashion to the number 11 (or 12, 13, or just about any other number or word). For example, the country code for Pakistan is 92 (9 + 2 = 11), a Boeing 757 holds about 11,000 gallons of fuel and the length is 155 feet (1 + 5 + 5 = 11), Nostradamus and Billy Graham have 11 letters, any word with 8 letters just put "the" in front of it; any word with 9 letters, just put 'ww' or any 2-letter word in front of it, any word with 10 letters, just put "a" in front of it or add "s" to the end of it, and so on. It is especially easy to do this since there is no specific guide before we begin our hunt for amazing facts as to what will and what won't count as being relevant. We have a nearly boundless array of items that could count as hits. Unfortunately for Geller and others who are impressed by these hits, there is also an even larger array of items that could count as misses. Geller doesn't see them because he isn't looking for them.

If we start hunting for items that seem relevant but don't fit the pattern, we will soon see that there is nothing special about Geller's list or the number 11. Only by focusing on anything that we can fit to our belief and ignoring everything that doesn't fit (**confirmation bias**) can we make these coincidences seem meaningful.

A List of Meaningless Coincidences

- the planes hit a little before and a little after 9 A.M.
- there were four flights that crashed with 266 people in them; 2 + 6 + 6 = 14
- one plane was a 767 (7 + 6 + 7 = 20)
- the other a 757 (7 + 5 + 7 = 19)
- the 767 has 20,000 gal fuel capacity
- the 757 has a 124-foot wingspan (1 + 2 + 4 = 7)
- a 767 has a 156-foot wingspan; 1 + 5 + 6 = 12
- one tower was 1,362 feet high; 1 + 3 + 6 + 2 = 12
- the other was 1,368; 1 + 3 + 6 + 8 = 18
- the supports are 39 inches apart; 3 + 9 = 12
- the buildings were leased for $3.25 billion; 3 + 2 + 5 = 10
- the other flights were AA 77 (7 + 7 = 14) and UA 93 (9 + 3 = 12) and UA 175 (1 + 7 + 5 = 13)
- flight 11 had 81 passengers (8 + 1 = 9)
- flight 77 had 58 passengers (5 + 8 = 13), 4 crew members, and 2 pilots (total = 64)
- Boston has 6 letters
- Massachusetts has 13 letters
- Pennsylvania has 12 letters
- Washington, D.C., has 12 letters
- Los Angeles has 10 letters
- The number of hijackers was 19 (9 + 1 = 10)

- Names and words that don't have 11 letters in them: Osama bin Laden, Khalid Al-Midhar, Majed Moqed, Nawaq Alhamzi, Salem Alhamzi, Satam Al Suqami, Waleed M. Alshehri, Wail Alshehri, Abdulaziz Alomari, Dick Cheney, Laura Bush, Iraq, Iran, Pakistan, Muslim, Islam, Pentagon, World Trade Center, terrorism, jihad, Taliban, Koran, United States, anti-American, murder, fire, hell, stupid.

What Geller and other numerologists are doing is a game, a game played with numbers and with people's minds. Sometimes it is amusing. Sometimes it is pathetic.

Further reading: Martin 1998; Paulos 1990, 1996.

Lenz, Frederick (1950–1998)

See **Rama**.

levitation

Levitation is the act of ascending into the air and floating in apparent defiance of gravity. Spiritual masters or **fakirs** are often depicted as levitating. Some take the ability to levitate as a sign of blessedness. Others see levitation as a **conjurer's** trick. No one really levitates; they just appear to do so. Clever people can use illusion, "invisible" string, and magnets to make things appear to levitate.

There are people in **Transcendental Meditation** who will sit cross-legged and hop up and down on their butts, claiming that they are flying. Perhaps they are—for one-millionth of a second—one millimeter above the ground. They say they feel lighter than air and are quite proud of their butt-hopping achievements.

See also **Indian rope trick**.

Further reading: Keene 1997; Randi 1982a; Rawcliffe 1988.

ley lines

Alleged alignments of ancient sites or holy places, such as stone circles, standing stones, cairns, and churches. Interest in ley lines began with the publication in 1922 of *Early British Trackways* by Alfred Watkins (1855–1935), a self-taught amateur archaeologist and antiquarian. Based on the fact that on a map of Blackwardine, near Leominster, England, he could link a number of ancient landmarks by a series of straight lines, Watkins became convinced that he had discovered an ancient trade route. Interest in these alleged trade routes as sources of mystical **energy** has become very popular among New Agers in Great Britain. Watkins called the lines "ley" (from an Old English word for "grasslands") because several of the aligned sites are named after variants (e.g., lee, leigh, lea).

Today, ley lines have been adopted by New Age **occultists** everywhere as sources of power or energy, attracting not only curious New Agers, but also aliens in their **UFOs** and locals with their **dowsing** rods. These New Age occultists believe that there are certain sites on the earth that are filled with special "energy," such as Stonehenge, Mt. Everest, Uluru (Ayers Rock) in Australia, **Nazca** in Peru, the Great Pyramid at Giza, Sedona (Arizona), and Mutiny Bay (Washington). Advocates claim that the alleged energy is connected to changes in magnetic fields, a claim that has not been scientifically verified. Maps have been produced, however, with lines on them that allegedly mark off special energy spots on earth. For example, the Seattle Arts Commission gave $5,000 to a group of New Age dowsers, the Geo Group, to do a ley line map of Seattle. Photographs of the result, which looks like a defaced satellite photo of the Seattle area, can be purchased for $7 from the group. It proudly proclaims that the "project made Seattle the first city on Earth to balance and tune its ley-line system." Skeptical citizens have criticized the Arts Commission for funding a New Age, **pagan** sect, but the artwork continues to be displayed on a rotating basis in city-owned buildings within Seattle.

Citizens had every right to be skeptical. Here is what the Geo Group has to say about their project:

> The vision of the Seattle Ley-Line Project is to heal the Earth energies within the Seattle city limits by identifying ley-line power centers in Seattle, neutralizing negative energies and then amplifying the positive potential of the ley-line power centers. We believe the result will be a decrease in disease and anxiety, an increased sense of wholeness and well-being and the achievement of Seattle's potential as a center of power for good on Spaceship Earth.

Uluru (Ayers Rock), a place of high spiritual "energy," according to aborigines and New Agers.

The Geo Group's vision is little more than a profession of faith. It is reminiscent of the claim of **Transcendental Meditation** that group meditation could reduce local crime rates. The Geo Group's methods have been just as effective.

lie detector

See **polygraph**.

Loch Ness monster (Nessie)

An alleged prehistoric creature living in Loch Ness, a long, deep lake near Inverness, Scotland. Many sightings of the "monster" have been recorded, going back at least as far as St. Columba, the Irish monk who converted most of Scotland to Christianity in the sixth century. Columba apparently converted Nessie, too; for it is said that until he went out on the waters and soothed the beast, she had been a murderess.

The modern Nessie legend began in 1934 with Dr. Robert Kenneth Wilson, a London physician, who allegedly photographed a plesiosaur-like beast with a long neck emerging out of the murky waters. Plesiosaurs have been extinct for 65 million years and the photo was a fake, but it created quite a fuss. Before the photo, Loch Ness was the stuff of legend and myth. The locals knew the ancient history of the sea serpent. But people came to the lake more to relax than to go on expeditions looking for mythical beasts. Scientific experts were called in to examine the photo and they declared that it could be a plesiosaur. Later, there would be explorations by a submarine with high-tech sensing devices. There have been many more photographs and many **testimonials** of eyewitnesses attesting to the existence of Nessie. Today, we have a full-blown tourist industry generating millions of dol-

lars a year, complete with submarine rides and a multimedia tourist center. For those who can't go to Scotland, there is a web camera site (www.lochness.scotland.net/camera.cfm) for your viewing pleasure with cameras placed both above and below the water.

Could anyone look at all the photographs and testimonials and still dismiss Nessie as a figment of people's imagination? As just another case of **pareidolia**?

As noted above, the famous photo of Nessie as a relative of the long-extinct plesiosaurs was faked. David Martin, a zoologist, and Alastair Boyd were members of a scientific project to find Nessie. They are credited by the *London Sunday Telegraph* (March 12, 1994) as having dug up the story of the faked photo. Christian Spurling, who died in the fall of 1993, was said to have made a deathbed confession of his role in the prank. The fake photo was not taken by Dr. Wilson—his name was used to give the photo stature and integrity—but by Spurling's stepbrother, Ian Wetherell. Ian's father, Marmaduke ("Duke") Wetherell, had been hired by the *London Daily Mail* to find the monster. Wetherell was a filmmaker who described himself as a "big game hunter." What bigger game could there be than Nessie? Except that the big game was in fact a small model of a sea serpent made of plastic wood attached to a 14-inch toy submarine! Actually, the game did get big, as the little hoax created such a huge fuss that the pranksters decided that the best thing for them to do was to keep quiet.

Alastair Boyd, mentioned above as one of the researchers who uncovered the photo hoax, claims he made a genuine sighting of Nessie in 1979. His Nessie didn't look like a plesiosaur, though. More like a whale, he said. It was at least 20 feet long and he says he saw it roll around in the water. Although it is possible he misinterpreted what he saw, Boyd is convinced there are 20-foot-long creatures in the loch.

Since the Loch Ness monster story has been around for more than 1,500 years, if there is a monster it is not likely that it is the same monster seen by St. Columba. In short, there must be more than one monster. Some zoologists calculate that a minimum population of ten creatures would be needed to sustain the population. However, Loch Ness is incapable of sustaining a predator weighing more than about 300 kilograms (about 660 pounds; *The Naturalist,* winter 1993/94, reported in the *Daily Telegraph*). Adrian Shine, head of the Loch Ness Project, once said the monster could be a Baltic sturgeon, a primitive fish with a snout and spines that can grow up to 9 feet long and weigh around 450 pounds. This may sound like just another fish story to some, but there is scientific evidence that Nessie is, at best, a big fish in a big lake, or a big wake in a big lake. Shine, who has been studying the Loch Ness story for some 25 years, now thinks that what people see when they think they see the "monster" is actually an underwater wave. A similar view has been presented by Luigi Piccardi, an Italian geologist "who is convinced that seismic rumblings far below the famous Scottish lake cause the roiling waves, deep groans and explosive blasts that have for centuries led people to believe that a giant beast lurks below the loch's murky surface" (Squatriglia 2001).

The Naturalist reported on extensive studies of the lake's ecology that indicate that the lake is capable of supporting no more than 30 metric tons of fish. (The food chain of the lake is driven by bacteria, which break down vegetation, rather than algae, as in most lakes.) Estimating that a group of predators would weigh no more than 10 percent of the total weight of the fish available for them to consume, researchers arrived at the 300 kilogram statistic. It seems extremely odd that with all the sophisticated technology, the submarines, and the thousands of observers, after all these years we still don't have a single specimen. We don't have a carcass; we don't even have a bone to examine. With at least ten of these huge monsters swimming around in the lake at any given time, you'd think that there would be at least one unambiguous sighting by now. You would think so, that is, unless you want to keep the legend alive. Of course, there are good economic reasons for keeping the Loch Ness monster myth alive. It's good for tourism.

Besides the photo that Mr. Boyd and others have exposed as a fake, there are many other photos of Nessie to consider. Not all photos of Nessie are fakes. Some are genuine photos of the lake. These photos are always very gray and grainy, taken of murky waters with lots of shadows and outlines. Anyone who has traveled around Loch Ness will not be disappointed in the variety of forms that one will see when looking out on the waters. The lake is very long and often very turbulent. There is no question that in some of these photos there does appear to be a form that could be taken for a sea serpent. The form could also be taken for a log, a shadow on a wave, or a wave itself. Or it could be a huge animal that has lived for thousands of years without leaving any evidence of its existence except questionable photos and the stories of eyewitnesses.

Further reading: Bauer 1986; Binns 1985; Brugioni 1999; Ellis 1994; Razdan and Kielar 1986.

lucid dreaming

Dreaming while being aware that you are dreaming. Lucid dreaming advocates strive to control and guide their **dreams**. Some desire to avoid recurring nightmares. Others desire fun. Some New Age lucid dreamers, however, believe that lucid dreaming is essential for self-improvement and personal growth. Stephen LaBerge, Ph.D., for example, claims that lucid dreaming is

> a priceless treasure that belongs to each of us. This treasure, the ability to dream lucidly, gives us the opportunity to experience anything imaginable—to overcome limitations, fears, and nightmares, to explore our minds, to enjoy incredible adventure, and to discover transcendent consciousness.
>
> With lucidity comes an astonishing, exhilarating feeling of freedom—the knowledge that you can do anything, unbound by any laws of physics or society. One of the first joys of many lucid dreamers is flying: soaring like a bird, freed from the restraints of gravity. From there, people can go on to discover the vast power of lucid dreaming for transforming their lives. [www.lucidity.com]

If you need help with your lucid dreaming, you can purchase books, tapes, scientific publications, and induction devices, such as the DreamLight, the DreamSpeaker, or the NovaDreamer, from LaBerge's Lucidity Institute. For $2,000 you can attend a seminar at a beautiful tropical resort where you can learn all the latest techniques to help you tap into your **unconscious mind**, an absolute necessity for living the good life. For a small additional charge, you can even get two nursing continuing education credits through the Institute of Transpersonal Psychology.

Why Dr. LaBerge doesn't just advocate *daydreaming* to do all this wonderful transcendent discovery is explained by Frederik van Eeden in *A Study of Dreams* (1913). When we're awake, we are logical and feel restricted by conventional social rules and oppressive laws of nature. Our imaginations would be too repressed by our waking consciousness to allow us to let go and dream of such things as flying with **spirits**.

For some, the main goal of lucid dreaming is to have lucid dreams that are indistinguishable from **out-of-body experiences**. Flying free from the restraints of gravity in one's dreams takes some people out of their bodies to hover and watch themselves dreaming lucidly.

Some skeptics do not believe that there is such a state as lucid dreaming (Malcolm 1959). Skeptics don't deny that sometimes in our dreams we dream that we are aware that we are dreaming. What they deny is that there is special dream state called the "lucid state." The lucid dream is therefore not a gateway to "transcendent consciousness" any more than nightmares are.

Self-awareness resides in the prefrontal cortex, which shows reduced activity during sleep for most people most of the time. This reduced activity may well be why we can dream of the most bizarre things without being aware of how bizarre they are until we wake up and remember them. Perhaps lucid dreaming is possible for some people because their frontal lobes don't rest during sleep.

Further reading: Blackmore 1991a; LaBerge 1990; Malcolm 1959.

lunar effects

The full moon has been linked to crime, suicide, mental illness, disasters, accidents, birthrates, fertility, and **werewolves**, among other things. Some people even buy and sell stocks according to phases of the moon, a method probably as successful as many others. Numerous studies have tried to find lunar effects. So far, the stud-

Full moon, April 19, 1999. Lick Observatory.

ies have failed to establish much of interest. Lunar effects that have been found have little or nothing to do with human behavior, for example, the discovery of a slight effect of the moon on global temperature, which in turn might have an effect on the growth of plants.

Ivan Kelly, James Rotton, and Roger Culver (1996) examined over 100 studies on lunar effects and concluded that the studies have failed to show a reliable and significant correlation (i.e., one not likely due to chance) between the full moon or any other phase of the moon and each of the following: the homicide rate, traffic accidents, crisis calls to police or fire stations, domestic violence, births of babies, suicide, major disasters, casino payout rates, assassinations, kidnappings, aggression by professional hockey players, violence in prisons, psychiatric admissions, agitated behavior by nursing home residents, assaults, gunshot wounds, stabbings, emergency room admissions, behavioral outbursts of psychologically

challenged rural adults, **lycanthropy**, **vampirism**, alcoholism, sleep walking, or epilepsy.

If so many studies have failed to prove a significant correlation between the full moon and anything of interest, why do so many people believe in these lunar myths? Kelly et al. suspect four factors: media effects, folklore and tradition, misconceptions, and cognitive biases. A fifth factor should be considered, as well: **communal reinforcement**.

Lunar myths are frequently presented in films and works of fiction. "With the constant media repetition of an association between the full moon and human behavior it is not surprising that such beliefs are widespread in the general public" (ibid.). Reporters also "favor those who claim that the full moon influences behavior." It wouldn't be much of a story if the moon was full and nothing happened. Anecdotal evidence for lunar effects is not hard to find and reporters lap it up, even though such evidence is unreliable for establishing significant correlations. Relying on personal experience ignores the possibility of **self-deception** and **confirmation bias**. Such evidence may be unreliable, but it is nonetheless persuasive to the uncritical mind.

Many lunar myths are rooted in folklore. For example, an ancient Assyrian/Babylonian fragment stated, "A woman is fertile according to the moon." Such notions have been turned into widespread misconceptions about fertility and birthrates. For example, Eugen Jonas, a Slovakian psychiatrist, was inspired by this bit of folklore to create a method of birth control and fertility largely rooted in **astrological** superstitions. The belief that there are more births during a full moon persists today among many educated people. Scientific studies, however, have failed to find any significant correlation between the full moon and number of births (Kelly and Martens 1994; Martens et al. 1988). In 1991, Benski and Gerin reported that they had analyzed birthdays of 4,256 babies born in a clinic in France and "found them equally distributed throughout the synodic (phase) lunar cycle" (Kelly et al. 1996: 19). In 1994, Italian researchers Periti and Biagiotti reported on their study of 7,842 spontaneous deliveries over a 5-year period at a clinic in Florence. They found "no relationship between moon phase and number of spontaneous deliveries" (ibid.: 19).

Despite the fact that there is no evidence of a significant correlation between phases of the moon and fertility, not only do some people maintain that there is, they have a "scientific" explanation for the nonexistent correlation. According to "Angela" of AstraConceptions at fertilityrhythms.com,

> photic (light) signals sent by the lens and retina of the eyes are converted into hormone signals by the pineal gland. It is the pineal gland which signals the onset of puberty in humans and plays a part in the fertility rhythms of all species.
>
> In animals which reproduce seasonally, it is the changing light patterns which trigger the fertility cycle. The gradual change in both the length of day and the changing angle of the sun in the sky (caused by earth's motion) is interpreted by the pineal gland as a signal to commence the fertility season.
>
> Of course, humans do not reproduce seasonally. Our fertility cycles exhibit an obvious monthly rhythm. The light source which has a monthly periodicity is, of course, the Moon.
>
> It is interesting to note that menstruation is actually a shedding process. Just as the average menstrual cycle is 28 days in length, the human body sheds a layer of skin approximately every 28 days.

Yes, that is very interesting to note—if you are interested in **sympathetic magic** and aren't bothered that approximations aren't equals. Angela continues:

> it is not only the changing day length but also the changing angular position of the sun which triggers this process; the pineal gland receives photic (light) impressions and converts these into hormonal messages which signal the onset of these cycles.
>
> With humans the cycles of fertility (and shedding) are triggered by photic impressions as well. Yet our cycles have a monthly periodicity which is obviously synchronized with fluctuations of the lunar light.

Obviously. However, the light of the moon is a very minor source of light in most women's lives, and is no more likely than the moon's gravitational force to have a significant effect on a woman's ovulation. Furthermore, the average menstrual cycle is 28 days but varies from woman to woman and month to month, while the length of the lunar month is a consistent 29.53 days. Some have noticed that these cycles are not identical. Furthermore, it would seem odd that natural selection would favor a method of reproduction for a species like ours that depended on the weather. Clouds are bound to be irregularly and frequently blocking moonlight, which would seem to hinder rather than enhance our species' chances of survival.

Some mythmakers believe that long ago women all bled in sync with the moon, but civilization and indoor electric lighting (or even the discovery of fire by primitive humans) have disturbed their rhythmic cycles. This theory may seem plausible until one remembers that there are quite a few other mammals on the planet that have not been affected by firelight or civilization's indoor lighting and whose cycles aren't in harmony with the moon. In short, given the large number of types of mammals on our planet, one would expect that by chance some species' estrus and menstrual cycles would harmonize with lunar cycles (e.g., the lemur). It is doubtful that there is anything of metaphysical significance in this.

What we do know is that there has been very little research on hormonal or neurochemical changes during lunar phases. James Rotton's search of the literature "failed to uncover any studies linking lunar cycles to substances that have been implicated as possible correlates of stress and aggression (e.g., serotonin, melatonin, epinephrine, norepinephrine, testosterone, cortisol, vasopressin [directly relevant to fluid content], growth hormone, pH, 17-OHCS, adrenocrotropic hormone)" (Rotton 1997). One would think that this area would be well studied, since hormones and neurochemicals are known to affect menstruation and behavior.

Misconceptions about things such as the moon's effect on tides have contributed to lunar mythology. Many people seem to think that since the moon affects the ocean's tides, it must be so powerful that it affects the human body as well. The lunar force is actually a very weak tidal force. A mother holding her child "will exert 12 million times as much tidal force on her child as the moon" (Kelly et al. 1996: 25). Astronomer George O. Abell claims that the moon's gravitational pull is less than that of a mosquito (Abell 1979). Despite these physical facts, there is still widespread belief that the moon can cause earthquakes. It doesn't; nor does the sun, which exerts much less tidal force on the earth than the moon.

The fact that the human body is mostly water largely contributes to the notion that the moon should have a powerful effect on the human body and therefore an effect on behavior. It is claimed by many that the earth and the human body

both are 80% water. This is false. Eighty percent of the *surface* of the earth is water. Furthermore, the moon only affects *unbounded* bodies of water, while the water in the human body is bounded.

Also, the tidal force of the moon on the earth depends on its distance from earth, not its phase. Whereas the synodic period is 29.53 days, it takes 27.5 days for the moon to move in its elliptical orbit from perigee to perigee (or apogee to apogee). Perigee (when the moon is closest to earth) "can occur at any phase of the synodic cycle" (Kelly et al. 1990: 989). Higher tides do occur at new and full moons, but not because the moon's gravitational pull is stronger at those times. Rather, the tides are higher then because "the sun, earth, and moon are in a line and the tidal force of the sun joins that of the moon at those times to produce higher tides" (ibid.: 989).

Many of the misconceptions about the moon's gravitational effect on the tides, as well as several other lunar misconceptions, seem to have been generated by Arnold Lieber in *The Lunar Effect* (1978), republished in 1996 as *How the Moon Affects You*. In *The Lunar Effect,* Leiber incorrectly predicted a catastrophic earthquake would hit California in 1982 due to the coincidental alignment of the moon and planets. Undeterred by the fact that no such earthquake had occurred, Leiber did not admit his error in the later book. In fact, he repeated his belief about the dangers of planet alignments and wrote that they "may trigger another great California earthquake." This time he didn't predict when.

Many believe in lunar myths because they have heard them repeated many times by members of the mass media, police officers, nurses, doctors, social workers, and other people with influence. Once many people believe something and enjoy a significant amount of **communal**

reinforcement, they get very selective about the type of data they pay attention to in the future. If one believes that during a full moon there is an increase in accidents, one will notice when accidents occur during a full moon, but be inattentive to the moon when accidents occur at other times. If something strange happens and there is a full moon at the time, a causal connection will be assumed. If something strange happens and there is no full moon, no connection is made, but the event is not seen as counterevidence to the belief in full moon causality. Memories get selective, and perhaps even distorted, to favor a full moon hypothesis. A tendency to do this over time strengthens one's belief in the relationship between the full moon and a host of unrelated effects.

Probably the most widely believed myth about the full moon is that it is associated with madness. However, in examining over 100 studies, Kelly et al. found that "phases of the moon accounted for no more than 3/100 of 1 percent of the variability in activities usually termed lunacy" (1996: 18). According to James Rotton, "such a small percentage is too close to zero to be of any theoretical, practical, or statistical interest or significance" (Rotton 1997).

Finally, the notion that there is a lunar influence on suicide is also unsubstantiated. Martin et al. (1992) reviewed numerous studies done over nearly three decades and found no significant association between phases of the moon and suicide deaths, attempted suicides, or suicide threats. In 1997, Gutiérrez-García and Tusell studied 897 suicide deaths in Madrid and found "no significant relationship between the synodic cycle and the suicide rate" (p. 248). These studies, like others that have failed to find anything interesting happening during the full moon, have gone largely unreported in the press.

Further reading: Abell 1986; Byrnes and Kelly 1992; Hines 1990; Jamison 1999; Plait 2002.

lycanthropy

The delusional belief that one has turned into an animal, especially a **werewolf**. In Europe during the Middle Ages, lycanthropy was commonly believed to occur due to **witchcraft** or magic. One modern theory is that the rye bread of the poor was often contaminated with the fungus ergot, which caused hallucinations and delusions.

Stories of humans turning into animals such as tigers, swans, or monkeys are widespread and seem to occur in all cultures, indicating shared human fears (e.g., fear of the wildest local beast), desires (e.g., wishing for powers such as great strength or the power of flight), or common brain disorders.

Further reading: Eisler 1951; Noll 1992.

Lysenkoism

Lysenkoism is an episode in Russian science featuring a nonscientific peasant plant breeder named Trofim Denisovich Lysenko (1898–1976). Lysenko was the leading proponent of Michurinism during the Lenin/Stalin years. I. V. Michurin, in turn, was a proponent of Lamarckism. Lamarck was an 18th-century French scientist who argued for a theory of evolution long before Darwin. Evolutionary scientists, however, have rejected Lamarck's theory, because it is not nearly as powerful an explanation of evolution as natural selection.

According to Lamarck, evolution occurs because organisms can inherit traits that have been acquired by their ancestors. For example, giraffes find themselves in a changing environment in which they can survive only by eating leaves high up on trees. So, they stretch their necks to reach the leaves, and this stretching and the desire to stretch gets passed on to later generations. As a result, a species of animal that originally had short necks evolved into a species with long necks.

Natural selection explains the long necks of the giraffes as a result of the workings of nature that allowed the species to feed on high leaves rather than feed on the ground, as do short-legged, short-necked animals. There was no purposive behavior that was a response to the environment that was then passed on to later generations. There was simply an environment that included trees with leaves up high, which was a favorable food source to long-legged, long-necked animals such as the giraffe. In fact, according to natural selection, if that were the only food source available, only animals with long necks, or animals that can climb or fly, would survive. All others would become extinct. There is no plan here, divine or otherwise, according to natural selection. Furthermore, there is nothing special signified by the fact that a species has survived. Survival of the fittest means only that those who have survived were fit to survive. It doesn't mean that those who survive are morally superior to those species that don't. They've survived because they were fit to adapt to their environment, for example, they had long necks when there was a good supply of food readily available high up in the trees and there were no other catastrophic disadvantages to their height. If a species got so tall that it became impossible to mate, it would become extinct. If the only food source on high happened to have a substance in it which rendered giraffes sterile, there would be no more giraffes, no matter how hard they tried to will themselves potent.

Lamarckism is favored by those who see *will* as the primary driving force of life (**vitalism**). Darwinism, or natural selection, is hated by many of those who believe that God created everything and that everything has a purpose. One might think that Marxists would prefer Darwinism with its mechanical, materialistic, deterministic, nonpurposive concept of natural selection. Lamarckism looks like it might be preferred by free market advocates with their emphasis on will, effort, hard work, and choice. In any case, Michurin's views on evolution, echoing Lamarck's, found favor with the party leadership in the Soviet Union. When the rest of the scientific world pursued the ideas of Mendel and developed the new science of genetics, Russia led the way in the effort to prevent the new science from being developed in the Soviet Union. Thus, while the rest of the scientific world could not conceive of understanding evolution without genetics, the Soviet Union used its political power to make sure that none of their scientists would advocate a genetic role in evolution.

Lysenko rose to dominance at a 1948 conference in Russia where he delivered a passionate address denouncing Mendelian thought as "reactionary and decadent" and declared such thinkers to be "enemies of the Soviet people" (Gardner 1957). He also announced that the Central Committee of the Communist Party had approved his speech. Scientists either groveled, writing public letters confessing the errors of their way and the righteousness of the wisdom of the Party, or they were dismissed. Some were sent to labor camps. Some were never heard from again.

Under Lysenko, **science** was practiced in the service of the state, or more precisely, in the service of ideology. The results were predictable: the steady deterioration of Soviet biology. The Soviet scientific community did not condemn

Lysenko's methods until 1965, more than a decade after Stalin's death.

Could something similar happen in the United States? Some might argue that it already has. First, there is the **creationist** movement that has had some success in preventing evolution from being taught in public schools. With creationist Duane Gish leading the way, who knows what would happen if Pat Robertson became President of the United States and Jerry Falwell his secretary of education? Then, of course, there are several well-known and well-financed scientists in America who also seem to be doing science in the name of ideology, not the ideology of fundamentalist Christianity but the ideology of racial superiority. Lysenko was opposed to the use of statistics, but had he been clever enough to see how useful statistics can be in the service of ideology, he might have changed his mind. Had he seen what J. Philippe Rushton, Arthur Jensen, Richard Lynn, Richard Herrnstein, or Charles Murray have done with statistical data to support an ideology of racial superiority, Lysenko might have created a department of Supreme Soviet Statistics and proven with the magic of numbers the superiority of Lamarckism to natural selection and genetics.

Further reading: Gardner 1957; Levins and Lewontin 1985.

M

macrobiotics

A way of life characterized by a special diet said to optimize the balance of **yin and yang**. George Ohsawa (1893–1966) started the macrobiotics movement with the pub-

lication of his *Zen Macrobiotics* in 1965. Michio Kushi popularized the movement in the United States. Sagen Ishizuka, a 19th-century Japanese army doctor, established the basics of the diet, which consists mainly of whole grains, vegetables, and beans.

Ishizuka claimed that foods have characteristics of yin or yang, and that a proper diet balances each. Ohsawa makes claims such as schizophrenia is a yin disease and one who is so afflicted should drink yang fluids. Kushi makes claims such as cancer "is the body's own defense mechanism to protect itself against long-term dietary and environmental abuse." How he knows this is a mystery. There is no reputable evidence that a macrobiotic diet is beneficial for cancer patients. The only reports of efficacy are **testimonials** by patients, many of whom received traditional medical treatment, according to the American Cancer Society.

If a macrobiotic diet is healthy, it is by accident, since foods are selected not for their physical or nutritional qualities, but for their **metaphysical** properties. It is possible that many people, like its founders, improve when on the diet not because of what they eat but because of what they quit eating, such as processed foods, meat, milk, and other animal products. All assignment of metaphysical properties to foods is arbitrary and seems to be based on **sympathetic magic**.

Further reading: Barrett and Jarvis 1993; Raso 1993.

magick

The alleged art and science of causing change in accordance with the *will* by nonphysical means. Magick is associated with all kinds of **paranormal** and **occult** phenomena, including but not limited to: **astral projection**, **chakras**, **ESP**, and

psychic healing. Magick uses various symbols, such as the **pentagram**, as well as a variety of symbolic ritual behaviors aimed at achieving powers that allow one to contravene the laws of physics, chemistry, and so on. Magick should not be confused with magic, which is the art of **conjuring** and legerdemain.

The religions based on the Old and New Testaments have long associated magick with false prophets, based on the belief that **Satan** regularly shares his powers with humans. Using powers that contravene natural forces is good if done by or through God (white magick or **miracles**), according to this view. If done by diabolical forces, it is evil (black magick).

The idea of being able to control things such as the weather or one's health by an act of will is very appealing. So is the idea of being able to wreak havoc on one's enemies without having to lift a finger: Just think it and thy will will be done. Stories of people with special powers are appealing, but for those contemplating becoming a magus, consider this warning from an authority on the subject:

> [M]agick ritual (or any magick or occultism) is very dangerous for the mentally unstable. If you should somehow "get out too far," eat "heavy foods" . . . and use your religious background or old belief system for support. But remember too, that weird experiences are not necessarily bad experiences. (Phil Hansford, *Ceremonial Magick:* www.ecauldron.com/cmagick00 .php)

On the other hand, weird experiences are not necessarily good, either.

The magic of performing magicians is related to magick in that performers use tricks and deception to make audiences think they have done things that, if real, would require supernatural or paranormal powers, for example, materializing objects such as rings or ashes, doves or rabbits, or bending spoons with one's mind.

See also **Wicca** and **witches**.

Further reading: Carus 1996; Randi 1989b; Sagan 1995.

magnet therapy

A type of **alternative health practice** that claims that magnetic fields have healing powers. Some claim that magnets can help broken bones heal faster, but most of the advocacy comes from those who claim that magnets relieve pain. Most of the support for these notions is in the form of **testimonials** and anecdotes, and can be attributed to "**placebo effects** and other effects accompanying their use" (Livingston 1998). There is almost no scientific evidence supporting magnet therapy. One highly publicized exception is a double-blind study done at Baylor College of Medicine that compared the effects of magnets and sham magnets on the knee pain of 50 post-polio patients. The experimental group reported a significantly greater reduction in pain than the control group. No replication of the study has yet been done.

A less publicized study at the New York College of Podiatric Medicine found that magnets did not have any effect on healing heel pain. Over a 4-week period, 19 patients wore a molded insole containing a magnetic foil, while 15 patients wore the same type of insole with no magnetic foil. In this randomized controlled trial, 60% of patients in both groups, reported improvement.

Despite the fact that there has been virtually no scientific testing of magnet therapy, a growing industry is producing magnetic bracelets, bands, insoles, back braces, mattresses, and so on and claiming miraculous powers for their products. The magnet market may be approaching $150 million annually (Collie 1999) or it may be much more (Lerner 2001). Magnets are becoming the gimmick of choice of **chiropractors** and other pain specialists.

The claim that magnets help circulate blood is a common one among supporters of magnet therapy, but there is no scientific evidence that magnets do anything to the blood. Even though the evidence is lacking that magnets have anything other than a placebo effect, theories abound as to how they work. Some say magnets are like a shiatsu massage; some claim magnets affect the iron in red blood cells; still others claim that magnets create an alkaline reaction in the body (Collie 1999). Bill Roper, head of Magnetherapy, claims, "Magnets don't cure or heal anything. All they do is set your body back to normal so the healing process can begin" (ibid.). How he knows this is not clear.

Some supporters of magnetic therapy seem to base their belief on a **metaphysical** assumption that all illness is due to some sort of imbalance or disharmony in **energy**. The balance or flow of electromagnetic energy must be restored to restore health, and magnets are thought to be able to do this.

The most rabid advocates of magnet therapy are athletes such as Jim Colbert and John Huston (golfers), Dan Marino (football), and Lindsay Davenport (tennis). Their beliefs are based on little more than **post hoc reasoning**. It is possible that the relief a magnetic belt gives to a golfer with a back problem, however, is not simply a function of the placebo effect or the **regressive fallacy**. It may well be due to the support or added heat the belt provides. The product might work just as well without the magnets. However, athletes are not given to scientific testing any more than are the manufacturers of magnetic gimmickry.

Athletes aren't the only ones enam-ored of the power of magnets to heal. Dr. Richard Rogachefsky, an orthopedic surgeon at the University of Miami, claims to have used magnets on about 600 patients, including people who have been shot. He says that the magnets "accelerate the healing process." His evidence? He can tell by looking at x-rays. Dr. William Jarvis is skeptical: "Any doctor who relies on clinical impressions, on what they think they see, is a fool" (ibid.). There is a good reason scientists do **controlled group studies** to test causal efficacy: to prevent **self-deception**.

While sales of magnetic products keep rising, there are few scientific studies of magnet therapy going on. The University of Virginia is testing magnets on sufferers of fibromyalgia. The universities of Miami and Kentucky are testing magnets on people with carpal tunnel syndrome (ibid.). At present, however, we have no good reason to believe that magnets have any more healing power than **crystals** or copper bracelets.

Further reading: B Beyerstein 1997; Kasler 1998; Livingston 1997, 1998.

mahasamadhi

A god-illumined master's conscious exit from the body at the time of physical death, according to Hinduism.

See also **incorruptible bodies**.

Malachy O'Morgair, Bishop (St. Malachy; 1094–1148)

Malachy was born in Armagh (in what is now Northern Ireland) and is believed by many to have had the gift of **prophecy**. He predicted British oppression for the Irish (good call) and conversion of the English back to Catholicism (bad call, but maybe next millennium). According to the Abbé Cucherat, Malachy had strange visions of the future, including a list of the popes until the end of time.

Some Roman Catholics think Malachy has predicted that Armageddon is just around the corner and that after John Paul II there will be only two more popes before the end of the world. Of course Malachy didn't name the popes by name—otherwise we'd all be believers in his prophetic skills. He gave them descriptive names. John Paul II, the current pope, is number 110 and he was christened De Labore Solis, ("from the labor of the sun"). Those who have the gifts of interpretation and **shoehorning** tell us that this is an accurate prophecy because John Paul II's father was a laborer and he has traveled around the earth (like the sun? Well, remember, the prediction was pre-Copernicus). Some think the name refers to the fact that there was a total eclipse of the sun when John Paul II was born.

Malachy's prophecies are said to have been locked away for 400 years before they were allegedly discovered in 1590 in the Roman Archives. Arnold de Wyon first published them. The debate has raged ever since among certain Catholics as to whether they are forgeries or genuine predictions of St. Malachy (www.newadvent.org/cathen/12473a.htm#malachy).

manifesting

Allegedly a way for the average person, without need of **paranormal** or divine powers, to do **magick** and perform **miracles**. All one needs is the will to exercise one's magic on the universe. "Manifesting is the art of creating what you want at the time that you want it," says John Payne (a.k.a. Omni, a "being of light" **channeled** by Mr. Payne).

Manifesting is an eclectic hodgepodge of CYOR (create your own reality), visual-

ization techniques, positive thinking, goal setting, self-analysis, **selective thinking**, and **post hoc** reasoning, supported by **anecdotes**. The purpose of manifesting is to get what you want by actively making your dreams come true, rather than passively waiting for someone to fulfill them. For example, Anne Marie Evers recommends "affirmation" as the best way to manifest one's desires. She has written a book titled *Affirmations: Your Passport to Happiness*. She writes:

> What Is an Affirmation: An Affirmation is a declaration of acceptance used to fill oneself with an abundance of freedom, prosperity and peace. An Affirmation is the vehicle of the manifestation of your desires. Affirmations are powerful, positive statements of belief recited consistently out loud and sent out into the Universe. The spoken word drives thoughts and images deep into both our conscious and subconscious minds. Slowly, firmly, concentrate on each word, phrase and the idea behind it. We know repetition is the Mother of Learning.

According to Ms. Evers, the first step to getting what you want is to "prepare the soil of your subconscious mind by forgiving everyone and everything that has *ever* hurt you, then forgive yourself." This may seem to be a bit too dramatic if all one wants to do is, say, fix a broken garage door. Jeannine, for example, didn't seem too concerned about forgiveness when her garage door was broken. She followed the advice of self-proclaimed expert manifesters Fred Fengler and Todd Varnum, authors of *Manifesting Your Heart's Desire*:

> I remembered reading your book and decided to manifest a fix. I started talking to the door and asking it to work. I . . . used to talk to plants and they tended to

grow better so I talked to the door. After a few minutes of communicating with the door I pushed the button and the door worked perfectly.

Fengler and Varnum give other examples of successful manifesters. For example, an anonymous writer told them how he or she sold a business:

> I decided to manifest using my will power. As I went to sleep, I said out loud, "OK universe, this is what I want. I want an offer. I want a good offer. In fact I want TWO offers. In fact I want them TOMORROW!"
>
> The next day was perfectly normal. I "reminded" the universe it was 4 PM and the office would close at 5:30. I felt confident that the universe would take care of me no matter what happened. Within ten minutes, I had a call from one prospect who said he had an offer and would be right over. Ten minutes after he left the offer off, I got a call from my business consultant. He told me that a second offer was being written and it would be on my desk in 24 hours, which it was.
>
> I accepted the first offer, and we flawlessly closed the deal in less than two weeks.

That's all there is to it. You let the universe know what you want and you'll get it! This should be good news to those superstitious folks who try to sell real estate by burying a statue of St. Joseph on the property. There is an easier way: manifesting!

Varnum explains that by asserting yourself to the universe you express extra **energy** in your emotion. The universe listens to people with extra energy as long as one has no fear and is willing to accept whatever the universe hands out. Varnum's caveat is akin to the warnings of faith healers who tell those who can't get rid of their cancer by faith that they don't

really have **faith**. If the universe fails to give you what you demand, it is because your desire is not coming from the right place. If you get what you desire, then your desire came from the right place.

Manifesting is another New Age technique that denies there is any such thing as **coincidence**. For example, Fengler and Varnum, in recommending a book on manifesting, write:

> Some people call it luck or coincidence—or just plain magic. It is the gift of being in the right place at the right time, of having opportunity fall into your lap. But what if you could create your own luck, make "coincidences" happen, even bring a few miracles into your daily life? Drawing on over twenty years of teaching the art of manifestation, David Spangler shows you how to do just that. Called a "strikingly new, spiritually aware approach to personal power and the fulfillment of your dreams," this new book [*Everyday Miracles: The Inner Art of Manifestation*] is a complete rewriting and updating of David's classic book, *Manifestation*.

Fengler and Varnum's book is hailed by the authors as a three-year "study," but it is little more than a collection of **testimonials** from a group of people who met regularly to learn a variety of manifesting techniques.

One of the more popular manifesting techniques is *visualization*. One of Fengler and Varnum's anecdotes involves a girl who was having trouble learning to ride a horse. She visualized riding the way her instructor told her to ride and at the next lesson she was riding well. Visualization seems quite different, however, from talking to your garage door or vocalizing your wishes so the whole universe can hear them. Yet the practices share much in common. Visualization is mental practice. It is a way to boost confidence. It requires clarifying goals. All of these can help a person who is trying to accomplish some physical feat, such as riding a horse or hitting a golf ball. But no amount of visualization will create reality. A golfer can visualize hitting a hole-in-one from now until doomsday without it ever happening. There are some people who believe they can fight cancer by visualizing little cellular warriors killing off cancer cells. The likelihood of such visualization creating the reality desired is near zero. You might as well visualize yourself flying or being in six places at once. If anyone could fix a flat tire by visualization, he or she would be collecting $1 million from James **Randi**. But manifesters don't need Randi; they can get $1 million just by visualizing it or letting the universe know that's what they want.

Despite the obvious falsity of many of the claims of manifesters, some of the techniques they recommend are quite good. For example, if you do not specify a goal, but merely express some vague wish like "someday I'm going to go to New Zealand," then you probably won't ever get your wish. But if you specify your desire, insist on having it satisfied, clearly identify the obstacles in the way of having it satisfied, determine what is needed to have your will be done, and create a plan for achieving your goal, then you have a very good chance of getting what you want. On the other hand, a lot of manifesting seems to be little more than refusing to accept coincidence as a fact of life, peppered with a lot of post hoc reasoning and selective thinking.

One good thing about manifesting is that it could take a person's attention away from the many bad things in life over which we have no control. By focusing on what you want, you may not dwell so much on the bad hand life has dealt you. By specifying your goals, you will be more likely to see troubles as obstacles to be overcome rather than as hindrances blocking your chance of success.

On the other hand, it could also be very depressing to think that the only reason you are not getting what you want is because your desires are not coming from the right place.

Marfa lights

Visible from a viewing area about 10 miles east of the town of Marfa, Texas, Marfa lights are the main tourist attraction in the area. The lights are said to bounce around in the sky, vanish and reappear, and thus are considered a mystery by some. To others, the lights are not a mystery. They are **ghosts**, swamp gas, radioactive bursts, **ball lightning**, or navigational lights for space aliens.

Skeptics who view the lights with strong binoculars claim that they are nothing more than the headlights and taillights of cars in the Chinati Mountains on U.S. highway 67.

Mars effect

The name given to Michel Gauquelin's "astrobiological" claim that when Mars is in certain sectors of the sky, great athletes are born in numbers indicative of a non-chance correlation. If this were true, **astrologers** believe it would provide support for their theories that heavenly bodies actively influence who and what we become. However, not only does correlation not prove causality, but such correlations are notoriously slippery. They are ambiguous (who counts as a "great" athlete?), and significant correlations between variables that are *not* significantly related are *expected* to occur occasionally.

In any case, what Gauquelin claims about Mars and athletes isn't true, according to a study by seven French scientists. They took a sample of 1,066 French athletes and compared them with 85,280 others for birth times, dates, and location of Mars at birth. The study didn't support the Mars effect.

Gauquelin also claimed to have found a significant correlation between Jupiter and military prowess, and between Venus and artists.

See also **law of truly large numbers**.

Further reading: Benski et al. 1996; Dean et al. 1996; Martin 1989; Nienhuys 1997; Ruscio 1998.

massage therapy

A massage is the rubbing or kneading of parts of the body to aid circulation, to stimulate nerves, or to relax the muscles. Massage therapy is often a massage plus a **metaphysical** explanation about some sort of **energy** or "structure" being balanced, unblocked, transferred, harmonized, tuned up, or aligned.

A massage is usually relaxing and usually feels good. Most of us, however, could not explain the physical and physiological mechanisms causing the relaxation and pleasure. Most of us probably suspect it has something to do with the enjoyment of being touched by another person, and with the physical movement of muscles and other body parts.

Massage therapists often claim to understand the metaphysical reasons for the uplifting and relaxing effect of massage. Their explanations vary. Karen Khamashta uses Ortho-Bionomy, Polarity therapy, and **reflexology**.

Ortho-Bionomy works by contacting the body's "trigger points." According to this theory, when a trigger point is contacted, you "immediately relieve pain and restore the body's natural balance and rhythm" (*The Davis Enterprise*, January 10, 1993).

Reflexology works by allegedly unblocking the 7,200 nerve endings in each foot so that they can respond to all of the glands, organs, and other parts of the body and improve the blood supply as well. This supposedly helps the body reach a "balanced state."

Polarity therapy is based on "balancing the life energy that moves through every part of the body . . . and . . . moves in currents, or channels within and around the body." Polarity therapy "attempts to eliminate blockages in these channels which can cause imbalance and illness." The theory is that "if the body's currents are balanced, then the person relaxes and is able to heal more efficiently" (ibid.).

Massage therapist Christy Freidrich says, "A lot of what I do is to try to help people with their structural balance. Over a period of time, people end up learning more about structure and how it works" (ibid.).

Massage therapy sounds as if it has as its goal something similar to **therapeutic touch**: restoring harmony and balance to one's life energy. But the massage therapist uses "palpation for assessment of . . . energy blockages," while the therapeutic touch practitioner allegedly manipulates the energy in your **aura**.

Massage therapists who are certified by the National Certification Board for Therapeutic Massage and Bodywork (NCBTMB) must take 500 hours of education classes and pass an examination. They must know some basic anatomy and physiology, as well as some first aid. Despite the emphasis on balancing energy, none of the practice questions provided by the NCBTMB involve metaphysics.

The American Massage Therapy Association (AMTA) claims:

Research shows [massage] reduces the heart rate, lowers blood pressure, increases blood circulation and lymph flow, relaxes muscles, improves range of motion, and increases endorphins, the body's natural painkillers. Therapeutic massage enhances medical treatment and helps people feel less anxious and stressed, relaxed yet more alert.

They don't mention who did the research and where one might verify these claims. Nor do they mention that these effects are likely to be temporary or that similar results might be achieved by meditating, walking, swimming, having sex, or reading a good book—not necessarily in that order.

The AMTA also claims that therapeutic massage "can help with"

allergies, anxiety, arthritis (both osteoarthritis and rheumatoid arthritis), asthma and bronchitis, carpal tunnel syndrome, chronic and temporary pain, circulatory problems, depression, digestive disorders (including spastic colon, constipation, and diarrhea), headache (especially when due to muscle tension), insomnia, myofascial pain (a condition of the tissue connecting the muscles), reduced range of motion, sinusitis, sports injuries (including pulled or strained muscles and sprained liga ments), stress, and temporomandibular joint dysfunction (TMJ).

Something that "can help with" so many disorders and dysfunctions should be very popular. According to the AMTA, Americans spend from $2 billion to $4 billion per year on massage therapy. However, "can help with" is an empty claim, and those with serious medical problems such as cardiac problems, depression, or sinusitis would do well to consult a physician.

Since massage therapy is essentially an unregulated profession, making claims that massage therapists are qualified to treat medical conditions such as allergies, infectious diseases, or phlebitis seems like **quackery**. This has not stopped the profession from expanding to the point where even dogs and horses can get a healing massage. One of the more popular animal therapies is "Tellington Touch," the creation of **animal quacker** Linda Tellington-Jones, who offers **holistic** treatment for pets.

So far no studies have been done to determine whether massage *therapists* are less likely than massage *parlors* to receive visits from the vice squad.

Further reading: Barrett and Jarvis 1993.

medium

In **spiritualism**, a medium is one with whom **spirits** communicate directly. In an earlier, simpler, but more dramatic age, a good medium would produce voices or **apports**, ring bells, float or move things across a darkened room, produce **automatic writing** or **ectoplasm**, and, in short, provide good entertainment value for the money.

Today, a medium is likely to write trite inspirational books and say he or she is **channeling**, such as J. Z. Knight and the *White Book* (1999) of her **Ramtha** from **Atlantis**. Today's most successful mediums simply claim the dead communicate through them. Under a thin guise of doing "spiritual healing" and "grief counseling," they use traditional techniques of **cold reading** and surreptitiously gathered information about their subjects to give the appearance of transmitting comforting messages from the dead. Using information they have gathered during the cold reading or in conversations with the subjects before the readings or during breaks from studio sessions, they claim to hear messages from the dead loved ones of those responding. The medium then passes on messages from the dead such as "he forgives you." In the good old days of **séances** and elaborate trickery, a spiritualist fraud would be more likely to pass on the message "give more money to me and my group" (Keene 1997).

Today, it is unnecessary to be so crude as to directly ask for money or to prey on elderly persons who have lots of cash and little time left. People are literally waiting for years to give money to those who give hope that a dead loved one will communicate with them via the medium. There is also a lucrative book business for those who have messages from the dead, and there is good money to be made by doing live shows for hundreds or thousands of people, each of whom pays $25 to $50 for the chance to connect with a lost child, spouse, or parent. George Anderson, a former switchboard operator and author of *Lessons from the Light: Extraordinary Messages of Comfort and Hope from the Other Side* (2000), got his own ABC special featuring celebrities who wanted to contact the dead. Some mediums even get their own syndicated television programs, such as John Edward and James Van Praagh, although the latter's show was canned by Tribune Media Services after only a few episodes.

Further reading: Hyman 1977; Nickell and Fischer 1991; Posner 1998; Randi 1982a, Rowland 2002.

memory

The retention of, and ability to recall, information, experiences, feelings, and procedures (skills and habits).

There is no universally agreed-on model of the mind/brain, and no universally agreed-on model of how memory works. Nevertheless, a good model for how memory works must be consistent with the subjective nature of consciousness and with what is known from scientific studies (Schacter 1996). Subjectivity in remembering involves at least three important factors:

1. Memories are constructions made in accordance with present needs, desires, influences, and so on.
2. Memories are often accompanied by feelings and emotions (Damasio 1999).
3. Memory usually involves awareness of the memory (Schacter 1996).

Two popular models of thinking are the behaviorist model (thinking is a set of behaviors) and the cognitive psychology model (the brain is like a computer). Neither can account for the subjective and present-need basis of memory (ibid.). The Freudian model posits an area of the **unconscious** where memories of traumatic experiences are stored. Though we are unconscious of them, such memories are claimed to be significant causal factors in shaping conscious thought and behavior. This model is not consistent with what is known about the memory of traumatic experiences. There is a great deal of supportive evidence for the claim that the more traumatic an experience, the more *likely* one is to remember it. Novel visual images, which would frequently accompany traumas, stimulate the hippocampus and left inferior prefrontal cavity and generally become part of long-term memory.

Current studies in neuroscience strongly support the notion that a memory is a set of encoded neural connections that are likely to go across various parts of the brain. The stronger the connections, the stronger the memory. Recollection of an event can occur by a stimulus to any parts of the brain where a neural connection for the memory occurs. If part of the brain is damaged, access to any neural data that were there is lost. On the other hand, if the brain is healthy and a person is fully conscious when experiencing some trauma, the likelihood that they will forget the event is nearly zero, unless either they are very young or they experience a brain injury.

Furthermore, the Freudian model often assumes that childhood sexual abuse is *unconsciously* repressed and that psychological problems in adulthood are caused by the unconscious memory of childhood abuse. There is, however, no body of scientific evidence to support the view that such abuse is *unconsciously* repressed.

Finally, the model of memory that sees the brain recording everything one experiences is a model that contradicts what is known about how memories are constructed. Even so, in a survey of psychologists by Loftus and Loftus, 84% said they believe every experience is permanently stored in the mind (Schacter 1996: 76).

One of the most popular models sees memory as a present act of consciousness, reconstructive of the past, stimulated by an analogue of an engram called the *retrieval cue*. The engram is the neural network representing fragments of past experiences that have been encoded. The evidence strongly indicates that there are distinct types and elements of memory that involve different parts of the brain, for example, the hippocampus and ongoing incidents of day-to-day living (short term or working memory), or the amygdala and emotional memories (ibid.: 213). Memories might better be thought of as a collage or a jigsaw puzzle than as tape recordings, pictures, or video clips stored as wholes. With this model, perceptual or conscious experience does not record all sense data experienced. Most sense data are not stored at all. What is stored are bits and fragments of experience that are encoded in engrams. Exactly how they are encoded is not completely understood.

This popular model of memory rejects the idea that individual memories are stored in distinct locations in the brain. That idea seems to have become solidified by Wilder Graves Penfield's experiments done in the 1950s. He placed electrodes on the surface of the exposed temporal lobes of patients and was able to elicit memories in 40 of 520 patients. Many psychologists (and lay people) refer to these experiments as proof that memories are just waiting for the right stimulus to be evoked. Schacter points out that the Penfield experiments are not very good evidence for this belief. Not only could Penfield elicit memories only in about one out of every thirteen patients, he did not provide support for the claim that what was elicited was actually a memory and not a hallucination, fantasy, or **confabulation**.

On the model described in the previous two paragraphs, forgetting is due to one of the following:

1. weak encoding (why we forget most things, including our nightly dreams)
2. lack of a retrieval cue (we seem to need something to stimulate memory)
3. time, and the replacement in the neural network by later experiences (how many experiences do you remember from many, many years ago?)
4. repetitive experiences (you'll remember the one special meal you had at a special restaurant, but you won't remember what you had for lunch a year ago Tuesday), or
5. a drive to keep us sane. (Imagine the brain overload that would occur if we were to never forget anything, the stated goal of L. Ron Hubbard in his book *Dianetics*, whose notion of engrams, by the way, is not the one scientists refer to. His followers should read the Jorge Luis Borges story

"Funes, the Memorious," a tale about a being who can forget nothing.)

The chances of remembering something improve by *consolidation,* which creates strong encoding. Thinking and talking about an experience enhance the chances of remembering it. One of the better-known techniques of remembering involves the process of association.

Source Memory. Many people have vivid and substantially accurate memories of events that are erroneous in one key aspect: the *source* of the memory. For example:

> In the 1980 presidential campaign, Ronald Reagan repeatedly told a heartbreaking story of a World War II bomber pilot who ordered his crew to bail out after his plane had been seriously damaged by an enemy hit. His young belly gunner was wounded so seriously that he was unable to evacuate the bomber. Reagan could barely hold back his tears as he uttered the pilot's heroic response: "Never mind. We'll ride it down together." . . . [T]his story was an almost exact duplicate of a scene in the 1944 film *A Wing and a Prayer.* Reagan had apparently retained the facts but forgotten their source. (Schacter 1996: 287)

An even more dramatic case of source amnesia (also called *memory misattribution*) is that of the woman who accused memory expert Dr. Donald Thompson of being the man who had raped her. Thompson was doing a live interview for a television program just before the rape occurred. The woman had seen the program and "apparently confused her memory of him from the television screen with her memory of the rapist" (ibid.: 114). Studies by Marcia Johnson et al. have shown that the ability to distinguish memory from imagination depends on the recall of source information.

Tom Kessinger, a mechanic at Elliott's Body Shop in Junction City, Kansas, gave a detailed description of two men who had rented a Ryder truck like the one used in the Oklahoma City bombing of the Alfred P. Murrah Federal Building on April 19, 1995. One looked just like Timothy McVeigh. The other wore a baseball cap and a T-shirt, and had a tattoo above the elbow on his left arm. The latter was Todd Bunting, who had rented a truck the day before McVeigh had rented his truck. Kessinger mixed the two memories, but was absolutely certain the two came in together.

Amnesia and Implicit Memory. Though all forgetting is a type of amnesia, we usually reserve that term for forgetting that is caused by the effects of drugs or alcohol, brain injuries, or physical or psychological traumas. One of the more interesting types of amnesia is what psychiatrists call the *fugue state.* An otherwise healthy person travels a good distance from his home, and when found has no memory of how he got there or who he is. The fugue state is usually attributed to recent emotional trauma. It is rare and is typically neither permanent nor recurring.

Limited amnesia, however, is quite common. Limited amnesia occurs in people who suffer a severe physical or psychological trauma. Football players who suffer concussions and accident victims who are rendered unconscious typically do not remember what happened immediately before the event. The scientific evidence indicates, however, that some sort of *implicit* memory may exist, which can be troubling to one whose amnesia is due to having been rendered unconscious by an assailant.

Implicit memory is memory without awareness. It differs substantially from **repressed memory**. Implicit memories are not necessarily repressed, nor are they necessarily the result of trauma. They are weakly encoded memories that can affect conscious thought and behavior. Retrieval cues do not bring about a complete memory of some events because parts, even most, of the event were not encoded.

Daniel Schacter and Endel Tulving introduced the terms "implicit memory" and "explicit memory" in their attempt to find a common language for those who believe there are several distinct memory systems and those who maintain there is only one such system. Schacter writes:

> The nonconscious world of implicit memory revealed by cognitive neuroscience differs markedly from the Freudian unconscious. In Freud's vision, unconscious memories are dynamic entities embroiled in a fight against the forces of repression; they result from special experiences that relate to our deepest conflicts and desires. . . . [I]mplicit memories . . . arise as a natural consequence of such everyday activities as perceiving, understanding, and acting. (ibid.: 190–191)

Perception is mostly a filtering and defragmenting process. Our interests and needs affect perception, but most of what is available to us as potential sense data will never be processed. And most of what is processed will be forgotten. Amnesia is not rare, but is the standard condition of the human species. We do not forget simply to avoid being reminded of unpleasant things. We forget either because we did not perceive closely in the first place or we did not encode the experience either in the parietal lobes of the cortical surface (for short-term or working memory) or in the prefrontal lobe (for long-term memory).

Semantic, Procedural, and Episodic Memory. Memory researchers distinguish several types of memory systems. *Semantic memory* contains conceptual and factual knowledge. *Procedural memory* allows us to learn new skills and acquire habits. *Episodic memory* allows us to recall personal incidents that uniquely define our lives (ibid.: 17). Another important distinction is that

between field and observer memory. *Field memories* are those where one sees oneself in the scene. *Observer memories* are those seen through one's own eyes. The fact that many memories are field memories is evidence, as Freud noted, of the reconstructive nature of memories (ibid.: 21).

How accurate and reliable is memory? Studies on memory have shown that we often construct our memories after the fact, that we are susceptible to suggestions from others that help us fill in the gaps in our memories. That is why, for example, a police officer investigating a crime should not show a picture of a single individual to a victim and ask if the victim recognizes the assailant. If the victim is then presented with a lineup and picks out the individual whose picture the victim had been shown, there is no way of knowing whether the victim is remembering the assailant or the picture.

Another interesting fact about memory is that studies have shown that there is no significant correlation between the *subjective feeling of certainty* about a memory and its *accuracy*. Also, contrary to what many people believe, **hypnosis** does not aid memory's accuracy. Because subjects are extremely suggestible while hypnotized, most states do not allow as evidence in a court of law testimony made while under hypnosis (Loftus 1979).

Furthermore, it is possible to create **false memories** in people's minds by suggestion, even false memories of **past lives**. Memory is so malleable that we should be very cautious in claiming certainty about any given memory without corroborative evidence.

See also Bridey **Murphy**, **reincarnation**, and **repressed memory therapy**.

Further reading: Ashcraft 1994; Baddeley 1998; Baker 1996a; Kandel and Schwartz 2000; Loftus 1980; Loftus and Ketcham 1991; Pinker 1997; Schacter 1997, 2001.

Men in Black (MIB)

Aliens or government agents, who visit **UFO** witnesses and warn them not to tell anyone about their UFO experiences.

John Keel in *The Mothman Prophecies* (1975) talks about the MIB, as did Gray Barker in *They Knew Too Much About Flying Saucers* (1956). The MIB are said to favor older model dark Cadillacs. They don't like to be photographed, though they have been reported to dress in black. Their mystique may soon wear off, however, as films, a TV series, a magazine, and a video game have been developed around their strange antics.

John Sherwood (a.k.a. Dr. Richard H. Pratt) has come clean about the role Gray Barker, head of Saucerian Publications, played in perpetuating the myth of the MIB. *"They Knew Too Much About Flying Saucers* made the Men in Black feared within UFO circles during the late 1950s and 1960s," claims Sherwood, but the book was impure fiction, written purely to make money with little concern for facts. It was Barker who published a 17-year-old Sherwood's *Flying Saucers Are Watching You,* which the author now claims was his "only corrupt journalistic experience"; that is, he wrote with little concern for the difference between fact and fiction. Sherwood admits that he encouraged the hoax by playing along when UFOers tagged him as having been silenced by the MIB.

Further reading: Rojecewicz 1987; Sherwood 1998.

mentalist

A performer who uses trickery and deception to create the illusion of having **paranormal** or supernatural powers.

Mentalists and **psychics** rely upon **selective thinking**. For example, James Randi tells the story of Peter Hurkos, who was astonishing people with his ability to recite intimate details about their homes and their lives. Two of the persons who had their minds read by Hurkos, and who were amazed at his accuracy, were invited by Randi to watch a tape of the mind readings. It was "discovered by *actual count* that this so-called psychic had, on the average, been correct in *one out of fourteen* of his statements. . . . Selective thinking had led them to dismiss all the apparent misses and the obviously wrong guesses and remember only the 'hits.' They were believers who *needed* this man to be the genuine article, and in spite of the results of this experiment they are still devoted fans of this charlatan" (Randi 1982a: 7).

mesmerism

A bit of medical **quackery** developed in the 18th century by Dr. Franz Anton Mesmer. It involves some social role-playing, with the mesmerizer making suggestions and his clients becoming absolutely mesmerized by him. Mesmer used his extraordinary powers of suggestion to send people into frenzied convulsions or sleeplike trances. He was so successful that to this day we use his name to describe the exercise of such powers over others.

In the early 1770s, Mesmer, a Viennese physician who got his doctorate on the basis of a plagiarized dissertation on how the planets affect health, met Maximillian Hell, a Viennese Jesuit and healer. The rest, as they say, is history. Father Hell cured people with a magnetic steel plate. Hell's "proof" of magnetic healing was that it worked; that is, he had a lot of satisfied customers. Mesmer adopted Hell's magnetic therapy and posited that it works because there is a very subtle magnetic fluid flowing through everything, which sometimes gets disturbed and needs to be restored to its proper flow. Hell, Mesmer theorized, was unblocking the flow of this magnetic fluid with his magnets. Mesmer eventually discovered that he got the same results *without* the magnets. Rather than attribute this to the **placebo effect** or admit that he wasn't really curing anybody, he posited that "animal magnetism" accounted for his ability to correct the flow of the universal magnetic fluid. (Today, the term "animal magnetism" means mesmerism or hypnotic power but also describes the attraction of men and women to one another).

Mesmer also discovered that even though he didn't need magnets to get results, the dramatic effect of waving a magnetized pole over a person, or having his subjects sit in magnetized water or hold magnetized poles, and so on, while he moved around in brightly colored robes

Depiction of mesmerism, from Ebenezer Sibly (1750–1800), *Key to Physic and the Occult Sciences* (1810).

playing the scientific faith healer, made for better drama and for larger audiences. He was able to evoke from a number of his clients entertaining behaviors ranging from sleeping to dancing to having convulsions. Mesmer did basically what today's **hypnotists** do in the showroom and the clinic, and what faith healers do in the circus tents and churches, only he did them together, making a great show out of his magnetic cures. With the help of Louis XVI and Marie Antoinette, Mesmer set up a Magnetic Institute where he had his patients do such things as sit with their feet in a fountain of magnetized water while holding cables attached to magnetized trees. He was later denounced as a fraud by the French medical establishment and by a commission that included Benjamin Franklin.

See also **exorcism** and **multiple personality disorder**.

Further reading: Baker 1990; Randi 1995; Spanos 1996; Spanos and Chaves 1989.

metaphysics

Metaphysics is a branch of philosophy consisting of ontology and cosmology. In the "weak" sense, metaphysics is used loosely to refer to New Age and nonempirical notions such as **energy** (**chi**, **prana**) being balanced, harmonized, tuned, aligned, unblocked, and so on. Although "metaphysics" in the weak sense is the most common usage in *The Skeptic's Dictionary*, here we are concerned with "metaphysics" in the strong sense.

The term "metaphysics" is often used to entail ideas and theories as to what kinds of beings are real, the nature of those beings, and the concepts used to think about those beings. For example, a theory of mind would be a metaphysical theory concerned with mental phenomena and related concepts such as perception, idea, consciousness, **memory**, intention, motive and reasoning.

However, typically, "metaphysics" refers to broad theories of reality, such as materialism and **dualism**, and to broad issues regarding the nature of reality. Why is there something rather than nothing? Is there free will or is every action determined by causes? Was the universe created or has it always existed? Are there spiritual beings, or **souls**? Is there life after death? What is the nature of the universe, of substance, of causality?

Most philosophers would agree that metaphysical claims are not **scientific** and that inconsistent or contradictory metaphysical positions cannot be tested empirically to determine which is false. For example, materialism and dualism are inconsistent with each other, but both theories are coherent and consistent with experience, and there is no empirical event that could falsify either theory.

Philosophers give various reasons for preferring one metaphysical belief to another. One thinks one's own theory is more coherent than a rival theory, or that one's own belief has more explanatory power or requires fewer assumptions. Some argue that their metaphysical beliefs fit better with what is known from other disciplines such as science, history, or psychology. Some criticize rival theories for being too farfetched: possible but implausible.

Some defend their metaphysical beliefs by appealing to the consequences of belief; for example, it gives hope for an afterlife or meaning to existence. Others maintain that such considerations are irrelevant to the truth of the claims, and indicate the belief is based more on desire than on good logic.

Since coherent metaphysical beliefs cannot be refuted, it is sometimes maintained that philosophers adhere to their metaphysical theories more out of personal disposition and temperament than evidence and proof.

Some consider metaphysics to represent what is highest in human nature, the drive to know and understand the nature of the universe and our place in it. Others consider metaphysics, specifically speculative metaphysics about nonempirical and transcendent realities, to be, more or less, bunk. Perhaps Kant was correct when he said that although we can never hope to answer our metaphysical questions, we cannot help asking them anyway.

Further reading: Gale 2002.

metempsychosis

The belief that at death the **soul** passes into another human or animal body.

See also **reincarnation**.

metoposcopy

The interpretation of facial wrinkles, especially those on the forehead, to determine the character of a person. Metoposcopy is also used as a type of **divination** in conjunction with **astrology**. Metoposcopy was invented by the 16th-century mathematician, physician, and astrologer Girolamo Cardano (1501–1576). Legend has it that Cardano starved himself to death at the age of 75 rather than live and run the risk of falsifying his horoscope and thereby discredit his beloved astrology.

The drawing reproduced here is from Cardano's *Metoposcopia* and shows the position of the planets on the wrinkles of the forehead. Cardano's science of forehead reading did not catch on, unlike the typhus fever of which he gave the first clinical description.

In all, Cardano worked up about 800 facial figures, each associated with astrological signs and qualities of temperament and character. He declared that one could

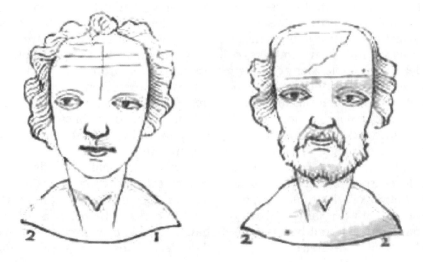

Metoposcopy illustrations, from Girolamo Cardano (1501–1576), *Metoposcopia* (Paris, 1658).

tell by the lines on her face which woman is an adulteress and which has a hatred of any lewdness. Long, straight furrows indicate nobility of character. He claimed to be able to tell the generous from the trickster by their distinct lines and noted that having three curved furrows on the forehead proves one is a dissolute simpleton. The strongest feature of metoposcopy is that it is noninvasive. Its weakest feature is that it has no scientific merit.

See also **personology**, **phrenology**, and **physiognomy**.

mind control (brainwashing)

The control of the thoughts and actions of another without his or her consent. Generally, the term implies that the victim has given up some basic political, social, or religious beliefs and attitudes, and has been made to accept contrasting ideas. "Brainwashing" is often used loosely to refer to being persuaded by propaganda.

There are many misconceptions about mind control. Some people consider mind control to include the efforts of parents to raise their children according to social, cultural, moral, and personal standards. Some think it is mind control to use behavior modification techniques to change one's own behavior, whether by self-discipline and autosuggestion, or through workshops and clinics. Others think that advertising and sexual seduction are examples of mind control. Still others consider it mind control to give debilitating drugs to a woman in order to take advantage of her.

Some of the tactics of recruiters for religious, spiritual, or New Age human potential groups are called mind control tactics. Many believe that terrorist kidnap victims who become sympathetic with their kidnapper's ideology are victims of mind control (the so-called Stockholm syndrome). Similarly, women who stay with abusive men are often seen as victims of mind control. Many consider **subliminal** messaging in Muzak, in advertising, or on self-help tapes to be a form of mind control. Many believe that it is mind control to use laser weapons, isotropic radiators, infrasound, nonnuclear electromagnetic pulse generators, or high-power microwave emitters to confuse or debilitate people. Many consider the alleged creating of **zombies** in voodoo as mind control. The "brainwashing" tactics (torture, sensory deprivation, etc.) of the Chinese during the Korean War are often cited as the paradigm of mind control. Finally, no one would doubt that it would be a clear case of mind control to be able to **hypnotize** or electronically program a person so that he or she would carry out one's commands without being aware that he or she is being controlled.

A term with such slack in its denotation is nearly useless. The denotation of "mind control" should not include activities where a person *freely chooses* to engage in the behavior. Nor should it include cases where fear or force is used to manipulate or coerce people into doing some action. Inquisitions do not succeed in capturing the minds of their victims. As soon as the threat of punishment is lifted, the extorted beliefs vanish. You do not control the mind of someone who will escape from you the moment you turn your back.

Also, it should be obvious that to render a woman helpless by drugs so you can rape her is not mind control. Using a frequency generator to give people headaches or to disorient them is not the same as controlling them. An essential component of mind control is that it involves *controlling another person,* not just putting them *out of control* or doing things to them over which they have *no control.*

Some of the more popular misconceptions of mind control originated in fiction, such as the film *The Manchurian Candidate* (1962). In that film, an assassin is programmed so that he will respond to a posthypnotic trigger, commit a murder,

and not remember it later. Other books and films portray hypnosis as a powerful tool, allowing the hypnotist to have his sexual way with beautiful women or to program her to become a robotic courier, assassin, and so on. One such book even claims to be based on a true story: *The Control of Candy Jones* (Playboy Press, 1976) by Donald Bain. To be able to use hypnosis in this powerful way is little more than **wishful thinking**.

Other fictional fantasies have been created that show that drugs or electronic devices, including brain implants, can be used to control the behavior of people. It has, of course, been established that brain damage, hypnosis, drugs, or electric stimulation to the brain or neural network can have a causal effect on thought, bodily movement, and behavior. However, the state of human knowledge on the effects of various chemical or electrical stimulation to the brain is so impoverished that it would be impossible using today's knowledge and technology to do anything approaching the kind of mind control accomplished in fantasy. It is certainly conceivable that some day we may be able to build a device that, if implanted in the brain, would allow us to control thoughts and actions by controlling specific chemical or electrical stimuli. Such a device does not now exist nor could it exist given today's state of knowledge in the neurosciences. (However, two Emory University neuroscientists, Dr. Roy Bakay and Dr. Philip Kennedy, have developed an electronic brain implant that can be activated by thoughts and in turn can move a computer cursor. Their goal is to help paralyzed patients move limbs or prosthetic devices, but military minds are probably drooling at the thought of other uses for this technology.)

There also seems to be a growing belief that the U.S. government, through its military branches or agencies such as the CIA, is using a number of horrible devices aimed at disrupting the brain. It is known that government agencies have experimented on humans in mind control studies with and without the knowledge of their subjects (Scheflin and Opton 1978). The claims of those who believe they have been unwilling victims of mind control experiments should not be dismissed as impossible or even as improbable. Given past practice and the amoral nature of our military and intelligence agencies, such experiments are not implausible. Nevertheless, it is a near certainty that our government is not capable of controlling anyone's mind—except through the usual methods of propaganda and censorship—though it is clear that many people in many governments lust after such power.

Some of the claims made by those who believe they are being controlled by electronic weapons do not seem plausible. For example, the belief that radio waves or microwaves can be used to cause a person to hear voices transmitted to him seems unlikely. We know that waves of all kinds of frequencies are constantly going through our bodies. The reason we have to turn on the radio or television to hear the sounds or see the pictures being transmitted through the air is because those devices have receivers that "translate" the waves into forms we can hear and see. What we know about hearing and vision makes it very unlikely that simply sending a signal to the brain that can be "translated" into sounds or pictures would cause a person to hear or see anything. Someday it may be possible to stimulate electronically or chemically a specific network of neurons to cause specific sounds or sights of the experimenter's choosing to emerge in a person's consciousness. But this is not possible today. Even if it were possible, it would not necessarily follow that a person

would obey a command, say, to assassinate the president just because he heard a voice telling him to do so. Hearing voices is one thing. Feeling compelled to obey them is quite another.

The above considerations should make it clear that what many people consider mind control would best be described by some other term, such as "behavior modification," "thought disruption," "brain disabling," "behavior manipulation," "mind-coercion," or "electronic harassment." People are not now being turned into robots by hypnosis or brain implants. Furthermore, given the state of knowledge in the neurosciences, the techniques for effective mind control are likely to be crude, and their mechanisms imperfectly understood.

We can also dismiss subliminal advertising as mind control. Despite widespread belief in the power of subliminal advertising and messaging, the evidence of its effectiveness is anecdotal and comes from interested parties. You will search in vain for the scientific studies that demonstrate that playing inaudible messages such as "do not steal" or "put that back" in Muzak significantly reduces employee or customer theft, or that subliminal messages increase sales of snacks at movie theaters.

We can also dismiss the tactics of husbands who control their wives and the alleged creation of zombies in voodoo. Wives who are terrorized by their husbands or boyfriends are victims not of mind control but of fear and violence. Zombies can be dismissed as cases of drugs being used to render people helpless or of passing on fraudulent stories.

Thus, if we restrict the term "mind control" to those cases where a person successfully controls another person's thoughts or actions without their consent, our initial list of examples of what people consider to be mind control will be pared down to just three items: the Stockholm syndrome and kidnap victims; the tactics of religious, spiritual, and other New Age recruiters; and the so-called brainwashing tactics of the Chinese inquisitors of American prisoners during the Korean War.

The tactics of **cult** recruiters differ substantially from those of kidnappers or inquisitors. Recruiters generally do not kidnap or capture their recruits, and they are not known to use torture as a typical conversion method. This raises the question of whether their victims are controlled without their consent. Many recruits are not truly victims of mind control and are willing members of their communities. Similarly, many recruits into mainstream religions should not be considered victims of mind control. To change a person's basic personality and character, to get them to behave in contradictory ways to lifelong patterns of behavior, to get them to alter their basic beliefs and values, would not necessarily count as mind control. It depends on how actively a person participates in their own transformation. Many may think that a person is out of his mind for joining **Scientology**, Jehovah's Witnesses, or Jim Roberts's The Brethren, but their crazy beliefs and behaviors are no wilder than the ones that millions of mainstream religious believers have chosen to accept and engage in.

Some recruits into nonmainstream religions seem to be brainwashed and controlled to the point that they will do great evil to themselves or others at the behest of their leader, including murder and suicide. These recruits are often in a state of extreme vulnerability when they are recruited, and their recruiter takes advantage of that vulnerability. Such recruits may be confused or rootless due to tragic life circumstances. Some may be people who are mentally ill or brain damaged, emotionally disturbed, greatly depressed, traumatized by self-abuse with drugs or abuse at the hands of others, and so on. But it would not be wise to actively recruit the emotionally disturbed. Such people are difficult to control. Recruiters are likely to look for people they can *make* vulnerable. One former cult recruiter told me: "Cults seek out strong, intelligent, idealistic people. They also seek out the rich, no matter what their mental status is." Recruiters and other manipulators are not using mind control unless they are depriving their victims of their free will. A person can be said to be deprived of his free will by another only if that other has introduced a causal agent that is irresistible. How could we ever demonstrate that a person's behavior is the result of irresistible commands given by a religious, spiritual, or personal growth leader? It is not enough to say that irrational behavior proves a person's free will has been taken from them. It may be irrational to give away all one's property, to devote all one's time and powers to satisfying the desires of one's divine leader, or to commit suicide or plant poison bombs in subways because ordered to do so, but how can we justify claiming such irrational acts are the acts of mindless robots? For all we know, the most bizarre, inhumane, and irrational acts done by the recruits are done freely, knowingly, and joyfully. Perhaps they are done by brain-damaged or insane people. In either case, such people would not be victims of mind control.

That leaves for consideration the acts of kidnappers and inquisitors, the acts of systematic isolation, control of sensory input, and torture. Do these methods allow us to wipe the cortical slate clean and write our own messages to it? That is, can we delete the old and implant new patterns of thought and behavior in our victims? First, it should be noted that not everybody who has been kidnapped comes to feel love or affection for his or her kidnappers. It may be that their tormentors reduce some kidnapped or captured people to a state of total dependency. They are put in a position similar to that of infancy and begin to bond with their tormentors much as an infant does with the one who feeds and comforts it. There is also the strange fascination some of us have with bullies. We fear them, even hate them, but often want to join their gang and be protected by them. It does not seem likely that people who fall in love with their kidnappers are victims of mind control. There is certainly some explanation why some people act as Patricia Hearst did and why others under similar circumstances would not have become Tanya had they been kidnapped by a rag-tag band of rebels calling themselves the Symbionese Liberation Army. It is doubtful that mind control should play much of a role in the explanation. Some women are attracted to gangsters, but have few opportunities to interact with them. We do not need to revert to mind control to explain why Hearst became intimate with one of her terrorist captors. She may have thought she had to in order to survive. She may have been genuinely attracted to him. Who knows? Mind control is a better defense than "changed my mind about a life of crime" when facing bank robbery and murder charges.

Finally, it is widely believed that the Chinese were successful in brainwashing American prisoners of war during the Korean War. The evidence that their tactics were successfully used to control the minds of their captives is slim. Very few (22 of 4,500, or 0.5%) of those captured by the Chinese went over to the other side (Sutherland 1979: 114). The myth of success by the Chinese is primarily due to the work of Edward Hunter, whose *Brainwashing in Red China: The Calculated Destruction of Men's Minds* (1951) is still referred to by those who see mind control tactics as a major menace

today. The CIA promoted the myth in 1950, however, to inspire hatred of the North Koreans and communism, to explain why some American soldiers said good things about their captors, and "to aggrandize their own role by arguing that they themselves must investigate brainwashing techniques in order to keep up with the enemy" (Sutherland 1979: 114).

If we define mind control as the successful control of the thoughts and actions of another without his or her consent, mind control exists only in fantasy. Unfortunately, that does not mean that it will always be thus.

Further reading: Delgado 1969; Kandel and Schwartz 2000; Lifton 1989; Sargant 1957; Valenstein 1973; Weinstein 1990.

miracle

A "transgression of a law of nature by a particular volition of the Deity, or by the interposition of some invisible agent" (Hume 1748: 123n). Theologians of the Old and New Testament religions consider only God-willed contravention of the laws of nature to be true miracles. However, they admit others can do and have done things that contravene the laws of nature; such acts are attributed to diabolical powers and are called "false miracles." Many outside of the Bible-based religions believe in the ability to transgress laws of nature through acts of will in consort with **paranormal** or **occult** powers. They generally refer to these transgressions not as miracles, but as **magick**.

All religions report numerous and equally credible miracles, said David Hume in his famous argument against miracles (1748: ch. X). Hume compares deciding among religions on the basis of their miracles to the task of a judge who must evaluate contradictory, but equally reliable, testimonies. Each religion establishes itself as solidly as the next, thereby overthrowing and destroying its rivals. Furthermore, the more ancient and barbarous a people is, the greater the tendency for miracles and marvels of all kinds to flourish. If a civilized people admit to miracles, "that people will be found to have received them from ignorant and barbarous ancestors, who transmitted them with that inviolable sanction and authority which always attend received opinions" (ibid.: 126). While there are still many people who believe in miracles, few scholars today would dare fill his or her writings with accounts of miraculous events. Indeed, only those who cater to the masses of superstitious and credulous consumers, such as the *National Enquirer* and the Fox TV network, would even think of reporting an alleged miracle without taking a very skeptical attitude toward it. No scholarly journal today would consider an author rational if he or she were to sprinkle reports of miracles throughout a treatise. The modern scholar generally dismisses such reports either as errors or frauds or as cases of **collective hallucination**.

Hume was aware that no matter how scientific or rational a civilization became, belief in miracles would never be eradicated. Human nature is such that we love the marvelous and the wondrous. Human nature is also such that we love even more to be the bearer of a story of the marvelous and the wondrous. The more wondrous our story, the more merit both we and it attain. Vanity, delusion, and zealotry have led to more than one **pious fraud** supporting a holy and meritorious cause with gross embellishments, delusions, and even outright lies about witnessing miraculous events (ibid.: 136).

Hume's greatest argument against belief in miracles, however, was modeled after an argument made by John Tillotson, Archbishop of Canterbury. Tillotson and others, such as William Chillingworth before him and his contemporary Bishop Edward Stillingfleet, had argued for what they called a "commonsense" defense of Christianity, that is, Anglicanism. Tillotson's argument against the Catholic doctrine of transubstantiation or "the real presence" was simple and direct. The idea contradicts sense, he said. The doctrine claims that the bread and wine used in the communion ceremony is changed in substance so that what is bread and wine to all the senses is really the body, blood, soul, and divinity of Christ. If it looks like bread, smells like bread, tastes like bread, then it is bread. *To believe otherwise is to give up the basis for all knowledge based on sense experience.* Anything could be other than it appears to the senses, and reasonable belief for anything would be impossible. This argument has nothing to do with the skeptical argument about the uncertainty of sense knowledge. This is an argument not about certainty but about reasonable belief. If the Catholics are right about transubstantiation, then a book might really be a bishop, for example, or a pear might actually be Westminster Cathedral. The accidents of a thing would be no clue as to its substance. Everything we perceive could be completely unrelated to what it appears to be. Such a world would be unreasonable and unworthy of God. If commonsense can't be trusted in this case, it can't be trusted in any. To believe in transubstantiation is to abandon the basis of all reasonable belief: the general trustworthiness of sense experience as a guide to the nature of things.

Hume begins his essay on miracles by praising Tillotson's argument as being "as concise and elegant and strong as any argument can possibly be supposed against a doctrine so little worthy of a serious refutation." He then goes on to say that he fancies that he has discovered an argument of a like nature which, if just, will, with the wise and learned, be an everlasting check to all kinds of superstitious delusion, and consequently will be useful as long as the world endures; for so long, I presume, will the accounts of miracles and prodigies be found in all history, sacred and profane. [Ibid.: 118.]

His argument is a paradigm of simplicity and elegance:

A miracle is a violation of the laws of nature; and as a firm and unalterable experience has established these laws, the proof against a miracle, from the very nature of the fact, is as entire as any argument from experience can possibly be imagined. [122]

Or put even more succinctly: "There must . . . be a uniform experience against every miraculous event, otherwise the event would not merit that appellation." The logical implication of this argument is that "no testimony is sufficient to establish a miracle unless the testimony be of such a kind that its falsehood would be more miraculous than the fact which it endeavors to establish."

Hume has taken the commonsense Anglican argument against the Catholic doctrine of transubstantiation and applied it to miracles, the basis of most, if not all, religious sects. The laws of nature have not been established by occasional or frequent experiences of a similar kind, but of *uniform* experience. It is "more than probable," says Hume, that all men must die, that lead can't remain suspended in air by itself, and that fire consumes wood and is extinguished by water. If someone were to report that a man could suspend lead in the air by an act of will, Hume would ask himself whether "the falsehood of his testimony would be more miraculous than the event

which he relates." If so, then he would believe the testimony. However, he does not believe there ever was a miraculous event established "on so full an evidence."

Consider the fact that the uniformity of experience of people around the world has been that once a human limb has been amputated, it does not grow back. What would you think if a friend of yours, a scientist of the highest integrity with a Ph.D. in physics from Harvard, were to tell you that she was in Spain last summer and met a man who used to have no legs but now walks on two fine, healthy limbs. She tells you that a holy man rubbed oil on his stumps and his legs grew back. He lives in a small village and all the villagers attest to this miracle. Your friend is convinced a miracle occurred. What would you believe? To believe in this miracle would be to reject the principle of the uniformity of experience, on which laws of nature are based. It would be to reject a fundamental assumption of all science, that the laws of nature are inviolate. The miracle cannot be believed without abandoning a basic principle of empirical knowledge: that like things under like circumstances produce like results.

Of course there is another constant, another product of uniform experience that should not be forgotten: the tendency of people at all times in all ages to desire wondrous events, to be deluded about them, to fabricate them, create them, embellish them, enhance them, and come to believe in the absolute truth of the creations of their own passions and heated imaginations. Does this mean that miracles cannot occur? Of course not. It means, however, that when a miracle is reported, the probability will always be greater that the person doing the reporting is mistaken, deluded, or a fraud than that the miracle really occurred. To believe in a miracle, as Hume said, is not an act of reason but of **faith**.

Further reading: Carroll 1974; Nickell 1993.

mokele-mbembe

Mokele-mbembe is an alleged living sauropod dinosaur now living in the Likouala swamp region of the Republic of Congo (Zaire). Mokele-mbembe means, depending on your source, "rainbow," "one that stops the flow of rivers," or "monstrous animal." Local pygmies, who have given the creature its name, have allegedly encountered the mokele-mbembe, which they say is the size of an elephant (the pygmies' favorite prey) with a very long reptilian neck. The creature is said to be hairless and reddish-brown, brown, or gray, with a tail 5 to 10 feet long. It apparently spends most of its time in the water, but the pygmies claim they've seen prints left on land of its three-clawed foot.

Reports of the mokele-membe have been circulating for the past 200 years, yet no one has photographed it or produced any physical evidence of its existence. Enthusiastic **cryptozoologists** such as Roy Mackal (*A Living Dinosaur: In Search of Mokele-Mbembe,* 1987) think we should give as much credence to the mokele-mbembe as to the **Loch Ness monster**. True, and it therefore seems unlikely the creature exists, since there would have to be a significant number of the huge creatures to produce descendents after all other dinosaurs were extinguished some 65 million years ago. One wonders how such creatures have flourished for millions of years without leaving a single carcass, bone fragment, or fossil.

Cryptozoologists argue that since a coelacanth was caught off the coast of South Africa, it is reasonable to think that a dinosaur might also have avoided detection for a few million years. However, there is a big difference between hiding a fish in the ocean and hiding a dinosaur on land. The fish is small and oceans cover two-thirds of the earth and have depths of up to 35,000 feet. Most of what goes on in the ocean is not visible to us. Dinosaurs, on the other hand, living on the visible part of the earth, would be much more likely to be detected than the coelacanth.

Moody, Raymond

A **parapsychologist** with a medical degree (from the Medical College of Georgia) and Ph.D.s in philosophy and psychology (from the University of Virginia), Raymond Moody has written several books on the subject of "life after life." He compiled a list of features many consider typical of the **near-death experience**: a buzzing or ringing noise; a sense of blissful peace; a feeling of floating out of one's body and observing it from above; moving through a tunnel into a bright light; meeting dead people, saints, Christ, angels, and so on; seeing one's life pass before one's eyes; and finding it all so wonderful that one doesn't want to return to one's body.

Moody conducts his **paranormal** studies at his private research institute in rural Alabama, which he calls the John Dee Memorial Theater of the Mind. Dee popularized **crystal** gazing in 16th-century England. Moody is continuing in the spirit of Dee, trying to evoke apparitions of the dead under controlled conditions. Moody has a mirrored room where guests come to **scry**, hoping for a visit from a dead loved one. ABC reporter Diane Sawyer tried it out for about 45 minutes but didn't have any visitors. Maybe she didn't have a strong enough desire to see a dead loved one. Maybe she didn't have a strong enough belief that gazing into mirrors can induce an **altered state of consciousness**. Maybe she should have stayed in the room for a day or two.

There are many frauds who claim to be able to see into the past or the future by various means of **divination** and there are many who hallucinate due to sensory deprivation, extreme concentration on a single item, or lengthy gazing at a uniform or kaleidoscopic surface. Nevertheless, Moody and many of his guests claim success at having spirits visit them in the mirrored room. Moody, like Charles **Tart**, is convinced that an altered state of consciousness is the gateway to the other world. Mirror gazing is just one of many methods Moody uses to try to induce an altered state.

Moody is also an advocate of **past life regression**. He claims that he was skeptical about **reincarnation** until undergoing hypnotherapy, during which he discovered that he had had nine past lives. He claims that just about anyone can experience a "past-life journey" and that such trips help one overcome phobias, compulsions, addictions, and depression, among other things.

On May 10, 1998, Moody succeeded Tart to the now-defunct Bigelow Chair in Consciousness Studies at the University of Nevada, Las Vegas (UNLV). On his appointment, Moody was quoted as saying:

> I am thrilled to have the opportunity to teach again, and believe that UNLV should be applauded for its determination to adhere to the strictest standards of scientific rigor regarding claims of rational "evidence" or "proof" of the continuation of consciousness upon bodily death. . . . [T]he extraordinary states of consciousness commonly deemed paranormal are an enduring human concern that will not go away, and I have been hopeful that students with a serious interest in these topics would have a setting within which they could learn about paranormal phenomena from a non-ideological perspective.

He gave some indication of his nonideological rigor by announcing that he had invited Brian Weiss, M.D., "expert in past life regression," to conduct a community forum at UNLV. He also invited to UNLV Dianne Arcangel, who, says Moody, is "an expert in the field of facilitated apparitions."

See also **parapsychology**.

moon, the

See **lunar effects**.

morphic resonance

"Morphic resonance" is a term coined by Rupert Sheldrake for what he thinks is "the basis of memory in nature . . . the idea of mysterious **telepathy**-type interconnections between organisms and of collective memories within species."

Sheldrake has been trained in 20th-century scientific models—he has a Ph.D. in biochemistry from Cambridge University (1967)—but he prefers Goethe, 19th-century **vitalism**, and teleological models to mechanistic models of reality. One of his books is entitled *Dogs That Know When Their Owners Are Coming Home: And Other Unexplained Powers of Animals*. One of his studies is on whether people can tell when someone is staring at them. He says they can; others have been unable to duplicate his results (Marks and Colwell 2000). He prefers a romantic vision of the past to the bleak picture of a world run by technocrats who want to control Nature and destroy much of the environment in the process. In short, he prefers **metaphysics** to **science**, though he seems to think he can do the former but call it the latter.

Morphic resonance (MR) is put forth as if it were an empirical term, but it is no more empirical than L. Ron Hubbard's "engram," the alleged source of all mental and physical illness. The term is on par more with the Stoic notion of the *Logos* or Plato's notion of the *eidos* than with any scientific notion of the laws of nature. What the rest of the scientific world terms lawfulness—the tendency of things to follow patterns we call laws of nature—Sheldrake explains by MR. He describes it as a kind of memory in things determined not by their inherent natures but by repetition. He also describes MR as something that is transmitted via "morphogenic fields." This gives him a conceptual framework wherein information is transmitted mysteriously and miraculously through any amount of space and time without loss of energy, and presumably without loss or change of content through something like mutation in DNA replication. Thus, room is made for **psychic** as well as physical transmission of information. Thus,

> it is not at all necessary for us to assume that the physical characteristics of organisms are contained inside the genes, which may in fact be analogous to transistors tuned in to the proper frequencies for translating invisible information into visible form. Thus, morphogenic fields are located invisibly in and around organisms, and may account for such hitherto unexplainable phenomena as the regeneration of severed limbs by worms and salamanders, phantom limbs, the holographic properties of memory, telepathy, and the increasing ease with which new skills are learned as greater quantities of a population acquire them. [www.sheldrake.org/interviews/quest_interview.html]

While this metaphysical proposition does seem to make room for telepathy, it does so at the expense of ignoring **Occam's razor**. Telepathy and such things as phantom limbs, for example, can be explained without adding the metaphysical baggage of morphic resonance (Ramachandran 1998).

So can **memory**, which does not require a holographic paradigm. The notion that new skills are learned with increasing ease as greater quantities of a population acquire them, known as the **hundredth monkey phenomenon**, is bogus.

In short, although Sheldrake commands some respect as a scientist because of his education and degree, he has clearly abandoned science in favor of theology and philosophy. This is his right, of course. However, his continued pose as a scientist is unwarranted. He is one of a growing horde of "alternative" scientists whose resentment at the aspiritual nature of modern scientific paradigms, as well as the obviously harmful and seemingly indifferent applications of modern science, have led them to create their own paradigms.

Moses syndrome

(1) A delusion characterized by uncritical belief in the promises of others to lead one to the Promised Land (or to provide any of the following: beauty, youth, wealth, power, peace of mind, happiness) or (2) a delusion characterized by the belief that one has been chosen by God, destiny, or history to lead others to the Promised Land, such as Descartes' dream of putting the sciences on a firm foundation or Hitler's belief in an "eternal law of nature," which gave Germany, "as the stronger power, the right before history to subjugate these peoples of inferior race, to dominate them and to coerce them into performing useful labors" (1941: Speech before German officers).

The Moses syndrome should not be confused with the *baby Moses syndrome* (the hope-in-a-basket fallacy), a kind of defense mechanism where one deceives oneself into inaction by the wishful thought that somebody else will eventually come along to solve your problems for you and save you from disaster.

Mozart effect

A term coined by Alfred A. Tomatis for the alleged increase in brain development that occurs in children under age three when they listen to the music of Wolfgang Amadeus Mozart.

The idea for the Mozart effect originated in 1993 at the University of California, Irvine, with physicist Gordon Shaw and Frances Rauscher, a former concert cellist and an expert on cognitive development. They studied the effects on a few dozen college students of listening to the first 10 minutes of the Mozart "Sonata for Two Pianos in D Major" (K. 448). They found a temporary enhancement of spatial-temporal reasoning, as measured by the Stanford-Binet IQ test. No one else has been able to duplicate their results. One researcher says that the best thing one can say about the results of the Shaw/Rauscher study is "listening to bad Mozart enhances short-term IQ" (Linton: www.firstthings.com/ftissues/ft9903/linton.html). Rauscher has moved on to study the effects of Mozart on rats. Both Shaw and Rauscher have speculated that exposure to Mozart enhances spatial reasoning and memory in humans.

Shaw and Rauscher have stimulated an industry. They have also created their own institute: the Music Intelligence Neural Development Institute. There is so much research going on to prove the wondrous effects of music that a web site has been created just to keep track of all the new developments: MuSICA (www.musica.uci.edu).

Shaw and Rauscher claim that their work has been misrepresented. What they have shown is that there are patterns of neurons that fire in sequences, and that there appear to be preexisting sites in the brain that respond to specific frequencies (Halpern: www.lightworks.com/MonthlyAspectarian/1998/January/0198-20.htm). This is not quite the same as showing that listening to Mozart increases intelligence in children. Nevertheless, Shaw is not going to wait for the hard evidence to pour in before he cashes in on the desire of parents to enhance their children's intelligence. He has a book and CD called *Keeping Mozart in Mind* (1999). You can buy it from his institute. He and his colleagues are convinced that since spatio-temporal reasoning is essential for many higher order cognitive tasks, stimulating the area of the brain associated with spatio-temporal reasoning and doing spatio-temporal exercises will increase a person's intelligence for math, engineering, chess, and science. They even have a software program for sale, which uses no language and aims at exercising spatio-temporal skills with the help of an animated penguin.

Shaw and Rauscher may have spawned an industry, but the mass media and others have created a kind of "alternative" science that supports that industry. Exaggerated and false claims about music have become so commonplace that it is probably a waste of time to try to correct them. For example, Jamal Munshi, an associate professor of Business Administration at Sonoma State University, collects tidbits of misinformation and gullibility. He used to post them on his web site as "Weird but True," including the claim that Shaw and Rauscher showed that listening to Mozart's sonata "increased SAT scores of students by 51 points." Actually, Shaw and Rauscher gave 36 University of California at Irvine students a paper folding and cutting test and found the Mozart group showed a temporary eight- to nine-point increase over their scores when they took the test after either a period of silence or listening to a relaxation tape.

Don Campbell, however, has become the Carlos **Castaneda** and P. T. Barnum of the Mozart effect, exaggerating and distorting the work of Shaw, Rauscher, and others for his own benefit. He has trademarked the expression "the Mozart effect" and peddles himself and his products on the Internet. Campbell claims that he made a blood clot in his brain disappear by humming, praying, and envisioning a vibrating hand on the right side of his skull. Uncritical supporters of alternative medicine don't question this claim, though it is one of those safe claims that can't be proved or disproved. He might as well claim that angels took the clot away.

> The claims that Campbell makes for music are of an almost rococo flamboyance. And like the rococo, just about as substantive. [Campbell claims music can cure just about anything that ails you.] His evidence is usually anecdotal, and even this he misinterprets. Some things he gets completely wrong.
>
> And the whole structure of his argument collapses under simple common sense. If Mozart's music were able to improve health, why was Mozart himself so frequently sick? If listening to Mozart's music increases intelligence and encourages spirituality, why aren't the world's smartest and most spiritual people Mozart specialists? [Linton: www.firstthings.com/ftissues/ft9903/linton.html]

The lack of evidence for the Mozart effect has not deterred Campbell from becoming a favorite on the lecture circuit with the naive and uncritical:

> When *McCall's* wants advice on how to lose the blues with music, when PBS wants to interview an expert on how the voice can energize you, when IBM wants a consultant to use music to increase efficiency and harmony in the workplace, when the National Association of Cancer Survivors wants a speaker on the healing

powers of music, they turn to Campbell. [www.mozarteffect.com]

The governors of Tennessee and Georgia have started programs that give a Mozart CD to every newborn. And the National Academy of Recording Arts and Sciences Foundation gave free CDs of classical music to hundreds of hospitals in May 1999. These are well-intentioned gestures, but they are not based on solid research that classical music increases a child's intelligence or an adult's healing process.

According to Kenneth Steele, a psychology professor at Appalachian State University, and John Bruer, head of the James S. McDonnell Foundation in St. Louis, there is no real intelligence-enhancing or health benefit to listening to Mozart. Steele and his colleagues Karen Bass and Melissa Crook claim that they followed the protocols set forth by Shaw and Rauscher but could not "find any kind of effect at all," even though their study tested 125 students. They concluded that "there is little evidence to support intervention programs based on the existence of the Mozart effect" (Steele et al. 1999). Bruer's book, *The Myth of the First Three Years* (1999), attacks not only the Mozart effect but also several other related myths based on the misinterpretation of recent brain research. According to Bruer,

> Stories stressing that children's experiences during their early years of life will ultimately determine their scholastic ability, their future career paths, and their ability to form loving relationships have little basis in neuroscience.

The Mozart effect is an example of how science and the media mix in our world. A suggestion in a few paragraphs in a scientific journal becomes a universal truth in a matter of months, eventually believed even by the scientists who initially recognized how their work had been distorted and exaggerated by the media. Others, smelling the money, jump on the bandwagon and play to the crowd, adding their own myths, questionable claims, and distortions to the mix. In this case, many uncritical supporters line up to defend the faith because at stake here is the future of our children. We then have books, tapes, CDs, institutes, government programs, and more. Soon millions believe the myth is a scientific fact. In this case, the process met with little critical resistance because we already know that music can affect feelings and moods, so why shouldn't it affect intelligence and health? It's just common sense, right? Yes, and all the more reason to be skeptical.

Further reading: Kandel and Schwartz 2000.

multilevel marketing (MLM)

Multilevel marketing is also called *network marketing* and *referral marketing*. The idea behind MLM is simple. Imagine you have a product to sell. A common MLM product is some sort of panacea, such as a vitamin or mineral supplement. You could do what most businesses do: Either sell it directly to consumers or find others who will buy your product from you and sell it to other people. MLM schemes require that you recruit people not only to buy and sell your product, but also to recruit people who will not only buy and sell your product but also to recruit people—ad infinitum. Only there never is an infinitum to move toward. This process may seem unusual to traditional business people. Why, you might wonder, would you recruit people to compete with you? MLM magic will convince you that it is reasonable to recruit competitors because they won't *really* be competitors since you will get a cut of their profits. This will take your mind off the fact that no matter how big your town or market, it is finite. The

well will go dry soon enough. There will always be some distributors who will make money in an MLM scheme. The majority, however, must fail due to the intrinsic nature of all pyramid schemes. Those who supply the products and the motivational materials that seem to be an essential part of any MLM scheme make the real money.

MLM is a system of marketing that puts more emphasis on the recruiting of distributors than on the selling of products. As such, it is intrinsically flawed. MLM is very attractive, however, because it sells *hope* and appears to be outside the mainstream of business as usual. It promises wealth and independence. Unfortunately, no matter what the product, MLM is doomed to produce more failures than successes. For every MLM distributor who makes a decent living or even a decent supplemental income, there are at least ten who do little more than buy products and promotional materials, costing them more than they will ever earn as an MLM agent. The most successful MLM scheme is Amway. It has millions of distributors worldwide with sales in the billions. At the turn of the century, the average Amway distributor earned about $700 a year in sales, but spent about $1,000 a year on Amway products. Distributors also have other expenses related to the business, for example, telephone, gas, publicity material, and time spent at motivational meetings (Amway.com; Klebniov 1991). The reason MLM schemes cannot succeed is because MLM marketing is, in essence, a legal **pyramid scheme**.

What are the benefits of MLM membership? You get certain tax write-offs. You get to buy products wholesale, some of which you will be happy with. You get to go to inspirational meetings, some of which will make you feel good. You may meet new friends and you may even make a little money. But more than likely you will end up alienating some family and friends. You will probably end up buying more goods than you sell. And you will learn a lot about deceiving yourself and others. You won't be allowed to tell anyone how you are really doing, for example. You will have to think positive, even if that means lying. You will have to tell anyone who asks that you are doing great, that business is wonderful, that you've never seen anything go so fast and bring you income so quickly, even if it isn't true.

The dangers of MLM schemes have been well articulated by others. If you are thinking of joining an MLM program, first read Dean Van Druff's *What's Wrong With Multi-Level Marketing* (www.vandruff.com/mlm.html) or Robert Fitzpatrick's *False Profits* (1997).

multiple personality disorder (MPD)

A psychiatric disorder characterized by having at least one alter personality that controls behavior. The alters are said to occur spontaneously and involuntarily, and function more or less independently of each other. The unity of consciousness, by which we identify our selves, is said to be absent in MPD. Another symptom of MPD is significant amnesia that can't be explained by ordinary forgetfulness or brain damage. In 1994, the American Psychiatric Association's *Diagnostic and Statistical Manual of Mental Disorders-IV* replaced the designation of MPD with DID: *dissociative identity disorder*. The label may have changed, but the list of symptoms remained essentially the same.

Memory and other aspects of consciousness are said to be divided up among "alters" in the MPD. The number of alters identified by various therapists ranges from several to tens to hundreds. There are even some reports of several thousand identities dwelling in one person. No current model of the brain and memory can

account for such a state of affairs. Furthermore, there does not seem to be any consensus among therapists as to what an alter is. Yet there is general agreement that MPD is caused by **repressed memories** of childhood sexual abuse and that it is a defense mechanism. The evidence for this claim has been challenged, however, and there are very few reported cases of MPD afflicting children.

Psychologist Nicholas P. Spanos argues that most cases of multiple personality disorder are "rule-governed social constructions established, legitimated, and maintained through social interaction" (Spanos 1996). Most MPD cases have been created by therapists with the cooperation of their patients and the rest of society. The experts have created both the disease and the cure. This does not mean that MPD does not exist, but that its origin and development are often, if not most often, explicable without the model of separate but permeable ego-states or alters arising out of the ashes of a destroyed "original self."

Philosopher Daniel Dennett gives a rather common view of MPD:

the evidence is now voluminous that there are not a handful or a hundred but thousands of cases of MPD diagnosed today, and it almost invariably owes its existence to prolonged early childhood abuse, usually sexual, and of sickening severity. Nicholas Humphrey and I investigated MPD several years ago [1989: "Speaking for Our Selves: An Assessment of Multiple Personality Disorder," *Raritan* 9: 68–98] and found it to be a complex phenomenon that extends far beyond individual brains and the sufferers.

These children have often been kept in such extraordinary terrifying and confusing circumstances that I am more amazed that they survive psychologically at all than I am that they manage to preserve themselves by a desperate redrawing of their boundaries. What they do, when confronted with overwhelming conflict and pain, is this: They "leave." They create a boundary so that the horror doesn't happen to them; it either happens to no one, or to some other self, better able to sustain its organization under such an onslaught—at least that's what they say they did, as best they recall. (Dennett 1991)

Dennett exhibits minimal skepticism about the truth of the MPD accounts, and focuses on how they can be explained metaphysically and biologically. For all his brilliant exploration of the concept of the self, the one perspective he doesn't seem to give much weight to is the one Spanos takes: that the self and the multiple selves of the MPD patient are social constructs, not needing a metaphysical or biological explanation so much as a social-psychological one. That is not to say that our biology is not a significant determining factor in the development of our ideas about selves, including the idea of our own self. It is to say, however, that before we go off worrying about how to metaphysically explain one or a hundred selves in one body (or one self in a hundred bodies, for that matter), we might want to consider that a phenomenological analysis of behavior that takes behavior at face value, or that attributes it to nothing but brain structure and biochemistry, may be missing the most significant element in the creation of the self: the sociocognitive context in which our ideas of self, disease, personality, memory, and so on emerge. Being a social construct does not make the self any less real. And Spanos should not be taken as denying the existence either of the self or of MPD.

But if thinkers of Dennett's stature accept MPD as something that needs explaining in terms of *psychological* dynamics limited to the psyche of the abused rather than in terms of social constructs, the task of convincing therapists who treat MPD to accept Spanos's way of thinking is Herculean. How could it be possible that most MPD patients have been created in the therapist's laboratory, so to speak? How could it be possible that so many people, particularly female people (85% of MPD patients are female), could have so many **false memories** of childhood sexual abuse? How could so many people behave as if their bodies have been invaded by numerous entities or personalities, if they hadn't really been so invaded? How could the defense mechanism explanation for MPD, in terms of repression of childhood sexual trauma and dissociation, not be correct? One might as well ask, how could so many people actually experience **past lives** under **hypnosis**, a standard procedure of many therapists who treat MPD? How could so many people be so wrong about so much? Spanos's answer makes it sound almost too easy for such a massive amount of **self-deception** and delusion to develop: It's happened before and we all know about it. Remember demonic possession?

Most educated people today do not try to explain epilepsy, brain damage, genetic disorders, neurochemical imbalances, feverish hallucinations, or troublesome behavior by appealing to the idea of demonic possession. Yet at one time, all of Europe and America would have accepted such an explanation. Furthermore, we had our experts—the priests and theologians—to tell us how to identify the possessed and how to **exorcise** the demons. An elaborate theological framework bolstered this world view, and an elaborate set of social rituals and behaviors validated it on a continuous basis. In fact, every culture, no matter how primitive and prescientific, has had a belief in some form of demonic possession. It has had its shamans and witch doctors who performed rituals to rid the possessed of their demons. In their own sociocognitive contexts, such beliefs and behaviors went unquestioned and were constantly reinforced by traditional and customary social behaviors and expectations.

Most educated people today believe that the behaviors of **witches** and other possessed persons—as well as the behaviors of their tormentors, exorcists, and executioners—were enactments of social roles. With the exception of superstitious religious folks (who still live in the world of demons, **witches**, **exorcisms**, and supernatural magic), educated people do not believe that in those days there really were witches, that demons really did invade bodies, or that priests really did exorcize those demons by their ritualistic magic. Yet for those who lived in the time of witches and demons, these beings were as real as anything else they experienced. In Spanos's view, what is true of the world of demons and exorcists is true of the psychological world filled with phenomena such as repression of childhood sexual trauma and its manifestation as a psychological survival mechanism in MPD.

Spanos makes a very strong case for the claim that "patients learn to construe themselves as possessing multiple selves, learn to present themselves in terms of this construal, and learn to reorganize and elaborate on their personal biography so as to make it congruent with their understanding of what it means to be a multiple." Psychotherapists, according to Spanos, "play a particularly important part in the generation and maintenance of MPD." According to Spanos, most therapists never see a single case of MPD, but some therapists report seeing hundreds of cases each year. It should be distressing to those trying to defend the integrity of psychotherapy that a patient's

diagnosis depends on the preconceptions of the therapist. However, an MPD patient typically has no memory of sexual abuse on entering therapy. Only after the therapist encourages the patient do memories of sexual abuse emerge. Furthermore, the typical MPD patient does not begin manifesting alters until after treatment begins (Piper 1998). MPD therapists counter these charges by claiming that their methods are tried and true, which they know from experience, and that those therapists who never treat MPD don't know what to look for (Dr. Ralph Allison: www.dissociation .com).

Multiple selves exist, and have existed in other cultures, without being related to the notion of a mental disorder, as is the case in North America today. According to Spanos, "Multiple identities can develop in a wide variety of cultural contexts and serve numerous different social functions." Neither childhood sexual abuse nor mental disorder is a necessary condition for multiple personality to manifest itself. In a number of different historical and social contexts, people have learned to think of themselves as "possessing more than one identity or self, and can learn to behave as if they are first one identity and then a different identity." However, "people are unlikely to think of themselves in this way or to behave in this way unless their culture has provided models from whom the rules and characteristics of multiple identity enactments can be learned. Along with providing rules and models, the culture, through its socializing agents, must also provide legitimation for multiple self enactments." Again, Spanos is not saying that MPD does not exist, but that the standard model of abuse, withdrawal of original self, and then emergence of alters is not needed to explain MPD. Nor is the psychological baggage that goes with that model:

repression, recovered memory of childhood sexual abuse, integration of alters in therapy. Nor are the standard diagnostic techniques: **hypnosis**, including **past life regression**, and **Rorschach** tests.

The widespread belief in the standard paradigm of MPD has been heavily influenced by books and films, for example, *Sybil, The Three Faces of Eve, The Five of Me,* or *The Minds of Billy Milligan*. These mass media presentations not only influence the general public's beliefs about MPD, but they affect MPD patients as well. For example, Flora Rheta Schreiber's *Sybil* is the story of a woman with sixteen personalities allegedly created in response to having been abused as a child. Before the publication of *Sybil* in 1973 and the 1976 television movie starring Sally Field as Sybil, there had been only about 75 reported cases of MPD. Since *Sybil* there have been some 40,000 diagnoses of MPD, mostly in North America.

Sybil has been identified as Shirley Ardell Mason, who died of breast cancer in 1998 at the age of 75. Her therapist has been identified as Cornelia Wilbur, who died in 1992, leaving Mason $25,000 and all future royalties from *Sybil*. Schreiber died in 1988. It is now known that Mason had no MPD symptoms before therapy with Wilbur, who used hypnosis and other suggestive techniques to tease out the so-called personalities. *Newsweek* reported that, according to historian Peter M. Swales (who first identified Mason as Sybil), "there is strong evidence that [the worst abuse in the book] could not have happened" (January 25, 1999).

Dr. Herbert Spiegel, who also treated Sybil, believes Wilbur suggested the personalities as part of her therapy and that the patient adopted them with the help of hypnosis and sodium pentothal. He describes his patient as highly hypnotizable and extremely suggestible. Mason was so helpful that she read the literature on MPD, including *The Three Faces of Eve*. The Sybil episode seems clearly to be symptomatic of an iatrogenic disorder. Yet the Sybil case is the paradigm for the standard case of MPD. A defender of this model, Dr. Philip M. Coons, of the Department of Psychiatry at the Indiana University School of Medicine, claims that "the relationship of multiple personality to child abuse was not generally recognized until the publication of *Sybil*" (www.php.iupui.edu/~pcoons/home.html).

The MPD community suffered another serious attack on its credibility when Dr. Bennett Braun, the founder of the International Society for the Study of Disassociation, had his license suspended over allegations that he used drugs and hypnosis to convince a patient she killed scores of people in **satanic rituals**. The patient claims that Braun convinced her that she had 300 personalities, among them a child molester, a high priestess of a satanic cult, and a cannibal. The patient told the *Chicago Tribune:* "I began to add a few things up and realized there was no way I could come from a little town in Iowa, be eating 2,000 people a year, and nobody said anything about it." The patient won $10.6 million in a lawsuit against Braun, Rush-Presbyterian-St. Luke's Hospital, and another therapist.

The defenders of the MPD/DID standard model of genesis, diagnosis, and treatment argue that the disease is underdiagnosed because its complexity makes it very difficult to identify. Dr. Philip M. Coons claims that "there is a professional reluctance to diagnose multiple personality disorder." He thinks this "stems from a number of factors including the generally subtle presentation of the symptoms, the fearful reluctance of the patient to divulge important clinical information, professional ignorance concerning dissociative disorders, and the reluctance of the clinician to believe that incest actually occurs and is not the product of fantasy." Dr. Coons also claims that demonic possession was "a forerunner of multiple personality."

Another defender of the standard model of MPD, Dr. Ralph Allison, has posted on the Internet his diagnosis of Kenneth Bianchi, the so-called Hillside Strangler, in which the therapist admits he has changed his mind several times. Bianchi, now a convicted serial killer serving a life sentence, was diagnosed as having MPD by defense psychiatrist Jack G. Watkins. Dr. Watkins used hypnosis on Bianchi, and "Steve" emerged to an explicit suggestion from the therapist. "Steve" was allegedly Bianchi's alter who did the murders. Prosecution psychiatrist Martin T. Orne, an expert on hypnosis, argued successfully before the court that the hypnosis and the MPD symptoms were a sham.

Dr. Allison claims that the controversy over MPD is one between therapists, who defend the standard model, and teachers, who deny MPD exists (www.dissociation .com/index/Definition). The battle took place in committee when preparing the DSM-IV, he claims. The teachers won and MPD was removed and DID replaced it. The DSM-IV is the current version (1994) of the American Psychiatric Association's Diagnostic and Statistical Manual of Mental Disorders. It lists 410 mental disorders, up from 145 in DSM-II (1968). The first edition in 1952 listed 60 disorders. Some claim that this proliferation of disorders indicates an attempt of therapists to expand their market; others see the rise in disorders as evidence of better diagnostic tools. According to Dr. Allison, MPD was called "hysterical dissociative disorder" in DSM-II and did not have its own code number. MPD was listed and coded in DSM-III, but removed in DSM-IV and replaced with DID.

See also **hystero-epilepsy**, **New Age psychotherapies**, and **repressed memory therapy**.

Further reading: Coons 1986; Lilienfeld et al. 1999; Morris 1990; Piper 1997; Ross 1996.

Murphy, Bridey

The name of a 19th-century woman from Cork, Ireland, who allegedly began speaking through Virginia Tighe during **hypnosis**. This happened in Pueblo, Colorado, in 1952 when Morey Bernstein, a local businessman and amateur hypnotist encouraged **past life regression** in his subject. Bernstein hypnotized Tighe many times. While under hypnosis, she sang Irish songs and told Irish stories in an Irish brogue, always as Bridey Murphy. Bernstein's book, *The Search for Bridey Murphy,* became a best-seller. (Tighe is called Ruth Simmons in the book.) Recordings of the hypnotic sessions were made and translated into more than a dozen languages. The recordings sold well, too.

Newspapers sent reporters to Ireland to investigate. Was there a redheaded Bridey Murphy who lived in Ireland in the 19th century? Perhaps, but one paper—the *Chicago American*—found one in Wisconsin in the 20th century. Bridie Murphey Corkell lived in the house across the street from where Virginia Tighe grew up. What Virginia reported while hypnotized were not memories of a *previous* life but memories from her early childhood.

Whatever else the hypnotic state is, it is a state where one's fantasies are often energetically displayed. Many people were impressed with the details of Tighe's hypnotic memories, but the details were not evidence of **channeling** or **reincarnation**. They were evidence of a vivid imagination, a confused memory, fraud, or a combination of the three.

Almost any hypnotic subject capable of going into a deep trance will babble about a previous incarnation if the hypnotist asks him to. He will babble just as freely about his future incarnations. . . . In every case of this sort where there has been adequate checking on the subject's past, it has been found that the subject was weaving together long forgotten bits of information acquired during his early years. (Gardner 1957)

The standards of critical thinking are often lowered for **paranormal** beliefs. Defenders of preposterous stories such as this one find easily accessible information to be incontrovertible proof of their veracity. For example, Tighe talks about kissing the Blarney stone and knew that the act requires the assistance of someone who holds you as you lean backward and face up to kiss the stone. This is common knowledge and photos of this are available in hundreds of sources, yet this fact has been cited as strong evidence that Tighe really kissed the stone in a previous incarnation. These same proponents of the strange and occult are not concerned that the kind of reincarnation they are considering contradicts everything we know about how **memory** works, not to mention that it is impossible to explain without rejecting everything we know about human consciousness and the brain. Such beliefs are works of pure imagination, which we tolerate in cartoons and for entertainment, but which any rational creature should rebuke in those who claim to be seeking the truth.

Further reading: Ready 1956.

Myers-Briggs Type Indicator

An instrument for measuring a person's preferences, using four basic scales with opposite poles. The four scales are: (1) extra-version/introversion, (2) sensate/intuitive, (3) thinking/feeling, and (4) judging/perceiving. "The various combinations of these preferences result in 16 personality types," says Consulting Psychologists Press, Inc. (CPP), which owns the rights to the instrument. Types are typically denoted by four letters—for example, INTJ ("Introversion, Intuition with Thinking and Judging")—to represent one's tendencies on the four scales.

According to CPP, the MBTI is "the most widely used personality inventory in history." According to the Center for Applications of Psychological Type, approximately 2,000,000 people a year take the MBTI. CPP claims that it "helps you improve work and personal relationships, increase productivity, and identify leadership and interpersonal communication preferences for your clients" (www.cpp-db.com/products/mbti/index.asp) Many schools use the MBTI in career counseling. A profile for each of the sixteen types has been developed. Each profile consists of a list of "characteristics frequently associated with your type," according to CPP. The INTJ, for example, is frequently

- insightful, conceptual, and creative
- rational, detached, and objectively critical
- likely to have a clear vision of future possibilities
- apt to enjoy complex challenges
- likely to value knowledge and competence
- apt to apply high standards to themselves and others
- independent, trusting their own judgments and perceptions more than those of others
- seen by others as reserved and hard to know

The people at CPP aren't too concerned if the list doesn't seem to match your type. They advise such persons to see the one who administered the test and ask for help in finding a more suitable list by changing a letter or two in your four-letter type. (See the report CPP publishes on its website at www.cpp-db.com/images/reports/11.pdf.) Furthermore, no matter what your preferences, your behavior will still sometimes indicate contrasting behavior. Thus, no behavior can ever be used to falsify the type, and any behavior can be used to verify it.

Jung's Psychological Types. The MBTI is based upon Carl **Jung**'s notions of psychological types. The MBTI was first developed by Isabel Briggs Myers (1897–1979) and her mother, Katharine Cook Briggs. Isabel had a bachelor's degree in political science from Swarthmore College and no academic affiliation. Katharine's father was on the faculty of Michigan Agricultural College (now Michigan State University). Her husband was a research physicist and became director of the Bureau of Standards in Washington. Isabel's husband, Clarence Myers, was a lawyer. Because Clarence was so different from the rest of the family, Katherine became interested in types. She introduced Isabel to Jung's book, *Psychological Types.* Both became avid "type watchers." Their goal was a noble one: to help people understand themselves and each other so that they might work in vocations that matched their personality types. This would make people happier and make the world a more creative, productive, and peaceful place in which to live.

According to Jung, some of us are extraverts (McGuire and Hull 1997: 213). (The spelling of "extravert" is Jung's preference. All citations are to McGuire and Hull.) They are "more influenced by their surroundings than by their own intentions" (302). The extravert is the person "who goes by the influence of the external world—say society or sense perceptions" (303). Jung also claims that "the world in general, particularly America, is extraverted as hell, the introvert has no place, because he doesn't know that he beholds the world from within" (303). The introvert "goes by the subjective factor. . . . [H]e bases himself on the world from within . . . and . . . is always afraid of the external world. . . . He always has a resentment" (303). Jung knows these things because he is a careful observer of people. He did only one statistical study in his life, and that was in **astrology** (315). In fact, Jung disdained statistics. "You can prove anything with statistics," he said (306). He preferred interpreting anecdotes.

Jung also claimed that "there is no such thing as a pure extravert or a pure introvert. Such a man would be in the lunatic asylum. They are only terms to designate a certain penchant, a certain tendency . . . the tendency to be more influenced by environmental factors, or more influenced by the subjective factor, that's all. There are people who are fairly well balanced and are just as much influenced from within as from without, or just as little" (304). Jung's intuition turns out to be correct here and should be a red flag to those who have created a typology out of his preference categories. A typology should have a bimodal distribution, but the evidence shows that most people fall between the two extremes of introversion and extraversion. Thus, "although one person may score as an E, his or her test results may be very similar to those of another person's, who scores as an I" (Pittenger 1993).

Jung claimed that *thinking/feeling* is another dichotomy to be used in psychological typing. "Thinking[,] roughly speaking, tells you *what* [something] is. Feeling tells you whether it is agreeable or not, to be accepted or rejected" (306). The final dichotomy, according to Jung, is the *sensa-*

tion/intuition dichotomy. "Sensation tells you that there is something. . . . And intuition—now there is a difficulty. . . . There is something funny about intuition" (306). Even so, he defines intuition as "a perception via the **unconscious**" (307).

Jung claims that it took him a long time to discover that not everybody was a thinking (or intellectual) type like himself. He claims that he discovered there are "four aspects of conscious orientation" (341), or psychic functions. He claims he arrived at his typology "through the study of all sorts of human types" (342). These four orientations cover it all, he claims:

> I came to the conclusion that there must be as many different ways of viewing the world [as there are psychological types]. The aspect of the world is not one, it is many—at least 16, and you can just as well say 360. You can increase the number of principles, but I found the most simple way is the way I told you, the division by four, the simple and natural division of a circle. I didn't know the symbolism then of this particular classification. Only when I studied the archetypes did I become aware that this is a very important archetypal pattern that plays an enormous role. (342)

Jung's evidence, from his clinical observations, is merely anecdotal. He talks about the extravert and the introvert as types. He also talks about the thinking type, the feeling type, the sensation type, and the intuition type. His evidence for his claims is not based on any **controlled studies**. He said he "probably would have done them" if he had had the means (315). But as it was, he says, "I had to content myself with the observation of facts" (315).

Jung seems to have realized the limitations of his work and may not have approved of the MBTI had he lived to see it developed in his name. "My scheme of typology," he noted, "is only a scheme of orientation. There is such a factor as introversion, there is such a factor as extraversion. The classification of individuals means nothing, nothing at all. It is only the instrumentarium for the practical psychologist to explain for instance, the husband to a wife or vice versa" (305).

However, his typology seems to imply that science is just a point of view and that using intuition is just as valid a way of seeing and understanding the world and ourselves as is careful observation under controlled conditions. Never mind that that is the only way to systematically minimize **self-deception** or prevent identifying causes where there are none.

Isabel Briggs Myers made similar mistakes:

> In describing the writing of the Manual, she mentioned that she considered the criticisms a thinker would make, and then directed her own thinking to find an answer. An extravert to whom she was speaking said that if he wanted to know the criticisms of thinkers, he would not look into his own head. He would go find some thinkers, and ask them. Isabel looked startled, and then amused. [www.capt.org/About_the_MBTI/ Isabel%20Myers.cfm]

This anecdote typifies the dangers of self-validation. To think that you can anticipate and characterize criticisms of your views fairly and accurately is arrogant and unintelligent, even if it is typical of your personality type. Others will see things you don't. It is too easy to create straw men instead of facing up to the strongest challenges that can be made against your position. It is not because of type that one should send out one's views for critical appraisal by others. It is the only way to be open-minded and complete in one's thinking. To suggest that only people of a certain type can be open-minded or concerned with completeness is to encourage sloppy and imprecise thinking.

The Myers-Briggs Instrument. Isabel Briggs Myers learned test construction by studying the personnel tests of a local bank. She worked up her inventories with the help of family and friends, and she tried her early tests on thousands of schoolchildren in Pennsylvania. Her first longitudinal study was on medical students, who she followed up after 12 years and found that their occupations fit their types. She eventually became convinced that she knew what traits people in the health professions should have ("accurate perception and informed judgment"). She not only thought her tests could help select who would make good nurses and physicians, but "she hoped the use of the MBTI in training physicians and nurses would lead to programs during medical school for increasing command of perception and judgment for all types, and for helping students choose specialties most suited to their gifts."

Others eventually helped her modify and develop her test, which was taken over by CPP in 1975. CPP has turned it into the instrument it is today. "I know intuitive types will have to change the MBTI," she said. "That's in their nature. But I do hope that before they change it, they will first try to understand what I did. I did have my reasons" (The Center for Applications of Psychological Type: www.capt.org).

As noted above, the Myers-Briggs instrument generates sixteen distinct personality profiles based on which side of the four scales one tends toward. Technically, the instrument is not supposed to be used to spew out personality profiles and pigeonhole people, but the temptation to do so seems irresistible. Providing personality tests and profiles has become a kind of entertainment on the Internet. There is also a pernicious side to these profiles: they can lead to discrimination and poor career counseling. Employers may hire, fire, or assign personnel by personality type, despite the fact that the MBTI is not even reliable at identifying one's type (Pittenger 1993).

Here are some excerpts from Myers-Briggs profiles. Note how parts of each profile could fit most people.

1. Serious, quiet, earn success by concentration and thoroughness. Practical, orderly, matter-of-fact, logical, realistic and dependable. See to it that everything is well organized. Take responsibility. Make up their own minds as to what should be accomplished and work toward it steadily, regardless of protests or distractions.

2. Usually have original minds and great drive for their own ideas and purposes. In fields that appeal to them, they have a fine power to organize a job and carry it through with or without help. Skeptical, critical, independent, determined, sometimes stubborn. Must learn to yield less important points in order to win the most important.

The first profile is of an ISTJ (introversion, sensation, thinking, judgment), a.k.a. "The Trustee." The second is of INTJ (introversion, intuition, thinking, judgment), a.k.a. "The Scientist." The profiles read like something from Omar the astrologer and seem to exemplify the **Forer effect**.

See also **enneagram** and **testimonial evidence**.

Further reading: Dawes 1994; Forer 1949; Quenk 1999; Thiriart 1991.

N

natural

Something present in or produced by nature is natural, such as an earthquake, typhoon, or poisonous mushroom. Death is natural in the sense that to die is to conform to the ordinary course of living things in nature. For a diabetic to die from lack of insulin would be natural. It would be unnatural for a diabetic to inject natural or synthetic insulin, since injections are not natural. Rotting wood on your porch is natural in the sense that you have not used anything artificial to protect it. The smell of rotting garbage is natural. Meanness and cruelty are natural to some people; that is, they are inherent, nonacquired personal traits. Some people are apparently natural-born killers. Squishing bugs and kicking cats is natural for some people, in the sense that they do such things spontaneously, without reflecting on their actions. Nudity is the only natural state for animals, even humans. All clothing is artificial, that is, not natural. So are the fillings in your teeth. So is all make-up and jewelry. Bearded men are natural. To shave is to do something unnatural.

Just because something is natural does not mean that it is good, safe, or healthy. Herbs are natural but they are also drugs when used in the diagnosis, treatment, or prevention of a disease. The chemicals that comprise synthetic drugs are natural. St. John's Wort (*Hypericum perforatum*) is natural, but it is a drug. Why do some people say that they prefer St. John's Wort to drugs for depression? If someone said that he preferred Irish whiskey to alcohol, we'd think he was confused. St. John's Wort contains hypericin, which inhibits mono-amine oxidase, a chemical associated with depression. In other words, St. John's Wort (hypericin) is an MAO inhibitor. Medical doctors commonly prescribe MAO inhibitors to treat depression. Other types of antidepressants have become more popular because they have far fewer side effects. MAO inhibitors should not be used when a person eats substances containing the amino acid tyramine or bacteria with enzymes that can convert tyrosine to tyramine, for example, alcoholic beverages, products made with yeast, aged cheese, sour cream, liver, canned meats, salami, sausage, pickled herring, eggplant, and soy sauce. Otherwise, convulsions, extremely high fever, and death by natural causes may occur.

Some plants are lethal even though they are natural. But if you die from eating a lethal but natural plant, you will not be said to have died of natural causes. Ditto if you die from being bitten by a poisonous snake whose venom is quite natural. If you die from lung cancer caused by smoking tobacco, a natural plant, you will, however, be said to have died of natural causes.

Fleas on dogs are natural. Flea collars are unnatural. Mosquitoes and flies are natural, though most people find them to be a nuisance and prefer the unnatural comfort of mosquitoeless nights and flyless barbecues. Eating meat might be a natural act, but eating cooked meat is unnatural. Most sauces put on meat are made with both natural and artificial ingredients. Salt is natural, but some healthy people avoid salt like the plague.

Civilization is unnatural. Indoor plumbing is unnatural. Corrective lenses are unnatural. So are automobiles. Think about that the next time you drive to the garden shop to get some natural fertilizer for your garden or to your naturalist herb shop for a little pick-me-up.

To have a broken arm set by a physician is unnatural. To let it heal spontaneously would be natural, if debilitating for life. Getting a medical degree is unnatural. Foraging and experimenting by trial and error would be natural, if often lethal. Children born with no brains or other monstrous deformities are natural. Brain surgery to remove a tumor is unnatural.

Anything supernatural is unnatural but is usually considered to be good by those who believe in the supernatural. Reading and writing are unnatural. Urinating whenever one has the urge is natural, but uncivilized. Marijuana is natural, so it must be good, right? LSD is unnatural, though mescaline is natural. Ergot is natural. Mold and bacteria and viruses are natural. Arsenic is natural. To strike back when struck is natural, but considered unchristian. Turning the other cheek when struck is considered Christian, but it is unnatural.

Monogamy is natural among some mammals, but unnatural for most mammals. Reproduction is natural but marriage is unnatural. Using condoms is unnatural. Dying of AIDS is to die of natural causes. Herpes is natural. Raping women is natural to some men, but it is usually regarded as evil nonetheless. Pedophilia seems to be natural in some people, but does that make it good?

In fact, ultimately everything that is made is comprised of nothing but natural atoms, molecules, elements, or substances. So, if everything is basically natural, why do followers of **naturopathy** make such a big fuss about using only what is natural? Such an obsession seems unhealthy, but it helps one avoid having to ask difficult questions about whether something really is good, safe, or healthy. All you need to know is that something is "natural" and you don't have to think about its value. Not thinking comes naturally to some people.

naturalism

A **metaphysical** theory that all phenomena can be explained in terms of **natural** (as opposed to supernatural) causes and laws.

Naturalism neither denies nor affirms the existence of God, either as transcendent or immanent. However, naturalism makes God an unnecessary hypothesis and essentially superfluous to scientific investigation. Reference to moral or divine purposes has no place in scientific explanations. On the other hand, the scope of science is limited to explanation of empirical phenomena without reference to supernatural forces, powers, or influences.

The difference between naturalistic and supernaturalistic views might best be understood by noting that the former favors mechanistic explanations, while the latter favors teleological ones. Mechanistic explanations are dysteleological, that is, they make no reference to purposes or design, except metaphorically as in biology (e.g., the heart was designed to pump blood).

The difference between mechanistic and teleological views may best be understood by considering a couple of examples.

The Sex Drive. From a teleological point of view, the sex drive is designed to reproduce the species. The pleasure that accompanies sex is the main inducement to carry out the purpose of reproduction. If sex were generally painful, most members of the species would avoid it, and hence the species would become extinct. Some theologians maintain that to engage in sex for the purpose of reproduction is the only proper sexual motive. To frustrate the reproductive purpose of sex is to act contrary to divine purpose and is immoral. Birth control and homosexuality, therefore, are morally wrong because they are unnatural.

From a mechanistic point of view, the sexual urge is purposeless. It was not designed to motivate animals to reproduce. Rather, animals with a strong sexual drive reproduce, and hence flourish. A species with a weak sexual drive would be unlikely to survive. According to this view, the purpose of sex can't be frustrated, since sex, in general, has no purpose.

Bee Pollination. From a teleological point of view, bee pollination of orchards is purposive and part of a design. To the mechanist, bees just do their thing, and, as a result, orchards get pollinated. If no animals existed that do what bees do, orchards wouldn't exist. The world would be a different place, but it would still be a world. Different mechanisms mean different worlds. The choice is not between this world or none at all, but this world or some other one.

See also **science**.

Further reading: Brooke 1991.

naturopathy

A system of therapy and treatment that relies exclusively on **natural** remedies such as sunlight, supplemented with such things as diet and **massage**. However, some naturopaths have been known to prescribe unnatural treatments such as colon hydrotherapy for diseases like asthma and arthritis. Naturopathy is often, if not always, practiced in combination with other forms of "**alternative**" **health practices**. For example, Bastyr University, a leading school of naturopathy since 1978, offers instruction in such things as **acupuncture** and spirituality.

Naturopathy is based on the belief that the body is self-healing and will repair itself and recover from illness spontaneously if it is in a healthy environment. Naturopaths have many remedies and recommendations for creating a healthy environment in the body.

Naturopaths claim to be **holistic**, which means they believe that the natural body is joined to a supernatural **soul** and that both must be treated as a unit, whatever that means. Naturopathy is fond of such terms as "balance," "harmony," and **energy**. It is often rooted in mysticism and a metaphysical belief in **vitalism** (Barrett 2001a).

Naturopaths assume that many diseases, including cancer, are caused by faulty immune systems. Naturopaths are also prone to make grandiose claims about some herb or remedy that can *enhance* the immune system. Yet only medical doctors are competent to do the tests necessary to determine whether an individual's immune system is in any way depressed (Green: www.quackwatch.com/01QuackeryRelatedTopics/Cancer/immuneboost.html). The immune system, in simple terms, is the body's own set of mechanisms that attacks anything that isn't "self." But in some cases the immune response goes haywire, and rather than attack foreign bodies such as viruses, fungus, or bacteria, the body attacks its own cells, for example, in lupus and multiple sclerosis.

Naturopaths also promote the idea that the *mind* can be used to enhance the immune system and thereby improve one's health. Yet "there are no credible reports in the scientific literature to support the contention that 'alternative' methods operate—on cancer or on any other disease—through an immune mechanism" (Green: www.hcrc.org/contrib/green/immunol.html). Furthermore, the evidence that diseases such as cancer occur mainly in people with compromised immune systems is lacking. This is an assumption made by many naturopaths, but the scientific evidence does not support it. Immunologists have shown that the most common cancers flourish in hosts with fully functional and competent immune systems (ibid.). The notion that vitamins and colloidal minerals, herbs, coffee enemas, colonic irrigation, Laetrile, and meditation can enhance the immune system and thereby help restore health is completely bogus. On the one hand, it is not necessarily the case that a diseased person even has a compromised immune system. On the other hand, there is no scientific evidence that any of these remedies either enhance the immune system or make it possible for the body to heal itself.

Much of the advice of naturopaths is sound: exercise, quit smoking, eat lots of fresh fruits and vegetables, practice good nutrition. Claims that these and other practices such as colonic irrigation or coffee enemas "detoxify" the body or enhance the immune system or promote "homeostasis," "harmony," "balance," "vitality," and so on are exaggerated and not backed up by sound research.

See also **Ayurvedic medicine**, **homeopathy**, and Joel D. **Wallach**.

Nazca lines

Geoglyphs and geometric line clearings in the Peruvian desert, made by the Nazca people, who flourished between 200 B.C.E. and 600 C.E. along rivers and streams that flow from the Andes. The desert itself runs for over 1,400 miles along the Pacific Ocean. The area of the Nazca art is called the Pampa Colorada ("Red Plain"). It is 15 miles wide and runs some 37 miles parallel to the Andes and the sea. Dark red surface stones and soil have been cleared away, exposing the lighter colored subsoil, creating the lines. There is no sand in this desert. From the air, the "lines" include not only lines and geometric shapes, but also depictions of animals and plants in styl-

ized forms. Some of the forms, including images of humans, grace the steep hillsides at the edge of the desert.

The Nazca lines must have taken hundreds of years to create and required a large number of people working on the project. Their size and their purpose have led some to speculate that visitors from another planet either created or directed the project. Erich von Däniken thinks that the Nazca lines formed an airfield for **ancient astronauts**, an idea first proposed by James W. Moseley in the October 1955 issue of *Fate* and made popular in the early '60s by Louis Pauwels and Jacques Bergier in *The Morning of the Magicians*. If Nazca was an alien airfield, it must have been a very confusing one, consisting as it does of giant lizards, spiders, monkeys, llamas, dogs, and hummingbirds, not to mention zigzagging and crisscrossing lines and geometric designs.

It was very considerate of the aliens to depict plants and animals of interest to the locals, even though it must have meant that navigation would be more difficult than with a straight runway or large clearing. Also, the airport must have been a very busy place, needing 37 miles of runway to handle all the traffic. However, it is unlikely spacecraft could have landed in the area without disturbing some of the artwork or the soil. There is no evidence of such disturbance.

The alien theory is proposed mainly because some people find it difficult to believe that a race of "primitive Indians" could have had the intelligence to conceive of such a project, much less the technology to bring the concept to fruition. The evidence points elsewhere, however. Mesoamerican civilizations such as the Aztecs, the Toltecs, the Inca, and the Maya are proof enough that the Nazca did not need extraterrestrial help to create their art gallery in the desert.

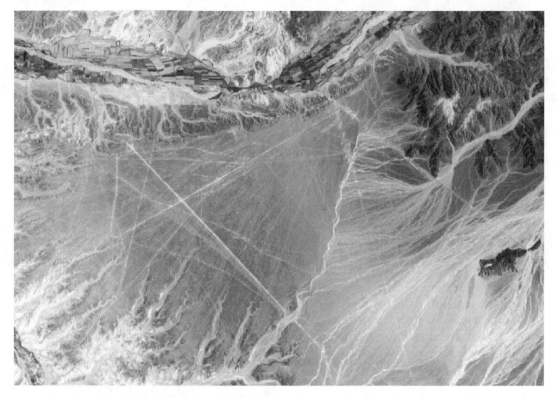

Nazca lines from the Terra satellite Dec. 22, 2000.

In any case, one does not need a very sophisticated technology to create large figures, geometrical shapes, and straight lines, as has been shown by the creators of so-called **crop circles**. The Nazca probably used grids for their giant geoglyphs, as their weavers did for their elaborate designs and patterns. The most difficult part of the project would have been moving all the stones and earth to reveal the lighter subsoil. There really is nothing mysterious about how the Nazca created their lines and figures.

Some think it is mysterious that the figures have remained intact for so many hundreds of years. However, the geology of the area solves that mystery:

> Stones (not sand) comprise the desert surface. Rusted by humidity, their dark-ened color increases heat absorption. The resulting cushion of warm surface air acts as a buffer against the wind; while minerals in the soil help to solidify the stones. On the "desert pavement" thus created in this dry, rainless environment, erosion is practically nil—making for remarkable preservation of the markings. [www.travelvantage.com/per_nazc.html]

The mystery is, why did the Nazca engage in such a project involving so many people for so many years?

G. von Breunig thinks the lines were used for running foot races. He examined the curved pathways and determined that they were partially shaped by continuous running. Anthropologist Paul Kosok briefly maintained that the lines were part of an irrigation system, but soon rejected the notion as impossible. He then speculated that the lines formed a gigantic calendar. Maria Reiche, a German immigrant and apprentice archaeologist to Julio Tello of the University of San Marcos, developed Kosok's theory and spent most of her life collecting data to show that the lines represent the Nazca's astronomical knowledge. Reiche identified many interesting astronomical alignments, which, had they been known to the Nazca, might have been useful in planning their planting and harvesting. However, there are so many lines going in so many different directions that *not* finding many with interesting astronomical alignments would have been miraculous.

The Nazca lines became of interest to anthropologists after they were seen from the air in the 1930s. It is likely that a project of this magnitude was religious in purpose. To involve the entire community for many centuries indicates the supreme significance of the site. Like pyramids, giant statues, and other monumental art, the Nazca art speaks of permanence. It says: We are here and we are not moving. These are not nomads, nor are they hunters and gatherers. This is an agricultural society. It is, of course, a prescientific agricultural society, that turned to magic and superstition (i.e., religion) to assist them with their crops. The Nazca had the knowledge to irrigate, plant, harvest, collect, and distribute. But weather is fickle. Things might go smoothly for years, or even centuries, and then, in a single generation, entire communities are forced to leave because of drought, floods, tidal waves, volcanic eruptions, earthquakes, fires, or whatever else Mother Nature might hurl their way. Hence, the theory of archaeologist David Johnson—that at least some of the lines mark out underground water supplies—has been gaining momentum in recent years (www.eggi.com/ken/peru).

Was this a site for worship? Was this the Mecca of the Nazca? A place of pilgrimage? Were the images used in rituals aimed at appeasing the gods or asking for help with the fertility of the people and the crops or for better weather or a good supply of water? That the figures could not be seen on the ground as those in the heavens might see them would not be that important for religious or magical purposes. In any case, similar figures to the giants at Nazca decorate the pottery found in nearby burial sites, and it is apparent from their cemeteries that the Nazca were preoccupied with death. Mummified remains litter the desert, discarded by grave robbers. Was this a place for rituals aimed at bringing immortality to the dead? We don't know, but if this mystery is ever to be cleared up it will be by serious scientists, not by alienated pseudoscientific speculators **shoehorning** the data to fit their extraterrestrial preconceptions.

Further reading: Davies 1998; Kroeber et al. 1999.

near-death experience (NDE)

There is a wide array of experiences reported by some people who have nearly died or who have thought they were going to die, all of which are described as near-death experiences. There is no single shared experience reported by those who have had NDEs. Even the experiences of most interest to **parapsychologists**—such as the "mystical experience," the "light at the end of the tunnel" experience, the "life review" experience, and the **out-of-body experience** (OBE)—rarely occur together in near-death experiences. However, the term NDE is most often used to refer to an OBE occurring while near death.

Two M.D.s who have popularized the idea that the NDE is proof of life after death are Elisabeth Kübler-Ross and Raymond **Moody**. The former is well known for her work on death and dying. The latter has written several books on the subject of life after life, and has comprised a list of features he considers to be typical of the NDE. According to Moody, the typical NDE includes a buzzing or ringing noise; a sense of blissful peace; a feeling of floating out of one's body and observing it from above; moving through a tunnel into a bright light; meeting dead people, saints, Christ, and angels; seeing one's life pass before one's eyes; and finding it all so wonderful that one doesn't want to return to one's body. This composite experience is based on interpretations of **testimonial evidence** from doctors, nurses, and patients. Characteristic of Moody's work is the glaring omission of cases that don't fit his hypothesis. If Moody is to be believed, no one near death has had a horrifying experience.

Yet there are numerous reports of bad NDE trips involving tortures by elves, giants, demons, and so on. Some parapsychologists take these good and bad NDE trips as evidence of heaven and hell. They believe that some souls actually leave their bodies and go to the other world for a time before returning to their bodies. If so, then what is one to conclude from the fact that most people near death experience neither the heavenly nor the diabolical? Is that fact good evidence that there is no afterlife or that most people end up in some sort of limbo? Such reasoning is on par with supposing that **dreams** in which one appears to oneself to be outside of one's bed are to be taken as evidence of the soul or mind actually leaving the body during sleep.

What little research there has been in this field indicates that the experiences Moody lists as typical of the NDE may be due to brain states triggered by cardiac arrest and anesthesia (Blackmore 1993). Furthermore, many people who have not been near death have had experiences that seem identical to NDEs. These mimicking experiences are often the result of psychosis (due to severe neurochemical imbalance) or usage of drugs such as hashish, LSD, or DMT (dimethyltryptamine; Strassman 2001).

Moody thinks that NDEs prove the existence of life after death. Skeptics believe that NDEs can be explained by neurochemistry and are the result of brain states that occur due to a dying, demented, or drugged brain. For example, neural noise and retino-cortical mapping explain the common experience of passage down a tunnel from darkness into a bright light. According to Susan Blackmore, the vision researcher Dr. Tomasz S. Troscianko of the University of Bristol speculated:

> If you started with very little neural noise and it gradually increased, the effect would be of a light at the centre getting larger and larger and hence closer and closer. . . . [T]he tunnel would appear to move as the noise levels increased and the central light got larger and larger. . . . If the whole cortex became so noisy that all the cells were firing fast, the whole area would appear light. (ibid.: 85)

Blackmore attributes the feelings of extreme peacefulness of the NDE to the release of endorphins in response to the extreme stress of the situation. The buzzing or ringing sound is attributed to cerebral **anoxia** and consequent effects on the connections between brain cells (ibid.: 64).

Dr. Karl Jansen (2000) has reproduced NDEs with ketamine, a short-acting, hallucinogenic, dissociative anaesthetic:

> The anaesthesia is the result of the patient being so "dissociated" and "removed from their body" that it is possible to carry out

surgical procedures. This is wholly different from the "unconsciousness" produced by conventional anaesthetics, although ketamine is also an excellent analgesic (pain killer) by a different route (i.e. not due to dissociation). Ketamine is related to phencyclidine (PCP). Both drugs are arylcyclohexylamines—they are not opioids and are not related to LSD. In contrast to PCP, ketamine is relatively safe, is much shorter acting, is an uncontrolled drug in most countries, and remains in use as an anaesthetic for children in industrialised countries and all ages in the third world as it is cheap and easy to use. Anaesthetists prevent patients from having NDE's ("emergence phenomena") by the co-administration of sedatives which produce "true" unconsciousness rather than dissociation.

According to Dr. Jansen, ketamine can reproduce all the main features of the NDE, including travel through a dark tunnel into the light, the feeling that one is dead, communing with God, hallucinations, OBEs, and strange noises. This does not prove that there is no life after death, but it does prove that an NDE is not proof of an afterlife. In any case, the so-called typical NDE is not typical of anything, except the tendency of parapsychologists to selectively isolate features of a wide array of experiences and **shoehorn** them to a **paranormal** or supernatural presupposition.

Finally, Raymond Quigg Lawrence (*Blinded by the Light,* 1996) thinks that NDEs are the work of **Satan**. That is at least as good a theory as the theory that they are due to real visits to other planes of reality.

See also **astral projection** and **remote viewing**.

Further reading: Blackmore 1991b; Jansen 2001.

neuro-linguistic programming (NLP)

A **Large Group Awareness Training** program. NLP is a competitor with **Landmark Forum**, Anthony Robbins & Associates, and legions of other enterprises that, like the Sophists of ancient Greece, travel from town to town to teach their wisdom for a fee. Robbins is probably the most successful graduate of NLP. He started his own empire after transforming from a self-described "fat slob" to a **firewalker** to "the nation's foremost authority on the psychology of peak performance and personal, professional and organizational turn-around" (www.tonyrobbins.com/bio .html). The founders of NLP, Richard Bandler and John Grinder, might disagree.

NLP has something for everybody, the sick and the healthy, individual or corporation. In addition to being an agent for change for healthy individuals taught en masse, NLP is also used for individual psychotherapy for problems as diverse as phobias and schizophrenia. NLP also aims at transforming corporations, showing them how to achieve their maximum potential and achieve great success.

NLP was begun in the mid-'70s by a linguist (Grinder) and a mathematician (Bandler) who had strong interests in successful people, psychology, language, and computer programming. It is difficult to define NLP because those who started it and those involved in it use such vague and ambiguous language that NLP means radically different things to different people. While it is difficult to find a consistent description of NLP among those who claim to be experts at it, one metaphor keeps recurring. NLP claims to help people change by teaching them to *program their brains*. We were given brains, we are told, but no instruction manual. NLP offers a user manual for the brain.

This brain-manual metaphor for NLP training is sometimes referred to as "software for the brain." Furthermore, NLP, consciously or unconsciously, relies heavily on the notion of the **unconscious** mind as a constant influence on conscious thought and action; metaphorical behavior and speech, especially building on the methods used in Freud's *Interpretation of Dreams;* and hypnotherapy as developed by Milton Erickson. NLP is also heavily influenced by the work of Gregory Bateson and Noam Chomsky.

One common thread in NLP is the emphasis on teaching a variety of communication and persuasion skills and using self-hypnosis to motivate and change oneself. Most NLP practitioners advertising on the Internet make grand claims about being able to help just about anybody become just about anything. The following is typical:

> NLP can enhance all aspects of your life by improving your relationships with loved ones, learning to teach effectively, gaining a stronger sense of self-esteem, greater motivation, better understanding of communication, enhancing your business or career . . . and an enormous amount of other things which involve your brain. [www.nlpinfo.com/intro/txintro.shtml]

Some advocates claim that they can teach an infallible method of telling when a person is lying (www.140.239.226.168/random/police-interrogation.htm). Some claim that people fail only because their teachers have not communicated with them in the right language. One NLP guru, Dale Kirby, informs us that one of the presuppositions of NLP is "No one is wrong or broken" (www.rain.org/~da5e/nlpfaq.html). So why seek remedial change? On the other hand, what Mr. Kirby does have to say about NLP that is intelligible does not make

it very attractive. For example, he says that, according to NLP, "There is no such thing as failure. There is only feedback." Was NLP invented by the U.S. military to explain their "incomplete successes"? When the space shuttle blew up within minutes of launch, killing everyone on board, was that "only feedback"? If I stab my neighbor and call it "performing non-elective surgery," am I practicing NLP? If I am arrested in a drunken state with a knife in my pocket for threatening an ex-girlfriend, am I just "trying to rekindle an old flame"?

Another NLP presupposition that is false is, "If someone can do something, anyone can learn it." This comes from people who claim they understand the brain and can help you reprogram yours. They want you to think that the only thing that separates the average person from Einstein or Pavarotti or the World Champion Log Lifter is NLP.

NLP is said to be the study of the structure of subjective experience, but a great deal of attention seems to be paid to observing behavior and teaching people how to read body language. However, there is no common structure to nonverbal communication, any more than there is a common structure to dream symbolism. There certainly are some well-defined culturally determined nonverbal ways of communicating, for example, giving someone "the finger" has a definite meaning in American culture. But when someone tells me that the way I wrinkle my nose during a conversation means that I am signaling that I think his idea stinks, how do we verify his interpretation? I deny it. He knows the sign, he says. He knows the meaning. I am not aware of my signal or of my feelings, he says, because the message is coming from my subconscious mind. While there is solid research (Eckman and Friesen 1975; Eckman and Rosenberg 1997) supporting the claim that certain identifiable facial expressions generally imply particular emotions, our knowledge of interpreting body language in general is about on par with our ability to interpret dreams. There are some cases where familiarity with a person and the context of their behavior makes it reasonable to infer what that person is *probably* thinking or feeling. But we are nowhere near a highly reliable method for the interpretation of nonverbal communication in general. Sitting cross-armed at a meeting might not mean that someone is blocking you out or getting defensive. She may just be cold or have a backache or simply feel comfortable sitting that way. It is dangerous to read too much into nonverbal behavior. Those splayed legs may simply indicate a relaxed person, not someone inviting you to have sex. At the same time, much of what NLP is teaching is how to do **cold reading**. This is valuable, but is an art, not a science, and should be used with caution.

Finally, NLP claims that each of us has a Primary Representational System (PRS), a tendency to think in specific modes: visual, auditory, kinesthetic, olfactory, or gustatory. A person's PRS can be determined by words the person tends to use or by the direction of one's eye movements. Supposedly, a therapist will have a better rapport with a client if they have a matching PRS. None of this has been supported by the scientific literature.

Richard Bandler's First Institute of Neuro-Linguistic Programming and Design Human Engineering has this to say about NLP:

> Neuro-Linguistic Programming (NLP) is defined as the study of the structure of subjective experience and what can be calculated from that and is predicated upon the belief that all behavior has structure. . . . Neuro-Linguistic Programming was specif-

ically created in order to allow us to do magic by creating new ways of understanding how verbal and non-verbal communication affect the human brain. As such it presents us all with the opportunity to not only communicate better with others, but also learn how to gain more control over what we considered to be automatic functions of our own neurology. [www.purenlp.com/whatsnlp.htm]

We are told that Bandler took as his first models Virginia Satir ("The Mother of Family System Therapy"), Milton Erickson ("The Father of Modern Hypnotherapy"), and Fritz Perls (an early advocate of Gestalt therapy) because they "had amazing results with their clients." The linguistic and behavioral patterns of such people were studied and used as models. These were therapists who liked expressions such as "self-esteem," "validate," "transformation," "harmony," "growth," "ecology," "self-realization," "unconscious mind," "nonverbal communication," "achieving one's highest potential"—expressions that serve as beacons to New Age transformational psychology. No neuroscientist or anyone who has studied the brain is mentioned as having had any influence on NLP. Also, someone who is not mentioned, but who certainly seems like the ideal model for NLP, is Werner Erhard. He started **est** in San Francisco just a couple of years before Bandler and Grinder started their training business a few miles to the south in Santa Cruz. Erhard seems to have set out to do just what Bandler and Grinder set out to do: help people transform themselves, and make a good living doing it. NLP and est also have in common the fact that they are built up from a hodgepodge of sources in psychology, philosophy, and other disciplines. Both have been brilliantly marketed as offering the key to success, happiness, and fulfillment to anyone willing to pay the price of admission. Best of all: No one who pays his fees flunks out of these schools.

When one reads what Bandler says, it may lead one to think that some people sign on just to get the translation from the Master Teacher of Communication Skills himself:

> One of the models that I built was called strategy elicitation which is something that people confuse with modeling to no end. They go out and elicit a strategy and they think they are modeling but they don't ask the question, "Where did the strategy elicitation model come from?" There are constraints inside this model since it was built by reducing things down. The strategy elicitation model is always looking for the most finite way of accomplishing a result. This model is based on sequential elicitation and simultaneous installation. [www.purenlp.com/rbnnn.htm]

Many would surely agree that with communication such as this, Bandler must have a very special code for programming his brain.

> I think the more you want to become more and more creative you have to not only elicit other peoples' strategies and replicate them yourself, but also modify others' strategies and have a strategy that creates new creativity strategies based on as many wonderful states as you can design for yourself. Therefore, in a way, the entire field of NLP is a creative tool, because I wanted to create something new. [www.purenlp.com/nlpfaqr.htm]

Bandler claims he keeps evolving. To some, however, he may seem mainly concerned with protecting his economic interests by trademarking his every burp. For someone bent on replicating other peoples' strategies, he seems incongruously concerned that some rogue therapist or trainer might steal his work and make money without him getting a cut. One might be charitable and see Bandler's obsession with trademarking as a way to protect the integrity of his brilliant new discoveries about human potential (such as "charisma enhancement") and how to sell it. To clarify, or perhaps to obscure matters, what Bandler calls "the real thing" can be identified by a license and the trademark from the Society of Neuro-Linguistic Programming™. He has sued former partner Grinder for millions of dollars over the issue. However, do not contact Bandler's organization if you want detailed, clear information about the nature of NLP or DHE (Design Human Engineering™, which will teach you to hallucinate designs like Tesla did), or PE (Persuasion Engineering™) or MetaMaster Track™, or Charisma Enhancement™, or Trancing™, or whatever else Mr. Bandler and associates are selling these days. Mostly what you will find on Bandler's page is information on how to sign up for one of his training sessions. For example, one can take a three-day seminar on Creativity Enhancement™ to learn why it's not creative to rely on other people's ideas, except for Bandler's, of course.

John Grinder, on the other hand, has gone on to try to do for the corporate world what Bandler is doing for the rest of us. He has joined Carmen Bostic St. Clair in an organization in Australia called Quantum Leap, "an international organisation dealing with the design and implementation of cross cultural communication systems." Like Bandler, Grinder claims he has evolved new and even more brilliant "codes":

> the New Code contains a series of gates which presuppose a certain and to my way of thinking appropriate relationship between the conscious and unconscious parts of a person purporting to train or

represent in some manner NLP. This goes a long way toward insisting on the presence of personal congruity in such a person. In other words, a person who fails to carry personal congruity will in general find themselves unable to use and/or teach the New Code patterns with any sort of consistent success. This is a design I like very much—it has the characteristic of a self-correcting system. [www .inspiritive.com.au/grinterv.htm]

It may strike some people that terms such as "personal congruity" are not very precise or scientific. This is probably because Grinder has created a "new paradigm." Or so he says. He denies that his and Bandler's work is an eclectic hodgepodge of philosophy and psychology, or that it even builds from the works of others. He believes that what he and Bandler did was "create a paradigm shift."

The following claim by Grinder provides some sense of what he thinks NLP is:

> My memories about what we thought at the time of discovery (with respect to the classic code we developed—that is, the years 1973 through 1978) are that we were quite explicit that we were out to overthrow a paradigm and that, for example, I, for one, found it very useful to plan this campaign using in part as a guide the excellent work of Thomas Kuhn *(The Structure of Scientific Revolutions)* in which he detailed some of the conditions which historically have obtained in the midst of paradigm shifts. For example, I believe it was very useful that neither one of us were qualified in the field we first went after—psychology and in particular, its therapeutic application; this being one of the conditions which Kuhn identified in his historical study of paradigm shifts. Who knows what Bandler was thinking?

One can only hope that Bandler wasn't thinking the same things that Grinder was thinking, at least with respect to Kuhn's classic text. Kuhn did not promote the notion that not being particularly qualified in a scientific field is a significant condition for contributing to the development of a new paradigm in science. Furthermore, Kuhn did not provide a model or blueprint for creating paradigm shifts. His is a *historical* work, describing what he believed to have occurred in the history of science. Nowhere does he indicate that a single person at any time did, or even could, create a paradigm shift in science. Individuals such as Newton or Einstein might provide theories that require paradigm shifts for their theories to be adequately understood, but they don't create the paradigm shifts themselves. Kuhn's work implies that such a notion is preposterous.

While I do not doubt that many people feel they benefit from NLP training sessions, there seem to be several false or questionable assumptions on which NLP is based. Their beliefs about **hypnosis** and the ability to influence people by appealing directly to the subconscious mind are unsubstantiated. All the scientific evidence that exists on such things indicates that what NLP claims is not true. You cannot learn to "speak directly to the unconscious mind," as Erickson and NLP claim, except in the most obvious way of using the power of suggestion and appealing to personal associations.

It seems that NLP develops models that can't be verified, from which it develops techniques that may have nothing to do with either the models or the sources of the models. NLP makes claims about thinking and perception that do not seem to be supported by neuroscience. This is not to say that the techniques won't work. They may work and work quite well, but there is no way to know whether the claims behind their origin are valid. Perhaps it doesn't matter. NLP itself proclaims that it is pragmatic in its approach: What matters is whether it works. However, how do you measure the claim "NLP works"? Anecdotes and **testimonial evidence** from satisfied customers seem to be the main measuring devices. Yet such measurements are notorious for misidentifying a *subjective feeling* of improvement with an *objective behavioral measurement* of improvement. Another criterion might be how persuasive the trainers are in getting clients to enroll in more training sessions and recruit others into the program. By this standard, NLP is an unqualified success.

See also **pragmatic fallacy**.

Further reading: Sacks 1984, 1985, 1995; Schacter 1996.

Noah's Ark

The boat built by the Biblical character Noah. At the command of God, according to the story, Noah was to build a boat that could accommodate his extended family, about 50,000 species of animals and about one million species of insects. The craft had to be constructed to endure a divinely planned universal flood aimed at destroying every other person and animal on earth (except, one supposes, those animals whose habitat is water). This was no problem, according to Dr. Max D. Younce, who says by his calculations from *Genesis* 6:15 that the ark was 450 feet long, 75 feet wide, and 45 feet deep. He says this is equivalent to "522 standard stock cars or 8 freight trains of 65 cars each" (www.heritagebbc.com/archive1/0012.html). By some divine calculation he figures that all the insect species and the worms could fit in 21 box cars. He could be right, though Dr. Younce does not address the issue of how the big boxcar filled with its cargo rose with the

rainwater level instead of staying put beneath the floodwaters.

Those not familiar with the story might wonder why God would destroy nearly all the descendants of all of the creatures he had created. The story is that God was displeased with all of his human creations, except for Noah and his family. Annihilating those with whom one is displeased seems to be a familiar tactic of the followers of this and many other gods, although some might see such behavior in God as a clear case of anthropomorphism.

Despite the bad example God set for Noah's descendants—imagine a human parent drowning his or her children because they were "not righteous"—the story remains a favorite among children. God likes good people. He lets them ride on a boat with a bunch of friendly animals. He shows them a great rainbow after the storm. And they all live happily ever after. Even adults like the story, though they might see it as an allegory with some sort of spiritual message, such as *God is all-powerful and we owe everything, even our very existence, to the Creator.* But there are many who take the story literally.

According to the story told in chapter 7 of *Genesis*, Noah, his crew, and the animals lived together for more than 6 months before the floodwaters receded. There are a few minor logistical problems with this arrangement. First, would this big boat float? Noah might have been given divine guidance here, so maybe this boat could float. This is all done before the discovery of metallurgy, so the boat would have been made of wood and other natural materials. How many forests would it take to provide the lumber for such a boat? How many people working how many years would be required? Building a pyramid would be peanuts compared with building the ark. But remember, people lived a lot longer in those days. Noah was

600 years old when he built his boat in the desert.

Let's say that, however implausible, such a boat could have been built using the wooden-boat technology of the earliest peoples. After all, Noah allegedly had God's help in building his boat. Thus, the need for **miracles** was no obstacle. There is still the problem of gathering the animals together from the various parts of the world that, as far as we know, Noah had no idea even existed. How did he get to the remote regions of the earth to collect exotic butterflies and Komodo dragons? How did he get all those species of dinosaurs to follow him home? How did he get just chimpanzees and bonobos to go to Africa, leave no trace behind, and never leave again? By the time he collected all his species, in twos and sevens, his boat would probably have rotted in the desert sun.

But let's grant that Noah was able to collect all the birds and mammals, reptiles and amphibians, and a couple of million insects that he is said to have gathered together on his boat. There is still the problem of keeping the animals from eating one another. Or are we to believe that the lion was sleeping with the lamb on the ark? Did the carnivores become vegetarians for the duration of the flood? How did he keep the birds from eating the insects? Perhaps the ark was stocked with foods for all the animals. After all, if Noah could engineer the building of a boat that could hold all those animals, it would have been a small feat to add room to store enough food to last for more than 6 months. Maybe he didn't need storage, since the carcasses of all the destroyed animals floating by would surely suffice for the journey. Of course, Noah would have to store enough food for himself and his family, too. But these would have been minor details to such a man with such a plan guided by God.

Still, it seems difficult to imagine how such a small crew could feed all these animals in a single day. There is just Noah, his wife, their three sons, and three daughters-in-law. The "daily" rounds would take years, it seems, not to mention the problems of the "clean-up" detail. Finally, if the noise of all those animals didn't drive Noah's crew insane (not to mention the insect bites), the smell should have killed them. At least they didn't have to worry about water to drink. God provided water in abundance.

As preposterous as this story seems, there are many people in the 21st century who believe it. Even more appalling: There are some who claim they have *found* Noah's ark. They call themselves "arkeologists." They say that when the flood receded, Noah and his zoo were perched on the top of Mt. Ararat in Turkey. Presumably, at that time, all the animals dispersed to the far recesses of the earth. How the animals got to the different continents, we are not told. Perhaps they floated there on debris. More problematic is how so many species survived when they had been reduced to just one pair or seven pairs of creatures. Also, you would think that the successful species that had the farthest to travel would have left a trail of offspring along the way. What evidence is there that all species originated in Turkey? None. But that's what the fossil record should look like if the ark landed on Mt. Ararat.

Still, none of this deters the true believer from maintaining that the story of Noah's ark is the God's truth. Nor does it deter those who think the ark has been found. For example, in 1977 a pseudodocumentary called *In Search of Noah's Ark* was played on numerous television stations. CBS showed a special in 1993 entitled *The Incredible Discovery of Noah's Ark*. The first is a work of fiction claiming to be a documentary. George Jammal masterminded the second, and has admitted that the story was a hoax. Jammal said he wanted to expose religious frauds. About 20 million people saw his hoax, most of whom probably still do not know that Jammal did not want them to take it seriously (Lippard 1994).

During his show, Jammal produced what he called "sacred wood" from the ark, which he later admitted was wood taken from railroad tracks in Long Beach, California, which he had hardened by cooking in an oven. He also prepared other fake wood by frying a piece of California pine on his kitchen stove in a mix of wine, iodine, and sweet-and-sour and teriyaki sauces. He also admitted that he had never been to Turkey (Lippard 1994). The program was produced by Sun International Pictures, based in Salt Lake City, which is responsible for several other pseudodocumentaries on the **Bermuda Triangle**, **Nostradamus**, the **Shroud of Turin**, and **UFOs**.

Further reading: Cerone 1993; Feder 2002; Moore 1993; Plimer 1994.

nocebo, nocebo effect

A nocebo (Latin for "I will harm") is something that should be ineffective but that causes symptoms of ill health. A nocebo effect is an ill effect caused by the suggestion or belief that something is harmful. The term "nocebo" became popular in the 1990s. Prior to that, both pleasant and harmful effects thought to be due to the power of suggestion were usually referred to as being due to the **placebo effect**.

Because of ethical concerns, nocebos are not commonly used in medical practice or research. Thus, it is not unexpected that the nocebo effect is not well established in the scientific literature. However, there are some anecdotes and some studies that are commonly appealed to in the literature to support its validity.

- More than two-thirds of 34 college students developed headaches when told that a nonexistent electrical current passing through their heads could produce a headache.
- "Japanese researchers tested 57 high school boys for their sensitivity to allergens. The boys filled out questionnaires about past experiences with plants, including lacquer trees, which can cause itchy rashes much as poison oak and poison ivy do. Boys who reported having severe reactions to the poisonous trees were blindfolded. Researchers brushed one arm with leaves from a lacquer tree but told the boys they were chestnut tree leaves. The scientists stroked the other arm with chestnut tree leaves but said the foliage came from a lacquer tree. Within minutes the arm the boys believed to have been exposed to the poisonous tree began to react, turning red and developing a bumpy, itchy rash. In most cases the arm that had contact with the actual poison did not react" (Morse 1999).
- In the Framingham Heart Study (begun in 1948), women who believed they are prone to heart disease were nearly four times as likely to die as women with similar risk factors who didn't believe (Reid 2002). Of course, one might argue that the women in both groups had good intuitions. The objective risk factors may have been the same, but subjectively the women knew their bodies better than the objective tests could reveal.
- C. K. Meador (1992) claimed that people who believe in voodoo may actually get sick and die because of their belief.
- "In one experiment, asthmatic patients breathed in a vapor that researchers told them was a chemical irritant or allergen. Nearly half of the patients experienced breathing problems, with a

dozen developing full-blown attacks. They were "treated" with a substance they believed to be a bronchodilating medicine, and recovered immediately. In actuality, both the 'irritant' and the 'medicine' were a nebulized saltwater solution." (Morse 1999)

Arthur Barsky, a psychiatrist at Boston's Brigham and Women's Hospital, found in a recent review of the nocebo literature that patient expectation of adverse effects of treatment or of possible harmful side-effects of a drug played a significant role in the outcome of treatment (Barsky et al. 2002).

Since patients' beliefs and fears may be generated by just about anything they come in contact with, it may well be that many things that are unattended to by many if not most physicians, such as the color of the pills they give, the type of uniform they wear, the words they use to give the patient information, and the kind of room they place a patient in for recovery, may be imbued with rich meaning for the patient and have profound effects for good or for ill on their response to treatment.

Nostradamus (1503–1566)

Michel Nostradamus was a 16th-century French physician and **astrologer**. His modern followers see him as a prophet. His prophecies have a magical quality for those who study them: They are muddled and obscure before the predicted event, but become crystal clear after the event has occurred.

Nostradamus wrote four-line verses (quatrains) in groups of a hundred (centuries). (Note: All quatrains below in modern French are translations from esoterism .com/nostradamus/bien2.htm; the translator prefers to remain anonymous.) Skeptics

consider the "prophecies" of Nostradamus to be mainly gibberish. For example:

L'an mil neuf cent nonante neuf sept mois,
Du ciel viendra un grand Roy deffrayeur:
Resusciter le grand Roy d'Angolmois,
Avant après Mars régner par bonheur.

The year 1999 seven months,
From the sky will come the great King of
    Terror:
To resuscitate the great king of the Mongols,
Before and after Mars reigns by good
    luck. (X.72)

Nobody, not even the most fanatical of Nostradamus's disciples, had a clue what this passage might have meant before July 1999. However, after John F. Kennedy Jr., his wife Carolyn Bessette, and her sister Lauren Bessette were killed in a plane crash on July 18, 1999, the retroprophets **shoehorned** the event to the "prophecy." Here is just one example culled from the Internet:

> Could the crash of John F. Kennedy Jr.'s airplane in July of 1999 fulfill the line "from the sky will come the great King of Terror"? Could the human fear of death and bodily injury be the intended definition of "the great King of Terror"? It might be possible!

"It might be possible"—now *there* is a precise bit of terminology. Other disciples were generous enough to think that Nostradamus was referring to a solar eclipse that would occur on August 11, 1999. Others feared a NASA space probe would come crashing down on earth.

Some claim that Nostradamus predicted the *Challenger* space shuttle disaster on January 28, 1986. Of course, they didn't recognize that he had predicted it until it was too late. Here is the passage:

D'humain troupeau neuf seront mis à part,
De jugement et conseil separés:
Leur sort sera divisé en départ,
Kappa, Thita, Lambda mors bannis égarés.

From the human flock nine will be sent
　away,
Separated from judgment and counsel:
Their fate will be sealed on departure
Kappa, Thita, Lambda the banished dead
　err. (I. 81)

Thiokol made the defective O-ring that is blamed for the disaster. The name has a "k," "th," and an "l." Never mind that there were seven who died, not nine. The rest is vague enough to retrofit many different scenarios.

　True believers, such as Erika Cheetham (*The Final Prophecies of Nostradamus,* 1989), believe that Nostradamus foresaw the invention of bombs, rockets, submarines, and airplanes. He predicted the Great Fire of London (1666) and the rise of Adolf Hitler and many other events.

　Skeptics cast doubt on the interpretation of Nostradamus's quatrains (Randi 1993). Here is how James Randi and Cheetham read one of the more famous quatrains, allegedly predicting the rise of Adolf Hitler to power in Germany:

　Bêtes farouches de faim fleuves tranner;
　Plus part du champ encore Hister sera,
　En caige de fer le grand sera treisner,
　Quand rien enfant de Germain observa.
　　(II. 24)

Cheetham's version:

　Beasts wild with hunger will cross the
　　rivers;
　The greater part of the battle will be
　　against Hitler,

He will cause great men to be dragged in
　a cage of iron,
When the son of Germany obeys no law.

Randi's version:

　Beasts mad with hunger will swim across
　　rivers;
　Most of the army will be against the
　　Lower Danube,
　The great one shall be dragged in an iron
　　cage,
　When the child brother will observe
　　nothing.

Neither translation seems to make much sense, but at least Randi's recognizes that "Hister" refers to a geographical region, not a person. So does "Germania," by the way; it refers to an ancient region of Europe, north of the Danube and east of the Rhine. It may also refer to a part of the Roman Empire, corresponding to present-day northeastern France and part of Belgium and the Netherlands. (Because Hister is an ancient name for the Danube region near Hitler's childhood home, some think the reference is clearly to him.)

　After the terrorist skyjackings and attacks in the United States on September 11, 2001, a rumor was spread that Nostradamus had predicted it. The following quatrains were offered as proof:

　In the year of the new century and nine
　　months,
　From the sky will come a great King of
　　Terror . . .
　The sky will burn at forty-five degrees.
　Fire approaches the great new city . . .

　In the city of York there will be a great
　　collapse,
　2 twin brothers torn apart by chaos
　While the fortress falls the great leader
　　will succumb

Third big war will begin when the big
city is burning.

These quatrains are hoaxes. The first two
lines seem to be an alteration of Centuries
X, quatrain 72:

L'an mil neuf cent nonante neuf sept mois,
Du ciel viendra un grand Roi d'Angolmois.

Or in English,

The year 1999 seven months
From the sky will come the great King of
Terror.

There is no reference in Nostradamus to
"the new century and nine months."

The next two lines are from Centuries
VI, quatrain 97.

Cinq et quarante degrés ciel brûlera,
Feu approcher de la grand cité neuve.

Or in English,

At forty-five degrees the sky will burn,
Fire to approach the great new city.

Some rumormongers speculated that 45
degrees refers to the latitude of New York,
but the latitude in Central Park is 40° 47′ N.
Any but the dimmest bulb should be able
to **shoehorn** "at 45 degrees the sky will
burn" with some aspect of the terrorist
attacks. The remainder of VI. 97 is

Instant grand flamme éparse sautera,
Quand on voudra des Normans faire preuue.

Or in English,

In an instant a great scattered flame will
leap up,
When one will want to demand proof of
the Normans.

The only thing in these lines that is even
vaguely close to what happened is the
mention of "a great scattered flame." Even
the dimmest should be bright enough to
find some way to connect "Normans" giving "proof" with what happened.

The only thing more detestable than
these hoaxes are the **psychics** such as Sil-
via Browne, Patricia Lane, and James Van
Praagh, who claimed *after the fact* that they
had predicted the attacks. Browne even
had the chutzpah to claim she couldn't tell
us the details in advance because she's not
"omniscient." One need not be a psychic
to know that.

According to Barbara and David P.
Mikkelson's Urban Legends page (www
.snopes2.com/rumors/predict.htm), one of
the hoaxed quatrains was written in 1997
by Neil Marshall, a Brock University
(Canada) student. Marshall wanted "to
demonstrate . . . that the writings of Nostradamus are so cryptic that they can be
interpreted to mean almost anything." If we
have some imagination, we can shoehorn
just about any event to some passage in
Nostradamus, or Bob Dylan, for that matter.
In 1981, Dylan wrote a song called *Angelina,*
which is as clear a prediction of September
11 as anything Nostradamus wrote:

There's a black Mercedes rolling through
the combat zone. . . .
Your servants are half-dead, you're down
to the bone. . . .
I see pieces of men marching, trying to
take heaven by force. . . .
In the valley of the giants where the stars
and stripes explode. . . .
Begging God for mercy and weeping in
unholy places.

Finally, there is the view of Jean-
Claude Pecker of the Collège de France in
Paris. He maintains that Nostradamus
described not *future* events but events of

his own and earlier times. According to Pecker, Nostradamus disguised "them in a sort of coded French" because "in his troubled period" he was "under constant threat" (*Skeptical Inquirer,* September/October 2001, p. 81).

One thing Nostradamus didn't predict was that he would become a one-man industry in the 20th century. Publishing houses will never go broke printing the latest predictions culled from the manuscripts of Nostradamus.

See also **Bible Code**, Edgar **Cayce**, Jeane **Dixon**, and **retroactive clairvoyance**.

Further reading: Hines 1990; Martin 1998; Prévost 1999.

numerology

The study of the **occult** meanings of numbers and their influence on human life.

According to an advertisement in *Parade* magazine (February 25, 1996), the definitive text on numerology was written by Matthew Goodwin, an MIT graduate who once worked in the personnel department of an architectural firm. He learned "this science of numbers" (as he calls it) from a clerk at the office. The ad is a pseudoarticle, a print "infomercial," allegedly authored by J. J. Leonard, who is probably Goodwin himself, since the ad is nothing more than an invitation to send him $9 for a numerological reading worth "$80 or more." In his advertisement, he explains how numerology works:

It all starts with your name and birth date. They are the database from which a numerologist is able to describe you, sight unseen. Number values are assigned to the letters in your name. By adding these—with the numbers in your birth date—in a multitude of combinations, a numerologist establishes your key numbers. He then interprets the meaning of these key numbers, which results in a complete description of your personal characteristics.

According to Mr. Goodwin, through numerology you can "see all the diverse parts of your personality and how they uniquely come together to make the person you are." This will enable you to "make the most of your strengths in a way that wasn't possible before."

Just what do you think are the numerical odds that a set of numbers associated with the letters of your name and your birth date will reveal who you are and what you should do with your life, *and* that someone in personnel has figured out how to read those numbers? I'd say the odds are about zero. Nevertheless, numerology shouldn't be brushed off without a thorough examination of its underlying theory. Unfortunately, there isn't any. We are just supposed to take Mr. Goodwin's word for it that numerology works, even though we have no idea how it works, much less do we have any clear idea of what it means for numerology to "work." Numerologists can produce a "reading" for you, just as **astrologers** can. You will be amazed at how

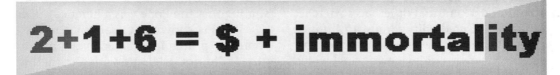

Numerological nonsense.

"accurate" the reading is. You may not even be aware at how selective your thinking has become as you are dazzled by the accuracy of your reading, unless, of course you know something about the **Forer effect**.

Some of the attractiveness of numerology comes from the desire to find somebody who will tell you that you are full of hidden strengths and powers and who will reinforce your deepest needs and emotions. Yet one must be desperate if one doesn't mind that the encouragement comes from a total stranger with no knowledge of who you are.

The attractiveness of numerology may be that numbers give an aura of scientific and mystical authoritativeness, especially if complex statistical analysis is involved. The ad for Mr. Goodwin's $9 numerological reading cites Pythagoras as the father of numerology. The Pythagoreans were a cult with esoteric notions about the universe and numbers, including the notion of the harmony of the spheres. No doubt they found something mystical about the relations of sides of triangles, which we have come to know as the Pythagorean theorem. But there is no evidence that Pythagoras thought he could analyze his disciples' personalities by assigning numbers to the letters of their names and their birth dates. For one thing, he would have realized the unreasonableness of such a notion. Different languages have different alphabets; different cultures use different calendars. It is unreasonable enough to think the universe is arranged according to numerical transcriptions of names, but to think that there are several equivalent transcriptions to accommodate cultural differences stretches the limits of credibility almost to infinity. Even if the universe were so unreasonably designed, how would we ever know which "reading" of a person's numbers is the "correct" one? Does the concept of "correct" reading" even have meaning in this so-called discipline?

It is a misrepresentation of history to cite mathematical mystics, or scientists who have been enamored of mathematics, as fellow travelers. In any case, even if Pythagoras, Plato, Kepler, Galileo, and Einstein were all numerologists, it would not make the theory of numerology one iota more plausible.

See also **Kabalarian philosophy** and the **law of truly large numbers**.

Further reading: Dudley 1998; Martin 1998; Paulos 1990.

O

OBE

See **out-of-body experience**

Occam's razor

Sometimes referred to as the *principle of parsimony* or the *principle of simplicity*. *Pluralitas non est ponenda sine neccesitate* (plurality should not be posited without necessity) are the words of the medieval English philosopher and Franciscan monk William of Ockham (ca. 1285–1349), known as Occam. Like many Franciscans, William was a minimalist, idealizing a life of poverty, and like St. Francis himself, he battled the Pope over the issue. Pope John XXII excommunicated William, who responded by writing a treatise demonstrating that the Pope was a heretic.

What is known as Occam's razor was a common principle in medieval philosophy. It did not originate with Occam, but his name has become indelibly attached to it because of his frequent usage of the principle. It is unlikely that Occam would appreciate what some of us have done in his name. For example, **atheists** often apply Occam's razor in arguing against the existence of God on the grounds that God is an unnecessary hypothesis.

William's use of the principle of unnecessary plurality occurs in debates over the medieval equivalent of **psi**. For example, in Book II of his *Commentary on the Sentences of Peter Abelard,* he is deep in thought about the question of "Whether a Higher Angel Knows Through Fewer Species than a Lower." Using the principle that "plurality should not be posited without necessity," he answers the question in the affirmative. He also cites Aristotle's notion that "the more perfect a nature is, the fewer means it requires for its operation." Some atheists have used this principle to reject the "God the Creator" hypothesis in favor of natural evolution: If a perfect God had created the universe, both the universe and its components would be much simpler. Again, William would not have approved.

He did argue, however, that natural theology is impossible. Natural theology uses reason alone to understand God, as contrasted with revealed theology, which is founded on scriptural revelations. According to Occam, the idea of God is not established by evident experience or evident reasoning. All we know about God we know from revelation. The foundation of all theology, therefore, is **faith**. It should be noted that while others might apply the razor to eliminate the entire spiritual world, Occam did not apply the principle of parsimony to the articles of faith. Had he done so, he might have become a Socinian such as John Toland (*Christianity Not Mysterious,* 1696) and pared down the Trinity to a Unity and the dual nature of Christ to a single nature.

William was somewhat of a minimal-ist in philosophy, advocating nominalism against the more popular view of realism. That is, he argued that universals (general notions) have no existence outside the mind; universals are just names we use to refer to groups of individuals and the properties of individuals. Realists claim that not only are there individual objects and our concepts of those objects, there are also universals. Occam thought that this was one too many pluralities. We don't need universals to explain anything. To nominalists and realists, there exist Socrates the individual and our concept of Socrates. To the realist there exist also such realities as the *humanity* of Socrates, the *animality* of Socrates, and so on. That is, every quality that may be attributed to Socrates has a corresponding reality—a universal, or *eidos,* as Plato called it. According to Occam, the realm of universals is not needed for logic, epistemology, or metaphysics, so why assume this unnecessary plurality? Plato and the realists *could be* right. Perhaps there is a realm of forms, of universal realities that are eternal, immutable models for individual objects. But we don't need to posit such a realm in order to explain individuals, our concepts, or our knowledge.

It might well be argued that Bishop George Berkeley applied Occam's razor to eliminate material substance as an unnecessary plurality. According to Berkeley, we need only minds and their ideas to explain everything. Berkeley was a bit selective in his use of the razor, however. He needed to posit God as the Mind who could hear the tree fall in the forest when nobody is present. Subjective Idealists might use the razor to get rid of God. All can be explained with just minds and their ideas. Of course this leads to solipsism, the view that *I and my ideas alone exist,* or at least they are all I *know* exist. Materialists, on the other hand, might be said to use the razor to eliminate minds altogether. We don't need to posit a plurality of minds as well as a plurality of brains.

Occam's razor is sometimes interpreted to mean something like "the simpler the explanation, the better," or "don't multiply hypotheses unnecessarily." In any case, Occam's razor is a principle that is frequently used outside ontology, for example, by philosophers of science to establish criteria for choosing among theories with equal explanatory power. When giving explanatory reasons for something, don't posit more than is necessary. For example, von Däniken *could* be right: Maybe **ancient astronauts** did teach ancient people art and engineering, but we don't need to posit alien visitations to explain the feats of ancient people. Why posit pluralities unnecessarily? Or as most would put it today, don't make any more assumptions than you have to. We can posit "ether" to explain action at a distance, but we don't need ether to explain it, so why assume it?

Oliver W. Holmes and Jerome Frank might be said to have applied Occam's razor in arguing that there is no such thing as "the Law." There are only judicial decisions and individual judgments, and the sum of them makes up the law. To confuse matters, these eminent jurists called their view legal *realism* instead of legal nominalism. So much for simplifying matters.

Because Occam's razor is sometimes called the principle of simplicity, some simpleminded **creationists** have argued that Occam's razor can be used to support creationism over evolution. After all, having God create everything is much simpler than evolution, which is a very complex mechanism. But Occam's razor does not say that the more simpleminded a hypothesis, the better. If it did, Occam's would be a dull razor for a dim populace.

Some have even found a use for Occam's razor to justify budget cuts, arguing that "what can be done with less is done in vain with more." This approach seems to apply Occam's razor to the principle itself, eliminating the word "assumptions." It also confuses matters by confusing "less" with "fewer." Occam was concerned with fewer assumptions, not less money.

The original principle seems to have been invoked within the context of a belief in the notion that perfection is simplicity itself. This seems to be a metaphysical bias that we share with the medievals and the ancient Greeks. For, like them, most of our disputes are not about this principle, but about what counts as necessary. To the materialist, dualists multiply pluralities unnecessarily. To the **dualist**, positing a mind as well as a body is necessary. To atheists, positing God and a supernatural realm is to posit pluralities unnecessarily. To the **theist**, positing God is necessary. And so on. To von Däniken, perhaps, the facts make it necessary to posit extraterrestrials. To others, these aliens are unnecessary pluralities. In the end, maybe Occam's razor says little more than that for atheists God is unnecessary, but for theists that is not true. If so, the principle is not very useful. On the other hand, if Occam's razor means that when confronted with two explanations, an implausible one and a probable one, a rational person should select the probable one, then the principle seems unnecessary because it's so obvious. But if the principle is truly a minimalist principle, then it seems to imply the more reductionism, the better. If so, then the principle of parsimony might better have been called Occam's chainsaw, for its main use seems to be for clear-cutting ontology.

Further reading: Hyman and Walsh 1973; Thorburn 1918.

occultism

The belief in hidden or mysterious powers that can be controlled by humans who have special knowledge of these powers. The occult includes such things as **alchemy**, **magick**, and other arts of **divination** that use **incantations** or magic formulae in an attempt to gain hidden knowledge or power.

Further reading: Nickell 1991; Seligmann 1997.

optional starting and stopping

A common practice among **psi** researchers who, while testing psychic powers, allow the subject to start or stop whenever he or she feels like it. For example, the subject may go through some warm-ups trying to psychically receive numbers or **Zener card** icons being psychically transmitted by another person. The responses of the warm-ups are recorded, however, and if they look good (i.e., seem to be better than would be expected by chance), then the responses are counted in the experimental data. If not, then the data are discarded. Likewise, if the psychic has had a good run at guessing numbers of card suits and starts to have a bad run, he can call it quits and the bad run is discarded.

This phenomenon seems to be related to another common factor in psi testing: optional keeping and optional disregarding of data. You get to keep all data favorable to your hypothesis and you get to disregard all unfavorable data. Some psi researchers consider this practice to be justified, since psychic powers may come and go. Yet any reasonable test of psychic powers should have a protocol that specifies exactly when the experiment will begin and when it will end.

Optional starting and stopping should not be confused with *displacement effects*, a practice of counting an event as a psychic hit not only if one guesses the target card, but also if one guesses either the one *before* or *after* the target card, thereby significantly increasing one's odds of a correct guess.

See also **parapsychology**.

Further reading: Alcock 1990; Gordon 1987; Hansel 1989; Hyman 1989, 1995; Reed 1988.

oracle

A shrine or temple sanctuary consecrated to the worship and consultation of a prophetic god. The person who transmits prophecies from a deity at such a shrine is also called an oracle, as is the prophecy or revelation itself.

Oracles are usually presented in the form of enigmatic or ambiguous statements or allegories. "Socrates is the wisest of men," "A great king will achieve victory." Such statements can have several meanings, thus affording a greater chance of being interpreted in such a way as to make them accurate than if they were more clear and precise, such as "Socrates has seven toes" or "Cyrus will defeat the Persians at Salamis on Tuesday."

orgone energy

An alleged type of "primordial cosmic energy" discovered by Freudian psychoanalyst Wilhelm Reich (1897–1957) in the late 1930s. Reich claimed that orgone **energy** is omnipresent and accounts for such things as the color of the sky, gravity, galaxies, the failure of most political revolutions, and a good orgasm. In living beings, orgone is called bio-energy, or life energy. Reich believed that orgone energy is "demonstrable visually, thermically, electroscopically and by means of Geiger-Mueller counters" (www.orgone.org). However, only true believers in orgone energy (i.e., *orgonomists* practicing the science of *orgonomy*) have been able to find success with the demonstrations.

Reich claimed to have created a new science (orgonomy) and to have discovered other entities, such as *bions,* which to this day only orgonomists can detect. Bions are alleged vesicles of orgone energy that are neither living nor nonliving, but transitional beings.

Reich died on November 3, 1957, in the federal penitentiary at Lewisburg, Pennsylvania, where he was sent for criminal contempt. To the end, Reich saw himself as a persecuted genius and considered the critics who ridiculed him to be ignorant fools. The criminal charge was levied because Reich refused to obey an injunction against selling quack medical devices such as the Orgone Accumulator and *orgone shooters,* devices that allegedly could collect and distribute orgone energy, thereby making possible the cure for just about any medical disorder except, perhaps, megalomania and self-delusion.

The Food and Drug Administration not only declared that there is no such thing as orgone energy, they had some of Reich's books burned—a sure-fire way to ignite interest in somebody. If the government burned his books, Reich must have been on to something big, so one theory goes. There is another theory: Don't ascribe to conspiracy what is probably due to stupidity.

Despite having no status in the scientific community, Reich's ideas have been passed on by a number of devoted followers led by Elsworth F. Baker, M.D., founder of the American College of Orgonomy, and Dr. James DeMeo of the Orgone Biophysical Research Laboratory, Inc., located in Ashland, Oregon. Baker's successors (he died in 1985) and DeMeo continue to defend both Reich the scientist and orgonomy.

See also **pathological science** and **pseudoscience**.

Further reading: Carlinsky 1994; Gardner 1957.

osteopathy

A medical practice based on the theory that diseases are due chiefly to loss of structural integrity, which can be restored to harmony or equilibrium by manipulation. The manipulation allegedly allows the body to heal itself. Osteopaths use manipulation for diagnosis, treatment, and prevention of disease.

Andrew Taylor Still (1828–1917), a Civil War surgeon in the Union Army, is credited with discovering osteopathy as an alternative to the medical practices common in his day, practices that failed to save his three children from spinal meningitis. Dr. Still became convinced that he could cure diseases by shaking the body or manipulating the spine. In his autobiography, he says he could "shake a child and stop scarlet fever, croup, diphtheria, and cure whooping cough in three days by a wring of its neck" (Barrett 1997). He also advocated clean living, including abstinence from alcohol and medically prescribed drugs. Surgery was to be avoided, if possible. Today, D.O.s (doctors of osteopathy) complement manipulation with standard medical methods of diagnosis and treatment, including recommending drug therapy and surgery if appropriate. D.O.s have four years of medical training at a college of osteopathic medicine and do a one-year internship in primary care. Some continue their education in an area of osteopathic specialization. Nevertheless, there has been no scientific validation of Still's theory of shaking and manipulating to remove obstructions.

See also **chiropractic** and **craniosacral therapy**.

Ouija board

Used in **divination** and **spiritualism**, the Ouija board usually has the letters of the alphabet inscribed on it, along with words such as "yes," "no," "good-bye," and "maybe." A planchette (a slidable three-legged device) or pointer of some sort is manipulated by those using the board. The users ask the board a question and together, or one of them singly, moves the pointer or the board until a letter is selected by the pointer. The selections spell out an answer to the question asked.

Some users believe that **paranormal** forces are at work in spelling out Ouija board answers. Skeptics believe that those using the board either consciously or unconsciously select what is read. They recommend using the board while blindfolded, having an innocent bystander take notes on the letters selected. Usually, the result will be unintelligible, indicating that the meaningful spellings issue from the players rather than some unseen force.

The movement of the planchette is not due to paranormal forces but to unnoticeable movements by those controlling the pointer, known as the **ideomotor effect**. The same kind of unnoticeable movement is at work in **dowsing**.

The Ouija board was first introduced to the American public in 1890 as a parlor game sold in novelty shops:

> E. C. Reiche, Elijah Bond, and Charles Kennard . . . created an all new alphanumeric design. They spread the letters of the alphabet in twin arcs across the middle of the board. Below the letters were the numbers one to ten. In the corners were "YES" and "NO."

Kennard called the new board Ouija (pronounced wE-ja) after the Egyptian word for good luck. Ouija is not really Egyptian for good luck, but since the board reportedly told him it was during a session, the name stuck. [www.museumoftalkingboards.com/history.html]

Kennard lost his company and it was taken over by his former foreman, William Fuld, in 1892.

> One of William Fuld's first public relations gimmicks, as master of his new company, was to reinvent the history of the Ouija board. He said that he himself had invented the board and that the name Ouija was a fusion of the French word "oui" for yes, and the German "ja" for yes. (ibid.)

Although Ouija boards are usually sold in the novelty or game section of stores, many people swear that there is something **occult** about them. For example, Susy Smith in *Confessions of a Psychic* (1971) claims that using the Ouija board caused her to become mentally disturbed. In *Thirty Years Among the Dead* (1924), American psychiatrist Dr. Carl Wickland claims that using the Ouija board "resulted in such wild insanity that commitment to asylums was necessitated." Is this what happens when amateurs try to dabble in the occult? Maybe it is, if they are suggestible, not very skeptical, and disturbed to begin with. However, even very intelligent people who have not gone insane are impressed by Ouija board sessions. They often find it difficult to explain the "communication" as simply the ideomotor effect reflecting unconscious thoughts. One reason they find such an explanation difficult to accept is that the "communications" are sometimes very vile and unpleasant. It is more psychologically pleasing to attribute vile pronouncements to evil spirits than to admit that any among you is harboring vile thoughts. Also, some of the "communications" express fears rather than wishes, such as the fear of death, and such notions can have a very visible and significant effect on some people.

Observing powerful messages and the powerful effect of messages on impressionable people can be impressive. Yet, as experiences with **facilitated communication** have shown, decent people often harbor indecent thoughts of which they are unaware. The fact that a person takes a "communication" seriously enough to have it significantly interfere with the enjoyment of life might be a sufficient reason for avoiding the Ouija board as being more than a "harmless bit of entertainment," but it is hardly a sufficient reason for concluding that the messages issue from anything but our own minds.

Ouspensky, Petyr Demianovich (1878–1947)

Petyr Demianovich Ouspensky was a mathematician and mystic who played Paul to **Gurdjieff**'s Christ, taking the occult and often unintelligible notions of the master and making them palatable, if no more comprehensible, in works such as *In Search of the Miraculous—Fragments of an Unknown Teaching* and *The Fourth Way—A Record of Talks and Answers to Questions Based on the Teaching of G. I. Gurdjieff.*

Unlike Paul, however, Ouspensky eventually lost faith in his master. He left Gurdjieff's Institute for the Harmonious Development of Man and founded the Society for the Study of Normal Man. Ouspensky is likely to remain a favorite among New Agers since he wrote books with titles such as *The Symbolism of the Tarot: Philosophy of Occultism in Pictures and Numbers* and *Tertium Organum: The Third Canon of*

Thought: A Key to the Enigmas of the World, an attempt to reconcile the mysticism of the East with the rationalism of the West.

out-of-body experience (OBE)

A feeling of departing from one's physical body and observing both one's self and the world from outside one's body. The experience is quite common in daydreams, **dreams**, and **memories**, where we quite often take the external perspective. Some people experience an OBE while under the influence of an anesthetic or while semiconscious due to trauma. Some people have an OBE while under the influence of drugs. OBEs have been induced by electrically stimulating the right angular gyrus (located at the juncture of the temporal and parietal lobes). Finally, some people experience an OBE when they are near death (**near-death experiences**, NDEs).

Susan Blackmore, a former **parapsychologist** with heavy skeptical leanings, is considered one of the world's leading authorities on OBEs and NDEs. By her own admission, while attending Oxford University during the early 1970s, she "spent much of the time stoned, experimenting with different drugs" (Shermer 1998). During her first year there, she had an OBE after several hours on the **Ouija board** while stoned on marijuana. The experience also occurred during a period of her life when sleep deprivation was common for her. She describes herself as having been in "a fairly peculiar state of mind" when she had the OBE (ibid.).

In her OBE, Blackmore went down a tunnel of trees toward a light, floated on the ceiling and observed her body below, saw a silver cord connecting her floating **astral body**, floated out of the building around Oxford and then over England, and finally across the Atlantic to New York. (In *An Unquiet Mind,* Kay Redford Jamison, who suffers from bipolar disorder, describes a similar voyage to Jupiter while she was enjoying the manic phase of her mental illness.)

After hovering around New York, Blackmore floated back to her room in Oxford where she became very small and entered her body's toes. Then she grew very big, as big as a planet at first, and then she filled the solar system, and finally she became as large as the universe.

Blackmore attributes her experience to peculiar brain processes such as might cause "neuronal disinhibition in the visual cortex," which is her explanation for hallucinations and NDEs. She did not consider investigating abnormal psychology, where she would find many similar cases of Alice-in-Wonderland voyagers. Instead, she says that she devoted her study to **astral projection** and **theosophy**, hoping to find an answer. Her experience with the silver cord is right out of traditional occult literature on astral projection.

One explanation of the OBE is that consciousness is a separate entity from the body (**dualism**) and can exist without the body and the body without it. The disembodied consciousness can "see," "hear," "feel," "taste," and "smell." Some speculate that "mind," "**soul**," or "consciousness" can operate over vast distances and perceive objects by some mysterious power not yet discovered. Others think that OBEs are due to brain states triggered by disease or stress.

If minds were leaving bodies, one would expect that there would be minds out of their bodies everywhere. You'd think that there'd be a mix-up occasionally and one or two souls or astral bodies would come back to the wrong physical bodies, or at least get their silver cords tangled up. One would expect some minds to get lost and never find their way back to their bodies. There should be at least a few mindless bodies wandering or lying around, abandoned by their souls as unnecessary baggage. There should also be a few confused souls who don't know who they are because they're in the wrong bodies.

The neuroscientists seem to be on the right track and someday we will understand the OBE in terms of brain pathology. That is not to say that OBEs are not real. For those who have them, they become turning points and defining factors in their lives.

See also **remote viewing**.

Further reading: Blackmore 1982, 1993; Grof 1975; Sacks 1985; Siegel 1992.

P

pagan

"Pagan" is the term Christians use for atheists or anyone who doesn't accept the God of the Bible. The word "pagan" is derived from the Latin word for "country dweller." Originally, pagans were the Greek and Roman polytheists who followed the cults of Mithras, Venus, Apollo, Demeter, and others. Today, however, Christians generally reserve the appellation for those who belong to no religion or to one of the New Age nature religions or anti-Christian cults.

See also **druids**, **Wicca**, and **witches**.

palmistry (chiromancy)

Palmistry is the practice of telling fortunes from the lines, marks, and patterns on the hands, particularly the palms.

Palmistry was practiced in many ancient cultures, such as India, China, and Egypt. The first book on the subject appeared in the 15th century. The term "chiromancy" comes from the Greek word for hand (*cheir*). The most famous

19th-century palmist went by the name of Cheiro.

Palmistry was used during the Middle Ages to detect witches. It was believed that certain spots on the hand indicate that one had made a pact with the Devil. The Catholic Church condemned palmistry, but in the 17th century it was taught at several German universities (Pickover 2001: 64). Britain outlawed palmistry in the 18th century. It is popular enough in America in the 20th century to deserve its own book in the *Complete Idiot's Guide* series.

According to Ann Fiery (*The Book of Divination*), if you are right handed, your left hand indicates inherited personality traits and your right hand indicates your individuality and fulfillment of potential. The various lines on the hand are given names like the "life line," the "head line," and the "heart line." The life line supposedly indicates physical vitality, the head line intellectual capacity, and the heart line emotional nature.

Some palmistry mimics **metoposcopy** or **physiognomy**. It claims that you can tell what a person is like by the *shape* of their hands. Creative people have fan-shaped hands and sensitive souls have narrow, pointy fingers and fleshy palms. There is about as much scientific support for such notions as there is for **personology** or **phrenology**. All such forms of **divination** seem to be based on **sympathetic magic**, intuition, and **confirmation bias**.

Like other forms of divination, palmistry can simplify the hard work of self-discovery. It can provide data that an indecisive person can use to make a decision.

Further reading: Fiery 1999; Marlock and Dowling 1994; Park 1986.

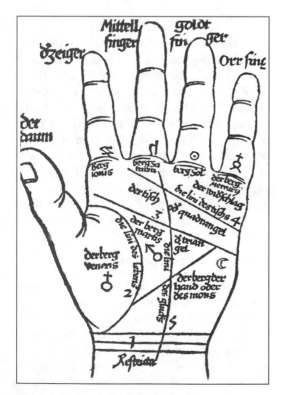

Lines of the hand. Barthélemy Coclès (*Physiognomonia* 1533).

paranormal

An empirical event or perception is said to be paranormal if it involves forces or agencies that are beyond scientific explanation. Some believe that only those with **psychic** powers, such as **ESP** or **psychokinesis**, can experience paranormal events. Some think everyone possesses paranormal abilities. Most skeptics believe that alleged paranormal events have **naturalistic** explanations and that believers in the paranormal are too quick to favor occult explanations for events that seem strange or beyond coincidence.

Parapsychologists study the paranormal, but unlike other sciences almost everyone involved in paranormal studies seeks to prove that their subject matter exists. Some skeptics also test paranormal claims because of the *possibility* that something truly paranormal might occur. James

Randi, for example, is willing to test any paranormal claim. So sure are he and his backers that everyone will fail the test, they offer more than $1,000,000 to anyone who can demonstrate a paranormal power.

Further reading: Alcock 1990; Frazier 1986; Gardner 1957, 1981; Gordon 1987; Hansel 1989; Hines 1990; Hyman 1989; Randi 1982a, 1989a; Stein 1996a.

parapsychology

The search for **paranormal** phenomena, such as **ESP** and **psychokinesis**. Most sciences try to explain observable phenomena. Parapsychologists try to observe unexplainable phenomena.

Scientific methodology in this field dates from at least 1882 at the founding of the Society for Psychical Research in London, which continues to flourish. Its initial members sought to distinguish **psychic** phenomena from **spiritualism**, and to investigate **mediums** and their activities. They studied **automatic writing**, **levitation**, and reports of **ectoplasmic** and poltergeist activity. In America, Joseph Banks Rhine (1895–1980) conducted **psi** experiments at Duke University in the 1930s. His work continues at the Rhine Research Center and at various labs across the country, where experiments have concentrated principally on **astral projection**, ESP, psychokinesis, and **remote viewing**. There are at least half a dozen peer-reviewed journals of parapsychology. However, research in this area has been characterized by deception, fraud, and incompetence in setting up properly controlled experiments and evaluating statistical data (Alcock 1990; Gardner 1981; Gordon 1987; Hansel 1989; Hines 1990; Hyman 1989; Park 2000; Randi 1982a). Psi researchers often find evidence for psi, but a year-long study done by the United States Air Force Research Laboratories (the VERITAC study, named after

the computer used) was unable to verify the existence of ESP. Some parapsychologists, such as Susan Blackmore, have abandoned the search for psi after years of failing to find any significant support for paranormal phenomena (Blackmore 1987).

Despite the fact that psychologists have been in the forefront of paranormal studies, a study of 1,100 college professors in the United States found that only 34% of psychologists believe that ESP is either an established fact or a likely possibility. Comparable figures for other disciplines are much higher: natural scientists (55%), social scientists (excluding psychologists; 66%), and academics in the arts, humanities, and education (77%). Of the psychologists surveyed, 34% believe psi is an impossibility, while only 2% of the other respondents maintained this position (Wagner and Monnet 1979).

Parapsychologists who claim to have found positive results often systematically ignore or rationalize their own studies if they don't support psi. Rhine discarded data that didn't support his beliefs, claiming subjects were intentionally getting answers wrong (**psi-missing**). Many, if not most, psi researchers allow **optional starting and optional stopping**. Most psi researchers limit their research to investigating parlor tricks (guessing the number or suit of a playing card, or "guess what **Zener card** I am looking at" or "try to influence this random number generator with your thoughts"). Any statistical strangeness is attributed to paranormal events.

From the standpoint of physics there seems to be a major problem with the assumption and alleged discovery by some parapsychologists that spatial distance is irrelevant to psi. Each of the four known forces in nature weakens with distance. Thus, as Einstein pointed out in a letter to Dr. Jan Ehrenwald, "This suggests . . . a very strong indication that a non-recognized source of systematic errors may have been involved [in ESP experiments]" (Gardner 1981: 153). The skeptic would rather believe that ESP doesn't exist than believe there is some very strong and powerful force that is undetectable and can act over vast distances—even though we're able to detect what must be a much weaker force—gravity, whose effect diminishes as distance increases—without any trouble at all.

See also **Ganzfeld experiment**, Raymond **Moody**, **pathological science**, and Charles **Tart**.

Further reading: Gardner 1957; Frazier 1986, 1991; Kurtz 1985; Randi 1989a; Stein 1996a.

paraskevidekatriaphobia

Paraskevidekatriaphobia is a morbid, irrational fear of Friday the 13th. Therapist Dr. Donald Dossey, whose specialty is treating people with irrational fears, coined the term. He claims that when you can pronounce the word you are cured.

Paraskevidekatriaphobia is related to *triskaidekaphobia,* the fear of the number 13.

Superstition about Friday the 13th may well be the number one superstition in America today. The number 13 is considered especially unfavorable though it was considered a lucky number in ancient Egypt and China. There were 13 people at the Last Supper. And several mass murderers have 13 letters in their names: Charles Manson, Jeffrey Dahmer, Theodore Bundy. Of course, millions of people who haven't committed any murders, such as Richard Cheney and Robert Redford, have 13 letters in their names, too. As far as I know, nobody has studied how many dinner parties with 13 present went off uneventfully. **Witches**, perhaps to clearly oppose themselves to a Christian superstition, sometimes collect in groups of 13, known as covens.

Some think 13 owes its bad reputation

to Loki, the Norse god of evil, who started a riot when he crashed a banquet at Valhalla attended by 12 gods.

Some cities may skip 13th Street, but not Sacramento, which has an intersection where 13th Street crosses 13th Avenue. Some buildings skip from the 12th to the 14th floor, which, of course, means that the 14th floor is actually the 13th floor.

The ancient Egyptians considered the 13th stage of life to be death, that is, the afterlife, which they thought was a good thing. The Death card in a **tarot** deck is numbered 13 and represents transformation. Those cultures with lunar calendars and 13 months don't associate 13 with anything sinister.

Friday may be considered unlucky because Christ is thought to have been crucified on a Friday, which was execution day among the Romans. Yet Christians don't call it Bad Friday. Friday was also Hangman's Day in Britain. Some even think that Friday was the day God threw Adam and Eve out of Eden.

Photo by R. Carroll.

Friday is Frigga's Day. Frigga was an ancient Nordic fertility and love goddess, equivalent to the Roman Venus, who had been worshipped on the sixth day of the week. The Norse worshipped Frigga on Friday and like the ancient Romans thought it a particularly good kind of day. Christians called Frigga a witch and Friday the witches' Sabbath; modern **Wiccans** are happy to oblige. Some call fear of Friday the 13th "friggatriskaidekaphobia."

Is Friday the 13th a particularly unlucky day? It could be—if you believe it is. Some prophecies are self-fulfilling.

Further reading: Dossey 1992; Radford 2000.

paratrinket

A trinket allegedly endowed with **occult**, **paranormal**, or supernatural powers, such as novelty **talismans** and **amulets**, **crystals**, and **takionic** beads.

pareidolia

A type of illusion or misperception involving a vague or obscure stimulus being perceived as something clear and distinct. For example, in the discolorations of a burnt tortilla one sees the face of Jesus Christ. Or one sees the image of Mother Teresa in a cinnamon bun, or the Virgin Mary in the bark of a tree.

Pareidolia provides a psychological explanation for many delusions based on sense perception. For example, it explains many **UFO** sightings, as well as the hearing of sinister messages on records played backward. Pareidolia explains Elvis, **Bigfoot**, and **Loch Ness monster** sightings. It explains numerous religious apparitions and visions. And it explains why some people see a **face on Mars** or the man in the moon.

Under clinical circumstances, some psychologists encourage pareidolia as a

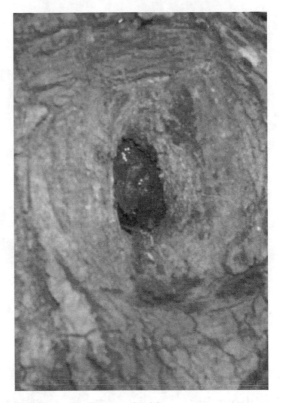

The Virgin Mary in tree bark?

means to understanding a patient, for example, in the **Rorschach inkblot test**.

Astronomer Carl Sagan believed that the human tendency to see faces in tortillas, clouds, and cinnamon buns is an evolutionary trait:

> As soon as the infant can see, it recognizes faces, and we now know that this skill is hardwired in our brains. Those infants who a million years ago were unable to recognize a face smiled back less, were less likely to win the hearts of their parents, and less likely to prosper. These days, nearly every infant is quick to identify a human face, and to respond with a goony [*sic*] grin. (Sagan 1995: 45)

I think Sagan is right about the tendency to recognize faces, but I don't see any reason to think there is an evolutionary advantage to seeing replicas of paintings, demons, ghosts, and so on in inanimate objects. It seems more likely that the mind is making associations with shapes, lines, shadows, and so on, and that these associations are rooted in desires, interests, hopes, and obsessions. Most people recognize illusions for what they are, but some become fixated on the reality of their perception and turn an illusion into a delusion. A little bit of critical thinking, however, should convince most reasonable persons that a cinnamon bun that looks like Mother Teresa or a burnt area on a tortilla that looks like Jesus are accidents and without significance. The Virgin Mary one sees in the reflection of a mirror or on the floor of an apartment complex or in the clouds is more likely to be generated from one's own imagination than be an actual manifestation of a person who has been dead for 2,000 years.

See also **apophenia** and Our Lady of **Watsonville**.

Further reading: Guthrie 1995; Schick and Vaughn 1998.

past life regression (PLR)

The alleged journeying into one's past lives while **hypnotized**. While it is true that many patients recall past lives, it is highly probable that their **memories** are **false memories**. The memories are from experiences in this life, pure products of the imagination, intentional or unintentional suggestions from the hypnotist, or **confabulations**.

Some New Age therapists do PLR therapy under the guise of personal growth; others under the guise of healing. As a tool for New Age explorers, there may be little harm in encouraging people to remember what are probably false memories about their living in earlier centuries or in encour-

aging them to go forward in time and glimpse into the future. But as a method of healing, it must be apparent even to the most superficial of therapists that there are great dangers in encouraging patients to create delusions. Some false memories may be harmless, but others can be devastating. They can increase a person's suffering, as well as destroy loving relationships with family members. The care with which hypnosis and PLR should be used seems obvious.

Some therapists think hypnosis opens a window to the **unconscious mind** where memories of past lives are stored. How memories of past lives get into the unconscious mind of a person is not known, but advocates loosely adhere to a doctrine of **reincarnation**, even though such a doctrine does not require a belief in the unconscious mind as a reservoir of memories of past lives.

PLR therapists claim that past life regression is essential to healing and helping their patients. Some therapists claim that past life therapy can help even those who don't believe in past lives. The practice is given credibility because of the credentials of some of its leading advocates, for example, Brian L. Weiss, M.D., who is a graduate of Columbia University and Yale Medical School and Chairman Emeritus of Psychiatry at the Mount Sinai Medical Center in Miami. There are no medical internships in PLR therapy, nor does being a medical doctor grant one special authority in **metaphysics**, the **occult**, or the supernatural.

Psychologist Robert Baker (1987–88) demonstrated that belief in reincarnation is the greatest predictor of whether a subject would have a past life memory while under past life regression hypnotherapy. Furthermore, Baker demonstrated that the subject's expectations significantly affect the past life regressive session. He divided a group of 60 students into three groups. He told the first group that they were about to experience an exciting new therapy that could help them uncover past lives. Eighty-five percent of this group were successful in remembering a past life. He told the second group that they were to learn about a therapy that may or may not work to engender past life memories. In this group, the success rate was 60%. He told the third group that the therapy was crazy and that normal people generally do not experience a past life. Only 10% of this group had a past life memory.

There are at least two attractive features of past life regression. Since therapists charge by the hour, the need to explore centuries instead of years will greatly extend the length of time a patient will need to be treated, thereby increasing the cost of therapy. Second, the therapist and patient can usually speculate wildly without much fear of being contradicted by the facts. However, this can backfire if anyone bothers to investigate the matter, as in the case of Bridey **Murphy**, the case that seems to have ignited this craze in 1952.

See also **channeling**, **cryptomnesia**, and **Dianetics**.

Further reading: Baker 1990, 1996a; Spanos 1987–88.

pathological science

A term coined by chemistry Nobel-laureate Irving Langmuir (1881–1957). Typical cases involve such things as barely detectable causal agents observed near the threshold of sensation that are nevertheless asserted to have been detected with great accuracy. The supporters offer fantastic theories that are contrary to experience and meet criticisms with **ad hoc** excuses. And, most telling, only supporters can reproduce the results. Critics can't duplicate the experiments.

Langmuir gave several examples, including **ESP** research and **Blondlot's N-Rays**, and stated:

> These are cases where there is no dishonesty involved but where people are tricked into false results by a lack of understanding about what human beings can do to themselves in the way of being led astray by subjective effects, **wishful thinking** or threshold interactions. These are examples of pathological science. These are things that attracted a great deal of attention. Usually hundreds of papers have been published on them. Sometimes they have lasted for 15 or 20 years and then gradually have died away. (Langmuir 1989)

Langumuir visited J. B. Rhine's lab at Duke University. Rhine had results from ESP experiments that could not be predicted by chance and that he claimed were probably due to **psychic** power. Langmuir found that Rhine was not counting all his data, however. He was leaving out the scores of those he believed were intentionally making wrong guesses of their **Zener cards**. "Rhine believed that persons who disliked him guessed wrong to spite him. Therefore, he felt it would be misleading to include their scores" (Park 2000: 42). Rhine determined that some of his subjects were deliberately guessing wrong because their scores were too low to have occurred by chance (**psi-missing**). "Indeed, he was convinced that abnormally low scores were as significant as abnormally high scores in proving the existence of ESP" (ibid.).

Cromer, commenting on Langmuir's characteristics of pathological science, noted that scientists are often not very good judges of the scientific process. Even the best intentions can be subverted by **self-deception**. Good science is not simply a matter of honesty or wisdom: "Real discoveries of phenomena contrary to all previous scientific experience are very rare, while fraud, fakery, foolishness, and error resulting from over-enthusiasm and delusion are all too common" (Cromer 1993).

Do Langmuir's observations imply that scientists should shy away from controversial topics such as prions, **facilitated communication**, cold fusion, ESP, **personology**, and zero-point energy? No. A research scientist must proceed with caution, tentativeness, a sense of the history of science, and an awareness of the tendencies in human nature that can easily lead the wisest of men and women astray. Also, to show little or no interest in allowing oneself and others to try to prove wrong one's fantastic theories, while immediately meeting every objection with ad hoc hypotheses, is a sign of pathological science, if not **pseudoscience**.

See also **confirmation bias**, **control group study**, **science**, and **orgone energy**.

penile plethysmograph (PPG)

A machine for measuring changes in the circumference of the penis. A stretchable band filled with mercury is fitted around the subject's penis. The band is connected to a machine with a video screen and data recorder. Any changes in penis size, even those not felt by the subject, are recorded while the subject views sexually suggestive or pornographic pictures, slides, or movies or listens to audio tapes with descriptions of such things as children being molested. Computer software is used to develop graphs showing "the degree of arousal to each stimulus." The machine cost about $8,000 in 1999 and was first developed in Czechoslovakia to prevent draft dodgers from claiming they were gay to avoid military duty. In addition to the United States, the device is being used in China, Norway, Britain, Brazil, and Spain.

Dr. Eugenia Gullick describes the theory behind the device:

> The plethysmograph . . . directly measures the outside evidence of sexual arousal. We know—it's established throughout the literature that when a man becomes sexually aroused—there is engorgement of the penis. It's a one-to-one relationship. . . . We know when the penis becomes engorged, we are measuring sexual arousal. (*State of North Carolina v. Robert Earl Spencer*, August 1, 1995, Mecklenburg County, No. 93 CRS 16225-26)

This much everyone seems to agree on: The device measures penile engorgement. Any male who has awakened with an erection knows, however, that penile engorgement is not always a measure of sexual arousal. On the other hand, most males would probably acknowledge that penile engorgement occurring while watching pornographic movies is due to sexual arousal.

What utility could such a device possibly have? In addition to identifying false gays, the machine is used to treat sex offenders and to identify potential sex offenders. Treatment is sometimes done in conjunction with aversion therapy, which involves subjecting patients to electric shocks or foul odors while being shown sexually suggestive pictures. The hope is that the treatment will prevent the patient from acting on his desire by associating something unpleasant with the desire.

Submission to a PPG has been made a condition of parole for certain sex offenders and has been used in sentencing decisions for sex offenders. The PPG has been used in child-custody cases to determine whether a father is likely to abuse his child. The PPG has even been given to children as young as 10 years old who had abused other children. This was done in Phoenix, Arizona, with no evidence either that the test was useful or that it would not be harmful when given to children. Not everyone submits quietly to the PPG requirements, however. Officials in Old Town, Maine, had to pay nearly $1 million to a policeman who was threatened with being fired for refusing to submit to a PPG.

Researchers at the University of Georgia have used the PPG to test the claim that homophobic men are latent homosexuals. In their study of 64 self-identified exclusively heterosexual men, 66% of the non-homophobic group showed no significant arousal while watching a male homosexual video, while only 20% of the homophobic men showed little or no evidence of arousal. This university study has not aroused much interest elsewhere, however.

Finally,

> there is an area where this device makes a valuable contribution: that of sorting out organic from psychogenic impotence. This is done by measuring changes in penile circumference during sleep, with increases expected during REM sleep. Men with psychogenic impotence still show erections, while those with an organic problem don't. (Dave Bunnell, personal correspondence)

Scientifically, what are we to make of such a device? The PPG can measure response time to a stimulus and it can measure change in penile girth over time. Apparently, it is assumed that the more quickly aroused and the greater the engorgement, the higher the "arousal level." Apparently, it is also assumed by many practitioners that any "arousal level" when viewing or listening to descriptions of naked children or adults having sex with children is "deviant." Yet according to studies done by the inventor of the PPG, Dr. Kurt Freund, "many so-called normal men who have not committed illegal sex acts show considerable arousal to stimuli depicting naked children or children involved in sexual activity" (www.delko .net/why.htm). In one court case, Dr. William Michael Tyson, a clinical and forensic psychologist specializing in the field of sexual criminal behavior, testified that "the vast majority of individuals who commit sexual offenses against children are not sexually aroused by stimulus material involving children." His expert adversary in that case, Dr. Gullick, claimed that "the plethysmograph has been extensively studied and recently shown to be ninety-five percent accurate in discriminating between individuals who had committed sexual offenses against children and a control group that was randomly drawn from the population." Yet other experts have claimed that there are "studies in which the devices have failed to detect nearly one out of three known sex offenders tested."

The device has been the subject of many scientific studies, and the results have been mixed. The reliability and utility of the device have been argued in court, and penile plethysmographic evidence has been declared inadmissible because of its "questionable reliability" (*State of North Carolina v. Robert Earl Spencer*, August 1, 1995, Mecklenburg County, No. 93 CRS 16225-26). Dr. Tyson testified that it was "generally accepted in the mental health community by both proponents and opponents of the plethysmograph that the plethysmograph data does not give any evidence that is useful in determining whether an individual did or did not commit a specific act." He also noted that "there is substantial disagreement as to the extent to which the penile response is subject to voluntary control and as to whether the penile response as measured by the plethysmograph can then be generalized to anything else pertaining to sexual behavior." In short, Dr. Tyson claims that the plethysmograph has very limited forensic utility.

Nevertheless, many therapists treating

sex offenders think the PPG will assist them "in determining whether someone who has committed a sex crime has a pattern of deviant sexual interests." Therapists use the PPG to help them devise treatment programs and to measure the success of their treatment. All this is done without any apparent concern that there is no compelling evidence that sexual arousal or nonarousal from pictures or sounds significantly correlates with criminal deviant behavior. There is no compelling evidence that a person who gets aroused by pictures or sounds is significantly *more* likely to commit sex crimes than one who does not get aroused. On the other hand, there is no compelling evidence that a person who does *not* get aroused by pictures or sounds is significantly *less* likely to commit sex crimes than one who does get aroused.

Still, the PPG can provide some information that might prove useful to therapists who work with sex offenders. The computer software used with the PPG enables the tester to develop graphs that indicate whether the subject is more aroused by males than by females, by children than by adults, by coerced than by consensual sex, and so on. The controversy begins, however, as soon as the therapist tries to convert "arousal levels" to anything meaningful, such as claims that a sex offender is "cured" or is "responding positively to treatment."

One glaring problem with the use of the PPG is the lack of standardized materials to use as stimuli for subjects, a factor that clearly biases the data. Therapists vary greatly in the kind of materials they use to arouse subjects. Some materials are rather tame, for example, nude adults or children in underwear or bathing suits. Others use hard-core pornography, including depictions of rape and pedophilia. Furthermore, there is no standard of "deviancy" for arousal. Worse, if therapists can define cer-

tain arousal as deviant, they can then suggest treatments for the deviancy and be the sole persons to determine when the "deviant" is "cured." Convicted sex offenders are in no position to protest either declarations that they have "deviant arousals" or treatments forced on them in the name of curing them of the "disease" of "deviant arousal."

More objectionable than the questionable scientific validity of the device, however, are the moral and legal questions its use raises. Some of the test materials would probably be illegal on the open market because they constitute child pornography. Much of the material is morally objectionable. Some of the uses of the device raise constitutional issues, for example, submission to the PPG test as a condition for employment, for enlistment into the armed forces, or for granting custody of children. As noted above, some penal institutions have made submission to the PPG a condition of parole, even though the device's usefulness as a predictor of behavior is unproven. The Seventh Circuit Court of Appeals has upheld the requirement. Parole boards have great latitude in establishing conditions for parole. These conditions do not have to meet the same standards of trial evidence. Nor do the normal liberties and constitutional protections of citizenship automatically apply to one being paroled.

From a scientific, moral, and legal point of view, what should matter is whether a person gives in to perverse desires and commits sex crimes. It is neither immoral nor a crime to get aroused. Furthermore, being aroused is not identical to having a desire. A man or woman may be aroused by the sight of animals copulating or be aroused by a film of a woman eating a banana and a man eating a fig in particularly provocative ways. Still, they may have no desire to engage in bestiality or to have sex with a bowl of fruit, much less engage in an act of molestation at the first opportunity.

Further reading: Adler 1993.

pentagram (pentacle, pentangle)

A five-pointed figure used as a magical or **occult** symbol by the **Freemasons**, Gnostics, Pythagoreans, **Satanists**, **Wiccans**, and others. There is apparently something attractive about the figure's geometry and proportions. In many symbolizations, the top point represents either the human head or a nonhuman **spirit**. To invert the figure is considered by some as a sign of relegating spirit to the bottom of the metaphysical heap. Others take inversion to be Satanic and on par with alleged mockeries such as inverting the cross or saying the Mass backward. Still others find nothing particularly diabolical about inversion and use the inverted pentagram without fear of accidentally invoking the forces of evil.

Some say the pentagram is mystical

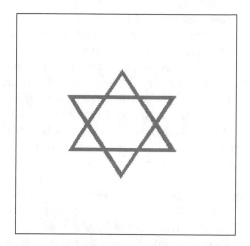

A hexagram star.

because the number 5 is mystical. It's a prime number, the sum of 2 and 3 (as well as of 1 and 4). Christ had five wounds, they say, if you don't count those inflicted by the crown of thorns, and he distributed five loaves of bread to 5,000 people. Most important, we have five fingers, toes, and senses.

Some Christian watchdogs apparently think the pentagram is the devil's hoof print. It can be bad for business if rumors are spread that one's company uses the pentagram or any other symbol deemed to be diabolical. Proctor and Gamble was once accused by some Amway competitors of being run by devil worshippers who flaunted their satanic religion with a diabolical logo. The logo consisted of an old man's bearded face in the crescent moon, facing 13 stars, all set within a circle. Some saw 666, the number of the Beast in *Revelation* (usually identified as Satan by the Christian watchdogs), lurking in the old man's beard and in the arrangement of the stars. Others saw a goat, another sign of the devil.

To the Wiccan, the five points of the pentagram represent air, fire, water, earth, and spirit. Wiccans usually put the symbol in a circle, which has traditionally represented the endless or eternity. The ancient

The pentagram, from Eliphas Levi (Alphonse-Louis Constant, 1910, *Transcendental Magic*).

Chinese believed there were five elements (wood, fire, earth, metal, and water), five planets, five seasons, five senses, as well as five basic colors, sounds, and tastes. However, the number 6 seems to have been more enchanting to them than five, for the *I Ching* uses 6 as its base number. So does the Star of David, which has six points and is made by overlapping two equilateral triangles. The Star of David is a hexagram but is not used to cast a "hex" on you. That kind of hex comes from the German word for **witch**, *Hexen,* which is related to the Old High German word *hagzissa,* a hag.

Occultists of all sorts wear pentagram **talismans** to protect them from evil or to help them get occult knowledge and power. They even draw pentagrams on the ground and stand within them to better call on occult powers. If the point is aimed north, they are not worshippers of Satan. However, if the point is aimed south, they are. So say the Christian watchdogs.

personology

A New Age variation of the ancient **pseudoscience** of **physiognomy**, which holds that outward appearance, especially the

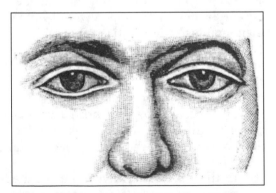

Licentious and unprincipled eyes, from M. O. Stanton, *The Encyclopedia of Face and Form Reading,* 6th revised ed. (Philadelphia: F. A. Davis, 1920), p. 950.

face, is the key to a person's predominant temper and character. The theory of personology, like physiognomy and **graphology**, seems to be based on **sympathetic magic** and intuition. For example, according to personology expert George Ronan:

> People with thin, soft, looser or porcelain-like skin tend to be more impressionable both emotionally and physically. . . . Those with thin, fine hair are refined emotionally. . . . A thick, full lower lip indicates spontaneous generosity to friends and strangers as well as talkativeness. . . . A ski-slope upturned nose person will usually be . . . a poor money manager. [biznet .maximizer.com/vedicastrology/msg1 .html]

According to Naomi Tickle, Los Angeles judge Edward Jones developed personology in the 1930s. Ms. Tickle is the author of *It's All in the Face: The Key to Finding Your Life Purpose* (1997). She is also the founder of the International Centre for Personology (not to be confused with the Personology Institute of San Diego or the Institute for Advanced Studies in Personology and Psychopathology in Coral Gables, Florida). According to Ms. Tickle, the judge "became fascinated by the relationship between facial features and behavior patterns of the people who appeared before him in court" (www.naomitickle.com). Then, like many naive people, he thought his personal observations were free of bias and constituted scientific data. Judge Jones even taught his "new science" to the public.

Judge Jones made the same mistake that Franz-Joseph Gall (**phrenology**), Ignatz Von Peczely (**iridology**), Edward Bach (**Bach's flower therapy**), and many others have made: He thought he observed a pattern and made no effort to scientifically test his thought. Jones thought he saw a pattern of facial similarities of people charged with similar crimes. He apparently did not realize that once he came to believe in this notion, he would find it easy to confirm his beliefs.

According to Ms. Tickle, the science was done by Robert L. Whiteside, a newspaper editor, who "used 1068 subjects and found the accuracy to be better than 90%" (www.naomitickle.com). Whiteside is the author of *Face Language* (1974). He became an advocate after watching Judge Jones do a **cold reading** of Whiteside's wife at a public lecture. Whiteside was amazed that Jones could know so much about his wife without knowing her. One searches in vain, however, for publications by Mr. Whiteside in scientific journals. Although Whiteside and his work have been universally ignored by the scientific community, the growth of personology has not been hindered.

Another Whiteside has added more science. Robert Whiteside trained Bill Whiteside. The latter notes that there is a scientific connection between genetics and behavior, and genetics and physical appearance. Therefore, he concludes in a lovely non sequitur, there must be a connection between behavior and physical appearance (he may as well have argued that since there is a connection between games and football, and between games and chess, there must be a connection between football and chess): "Over the years, [scientists] have conclusively proven that our genetic inheritance shows up in our structure and, therefore, so do our behavior patterns" (www.ireadfaces.com/ovrview7.htm). He might as well argue that since eye color is genetically determined, eye color is a key to understanding personality.

According to Bill Whiteside:

> There are 68 behavioral traits in Personology. A trained observer identifies each one with sight, measurement or touch. There are five trait areas: Physical, Automatic Expression, Action, Feeling and Emotion, and Thinking. The placement of each trait into an area develops logically from its location and relationship to a corresponding area of the brain. (ibid.)

This all sounds very scientific, but nothing we know about the brain supports these notions. Personology has failed to make any advancement over physiognomy.

Philadelphia Experiment

Allegedly conducted by the U.S. Navy (Project Rainbow) on October 28, 1943. According to legend, the destroyer USS *Eldridge* was made invisible, dematerialized, and **teleported** from the Philadelphia naval yard to Norfolk, Virginia, and back again. The experiment allegedly had terrible side effects, such as making sailors invisible and causing them to go mad, so the navy quit exploring this exciting new technology.

Acquisitive eyes, from M. O. Stanton, *The Encyclopedia of Face and Form Reading,* 6th revised ed. (Philadelphia: F. A. Davis, 1920), p. 951.

Dr. Franklin Reno allegedly did the experiment as an application of Einstein's unified field theory. The experiment supposedly demonstrated a successful connection between gravity and electromagnetism: *electromagnetic space-time warping.*

The Navy denies that it ever did such a test. The denial is taken as proof by the conspiratorially minded that the experiment must have occurred. The less gullible ask, Where did this story come from?

The story is a mixture of fact, fiction, speculation, and madness.

The facts are that the Navy does all kinds of experiments, many of them secret. Many of these experiments attempt to find military applications for the latest discoveries or theories in physics, such as Einstein's unified field theory. It seems to be a fact that the Navy was experimenting with "invisibility" in 1943, but not with making ships disappear. Edward Dudgeon, who says he was on the USS *Engstrom,* claims that they hoped to make our ships "invisible to magnetic torpedoes by de-Gaussing them." Dudgeon described the procedure to **UFO** investigator Jaques Vallee:

> They sent the crew ashore and they wrapped the vessel in big cables, then they sent high voltages through these cables to scramble the ship's magnetic signature. This operation involved contract workers, and of course there were also merchant ships around, so civilian sailors could well have heard Navy personnel saying something like, "they're going to make us invisible," meaning undetectable by magnetic torpedoes. (Vallee 1994)

The *Engstrom* and the *Eldridge* were harbored together. According to Dudgeon, crew members from both ships had parties together on shore, but "there was never any mention of anything unusual."

Dudgeon also thinks that some facts may have gotten distorted.

I was one of the two sailors who are said to have disappeared mysteriously. . . . [A] fight started when some of the sailors bragged about the secret equipment [radar, sonar, special screws, a new compass, etc.] and were told to keep their mouths shut. Two of us were minors. . . . The waitresses scooted us out the back door as soon as trouble began and later denied knowing anything about us. We were leaving at two in the morning. The *Eldridge* had already left at 11 P.M. Someone looking at the harbor that night might have noticed that the *Eldridge* wasn't there any more and it did appear in Norfolk. It was back in Philadelphia harbor the next morning, which seems like an impossible feat: if you look at the map you'll see that merchant ships would have taken two days to make the trip. They would have required pilots to go around the submarine nets, the mines and so on at the harbor entrances to the Atlantic. But the Navy used a special inland channel, the Chesapeake-Delaware Canal, that bypassed all that. We made the trip in about six hours. (ibid.)

Marshall Barnes, who identifies himself as a "Special Civilian Investigator," claims Dudgeon's story is disinformation and that Vallee is a hoaxer out to cover up the government's real activities (www .cbjd.net/orbit/text/ph1.html). Maybe so, but it is unlikely. In March 1999, sailors who had served on the *Eldridge* reunited and told a *Philadelphia Inquirer* reporter that they "find the story amusing—especially because the ship never docked in Philadelphia" (www.cpcn.com/articles/ 081999/news.cb.ship.shtml). Barnes also claims that he can prove that "optical invisibility" is possible through "the use of an intense electromagnetic field that would create a mirage effect of invisibility by refracting light" (ibid.). He claims he proved this to the cable network A&E for an episode of *The Unexplained,* but that they reneged on the deal and his "proof" was never shown. One would think A&E would have jumped at the chance to demonstrate something so wondrous.

In 1955 an auto parts salesman and amateur astronomer named Morris K. Jessup published a book called *The Case for the UFO.* In his book, Jessup speculated—among other things—that antigravity and electromagnetism would be better than rocket fuel for propelling space vehicles. The following year, Carl Allen (a.k.a. Carlos Miguel Allende), a somewhat brilliant but very disturbed human being, started the hoax by writing letters to Jessup telling him of the Philadelphia Experiment (Goerman 1980; Goerman is the source for the following material regarding Allen/Allende). Allende claims that he witnessed the disappearance of a ship while on board the SS *Andrew Furuseth,* a merchant ship. He also claims he saw some *Eldridge* crew members disappear into thin air during a fight. Allen sent an annotated copy of Jessup's book to the Office of Naval Research in Washington, D.C. Jessup was summoned to Washington, where he turned over the Allen letters. Later, the Varo Corporation, a firm that did research for the military, published the annotated version along with Allen's letters to Jessup. Jessup committed suicide in 1959. Allen continued sending strange annotations to relatives for many more years, as he drifted from place to place.

The speculations regarding the origin of Allen's story have run rampant. Some say that he was there and saw it all. Some say that Allen is an alien and channels information. Some claim that the Navy is covering up the experiment and their complicity with aliens. The simple truth is that Allen made it all up. According to Goerman, "If we are to believe Allen, our naval hierarchy abandoned sanity and historical precedent by conducting an experiment of enormous importance in broad daylight using a badly needed destroyer escort vessel. . . . If someone were to write a book telling the real story, its title might be *The Philadelphia Hoax: Project Gullibility."*

Allen's hoax has grown into a legend that has been spurred on by a number of books, some of them fictional, some nonfictional, and others fictional but claiming to be nonfictional. In 1965, Vincent H. Gaddis's *Invisible Horizons: True Mysteries of the Sea* was published. In addition to stories about various disappearing islands, aircraft, and ships, Gaddis presents the basics of the legend as created by Allen in his letters and published in the Varo edition of Jessup's work. In 1977, Charles Berlitz published *Without a Trace: New Information from the Triangle,* which included a chapter on the Philadelphia Experiment. Berlitz is a frequent source of stories on strange phenomena, such as **Atlantis**, the **Bermuda Triangle**, and **Noah's Ark**.

In the fictional category, *Thin Air* (1978) by George E. Burger and Neil R. Simpson stands out. It is about a Navy investigation of a cover-up of an experiment involving the USS *Eldridge* in 1943.

In 1979, *The Philadelphia Experiment: Project Invisibility* by William L. Moore and Charles Berlitz was published. This book is fiction but claims to be fact, and plagiarizes parts of *Thin Air.* In the Moore and Berlitz book, not only the ship, but also several crewmembers disappear into a new dimension, never to be seen again.

In 1984, a movie called *The Philadelphia Experiment* was produced. It was directed by Stewart Raffill and was based on a screenplay by William Gray and Michael Janover.

There have been other attempts to exploit the gullible with stories about this so-called experiment, but two stand out as more inane than the rest: *The Philadelphia Experiment, and Other UFO Conspiracies,* by

Brad Steiger, with Alfred Bielek and Sherry-Hanson Steiger (1990), and *The Philadelphia Experiment Part 1—Crossroads of History,* presented by Alfred Bielek. The former is a book that rehashes the usual stories of CIA plots, government conspiracies, secret meetings with aliens, trips to Mars, and visits from the **Men in Black**. The latter is a video featuring a man who claims he was a physicist on the USS *Eldridge* in 1943 and was part of the team that conducted the experiment. Bielek claims he time-traveled in 1943 to 1983 during the experiment and lived to tell the story, only to be harassed by the U.S. government for his troubles.

philosopher's stone

The magical substance in **alchemy** that can turn base metals into gold, cure all ills and ailments, and allow its possessor to achieve immortality. Unfortunately, like many wonderful things dreamed up by **occultists**, this substance exists only in the imagination.

phrenology (cranioscopy)

The study of the structure of the skull to determine a person's character and mental capacity. It is based on the false assumption that mental faculties are located in brain organs on the surface of the brain and can be detected by visible inspection of the skull. The Viennese physician Franz-Joseph Gall (1758–1828) claimed there are some 26 organs on the surface of the brain that affect the contour of the skull, including organs such as a murder organ present in murderers. Gall was an advocate of the "use it or lose it" school of thought. Brain organs that were used got bigger and those that were not used shrunk, causing the skull to rise and fall with organ development. These bumps and indentations on

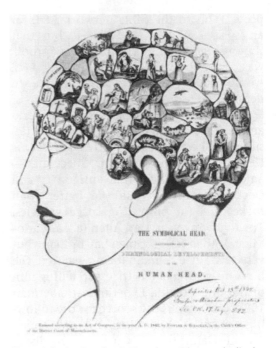

Copyright by Fowler & Strachan: The symbolical head, illustrating all the phrenological developments of the human head (1842).

the skull, according to Gall, reflect specific areas of the brain that determine a person's emotional and intellectual functions. Gall called the study of these cranial hills and valleys "cranioscopy." Others, such as Johann Kaspar Spurzheim (1776–1832), who spread the word in America, and George Combe (1788–1858), who founded the Edinburgh Phrenological Society, followed with even zanier and more specious divisions and designations of the brain and skull, such as "metaphysical spirit" and "wit." In 1815, Thomas Foster called the work of Gall and Spurzheim "phrenology" (*phrenos* is Greek for "mind"), and the name stuck.

Phrenology advanced the correct notion that the human brain is the seat of character, emotions, perception, intellect, and so on, and that different parts of the brain are responsible for different mental functions. However, in Gall's time it was possible to study the brains of only the dead. Thus, phrenologists associated the different structures in the brain with supposed mental functions that were in turn associated only with the contour of the skull, rather than actual behavior. Little was done to study the brains or the behavior of persons known to have had neurological problems, which might have helped in the process of locating parts of the brain responsible for specific neurological functioning. Instead, mental faculty localization was arbitrarily selected. Gall's early work was with criminals and the insane, and his brain organs reflected this interest. Spurzheim removed theft organs and murder organs from the phrenologist's map, but he added areas for benevolence, self-esteem, and conjugal love.

Although phrenology has been thoroughly discredited and has been recognized as having no scientific merit, it still has its advocates. It remained popular, especially in the United States, throughout the 19th century and gave rise to **anthropometry** and **craniometry**. Phrenology was highly praised by Thomas Edison, Ralph Waldo Emerson, Horace Mann, and Alfred Russell Wallace. The Boston Medical Society welcomed Spurzheim as a heroic figure when he arrived in 1832 for an American tour. The Fowler Brothers and Samuel Wells published the *American Phrenological Journal and Life Illustrated,* which lasted from 1838 until 1911. In Edinburgh, Combe's *Phrenological Journal* was published from 1823 until 1847, and his *The Constitution of Man* sold more than 300,000 copies between 1828 and 1868.

It may seem difficult to explain the early popularity of phrenology among scientists, since the empirical evidence for a direct relationship between the brain and character was scant. However, phrenology offered an explanation of human behavior that was materialistic and could be studied scientifically without reference to elusive **souls**. Also, an unplanned experiment provided some solid evidence for such a relationship in 1848, when Phineas Gage's moral character changed dramatically after an explosion blew a tamping iron through his head (Damasio 1995). Gage was leading a railroad construction crew near Cavendish, Vermont, when the accident occurred. "Before the accident he had been a most capable and efficient foreman, one with a well-balanced mind, and who was looked on as a shrewd smart business man." After the accident, he became "fitful, irreverent, and grossly profane, showing little deference for his fellows. He was also impatient and obstinate, yet capricious and vacillating, unable to settle on any of the plans he devised for future action" (www.deakin.edu.au/hbs/gagepage/pgstory.htm). On the other hand, one might conclude that the Gage incident blew a hole through the theory that bumps on the head were the key to brain function.

Bob McCoy wears the psycograph.

Phrenology gave rise to the invention of the psycograph by Lavery and White, a machine that could do a phrenological reading, complete with printout. It is said that this device netted its owners about $200,000 at the 1934 Century of Progress Exposition in Chicago (McCoy 1996). Phrenological readings are not unlike astrological readings, and many who have them done are satisfied that the results are uncannily accurate, as would be predicted by the **Forer effect**.

Further reading: Gardner 1957; McCoy 2000; Skrabanek and McCormick 1990; Stern 1982.

physiognomy

The interpretation of outward appearance, especially the features of the face, to discover a person's predominant temper and character. Physiognomy has also been used as a kind of **divination** and is often associated with **astrology**. The faces depicted here are from a German edition of *Physiognomy and Chiromancy* (1540), by Barthélemy Coclès (1467–1504).

Coclès, like others before and after him, tried to create a science out of something each of us does from time to time: judge a person by his or her facial characteristics. Physiognomists such as Coclès are wont to say such things as "People with snub noses are vain, untruthful, unstable, unfaithful and seducers." The snub-nosed of the world tend to snub their noses at such pseudoscientific drivel.

Three hundred years later, M. O. Stanton would write under the head of the *pug type nose*:

> the interpretation of character is in consonance with the peculiarities of the form [of the nose], whether it be rounded, blunt, pug or a sharpened narrow pug. In

Page from Barthélemy Coclès (1467–1504), *Phisionomi und Chiromanci* (Strassburg: M. Jacob Cammerlander von Mentz, ca. 1540).

regard to its meanings, it indicates lowness, coarseness, or commonplace mentality. If it be relatively sharp, the character is more acute and the subject quicker in his perceptions than where a blunt pug is exhibited, yet all of this class of noses have the same general meaning in absence of reasoning power, pugnacity, irritability, quarrelsomeness, and opposition. (*The Encyclopedia of Face and Form Reading*, 6th revised ed., 1920)

Stanton's musings are clearly based on **sympathetic magic**.

In the 18th and 19th centuries, physiognomy was used by some of its proponents as a method of detecting criminal tendencies. Many bigots and racists still use physiognomy to judge character and

Pug nose (*left*) and the blunt pug nose. Drawings from M. O. Stanton, *The Encyclopedia of Face and Form Reading,* 6th revised ed. (Philadelphia: F. A. Davis, 1920), p. 898.

personality. This is not to say that there are not certain physiognomic features associated with certain genetic disorders such as Down's syndrome or Williams syndrome.

See also **metoposcopy**, **personology**, and **phrenology**.

Piltdown hoax

Piltdown was an archaeological site in England where in 1908 and 1912 human, ape, and other mammal fossils were found. In 1913 at a nearby site, an ape's jaw and a canine tooth worn down like a human's were found. The general community of British paleoanthropologists came to accept the idea that the fossil remains belonged to a single creature that had had a human cranium and an ape's jaw. In 1953, Piltdown "man" was exposed as a forgery. The skull was modern and the teeth on the ape's jaw had been filed down.

To those who are skeptical of science, such episodes as Piltdown are taken to be proof that science is, more or less, bunk. Some **creationists**, for example, think that Piltdown demonstrates that scientists can't accurately date bones. But methods of dat-

ing such things have greatly improved since 1910 (Feder 2002; Gould 1982). To those who have a better understanding of the nature and limits of science, Piltdown is little more than a wrong turn down a road that, despite such detours, eventually arrives at the right destination. Because of the public nature of science and the universal application of its methods, and because of the fact that the majority of scientists are not crusaders for their own untested or untestable prejudices, whatever errors are made by scientists are likely to be discovered by other scientists. The same can't be said for the history of quacks and **pseudoscientists**, where errors do not get detected because their claims are not tested properly.

How had so many scientists been duped? Stephen Jay Gould (1982) offers several reasons, among them **wishful thinking** and cultural bias. The latter, no doubt, played a role in the lack of critical thinking among British paleoanthropologists. Above all, however, the Piltdown forgery demonstrates the fallibility of scientific knowledge. It demonstrates, too, the way that theories and facts are related in **science**. Theories are the filters through which facts are interpreted (Popper 1959). Theories try to explain and make sense of facts. On the other hand, facts are used to test theories. Gould notes that today a human cranium with an ape's jaw is considered to be extremely implausible and far-fetched. But in the early part of this century, anthropologists were imbued with the cultural prejudice that considered man's big brain as his ticket to rule, the main evolutionary feature that made it possible for man to develop all his other unique features. Since there was a preconceived notion that man's brain must have developed to its human size before other changes occurred in human structure, a human cranium with an ape's jaw didn't arouse as much suspicion as it would today. Fossil discoveries since Piltdown clearly show a progression from small-brained, upright, nonsimian hominids to larger-brained upright humans. Scientists "modeled the facts" and confirmed their theory, "another illustration," says Gould, "that information always reaches us through the strong filters of culture, hope, and expectation" (Gould 1982: 118). Once committed to a theory, we see what fits with the theory. Piltdown is a reminder that *confirmation* of a theory is not *proof* of the theory.

The main reason Piltdown was not spotted as a fraud much earlier was that scientists weren't allowed to see the evidence, which was locked away securely in the British Museum. Instead of focusing their attention on examining the facts more closely with an eye to discovering the fraud, scientists weren't even allowed to examine the physical evidence at all. They had to deal with plaster molds and be satisfied with a quick look at the originals to justify the claim that the models were accurate.

Another reason some scientists were duped was probably because it was not in their nature to consider someone would be so malicious as to intentionally engage in such deception. In any case, one of the main fallouts of Piltdown has been a small industry of detectives turned authors trying to identify the hoaxer. The list of suspects includes:

- Charles Dawson, an amateur archaeologist who brought in the first cranial fragments from Piltdown;
- Pierre Teilhard de Chardin, theologian and scientist who accompanied Dawson and Arthur Smith Woodward (Keeper of Geology at the British Museum [Natural History] in 1912) to Piltdown on expe-

ditions where they discovered the mandible;
- W. J. Solass, a professor of geology at Oxford;
- Grafton Elliot Smith, who wrote a paper on the find in 1913;
- Arthur Conan Doyle, the creator of Sherlock Holmes; and
- Martin A. C. Hinton, a curator of zoology at the time of the Piltdown hoax. A trunk with Hinton's initials on it was found in an attic of London's Natural History Museum. The trunk contained bones stained and carved in the same way as the Piltdown fossils.

The evidence in each case is circumstantial and not very strong. What is highly probable is that there will be more books and articles speculating on the identity of the Piltdown hoaxer.

Further reading: Anderson 1996; Edey and Johanson 1990; Gee 1996; Gould 1983; Johanson and Edey 1981.

pious fraud

Someone whose fraud is motivated by religious zeal. It is very difficult, if not impossible, to prove fraud in many cases regarding such things as a miraculous painting on a peasant's cloak or on a shroud, receiving the **stigmata**, or being instructed by an apparition of the Virgin Mary. Yet fraud seems most probable in some cases. For example, Catalina Rivas, originally from Bolivia but now residing in Mexico, claims to have the stigmata and that Jesus and the Virgin Mary have dictated messages to her in various languages. She is revered as the spiritual mother of international religious movements known as the Apostolate of the New Evangelization (in Spanish, Apostolado de la Nueva Evangelización) and the Great Crusade of

Love and Mercy (in Spanish, La Gran Cruzada del Amor y Misericordia). She has followers all over the world because of her stigmata and the fact that she can write books in languages she does not understand. Her followers believe her books and her wounds are miraculous. However, her books are not dictated by Jesus or Mary. They have been copied without authorization from the works of Catholic writers José Prado Flores and Salvador Gómez of Guadalajara, Mexico (Carroll, www.skepdic .com/rivas.html). Furthermore, her wounds seem to be self-inflicted (ibid.). A mental aberration may explain how she could wound herself and be deluded into thinking otherwise, but it seems very unlikely that she could plagiarize large sections of books and pass them off as messages from Jesus or Mary without knowing she was lying.

Other cases that seem also to be pious frauds include all those who claim to do **psychic surgery**, many faith healers, and anyone who has lied about performing or witnessing a **miracle**.

placebo effect

The measurable, observable, or felt improvement in health not attributable to treatment, which is believed to be inert or innocuous. This effect is believed by many people to be due to the placebo itself in some mysterious way. A placebo (Latin for "I shall please") is a substance or treatment believed by the administrator of the treatment to be inert or innocuous. Placebos may be sugar pills, saline solution, or starch pills. Even fake surgery and fake psychotherapy are considered placebos.

Researchers and medical doctors sometimes give placebos to patients without telling them. Anecdotal evidence for the placebo effect is garnered in this way,

though this method may often test the power of suggestion, which some researchers identify with the placebo effect (Ramachandran and Blakeslee 1998: 53). Those who believe there is scientific evidence for the placebo effect point to clinical studies, many of which use a **control group** treated with a placebo. It is not known how an inert substance or a fake surgery or therapy could be effective.

Some believe the placebo effect is psychological, due to a belief in the treatment or to a subjective feeling of improvement. Irving Kirsch, a psychologist at the University of Connecticut, believes that the effectiveness of Prozac and similar drugs may be attributed almost entirely to the placebo effect. He and Guy Sapirstein analyzed 19 clinical trials of antidepressants and concluded that the expectation of improvement, not adjustments in brain chemistry, accounted for 75% of the drugs' effectiveness (Kirsch 1998). "The critical factor," says Kirsch, "is our beliefs about what's going to happen to us. You don't have to rely on drugs to see profound transformation." In an earlier study, Sapirstein analyzed 39 studies, done between 1974 and 1995, of depressed patients treated with drugs, psychotherapy, or a combination of both. He found that 50% of the drug effect is due to the placebo response.

A person's beliefs and hopes about a treatment, combined with his or her suggestibility, may have a significant biochemical effect. Sensory experience and thoughts can affect neurochemistry. The body's neurochemical system affects and is affected by other biochemical systems, including the hormonal and immune systems. Thus, it is consistent with current knowledge that a person's hopeful attitude and beliefs may be very important to their physical well-being and recovery from injury or illness.

However, it may be that much of the

placebo effect is not a matter of brain over molecules, but of brain over behavior. A part of the behavior of a "sick" person is learned. So is part of the behavior of a person in pain. In short, there is a certain amount of role-playing by ill or hurt people. Role-playing is not the same as faking or malingering. The behavior of sick or injured persons is socially and culturally based to some extent. The placebo effect may be a measurement of changed behavior affected by a belief in the treatment. The changed behavior includes a change in attitude, in what one says about how one feels, and how one acts. It may also affect one's brain chemistry.

The psychological explanation seems to be the one most commonly believed. Perhaps this is why many people are dismayed when they are told that the effective drug they are taking is a placebo. This makes them think that their problem is "all in their head" and that there is really nothing wrong with them. Yet there are too many studies that have found objective improvements in health from placebos to support the notion that the placebo effect is entirely psychological.

> Doctors in one study successfully eliminated warts by painting them with a brightly colored, inert dye and promising patients the warts would be gone when the color wore off. In a study of asthmatics, researchers found that they could produce dilation of the airways by simply telling people they were inhaling a bronchiodilator, even when they weren't. Patients suffering pain after wisdom-tooth extraction got just as much relief from a fake application of ultrasound as from a real one, so long as both patient and therapist thought the machine was on. Fifty-two percent of the colitis patients treated with placebo in 11 different trials reported feeling better—and 50 percent of

the inflamed intestines actually looked better when assessed with a sigmoidoscope. (Talbot 2000)

It is unlikely that such effects are purely psychological. But it is not necessarily the case that the placebo is actually effective in such cases. Some believe that at least part of the placebo effect is due to an illness or injury taking its natural course. We often heal spontaneously if we do nothing at all to treat an illness or injury. That is, healing bodily mechanisms occur, the details of which we are ignorant. Furthermore, many disorders, pains, and illnesses wax and wane. What is measured as due to the placebo effect could be, in many cases, due to natural **regression**. In short, the placebo may be given credit that is due to nature.

However, spontaneous healing and spontaneous remission of disease cannot explain *all* the healing or improvement that takes place because of placebos. People who are given no treatment at all often do not do as well as those given placebos or real medicine and treatment.

Another theory gaining popularity is that a process of treatment that involves showing attention, care, and affection to the patient or subject, a process that is encouraging and hopeful, may itself trigger physical reactions in the body that promote healing. According to Dr. Walter A. Brown, a psychiatrist at Brown University,

> [T]here is certainly data that suggest that just being in the healing situation accomplishes something. Depressed patients who are merely put on a waiting list for treatment do not do as well as those given placebos. And—this is very telling, I think—when placebos are given for pain management, the course of pain relief follows what you would get with an active drug. The peak relief comes about an hour after it's administered, as it does with the real drug, and so on. If placebo analgesia

was the equivalent of giving nothing, you'd expect a more random pattern. (ibid.)

Dr. Brown and others believe that the placebo effect is mainly or purely physical and due to physical changes that promote healing or feeling better. Brown and others think that the touching, caring, attention, and other interpersonal communication that is part of the controlled study process (or the therapeutic setting), along with the hopefulness and encouragement provided by the experimenter or healer, affect the mood of the subject, which in turn triggers physical changes such as release of endorphins. The process reduces stress by providing hope or reducing uncertainty about what treatment to take or what the outcome will be. The reduction in stress prevents or slows down further harmful physical changes from occurring.

The process-of-treatment hypothesis would explain how inert **homeopathic** remedies and the questionable therapies of many **"alternative" health practitioners** are often effective or thought to be effective. It would also explain why pills or procedures used by conventional medicine work until they are shown to be worthless.

Forty years ago, a young Seattle cardiologist named Leonard Cobb conducted a unique trial of a procedure then commonly used for angina, in which doctors made small incisions in the chest and tied knots in two arteries to try to increase blood flow to the heart. It was a popular technique—90 percent of patients reported that it helped—but when Cobb compared it with placebo surgery in which he made incisions but did not tie off the arteries, the sham operations proved just as successful. The procedure, known as internal mammary ligation, was soon abandoned. (ibid.)

Of course, spontaneous healing or regression can also adequately explain why homeopathic remedies might appear to be effective. Whether the placebo effect is mainly psychological, misunderstood spontaneous healing, due to showing care and attention, or due to some combination of all three may not be known with complete confidence. A fourth possibility has been suggested: The placebo tricks the brain, which issues a healing response based on past experience (Ramachandran 1998).

One thing that does not seem to be in doubt is the powerful effect of the placebo. It should be, however, according to Danish researchers Asbjørn Hróbjartsson and Peter C. Götzsche. Their metastudy of 114 studies involving placebos found "little evidence in general that placebos had powerful clinical effects . . . [and] . . . compared with no treatment, placebo had no significant effect on binary outcomes, regardless of whether these outcomes were subjective or objective. For the trials with continuous outcomes, placebo had a beneficial effect, but the effect decreased with increasing sample size, indicating a possible bias related to the effects of small trials" (Hróbjartsson and Götzsche 2001).

According to Dr. Hróbjartsson, professor of medical philosophy and research methodology at University of Copenhagen, "The high levels of placebo effect which have been repeatedly reported in many articles, in our mind are the result of flawed research methodology." This claim flies in the face of more than 50 years of research. At the very least, we can expect to see more rigorously designed research projects trying to disprove Hróbjartsson and Götzsche.

The idea of the powerful placebo in modern times originated with H. K. Beecher. He evaluated over two dozen studies and calculated that about one-third of those in the studies improved due to the placebo effect ("The Powerful Placebo,"

1955). Other studies calculate the placebo effect as being even greater than Beecher claimed. For example, studies have shown that placebos are effective in 50 or 60% of subjects with certain conditions, for example, "pain, depression, some heart ailments, gastric ulcers and other stomach complaints" (Talbot 2000). As effective as the new psychotropic drugs seem to be in the treatment of various brain disorders, some researchers maintain that there is not adequate evidence from studies to prove that the new drugs are more effective than placebos (Fisher and Greenberg 1997).

Placebos have even been shown to cause unpleasant side effects. Dermatitis medicamentosa and angioneurotic edema have resulted from placebo therapy (Dodes 1997). There are even reports of people becoming addicted to placebos. How dangerous are placebos? Some ask, "What difference does it make why something works, as long as it seems to work?" They should consider that it is likely that there is something that works even better, something for the other two-thirds or one-half of humanity who, for whatever reason, cannot be cured or helped by placebos or spontaneous healing or natural regression of their pain. Furthermore, placebos may not always be beneficial or harmless.

> Patients can become dependent on nonscientific practitioners who employ placebo therapies. Such patients may be led to believe they're suffering from imagined "reactive" hypoglycemia, nonexistent allergies and yeast infections, dental filling amalgam "toxicity," or that they're under the power of Qi [**chi**] or extraterrestrials. And patients can be led to believe that diseases are only amenable to a specific type of treatment from a specific practitioner. (ibid.)

In other words, the placebo can be an open door to **quackery**.

Further reading: Fisher and Greenberg 1997; Harrington 1999; Hart 1999; Kirsch and Sapirstein 1998; Shapiro and Shapiro 1997; Sternberg and Gold 1997.

plant perception

Plant perception is the alleged ability of plants to feel and to read human minds, an idea first put forth in the 1960s by Cleve Backster, a **polygraph** school operator.

Plants are living things with cellulose cell walls, lacking nervous or sensory organs. Animals do not have cellulose cell walls but do have nervous or sensory organs. It would never occur to a plant or animal physiologist to test plants for consciousness or **ESP** because their knowledge would rule out the possibility of plants having feelings or perceptions on the order of human feeling or perception. In layman's terms, plants don't have brains or anything similar to brains.

However, Cleve Backster, a person apparently ignorant of plant and animal science, not only has tested plants for perception and feeling, but claims he has scientific proof that plants experience a wide range of emotions and thoughts. He also claims that plants can read human minds. Backster published his research in the *International Journal of Parapsychology* ("Evidence of a Primary Perception in Plant Life" [1968], vol. 10, no. 4, Winter, pp. 329–348). He tested his plants on a polygraph machine and inferred from the contours of their polygraph test that plants react to thoughts and threats.

Backster claims to have a D.Sc. in Complementary Medicine from Medicina Alternativa (1996). Dr. Backster is on the faculty of the California Institute for Human Science Graduate School and Research Center, an unaccredited institution founded by Dr. Hiroshi Motoyama for the study of "the human being as tridimensional." Dr.

Motoyama is said to be a scientist and Shinto priest who "has awakened to states of consciousness that enable him to see beyond the limits of space and time" (www.cihs.edu/whatsnew/motoyama_book .asp).

Backster's claims about plant perception were refuted by Horowitz, Lewis, and Gasteiger (1975) and Kmetz (1977, 1978). Kmetz summarized the case against Backster in an article for the *Skeptical Inquirer* in 1978. Backster had not used proper controls in doing his study. When controls were used, no detection of plant reaction to thoughts or threats could be found. Careful study found that the cause of the polygraph contours noted by Backster could have been due to static electricity, movement in the room, changes in humidity, or other natural causes.

Nevertheless, Backster has become the darling of several **parapsychological** and **pseudoscientific** notions. His work has been cited in defense of **dowsing**, various forms of **energy** healing, **remote viewing**, and the Silva mind control program (now known as the **Silva method**). In 1995, Backster was invited to address the Silva International Convention in Laredo, Texas. Nearly 30 years after his original discovery, he is telling the same story. It is a very revealing story and worth repeating. It shows his curious nature, as well as his apparent ignorance of the dangers of **confirmation bias** and **self-deception**. Backster clearly does not understand why scientists use controls in causal studies.

Backster tells us that it was on February 2, 1966, in his lab in New York City that he did his first plant experiment. His lab was not a science lab. In fact, it wasn't much of a lab at all in the beginning. It was a place where he conducted training in the use of the polygraph. There was a plant in the room. He recalls the following:

For whatever reason, it occurred to me that it would be interesting to see how long it took the water to get from the root area of this plant, all the way up this long trunk and out and down to the leaves.

After doing a saturation watering of the plant, I thought, "Well gee whiz, I've got a lot of polygraph equipment around; let me hook the galvanic skin response section of the polygraph onto the leaf." [www.avlispub.com/id340.htm]

Clearly, Backster is a very curious individual. A less inquisitive person would probably not care how long it would take water to get from the root to the leaves in an office plant. Not only did Backster care, but he cared enough to put his polygraph equipment to use as a measuring device. He reasoned as follows:

I felt that as the contaminated water came up the trunk and down into the leaf that the leaf becoming more saturated and a better conductor it would give me the rising time of the water. . . . I would be able to get that on the polygraph chart tracing.

Why would the polygraph indicate this? Because, he says, he was using a "whetstone bridge circuit that is designed to measure resistance changes." Presumably, resistance changes would be picked up by the polygraph as the water reached the leaf. He predicted that the resistance would slowly drop and the tracings on his polygraph paper would rise as the water reached the leaf. Instead, the opposite happened, which, he says, "amazed me a little bit."

Apparently, he moved the electrodes and saw that the contour of the polygraph chart was "the contour of a human being tested, reacting when you are asking a question that could get them in trouble." Backster claims that he then gave up his interest in measuring how long it takes water to get from the roots to the leaves of his plant. He says he believed that the plant was trying "to show me people-like reactions." He claims his next thought was: "What can I do that will be a *threat* to the well-being of the plant, similar to the fact that a relevant question regarding a crime could be a threat to a person taking a polygraph test if they're lying?" The contour of the graph triggered in him an immediate identification of the plant with one of his subjects. Until that moment, Backster had not suspected that the plants in his office were just like people and would respond similarly. Why he thought of threatening the plant isn't quite clear. Maybe this was his standard procedure with human subjects. Still, it is not quite clear why the response to a threat would result in the same kind of response as being caught in a lie.

Backster says he tried for 13 minutes and 55 seconds to get a reaction out of the plant by doing such things as dipping a leaf in warm coffee, but he got no response. A less devoted inquisitor might have given up and gone home at this point, but not Backster. He concluded that the plant seemed like it was *bored*. Then, he had his Eureka! experience: "I know what I am going to do: I am going to burn that plant leaf, that very leaf that's attached to the polygraph." Why he would burn the leaf isn't clear, since burning it would eliminate its moisture, making measurement of galvanic response impossible, and might damage his equipment attached to the leaf. Anyway, he tells us that there was a problem with carrying out his plan: He didn't have any matches. He claims, however, that while standing some 5 feet from the plant the polygraph "went into a wild agitation." Rather than conclude that maybe the water finally got to the leaf or some other natural event was causing the polygraph needle movements, Backster became convinced that the plant was *reading his mind* and was reacting to his

intent to burn it. He gives no indication that he even considered that there might be other possible explanations for the movement of his polygraph. This may strike some readers as a good thing, that a gifted mind immediately grasps the truth. But actually this is a bad thing, because one's intuition could be wrong. What is very curious is that after three decades of experiments, there is still no evidence that Backster and his many supporters see the importance of using controls in their studies of alleged plant perception.

Backster admits that he committed a bit of petty larceny in the name of science: He went to another office, went into a secretary's desk drawer, and retrieved some matches. When he got back to his experiment, he lit a match, but—careful and observant scientist that he was—he realized that the machine was so agitated he wouldn't be able to measure any additional agitation. So, he left the room. When he returned, "the thing just evened right out again, which really rounded it out and gave me a very, very high quality observation." What he meant by "a very, very high quality observation" is not clear.

Backster's true genius is exhibited in his final remark on the remarkable experiment:

> Now when my partner in the polygraph school we were running at the time came in, he was able to do the same thing also, as long as he intended to burn the plant leaf. If he pretended to burn the plant leaf, it wouldn't react.
>
> It could tell the difference between *pretending* you are going to, compared to when you actually *intend* to do it, which is quite interesting in itself from a plant psychology standpoint.

Plant psychology? Backster invented it that night. He is still at it 30 years later, at the

Backster Research Center in San Diego, California, where he claims to be able to demonstrate that plants respond to his loving thoughts and obey his thought commands, proving that a little ignorance, a nonskeptical mass media, and a lot of **communal reinforcement** can go a long way.

Pleiadians

Alien beings from the star cluster in the constellation Taurus known as The Pleiades. In her **channeled** book, *Bringers of the Dawn,* Barbara Marciniak claims that the Pleiadians chose her to be their messenger. According to Marciniak, the message is: "If you can clear people of their personal information, they can go cosmic."

Another message is: over 280,000 copies in print, a message not lost on Lia Shapiro, a.k.a. Lia Light. Lia claims the Pleiadians used her as a channel for *her* book, *Comes the Awakening—Realizing the Divine Nature of Who You Are.*

See also Bridey **Murphy**, *Celestine Prophecy, Course in Miracles,* Dr. **Fritz, Ica stones, Ramtha,** and *Urantia Book*.

Further reading: Korff 1995.

poltergeist

See **ghosts**

polygraph ("lie detector")

A polygraph is an instrument that simultaneously records changes in physiological processes such as heartbeat, blood pressure, and respiration. The polygraph is used as a lie detector by police departments, the FBI, the CIA, federal and state governments, and numerous private agencies. The underlying theory of the polygraph is that when people lie, they also get measurably nervous about lying. The heartbeat increases, blood pressure goes up, breathing rhythms change, perspiration increases, and so on. A baseline for these physiological characteristics is established by asking the subject questions whose answers the investigator knows. Deviation from the baseline for truthfulness is taken as sign of lying.

There are three basic approaches to the polygraph test:

1. *The Control Question Test (CQT).* This test compares the physiological response to relevant questions about a crime with the response to questions relating to possible prior misdeeds. "This test is often used to determine whether certain criminal suspects should be prosecuted or classified as uninvolved in the crime" (Iacono and Lykken 1997).

2. *The Directed Lie Test (DLT).* This test tries to detect lying by comparing physiological responses when the subject is told to deliberately lie with physiological responses when they tell the truth.

3. *The Guilty Knowledge Test (GKT).* This test compares physiological responses to multiple-choice type questions about the crime, one choice of which contains information only the crime investigators and the criminal would know about.

Psychologists do not think either the CQT or the DLT is scientifically sound, but a majority surveyed by the American Psychological Association thinks that the Guilty Knowledge Test is based on sound scientific theory and consider it "a promising forensic tool" (ibid.). However, they "would not advocate its admissibility [in court] in the absence of additional research with real-life criminal cases" (ibid.). One major problem with this test is that it has no controls. Also, unless the investigators have several pieces of insider information to use in their questioning, they run the risk of making a hasty conclusion based on just one or two deviant responses. There may be many reasons why a subject would select the insider choice. Furthermore, not responding differently to the insider choices for several questions should not be taken as proof the subject is innocent. He or she may be a sociopath, a psychopath, or simply a good liar.

Is there any evidence that the polygraph is really able to detect lies? The machine measures changes in heart rate, blood pressure, and respiration. When a person lies, it is assumed that these physiological changes occur in such a way that a trained expert can detect whether the person is lying. Is there a scientific formula or law that establishes a regular correlation between such physiological changes and lying? No. Is there any scientific evidence that polygraph experts can detect lies using their machine at a significantly better rate than nonexperts using other methods? No. There are no machines and no experts that can detect with a high degree of accuracy when people, selected randomly, are lying and when they are telling the truth.

Some people, such as Senator Oren Hatch, don't trust the polygraph machine, even if used by an expert such as Paul Minor, who trained FBI agents in their use. Anita Hill passed a polygraph test administered by Minor, who declared she was telling the truth about Clarence Thomas. Hatch declared that someone with a delusional disorder could pass the test if the liar really thought she was telling the truth. Hatch may be right, but the ability of the sociopath and the deluded to pass a polygraph test is not the reason such machines cannot accurately detect lies with accuracy any greater than other methods of lie detection.

The reason the polygraph is not a lie detector is that what it measures—changes in heartbeat, blood pressure, and respiration—can be caused by many things. Nervousness, anger, sadness, embarrassment and fear can all be causal factors in altering one's heart rate, blood pressure or respiration rate. Having to go to the bathroom can also be causative. There are also a number of medical conditions such as colds, headaches, constipation, or neurological and muscular problems that can cause the physiological changes measured by the polygraph. The claim that an expert can tell when the changes are due to a lie and when they are due to other factors has never been proven. Even if the device measures nervousness, one cannot be sure that the *cause* of the nervousness is fear of being caught in a lie. Some people may fear that the machine will indicate they are lying when they are telling the truth and that they will be falsely accused of lying. Furthermore, even the most ardent advocate of the polygraph must admit that liars can sometimes pass their tests. One need only remember the spy Aldrich Ames, who passed the polygraph test several times. This lesson was lost on the FBI, however, which started requiring polygraph tests of its employees after spy Robert Hanssen was caught. Heretofore, the FBI had used the polygraph only on suspected criminals. Apparently, the FBI thinks that they could have prevented Hanssen's betrayal if only he had been made to take the polygraph.

In California and many other states, the results of polygraph tests are inadmissible as evidence in a court of law. This may be because polygraph tests are known to be unreliable, or it may be because what little benefit may be derived from using the polygraph is far outweighed by the potential for significant abuse by the police. The test can easily be used to invade a person's privacy or to issue a high-tech browbeating of suspects. Skeptics consider evidence from polygraphs no more reliable than testimony evoked under **hypnosis**, which is also not allowed in a court of law in many states. Also, in 1998 the U.S. Supreme Court argued that Military Rule of Evidence 707, which makes polygraph evidence inadmissible in court-martial proceedings, does not unconstitutionally abridge the right of accused members of the military to present a defense (*United States, Petitioner v. Edward G. Scheffer*, 1998, No. 96-1133).

The American Civil Liberties Union strongly supported the passage of the Employee Polygraph Protection Act of 1988 (EPPA), which outlaws the use of the polygraph "for the purpose of rendering a diagnostic opinion regarding the honesty or dishonesty of an individual." The EPPA doesn't outlaw the polygraph across the board, however. Federal, state, and local governments can still use the polygraph. The federal government can give polygraph tests to government contractors involved in national security projects. In the private sector, security and pharmaceutical firms can still use the polygraph on current or prospective employees. Furthermore, any employer can administer polygraph tests

> in connection with an ongoing investigation of an economic loss or injury to his/her business on these conditions: The employee under suspicion must have had access to the property, and the employer must state in writing the basis for a reasonable suspicion that the employee was guilty. [ACLU: www.aclu.org/library/pbp4.html]

The ACLU supported the EPPA not only because of the lack of evidence for the accuracy of the polygraph, but also because of abuses related with its administration, including, but not limited to, invasion of privacy:

> For example, in order to establish "normal" physiological reactions of the person

being tested, "lie detector" examiners ask questions that purposely embarrass, frighten and humiliate workers. An ACLU lawsuit in 1987 revealed that state employees in North Carolina were routinely asked to answer such questions as "When was the last time you unintentionally exposed yourself after drinking?" and "Who was the last child that got you sexy?" Polygraphs have been used by unscrupulous employers to harass union organizers and whistle-blowers, to coerce employees into "confessing" infractions they did not commit, and to falsely implicate fellow employees. (ibid.)

Why would so many government and law enforcement agencies, and so many private sector employers, want to use the polygraph if the scientific community is not generally convinced of its validity? Is it just **wishful thinking**? Do the users of the polygraph want to believe there is a quick and dirty test to determine who's lying and who's not, so they blind themselves to the lack of evidence? Perhaps, but there are other factors as well, such as the "esoteric technology factor." The polygraph machine looks like a sophisticated, space-age device of modern technology. Only experts trained in its arcane ways can administer it correctly. Nonexperts are at the mercy of the high-tech, specially trained wizards who alone can deliver the prize: a decision as to who is lying and who is not.

Another reason for the polygraph's popularity is the **pragmatic fallacy**: It works! Case after case can be used to confirm that the polygraph works. There are the cases of those who failed the test and whose lying was corroborated by other evidence. There are the cases of those who, seeing they are failing the test, suddenly confess. What is the evidence that the rate of correct identification of lying corroborated by extrinsic evidence is greater than the rate of identification of lying where the polygraph is not used? There isn't any. The proofs are anecdotal or based on fallacious reasoning such as thinking that a correlation proves a causal connection.

On the other hand, it is possible that one of the main reasons so many government, law enforcement, and private sector employers want to use polygraphs is because they think the test will *frighten away* liars and cheats who are seeking jobs or will *frighten confessions* out of those accused of wrongdoing. In other words, the users of the machine don't really believe the machine can detect lies, but they believe that many of the people they administer it to *think* the machine can catch them in a lie. So, the result is the same as if the test really worked: They don't hire the liar/cheat and they catch the dishonest employee. As President Richard Nixon once said: "I don't know anything about lie detectors other than they scare the hell out of people."

Further reading: Lykken 1998; Shneour 1990; Zelicoff 2001.

Ponzi schemes

See **pyramid schemes**.

positive-outcome bias

The tendency of researchers and journals to publish research with positive outcomes much more frequently than research with negative outcomes. Positive outcomes are those that find a significant correlation or likely causal connection between at least two variables, such as a new medication and reduction of anxiety. Negative outcomes are not those that discover harmful effects of substances, but rather those that find no effects at all.

Positive-outcome bias also refers to the tendency of the *media* to publish medical research stories with positive outcomes much more frequently than such stories with negative outcomes. The media bias may be due to the scientific journal bias, but the latter seems to be due to researchers not submitting negative outcome studies for publication, rather than bias on the part of publication or peer review editors.

Further reading: Easterbrook et al. 1991; Koren and Klein 1991.

post hoc fallacy

The fallacy of *post hoc ergo propter hoc* ("after this therefore because of this") is based on the mistaken notion that because one thing happens after another, the first event caused the second event. Post hoc reasoning is the basis for many superstitions and erroneous causal beliefs.

Many events follow sequential patterns without being causally related. A solar eclipse occurs, so you beat your drums and the sun returns, but you're mistaken if you conclude that your drumming caused the sun to return. You use your **dowsing** stick and you find water. You have a vision that a body is going to be found near water or in a field and later a body is found near water or in a field. You have a dream that an airplane crashes and an airplane crashes the next day or crashed the night before. Just because one thing happened after another is not sufficient to establish a causal connection.

Coincidences happen. To establish the probability of a causal connection between two events, controls must be established to rule out other factors such as chance or some unknown causal factor. Anecdotes and **testimonial evidence** aren't sufficient because they rely on intuition and subjective interpretation. A **controlled study** is necessary to reduce the chance of error from **self-deception**.

See also **regressive fallacy**.

Further reading: Carroll 2000; Damer 2001; Giere 1998; Kahane 1997; Moore and Parker 2000.

pragmatic fallacy

Arguing that something is true because "it works." For example, **astrology** works, **numerology** works, **therapeutic touch** works. What "works" means here is imprecise. At the least, it means that one perceives some practical benefit in believing that it is true, despite the fact that the utility of a belief is independent of its truth-value. At this level, "works" seems to mean "I'm satisfied with it," which in turn might mean "I feel better" or "It explains things for me." At most, "works" means "has beneficial effects" even though the evidence may be very weak for establishing causality.

The pragmatic fallacy is common in "**alternative**" **health claims** and is often based on **post hoc** reasoning. For example, one has a sore back, wears the new **magnetic** or **takionic** belt, finds relief soon afterward, and declares that the magic belt caused the pain to go away. How does one know this? Because it works!

There is a common retort to the skeptic who points out that customer satisfaction is irrelevant to whether the device, medicine, or therapy in question really is a significant causal factor in some outcome: Who cares *why* it works, as long as it works? You can argue about why it works, but you can't argue about customer satisfaction. They feel better after using the product. That's all that matters.

It isn't all that matters. **Testimonial evidence** is not a substitute for scientific studies, which are done to make sure that we are not deceiving ourselves about what appears to be true. It is especially necessary to do **controlled studies** of alleged pain relievers to avoid **self-deception** due to the **placebo effect**, post hoc reasoning, or the **regressive fallacy**. We may not want to question too deeply the felt relief, but we should question the *cause* of that relief. Also, one's mood and feelings are subjective and do not necessarily correspond to objective, measurable changes, which are better indicators of whether something works.

It is easy to understand why someone diagnosed with terminal cancer who seeks out an "alternative" treatment and finds the cancer goes into remission soon afterward would attribute miraculous causal efficacy to the alternative treatment. However, if the alternative treatment is not really the cause of the remission, then others who seek the treatment will be filled with false hope. Of course, those patients who try the same treatment but who die anyway are not around to tell their story. Their surviving loved ones may even claim that the only reason the treatment did not work was because the patient came to it too late. The only way to know for sure whether the treatment has causal efficacy is to study its application under controlled conditions. Testimonials regarding how well the treatment works may be heartfelt, but they can be dangerously misleading, even deadly.

prana

The all-pervading vital **energy** of the universe; the Hindu version of **chi**.

prayer

Attempted communication with supernatural beings (SBs). The word derives from a 14th-century French word (*preiere*) meaning "to obtain by entreaty." The most common use of the word "prayer" is asking an SB for some favor. This type of prayer is called *intercessory prayer* because it is done to ask an SB to intercede on behalf of oneself or someone else. There are some people who believe that such prayers are effective in curing diseases, reducing crime, defeating enemies, and winning high school football games. Some religions require parents to ignore medical treatment for their children, even if to do so is likely to prove fatal, in favor of prayer. (These religions may not ban medical treatment altogether, but parents try healing prayer first, a practice that sometimes proves fatal to their children.) The prayer of such people, however, is not intercessory prayer, but the prayer of total submission to the will of an all-powerful, perfect God, and **faith** that whatever happens happens only because God wills it. Such was the belief of the founder of Christian Science, Mary Baker Eddy (1821–1910), who wrote *Science and Health with Key to the Scriptures* (1875), the Bible of faith healing. "If the sick recover because they pray or are prayed for audibly," said Eddy, "only petitioners should get well."

For an SB to intercede would be for a being from the supernatural world to cause things to happen in the natural world that would not happen naturally. This might sound like a good thing. After all, who wouldn't like to be able to contradict the laws of nature whenever it was convenient to do so? However, there are at least two reasons for believing that beseeching an SB to intervene in the natural course of events is absurd.

SBs, if they exist, would not be SBs if anything that mere humans or other earthly creatures did could please or displease them. Epicurus made a most elegant argument centuries ago demonstrating this point. He argued persuasively that men make their gods in their own image rather than the other way around (anthropomorphism) and that the gods would not be perfect if our antics or pleas could affect them in any way. Mary Baker Eddy obviously agreed with Epicurus. "God is not influenced by man," she said. "Do we expect to

change perfection?" she asked rhetorically.

Second, and more important, if SBs could contravene the laws of nature at will, human experience and **science** would be impossible. We are able to experience the world only because we perceive it to be an orderly and lawful world. If SBs could intervene in nature at will, then the order and lawfulness of the world of experience and of the world that science attempts to understand would be impossible. If that order and lawfulness were impossible, then so would be the experience and understanding of it.

David Hume gave an elegant argument on **miracles** that applies to intercessory prayer; for, by asking an SB to intervene in the ordinary course of natural events, one is requesting an SB to perform a miracle. As Hume argued, to believe miracles have been witnessed is to go against all one's experience that there is an inexorable order and lawfulness to our sense perceptions. All our rules of reasoning are based on this experience. We would have to abandon them to believe in miracles. Likewise, we would have to abandon any hope of experiencing, much less understanding, the world we perceive, if it were possible that any event could follow any other event based on the will of SBs. Only if our experience of events following other events is constant and consistent, can we perceive and understand the world. If you don't like Hume's approach, there is Kant's: Only if we experience events as causal can we have any experience at all.

Testing causal hypotheses would be impossible if SBs could interfere with the regular course of nature. Scientists test causal hypotheses. Thus, for a scientist to do a causal test on intercessory prayer would be absurd. Then what are we to make of those scientists who design controlled, double-blind studies to test the effectiveness of intercessory prayer? For example, what should we make of Elisabeth Targ's study on "distance healing" using prayer? The National Institutes of Health granted Targ hundreds of thousands of taxpayer dollars to investigate an absurdity (Gardner 2001). However, she died of brain cancer in 2002, before her study was completed.

Targ's study of the effectiveness of prayer on healing seems to be self-refuting. That is, if God or some other SB were to answer prayers and heal some patients but not others, depending on which patients had prayers said for them, then we could never know whether anything occurred due to natural causes or due to divine intervention. No causal study could rule out the possibility that its results were not due directly to an SB interfering with the course of nature. In short, it would be pointless to do causal studies and, hence, pointless to study whether prayer is effective in healing.

There are other problems, as well. Those who are not healed may not have died due to natural causes; it is always possible that some malevolent but powerful SB interfered with natural processes and caused the deaths. Once you introduce the possibility of SBs being the cause of events, there is no justification for assuming that only the Judeo-Christian God can be that cause or that God only interferes when prayers are involved.

Thus, there are logical, scientific, and metaphysical reasons for not seriously investigating such notions as the healing power of prayer. The idea is logically contradictory, scientifically preposterous, and metaphysically demeaning. It requires God to be perfect and imperfect, it makes a mockery of the notion of scientific tests of causality, and it belittles the Omnipotent Infinite God, if such exists, and ignores the possibility of lesser supernatural powers interfering with nature in untold ways.

See also **pious fraud** and **positive-outcome bias**.

Further reading: Barrett 2000; Humphrey 2000; Posner 1990, 2000a; Randi 1982a, 1989a, 1995.

precognition

Psychic knowledge of something in advance of its occurrence.

See also **clairvoyance**.
Further reading: Hyman 1989.

Princeton Engineering Anomalies Research (PEAR)

The brainchild of Robert G. Jahn, who, in 1979, when he was Dean of the School of Engineering and Applied Science at Princeton University, claimed he wanted "to pursue rigorous scientific study of the interaction of human consciousness with sensitive physical devices, systems, and processes common to contemporary engineering practice." In short, he wanted to be a **parapsychologist** and test **psychokinesis**. Not so unbelievably, he has found several others at Princeton who also were tired of humdrum work in the humanities, social sciences, engineering, and physics, and have joined the quest to prove that the mind alone can alter matter.

Jahn, six of his associates, and PEAR even have a patent (US5830064) on an "Apparatus and method for distinguishing events which collectively exceed chance expectations and thereby controlling an output." This patent is based on their experiments where human operators try to use their minds to influence a variety of mechanical, optical, acoustical, and fluid devices. In short, the PEAR people are doing what many drivers do when they try to use their thoughts to make a red light turn green.

PEAR claims to have attained results that can't be due to chance and *"can only be attributed to the influence of the human operators"* (emphasis added). This is an extraordinary claim, especially coming from such scholars at such a distinguished institution. I would think it would be impossible to rule out all of the following explanations for statistics not likely due to chance:

- deliberate fraud or cheating
- errors in calibration
- unconscious cheating
- errors in calculation
- software errors
- self-deception

Of course, they could be hedging here. After all, fraud, unconscious cheating, errors in calculation, software errors, and self-deception could be considered as "influence of human operators." So could the fact that "operator 10," believed to be a PEAR staff member, "has been involved in 15% of the 14 million trials, yet contributed to a full half of the total excess hits" (McCrone 1994).

The PEAR people are so convinced of the breakthrough nature of their work that they have incorporated as Mindsong Inc. They claim their corporation "is developing a range of breakthrough products and research tools based on a provocative new technology—proprietary microelectronics which are responsive to the inner states of living systems." One of their breakthrough products is some software "that allows you to influence, with your mind, which of two images will be displayed on your computer screen." They also sell a device for several hundred dollars that lets you do your own testing of mental influence on randomized outputs. Stan Jeffers, a physicist at York University, Ontario, replicated the Jahn experiments, but with chance results.

See also **law of truly large numbers**. Further reading: Hyman 1996b.

prophecy

See **oracle**.

Protocols of the Elders of Zion

A forgery made in Russia for the Okhrana (secret police) that blames the Jews for the country's ills. It was privately printed in 1897 and was made public in 1905. It is copied from a 19th-century novel by Hermann Goedsche (*Biarritz,* 1868) and claims that a secret Jewish cabal is plotting to take over the world.

Goedsche, a German novelist and anti-Semite who used the pseudonym of Sir John Retcliffe, composed the basic story. Goedsche stole the main story from another writer, Maurice Joly, whose *Dialogues in Hell Between Machiavelli and Montesquieu* (1864) involved a plot aimed at opposing Napoleon III. Goedsche's original contribution consists mainly of introducing Jews to do the plotting to take over the world.

The Russians used big chunks of a Russian translation of Goedsche's novel, published it separately as the *Protocols,* and claimed it was not fiction but evidence of a Zionist plot to take over the world. Their purpose was political: to strengthen Czar Nicholas II's position by exposing his opponents as allies with those who were part of a massive conspiracy. In short, the *Protocols* are a forgery of a plagiarized fiction.

Lucien Wolf exposed the *Protocols* as a forgery in *The Jewish Bogey and the Forged Protocols of the Learned Elders of Zion* (1920). In 1921, Philip Graves, a correspondent for the *London Times,* publicized the forgery. Herman Bernstein, in *The Truth About "The Protocols of Zion": A Complete Exposure*

(1935), also tried and failed to convince the world of the forgery.

The *Protocols* were published in 1920 in a Michigan newspaper, the *Dearborn Independent,* started by Henry Ford mainly to attack Jews and Communists. Even after it was exposed as a forgery, Ford's paper continued to cite the document. On February 17, 1921, Ford himself said: "The only statement I care to make about the Protocols is that they fit in with what is going on. They are sixteen years old, and they have fitted the world situation up to this time. They fit it now." Adolf Hitler later used the *Protocols* to help justify his attempt to exterminate Jews during World War II. In *Mein Kampf,* Hitler wrote, "To what extent the whole existence of this people [the Jews] is based on a continuous lie is shown incomparably by the *Protocols of the Wise Men of Zion.*"

See also **Holocaust denial** and **Illuminati**.

Further reading: Bronner 2000; Cohn 1967; Goldberg 1936; Segel 1995; Wolf 1921.

pseudohistory

Purported history that

- treats myths, legends, sagas, and similar literature as literal truth;
- is neither critical nor skeptical in its reading of ancient historians, taking their claims at face value and ignoring empirical or logical evidence contrary to the claims of the ancients;
- is on a mission, not a quest, seeking to support some contemporary political or religious agenda rather than find out the truth about the past;
- often denies that there is such a thing as historical truth, clinging to the extreme skeptical notion that only what is

absolutely certain can be called "true" and nothing is absolutely certain, so nothing is true;

- often maintains that history is nothing but mythmaking and that different histories are not to be compared on traditional academic standards such as accuracy, empirical probability, logical consistency, relevancy, completeness, fairness, and honesty but on moral or political grounds;
- is selective in its use of ancient documents, citing favorably those that fit with its agenda and ignoring or interpreting away those documents that don't fit;
- considers the *possibility* of something being true as sufficient to believe it is true if it fits with one's agenda; and
- often maintains that there is a conspiracy to suppress its claims because of racism, atheism, or ethnocentrism, or because of opposition to its political or religious agenda.

Examples of pseudohistory include **Afrocentrism**, **creationism**, various theories regarding aliens as the creators of human life on earth (e.g., the theories of **von Däniken**, **Raël**, and **Sitchin**), and the myth-based catastrophism of Immanuel **Velikovsky**.

One should also refer to writers such as the Abbé Jean Terrasson (1670–1750) as pseudohistorians. These are writers of historical fiction who intentionally falsify and invent ancient history, as Terrasson did in his *Sethos, a History or Biography, Based on Unpublished Memoirs of Ancient Egypt.* This technique of claiming to find an ancient document and publishing it in order to express one's own ideas is still used, e.g., *The Celestine Prophecy.* A variation on this theme is to claim that one is **channeling** a book from some ancient being, for example,

*The **Urantia Book,** A **Course in Miracles,*** and *Bringers of the Dawn* (see **Pleiadians**).

Further reading: Lefkowitz 1996; Shermer and Grobman 2000; Walker 2001.

pseudoscience

Nonscientific theories that are claimed to be scientific by their advocates. Scientific theories are characterized by such things as:

- being based on empirical observation rather than the authority of some sacred text;
- explaining a range of empirical phenomena;
- being empirically tested in some meaningful way, usually involving testing specific predictions deduced from the theory;
- being confirmed rather than falsified by empirical tests or with the discovery of new facts;
- being impersonal and therefore not dependent on faith or personal belief in order to be utilized or tested;
- being testable by anyone regardless of personal religious or metaphysical beliefs;
- being dynamic and fecund, leading investigators to new knowledge and understanding of the interrelatedness of the natural world, rather than being static and stagnant leading to no research or development of a better understanding of anything in the natural world;
- being approached with skepticism rather than gullibility, especially regarding paranormal forces or supernatural powers; and
- being fallible and put forth tentatively rather than being put forth dogmatically as infallible.

Some pseudoscientific theories are based on an authoritative text rather than observation or empirical investigation. **Creation scientists**, for example, make observations only to confirm infallible dogmas, not to discover the truth about the natural world. Such theories are static and lead to no new scientific discoveries or enhancement of our understanding of the natural world.

Some pseudoscientific theories explain what nonbelievers cannot even observe, for example, **orgone energy**.

Some can't be tested because they are consistent with every imaginable state of affairs in the empirical world, for example, L. Ron Hubbard's theory of engrams (see **Dianetics**).

Some pseudoscientific theories can't be tested because they are so vague and malleable that anything relevant can be **shoehorned** to fit the theory, for example, the **enneagram**, **iridology**, **reflexology**, the **Myers-Briggs Type Indicator**, and the theories behind much New Age **psychotherapy**.

Some theories have been empirically tested, and rather than being confirmed, they seem either to have been falsified or to require numerous **ad hoc hypotheses** to sustain them, for example, **astrology**, **biorhythms**, **ESP**, **facilitated communication**, and **plant perception**. Yet despite seemingly insurmountable contrary evidence, adherents won't give the theories up.

Some pseudoscientific theories rely on ancient myths and legends rather than on physical evidence, even when their interpretations of those legends requires a belief contrary to the known laws of nature or to established facts, for example, **Däniken's**, **Sitchen's**, and **Velikovsky's** theories.

Some pseudoscientific theories are supported mainly by selective use of anecdotes, intuition, and examples of confirming instances, for example, **anthropometry**, **aromatherapy**, **craniometry**, **graphology**, **metoposcopy**, **personology**, and **physiognomy**.

Some pseudoscientific theories confuse metaphysical claims with empirical claims, for example, the theories of **acupuncture**, **alchemy**, **cellular memory**, **Lysenkoism**, **naturopathy**, **reiki**, **Rolfing**, **therapeutic touch**, and **Ayurvedic medicine**.

Some pseudoscientific theories not only confuse metaphysical claims with empirical claims, but also maintain views that contradict known scientific laws and use ad hoc hypotheses to explain their belief, for example, **homeopathy**.

Pseudoscientists claim to base their theories on empirical evidence, and they may even use some scientific methods, though often their understanding of a **controlled experiment** is inadequate (e.g., Cleve Backster and his experiments on **plant perception**). Many pseudoscientists relish being able to point out the consistency of their theories with known facts or with predicted consequences, but they do not recognize that such consistency is not proof of anything. It is a *necessary* but not a *sufficient* condition that a good scientific theory be consistent with the facts. A theory that is contradicted by the facts is obviously not a very good scientific theory, but a theory that is consistent with the facts is not necessarily a good theory. For example, "the truth of the hypothesis that plague is due to evil spirits is not established by the correctness of the deduction that you can avoid the disease by keeping out of the reach of the evil spirits" (Beveridge 1957: 118).

Further reading: Carroll 1999; Dawes 1994; Friedlander 1995; Gardner 1957; Gilovich 1993; Glymour and Stalker 1990; Gould 1979; Radner and Radner 1982; Sagan 1979, 1995; Shermer 1997, 2001a; Singer and Lalich 1996; Spanos 1996.

psi

"Psi" (pronounced "sigh") is a term commonly used by **parapsychologists** to refer to **ESP** and **psychokinesis** taken together. The term was coined by R. H. Thouless and B. P. Weisner (1942).

psi-missing

An **ad hoc hypothesis** invented by **parapsychologists** to explain away failures to demonstrate **ESP**. The tests usually involve trying to use ESP to identify various targets, such as **Zener cards** or pictures, that are hidden from direct view of the subject. The failure to do better than would be expected by chance is explained away as due to unconscious direction to avoid the target. Parapsychologist J. B. Rhine even claimed that persons who didn't like him would consciously guess wrong to spite him, so he wouldn't count their results (Park 2000: 42).

psychic

As an adjective, "psychic" refers to forces or agencies of a **paranormal** nature. As a noun, "psychic" refers to a **medium** or a person who has paranormal powers.

Many skeptics accuse psychics of being frauds. Some are, but many psychics genuinely believe in their powers. However, they've never tested their powers in any meaningful way. James **Randi**, who has tested many people who think they have psychic abilities, has found that when he has tested the alleged paranormal powers of psychics, they had never before tested their powers under controlled conditions, and those who don't offer preposterous rationalizations for their inability to perform seem genuinely baffled at their failure.

To believe in the ability of a person to **channel** spirits; to "hear" or "feel" the voices or presence of the dead; to "see" the past, the future, or what is presently in another's mind; or to make contact with a realm of reality that transcends natural laws is to believe in something highly improbable. Psychics don't rely on psychics to warn them of impending disasters. Psychics don't predict their own deaths or diseases. When they get a toothache, they call the dentist; they don't call months in advance. They're as surprised and disturbed as the rest of us when they have to call a plumber or an electrician to fix some defect at home. They get stuck at airports when their planes are delayed; they don't stay in the hotel because they anticipate the delay. If they want to know something about Abraham Lincoln, they go to the library; they don't try to talk to Abe's spirit. In short, psychics live by the known laws of nature except when they are playing the psychic game with people. Psychics aren't overly worried about other psychics reading their minds and revealing their innermost secrets to the world. No casino has ever banned psychics from the gaming room. There is no need. "When confidential information leaks out of an organization," notes John Allen Paulos, "people suspect a spy, not a psychic" (1990).

The main reasons for belief in such paranormal powers as **clairaudience** and **clairvoyance** are the perceived accuracy of psychic predictions and readings; the seemingly uncanny premonitions which many people have, especially in **dreams**; and the seemingly fantastic odds against such premonitions or predictions being correct by **coincidence** or chance.

However, the accuracy of psychic predictions is grossly overrated. The belief in the accuracy of clairvoyants such as Edgar **Cayce** and Jeane **Dixon** is due to several factors, including mass media error and hype. For example, it has been repeatedly reported in the mass media that Jeane Dixon predicted the assassination of President Kennedy. She did not. The *New York Times* helped spread the myth that Edgar Cayce transformed from an illiterate into a

healer when hypnotized. One of the more egregious cases of mass media complicity in promoting belief in psychics is the case of Tamara Rand, Dick Maurice, and Gary Grecco of KNTV in Las Vegas. All conspired to deceive the public by claiming that a videotape of a *Dick Maurice Show,* on which Rand predicts the assassination attempt by John Hinkley on Ronald Reagan, was done on January 6, 1981. The tape was actually made on March 31, 1991, a day after Hinkley shot Reagan (Steiner 1996).

Another reason the accuracy of psychic predictions is grossly overrated is because many people do not understand how psychics use techniques such as warm and **cold reading**. Also, many people lack an understanding of **confirmation bias** and the **law of truly large numbers**. The accuracy of premonitions and prophecies is also exaggerated because of ignorance about how **memory** works, especially about how dreams and premonitions are often filled in after the fact.

The strongest kind of evidence for psychic power, however, comes from witnessing an alleged psychic perform. Some performers seem to be able to do things that require paranormal powers; they are masters of the art of **conjuring**. Others seem to be able to tell us things about ourselves and our departed loved ones that only we should know; these are the masters of cold and warm reading. Others surreptitiously gather information about us and deceive us into thinking they obtained their data by psychic means.

The success of numerous hoaxes by fraudulent psychics testifies to the difficulty of seeing through the performance. Psychologist Ray Hyman, who worked as a psychic to help pay his way through college, claims that the most common method used by psychics is cold reading. The following borrows heavily from his "Guide to Cold Reading" (2002):

You must act with confidence.

You must do your research. You have to be up on the latest statistics (e.g., most plane crashes are in April; most planes have something red on their tails). You have to know what people in general are like from polls and surveys. Also, you must pick up in casual conversation before a performance any information that might be useful later, like talking to a cameraman in the afternoon and then during the evening performance you are "contacted" by his dead father, whom he told you all about that afternoon.

You must convince the mark that he or she will be the reason for success or failure. This is actually true because it is the mark who will provide all the vital information that seems so shocking and revealing. It is human nature to find meaning, so this is not a difficult chore.

Be observant. Does the person have expensive jewelry on but worn out clothes? Is she wearing a pin with the letter "K" on it? (You better know that "Kevin" is a good guess here. But it doesn't matter, really. Since, when the mark tells you the name of the person, she'll think you are the one who told her the name!)

Use flattery and pretend you know more than you do.

What looks like psychic power to some may appear to be little more than a game of 20 questions or a fishing expedition to others. But the experienced cold reader doesn't just randomly throw chum at his audience. His questions have purpose and issue from a storehouse of facts and probabilities. He must know how to use effectively whatever information is provided by the client (Rowland 2002). The mark provides all the relevant details and connects all the dots, while the "psychic" appears to be getting messages from beyond. Of course, sometimes the "psychic" is simply an observant, thoughtful person, who says things appropriate for the age and gender of the subject. For example, one of my students—right out of high school, tall, handsome, strong, and athletic—was told by a psychic to stay away from sex or he'd soon be a father. The student became an immediate convert. He'd already gotten a girl pregnant and had a daughter. Good advice became proof of psychic power in this young man's mind. She also told him other things he thought nobody could have known, such as that he had once "thrown up all over himself and crapped in his pants." He apparently had done this as a young man and didn't realize that she was describing a nearly universal situation for babies and toddlers.

The deception can be more dramatic than cold reading, of course. According to Lamar Keene, a reformed psychic, some people seek psychic advice from professionals who exchange information on their marks. Some psychics do what is called a "hot reading," that is, they have done research on you and that's why they know things they shouldn't know. Still others are magicians who try to pass off their **conjuring** skills as paranormal powers.

It has also been argued that if psychic power existed, to use it would be "a gross and unethical violation of privacy" and that "professions that involve deception would be worthless" (B. Radford 2000). There wouldn't be any need for undercover work or spies. Every child molester would be identified immediately. No double agent could ever get away with it. Psychics would be on demand for high-paying jobs in banks, businesses, and government. "Most psychics would be very, very rich...." (ibid.) And since psychics are such altruistic persons, giving up their time to help others talk to the deceased or figure out what to do with their lives, they would be winning lotteries right and left and giving part of their winnings to help the needy. We wouldn't need trials of accused persons; psychics could tell us who is guilty and who is not. The **polygraph** would be a thing of the past. Of course, the operative word here is "if." *If* psychic power existed, the world would be very different.

It seems clear that psychics can be explained in one of three ways: They truly are psychic; they are frauds, taking advantage of people's gullibility and weaknesses; or they're deluded and self-deceived. Of the three options, the least probable is the first. Psychics who are honest about their deception call themselves **mentalists** and call their art "magic" or "conjuring." Yet it is the psychics, not the mentalists, who are the darlings of the mass media. Thus, when the mass media promote psychics for their entertainment or news value, they are either promoting fraud or encouraging delusions. Perhaps the media think that because most parties in the psychic game are consenting adults, that makes it acceptable. Perhaps the police agree and that is why telepsychics such as Miss Cleo can practice without fear of arrest for fraud—at least not for the fraud associated with claiming you can see into other people's pasts and futures over the telephone.

See also **Akashic record** and **séance**.

Further reading: Alcock 1990; Blackmore 1992, 1992; Frazier and Randi 1981; Gardner 1957, 1989; Gordon 1987; Hansel 1989; Hyman 1989; Marks and Kammann 1979; Paulos 1990; Randi 1982a; Steiner 1996; Stenger 1990; Wiseman 1997; Wiseman and Morris 1995.

psychic detectives (PDs)

Alleged **psychics** who offer to help law enforcement agencies solve crimes. In their book, *The Blue Sense: Psychic Detectives and Crime* (1991), Arthur Lyons and Marcello Truzzi list many reasons people without

any psychic powers gain a reputation for assisting in crime detection. In many cases, most of the evidence attesting to the psychic detective's prowess is provided to the mass media by the psychic rather than by an independent source. The mass media is rarely critical or skeptical of the claims of psychics. For example, alleged psychic detective Sylvia Browne has declared many times that she has used her psychic powers to solve crimes, yet it is rare to see her challenged as she was by *Brill's Content*:

> *Brill's Content* has examined ten recent Montel Williams programs that highlighted Browne's work as a psychic detective (as opposed to her ideas about "the afterlife," for example), spanning 35 cases. In 21, the details were too vague to be verified. Of the remaining 14, law-enforcement officials or family members involved in the investigations say that Browne had played no useful role.
>
> "These guys don't solve cases, and the media consistently gets it wrong," says Michael Corn, an investigative producer for "Inside Edition" who produced a story last May debunking psychic detectives. Moreover, the FBI and the National Center for Missing & Exploited Children maintain that to their knowledge, psychic detectives have never helped solve a single missing-person case.
>
> "Zero. They go on TV and I see how things go and what they claim but no, zero," says FBI agent Chris Whitcomb. "They may be remarkable in other ways, but the FBI does not use them." ("Prophet Motive," *Brill's Content*, November 27, 2000)

Browne has made many claims on the *Larry King Live* show about her crime-solving powers, including the claim that she solved the 1993 World Trade Center bombing. James **Randi** challenged another of Browne's claims made on *Larry King*, namely, that she was working with Stephen Xanthos of the Rumson, New Jersey, police department. She said she was getting ready to close a case.

> [N]o person named Xanthos ever worked with that police department, though there was a Stephen Xanthos who was canned from another New Jersey police department. Looking a little further into this mythical claim of Sylvia's, we discovered that Xanthos had a private investigator's license at one time, but it expired in 1994 [i.e., 6 years earlier]. It's interesting to note that if this man really had been working with Browne, as she stated . . . on the Larry King show, he would be subject to charges of a third degree felony, under New Jersey State law—that's on a par with burglary and car theft. Not that we ever believed Sylvia was telling the truth, but she should be a bit more clever with her mendacity. [www .randi.org/jr/092801.html]

There are other reasons for the undeserved reputations of psychic detectives besides blowing their own horns to an uncritical media. They do sometimes guess correctly. Everybody can have a 50% hit rate if we guess "dead" or "alive" about a missing person. The odds are good that by the time a psychic gets involved in a missing person case, the person is probably dead. The events predicted by PDs are commonplace events that are predicted by thousands of psychics every year. A missing person will be either dead or alive; if dead, probably buried; if buried, probably in a remote place such as the woods or a field. Shallow graves are likely to be pretty common, too. How many killers take the time to dig a deep grave? Water can be found nearly anywhere a person is likely to be buried. Yet predicting that a body will be found in a shallow grave in a wooded area near water is taken by some to be truly astounding if it turns out to be the case. In other words, some PDs' "visions" are bound to be correct often enough for the credulous to be duped. What seems like an accurate perception is due to its vagueness, commonness, and the latitude given to what will count as a psychic hit: for example, "I see water near the body" or "I see trees." Some PDs are very skillful in their use of vagueness and ambiguity, and provide "the verbal equivalent of a **Rorschach test**," according to Piet Hein Hoebens, one of Truzzi's collaborators in a "Psychic Sleuths" project.

Lyons and Truzzi note that, over time, reports of psychic achievements get exaggerated and distorted. Vague claims become specific. Errors become replaced with correct predictions. Events that never happened become "facts." Often, the PD herself or himself is the source of this historical reconstruction. Sometimes a psychic's "predictions" are made *after* an event, but are claimed to have been made before it, such as James Van Praagh's and Sylvia Browne's claims after the September 11, 2001, terrorist attacks.

Some of the undeserved reputation of PDs comes from their clients: the police or relatives of crime victims. The clients count misses and errors as hits. For example, Browne told a woman her husband died of a clot and, even though he died of a hemorrhage, the client agreed that Browne was right. The difference between the two is like the difference between a plugged drain and a leaky pipe.

Clients often take coincidences for hits. Sometimes, as Lyons and Truzzi point out, the information provided by the PD was garnered from another source, often from an unwitting law enforcement agent. The psychic just feeds back information initially provided by the client himself. Some psychic successes are merely self-fulfilling prophecies. Clients find ways to retrofit facts with the vague and ambiguous pronouncements of the psychic. Clients also often use **selective thinking**, remembering what seems accurate and forgetting what was clearly not on the mark. Furthermore, the mass media publish stories about alleged psychic successes, while generally ignoring stories about psychic failures and frauds. Reputations are thereby created and enhanced from trivial or paltry evidence of psychic detective powers.

According to Lyons and Truzzi, PDs often use **shotgunning** to provide information, that is, they provide a large quantity of information, some of which is bound to fit the case. Shotgunning relies on **confirmation bias** and **cold reading**: The cop tunes in to the information that is correct and ignores what isn't, and unknowingly gives cues to the psychic as he or she fires salvo after salvo.

Some PDs are simply frauds, according to Lyons and Truzzi. Some psychics even use accomplices to accomplish their frauds and deceptions. Some bribe informants, including police officers, for information they pass off as acquired by psychic means.

While it is true that some cops believe in psychics, many simply use them for their own purposes. Lyons and Truzzi tell the story of a cop who considered psychic Noreen Reiner's drawing of a circle to be a correct clue in a crime because the person arrested drove a cement mixer. Another cop considered Dorothy Allison's clues in a case to be on the money even though she predicted a missing person was dead who was not dead but living in a religious cult community. The cop admitted he was baffled by Allison's error about the person being dead, but *which way was he dead?* asked the cop, "Biologically? Clinically? Dead tired?" However, such **wishful thinking** and **self-deception** seem to be the exception rather than the rule among law enforcement officers. Cops are more likely to use psychics to cover up their real sources of information, to protect an informant, or to conceal the fact that information was obtained illegally. Finally, some cops use psychics, or even pretend to be psychic, to psych out superstitious suspects.

Further reading: Nickell 1994b; Seckel 1987a, 1987b; Steiner 1996; Wiseman et al. 1996.

psychic healers

See **intuitives**.

psychic photography

The alleged production on photographic media of images of **ghosts** or **astral bodies** or images of ordinary things like buildings by **paranormal** means such as **psychokinesis**.

The first psychic photographs appeared almost immediately after the first photographs. "As early as 1856, prints of ghostly looking ethereal figures sitting next to the person being photographed were being sold as joke novelties" (Williams 2000: 205). In 1862, William Mumler made a good living in Boston using double exposure to produce photographs with alleged spirits of dead people in them (ibid.: 326). Many have followed in Mumler's footsteps.

Psychic photograph is a piece of cake in the digital age.

Some paranormal researchers, apparently unaware of or unwilling to accept that spirit photos are faked or misinterpretations of ordinary phenomena, try to chase down and photograph spirits. Still others claim that they can transfer their thoughts to film directly, a trick known as "thoughtography."

Thoughtography was made popular by psychiatrist Dr. Jule Eisenbud. He wrote a book about a Chicago bellhop name Ted Serios, who claimed he could make images appear on Polaroid film just by thinking of an image. Since the publication of Eisenbud's *The World of Ted Serios: "Thoughtographic" Studies of an Extraordinary Mind* (1966), others have claimed to be able to perform this feat. Eisenbud claimed that Serios made his thoughtographs by psychokinesis, and that some of them were instigated during **out-of-body experiences**. Charlie Reynolds and David Eisendrath, both amateur magicians and professional photographers, exposed Serios as a fraud after spending a weekend with him and Eisenbud. Serios claimed he needed a little tube in front of the camera lens to help him concentrate, but he was spotted slipping something into the tube. Most likely it was a picture of something that the camera would take an image of, but which Serios would claim came from his mind rather than his hand. Their exposé appeared in the October 1967 issue of *Popular Photography*. Serios's psychokinetic powers began to fade after the exposure and he has remained virtually unheard from for the past 30 years.

Many years after Serios faded from the paranormal spotlight, Uri **Geller** began doing a trick in which he produced thoughtographs. Geller would leave the lens cap on a 35 mm camera and take pictures of his forehead. He claimed the developed film had pictures on it that came directly from his mind. There is no doubt that the images came from Geller's mind,

but perhaps they took a more circuitous route than he says. James Randi, magician and debunker of all things paranormal, claims that psychic photography is actually trickery done using a handheld optical device (Randi 1982a: 222ff.; 1995: 233) or by taking photos on already exposed film.

Many psychic photos are fanciful interpretations of flaws in a camera or film or effects due to various exposures, film-processing errors, lens flares (caused by interreflection between lens surfaces), the camera or lens strap hanging over the lens, the flash reflecting off mirrors or jewelry, or light patterns, polarization, chemical reactions, and so on (Nickell 1994c, 1997). As some pundit once noted: "Whoever thinks the camera doesn't lie, doesn't think."

It does seem strange that spirits and other paranormal forces have the power to appear on film or on electronic devices, or to communicate to a select few in cryptic noises that must be deciphered by **shotgunning** in a game of 20 questions. The spirits never simply sit down at the table and say directly what is on their minds. Perhaps this explains our love for hide and seek, the children's game that may hold the key to understanding human nature and the great secrets of the universe.

Further reading: Brugioni 1999; Randi 1982b.

psychic surgery

A type of fake surgery performed by a nonmedical healer. The healer fakes an incision by running a finger along the patient's body, apparently going through the skin without using any surgical instruments. The healer pretends to dig his hands into the patient's innards and to pull out tumors. Using trickery, the healer squirts animal blood from a hand-held balloon while discarding items such as chicken livers and hearts. The patient then goes home to die, if he or she was really dying, or to live if there was nothing seriously wrong in the first place.

Psychic surgery is big business around the world, but especially in the Philippines and Brazil. Tony Agpaoa put psychic surgery on the map in Manila, where there are now several hundred practicing psychic surgeons, many working out of hotels. In 1967, Agpaoa was indicted for fraud in the United States. He jumped bail and went home, forfeiting a $25,000 bond.

Some people find solace in psychic surgeons and other faith healers because they think the healers are divine agents. The practice is not restricted to third-world countries, either. Chris Cole practices psychic surgery in Sydney, Australia. One of the more popular psychic surgeons outside the Philippines is Stephen Turoff, who runs the Danbury Healing Clinic in Chelmsford, England. Turoff, a follower of **Sai Baba**, performs **therapeutic touch** at no extra charge.

Turoff has been performing for a quarter of a century and is popular enough to warrant a biographer, Grant Solomon (*Stephen Turoff, Psychic Surgeon: The Extraordinary Story of a Remarkable Healer*, 1999). Turoff describes himself as "a 16-stone, six-and-a-half foot, middle-aged, Jewish-Christian former carpenter from Brick Lane in London's East End whom many believe to be an instrument of God." To others, Turoff is a **pious fraud**.

In December 2002, the Rev. Alex Orbito, a psychic surgeon from the Philippines, was arrested in Padua, Italy, on charges of aggravated fraud, deception of incompetent people, and medical malpractice. Despite the fact that James **Randi** (1982a), the Italian skeptic's group Comitato Italiano per il Controllo delle Affermazioni sul Paranormale (CICAP), and others have exposed Orbito's frauds, he still manages to make a living around the world with his fake surgeries. (In one weekend, before his arrest, he had performed his operations on 150 to 170 people, who had each paid 485 euros [at the time of this writing $522].) Orbito is the healer who removed, without instruments, "negative energy clots" and "negative stress clots" from the body of actress and author Shirley MacLaine (*Going Within,* 1989).

Psychic dentistry is also available for those who prefer dentistry without anesthesia or dental drills performed by a faith healer. "Willard Fuller has supposedly healed more than 40,000 people since he began practicing in 1960. Those who flock to his healing ministry claim his magic touch can fill cavities, make bad teeth whole again, and even produce a new set of teeth in some elderly patients" (True: www.netasia.net/users/truehealth/Psychic%20Dentistry.htm). Many patients are afraid to admit they've been defrauded because that would imply that they lack true **faith**. According to George Nava True II, who operates the "only Philippine skeptical website to challenge the claims of alternative healers, psychics, and other quacks," psychic dentistry "has never been demonstrated under controlled laboratory conditions and most practitioners are simply sleight-of-hand artists who can't produce a shred of proof of their alleged powers." As those with faith are wont to say: For those who have no faith, proof is not possible; for those who have faith, proof is not needed.

Further reading: Barret and Butler 1992; Barret and Jarvis 1993; Brenneman 1990; True: www.netasia.net/users/truehealth/Psychic%20Surgery.htm; Randi 1982a; 1989a; Raso 1994.

psychoanalysis

Psychotherapy developed by Sigmund Freud (1856–1939) and rarely modified significantly by later practitioners. The therapy is often criticized because adherents have made false or misleading claims about the mind, mental health, and mental illness. For example, in psychoanalysis, schizophrenia and depression are not neurochemical disorders but *narcissistic* disorders. Autism and other brain disorders are not brain chemistry problems but *mothering* problems. These illnesses do not require pharmacological treatment. They require only talk therapy. Similar positions are taken for anorexia nervosa and Tourette's syndrome (Hines 1990: 136).

Freud thought he understood the nature of schizophrenia. It is not a brain disorder, but a disturbance in the **unconscious** caused by unresolved feelings of homosexuality. However, he maintained that psychoanalysis would not work with schizophrenics because such patients ignore their therapist's insights and are resistant to treatment (Dolnick 1998: 40). Later psychoanalysts would claim, with equal certainty and equal lack of scientific evidence, that schizophrenia is caused by smothering mothering. In 1948, Frieda Fromm-Reichmann, for example, gave birth to the term "schizophrenogenic mother," the mother whose bad mothering causes her child to become schizophrenic (ibid.: 94). Other analysts before her had supported the notion with anecdotes and intuitions, and over the next 20 years many more would follow her misguided lead.

Would you treat a broken leg or diabetes with "talk" therapy or by interpreting the patient's **dreams**? Of course not. Imagine the reaction if a diabetic were told that her illness was due to "masturbatory conflict" or "displaced eroticism." One might as well tell the patient she is possessed by demons as give her a psychoanalytic explanation of her physical disease or disorder.

The most fundamental concept of psychoanalysis is the notion of the unconscious mind as a reservoir for **repressed memories** of traumatic events that continuously influence conscious thought and behavior. The scientific evidence for this notion of unconscious repression is lacking, though there is ample evidence that conscious thought and behavior are influenced by unconscious memories.

Related to these questionable assumptions of psychoanalysis are two equally questionable methods of investigating the alleged memories repressed in the unconscious: *free association* and the *interpretation of dreams*. Neither method is capable of scientific formulation or empirical testing. Both are metaphysical blank checks to speculate at will without any check in reality.

Essentially connected to the psychoanalytic view of repression is the assumption that parental treatment of children, especially mothering, is the source of many, if not most, adult problems ranging from personality disorders to emotional problems to mental illnesses. There is little question that if children are treated cruelly throughout childhood, their lives as adults will be profoundly influenced by such treatment. It is a big conceptual leap from this fact to the notion that all sexual experiences in childhood will cause problems in later life, or that all problems in later life, including sexual problems, are due to childhood experiences. The scientific evidence for these notions is lacking.

In many ways, psychoanalytic therapy is based on a search for what probably does not exist (repressed childhood memories), on an assumption that is probably false (that childhood experiences caused the patient's problem), and on a therapeutic theory that has nearly no probability of being correct (that bringing repressed memories to consciousness is essential to the cure). Of course, this is just the foundation of an elaborate set of scientifically sounding concepts that pretend to explain the deep mysteries of consciousness and behavior. But if the foundation is illusory, what is the future of concepts built on it? Perhaps it would be good to remember that Freud once wrote:

> I am actually not at all a man of science, not an observer, not an experimenter, not a thinker. I am by temperament nothing but a conquistador—an adventurer, if you want it translated—with all the curiosity, daring, and tenacity characteristic of a man of this sort. (letter to Wilhelm Fliess, February 1, 1900)

There are some good things, however, that have resulted from the method of psychoanalysis developed by Freud a century ago in Vienna. Freud should be considered one of our greatest benefactors if only because he pioneered the desire to understand those whose behavior and thoughts cross the boundaries of convention set by civilization and cultures. That it is no longer fashionable to condemn and ridicule those with behavioral, mood, or thought disorders is due in no small part to the tolerance promoted by psychoanalysis. Furthermore, whatever intolerance, ignorance, hypocrisy, and prudishness remains regarding the understanding of our sexual natures and behaviors cannot be blamed on Freud. Psychoanalysts do Freud no honor by blindly adhering to the doctrines of their master in this or any other area.

Finally, as psychiatrist Anthony Storr put it: "Freud's technique of listening to distressed people over long periods rather than giving them orders or advice has formed the foundation of most modern forms of psychotherapy, with benefits to both patients and practitioners" (Storr 1996: 120).

See also Carl **Jung**, New Age **psychotherapies**, and **repressed memory therapy**.

Further reading: Dawes 1994; Dineen 1998; Feinberg and Farah 1997; Freud 1927, 1930; Gold 1995; Marquis 1996; Pincus and Tucker 1985; Schacter 1996; Torrey 1992.

psychokinesis

The production of motion in physical objects by the exercise of **psychic** or mental powers. Uri **Geller** claims he can bend spoons and stop watches using only his thoughts to control the external objects, yet he always handles these items while his mind is supposedly affecting them. Others claim to be able to make a pencil roll across a table by a mere act of will (they actually pull it by a thin thread). The variety of parlor tricks used to demonstrate psychokinetic powers is endless.

See also **telekinesis**.

Further reading: Gardner 1957; Randi 1982a, 1982b.

psychometry

An alleged **psychic** power that enables one to **divine** facts by handling objects. Commonly, the psychic handles some jewelry or clothing and begins **shotgunning**. Many **psychic detectives** claim to have psychometric powers. Skeptics explain this "power" as a matter of **cold reading** and **selective thinking**.

Further reading: Randi 1982a; Lyons and Truzzi 1991; Nickell 1994b.

psychotherapies, New Age

Psychotherapy is treatment for mental or emotional disorders. There are many types of psychotherapy. Some have been empirically tested and are known to be very effective, such as cognitive therapy. Many New Age therapies, however, are little more than a mixture of **metaphysics**, religion, and **pseudoscience**. They variously require belief in **alien abductions**, **channeling**, God, inner children, **miracles**, possession by entities, primal pains, and **reincarnation**, or are based on empirical beliefs not grounded in any known science. It is difficult to select the most egregious New Age therapy, but Neural Organization Technique (NOT), developed by **chiropractor** Carl Ferreri, is hard to top. Ferreri decided, without the slightest hint of scientific evidence, that all mental and physical problems are due to misaligned skulls. Ferreri believes that as you breathe, the bones in your skull move, causing misalignments that can be corrected by manipulation. This theory was put into practice without the slightest proof that cranial bones move or that there is any sense to the notion of "standard alignment" of the cranial bones. Ferreri was not stopped by logic or good science, however. He was stopped by lawsuits and criminal charges.

For in-depth descriptions of some of the latest New Age therapies, one should read *"Crazy" Therapies* (1996) by Margaret Thaler Singer and Janja Lalich, or view Ofra Bikel's "Divided Memories," first aired on PBS's *Frontline* on April 4, 1994.

Singer and Lalich attribute the popularity of bizarre therapies to the rise in irrationality and the demand for such items on talk shows and the book circuit. Some therapists, such as Sondra Ray, an advocate of "rebirthing therapy," consider themselves to be *spiritual guides*. They are proud of their lack of scientific support. Some claim that mental illness is caused by possession by spirit entities that must be placated. Others use **pastlife regression** to find the cause of the problem. Some treat alien abduction claims as nondelusional. There are several cathartic therapies that involve "primal screaming" or "reparenting." The support for these therapies comes mainly from the "insight" and observations of their founders and patient response, which is analyzed and evaluated by the therapists themselves.

The list of "crazy" therapies is too long to reproduce here, but Singer and Lalich describe the following:

- Leonard Orr developed *energy breathing and rebirthing theory*. According to Orr, if you learn how to breathe **energy** well, you can breathe away diseases and physical or emotional pain.
- Marguerite Sechehaye and John Rosen practice the theory of *regression and reparenting*. The therapist becomes the patient's surrogate parent to make up for the terrible job her real parents did.
- Jacqui Shiff's theory is that the patient must wear diapers, suck his thumb and drink from a baby bottle to be cured.
- Sondra Ray and Bob Mandel believe that your problems are due to the way you were born. They will help "rebirth" you, properly this time. Connell Watkins and Julie Ponder were sentenced to 16 years in prison for smothering to death a 10-year-old girl using this therapy, which is now illegal in Colorado.
- Richard Boylan, Edith Fiore, John Fuller, Bruce Goldberg, Budd Hopkins, David Jacobs, John Mack, and Brian Weiss use **hypnosis** to discover the patient's past or future lives as an alien abductee, in an effort to "help" them.
- John Bradshaw's theory is that you have an "inner child" you must nurture and be good to, if you are to be healthy.
- Arthur Janov practices primal therapy. According to Janov, the patient must rid herself of primal pain, which can be eradicated only by learning the proper way to scream.
- Daniel Casriel's New Identity Process involves screaming, which allegedly unblocks what's blocked. Casriel's scream is apparently a better kind of scream than Janov's.

- Nolan Saltzman practices Bio Scream Psychotherapy. His screaming is apparently better than both Casriel's and Janov's because it has more love in it.
- Finally, there is hypnotherapy, which is extremely popular and is practiced by thousands of therapists, many of whom got their training in a weekend seminar.

Singer and Lalich note that

> There are no licensing requirements, no prerequisites for training, and no professional organization to which those who hypnotize others are accountable. You can be a real estate agent, a graphic artist, an English teacher, or a hairdresser and also call yourself a hypnotherapist by hanging a certificate on your wall that states you took as few as eighteen hours of courses in hypnosis. (1996: 53)

They also note:

> To society's loss, there is an alarming laxity within the mental health professions when it comes to monitoring, commenting on, and educating the public about what is good therapy, what is negligent behavior by trained professionals, and what is or borders on **quackery**.

This lack of oversight leads to all sorts of abuses and malpractice.

For example, many hypnotherapists seem unaware that they are priming their patients. The dangers of this practice are stated by psychiatrist Martin Orne, considered one of the world's leading authorities on hypnosis:

> The cues as to what is expected may be unwittingly communicated before or during the hypnotic procedure, either by the hypnotist or by someone else; for example, a previous subject, a story, a movie, a stage show, etc. Further, the nature of these cues may be quite obscure to the hypnotist, to the subject, and even to the trained observer. (ibid.: 96)

Yet many hypnotherapists seem oblivious to the dangers and pitfalls of using hypnosis in a therapeutic session.

Many New Age therapists seem unaware of many things that any competent therapist should be concerned with. For example, they generally exclude the possibility that a patient might either have a *physical* problem or a *character flaw*. Seemingly, no patient is physically ill. No mental disorder is biochemical or brain-based. No patient is responsible for his or her problems. Patients apparently never lie, manipulate, deceive, cheat, distort, rationalize, or err. If a patient has a fault, it is that he or she is not completely trusting of the therapist. Patients have mental diseases, emotional problems, or syndromes, not character flaws. It would be an astounding fact to discover that emotionally disturbed or mentally troubled persons are completely without flaws in their moral character.

Bikel's documentary of therapists allows practitioners to confidently display their arrogance and incompetence. The therapists are oblivious to the fact that they are being used to demonstrate the monstrosity of their pseudoscientific and self-deceptive work. Therapist after therapist talks freely about how uninterested he or she is in the truth and how indifferent he or she is to the families they help destroy. They are uniform in their dismissal of critics as being in denial. Patient after patient is paraded forth by the therapists as evidence of their good work, yet none of the patients seems better for the therapy and many seem hopelessly ill.

Trying to find a meaningful common thread in the therapies examined by Bikel is not too difficult, but its meaningfulness does not enhance the position of those who think these therapies are scientific. One common thread is the belief that the cause of a problem is some traumatic past event, such as being stabbed in the stomach in a previous lifetime or being sexually abused as a child, the latter of which is the **repressed memory therapists'** one-size-fits-all explanation of nearly every emotional disorder. Childhood sexual abuse is not only the cause of most problems, according to these therapists; it is the cause around which their lives revolve. The repressed memory therapists are not bothered that most of their patients do not remember being abused. Repressed memory therapy will help them recall the trauma. Several therapists in "Divided Memories" claim to have been abused themselves; one discovered her abuse while treating a patient who was remembering her abuse. That a therapist would inject his or her own problems into treatment and consider the beliefs about a past life of a patient to be relevant to the patient's illness make these New Age therapies look more like **cults** than science.

Another common thread is the belief that the patient must discover the *cause* of his or her problem to be helped. This insight approach to psychotherapy goes back at least to Freud, but has never been scientifically tested or validated. Nor does there seem to be any clear idea as to what it means to be *helped* by psychotherapy. The only common thread regarding cure seems to be that the patient *believes* she knows what caused her problems. Believing you know who or what harmed you in the past *is* the cure, it seems. The quality of the patient's life, the interaction of the patient in significant social settings—such as with one's family, friends, and coworkers—is irrelevant. Having the patient trust the therapist is all-important. To gain this trust one of the common tactics of the therapists is to turn the patient against the patient's family. This is done by leading the patient to believe that the cause of the patient's problems is a family member or several family members. The family cannot

help the patient because the family is the *cause* of the patient's problems. One or more family members abused the patient and is now either a liar or in denial; the other family members are deluded or in conspiracy to protect the evil family member. The patient has been persecuted; the therapist is her savior.

The most appalling thread holding these therapies together is the profound lack of interest in objective truth or accuracy. Neither patient nor therapist is to be concerned with facts or tangible evidence that the believed cause actually happened. In fact, whether the believed cause is the real cause is irrelevant to the therapy. The patient creates truth, and it is as real to the patient as facts are to the skeptic. *We all live in a delusion,* proclaims one therapist. So it is of no concern to him that his patient's believed cause is pure delusion. Any first-year psychology student would recognize the projection in that claim. The viewer of Bikel's work, however, needs no training to see that this therapist is clearly deluded when he claims that he did not induce his patient's bizarre tale of **satanic ritual abuse** by her parents and grandparents. His total lack of interest in corroborating evidence to his patient's story, his lack of concern for the family he was helping to destroy, his disingenuous claims about needing to accept on faith everything his patient tells him, his apparent obliviousness to the absurdity and cruelty of encouraging his patient to file a $20 million lawsuit against her family, his deluded claim that *he can tell in the first session* with a patient whether she has been abused as a child, all add up to the self-labeled therapeutic package: delusion.

Most of the therapists discussed by Bikel, Singer, and Lalich seem oblivious or indifferent to their role in priming and prompting their patients. They condition their patients, prompt them, and in some cases, clearly plant notions in their minds. They give their patients books to read or videos to watch, not to help the patient understand a problem but to prime the patient for belief in some "crazy" therapy. They plant notions during hypnosis or group sessions and then these planted notions are "recovered" and offered as validation of their therapeutic techniques and theories. Rather than provide real therapy, these "crazy" therapists indoctrinate patients into their own worldviews. Perhaps most disturbing of all is that this surreal pseudoscience goes nearly unchallenged by professional mental health associations and the mass media.

Further reading: Dawes 1994; Gold 1995; Haley 1986; Kandel and Schwartz 2000.

publication bias

See **positive-outcome bias**.

pyramid schemes, chain letters, and Ponzi schemes

A pyramid scheme is a fraudulent system of making money that requires an endless stream of recruits for success. Recruits give money to recruiters and enlist fresh recruits to give them money.

Why it is called a pyramid scheme is a mystery, since a pyramid is build by putting *fewer* stones on each rising layer, whereas a pyramid scheme involves adding *more* people with each layer of the pyramid. Pyramids are built from the bottom up; pyramid schemes are built from the top down. If a pyramid scheme were started by a human being at the top with just 10 people beneath him, and 100 beneath them, and 1,000 beneath them, and so on, the pyramid would involve everyone on Earth in just 11 layers

of people with one con man on top. The human pyramid would be about 60 feet high and the bottom layer would have more than 5.4 billion people!

A diagram might help you see this:

1
10
100
1,000
10,000
100,000
1,000,000
10,000,000
100,000,000
1,000,000,000
10,000,000,000

Thus, in very short order, 10 recruiting 10 and so on would reach 10 billion, well in excess of the earth's population. If the entire population of earth were 6 billion and we all got involved in a pyramid scheme, the bottom layer would consist of about 90 percent of the planet, that is, about 5.4 billion people. Thus, for 600 million people to be winners, 5.4 billion must be losers, for whom would the bottom layer sell to, or recruit?

In a straightforward pyramid scheme, a recruit is asked to give a sum of money, say, $100, to a recruiter. The new recruit then enlists, say, 10 more recruits, to give up $100 each. In the simplest example, the recruiter keeps all the money he gets from his recruits. In our example, each recruit gives up $100 in exchange for $900 ($100 from each of his 10 recruits minus the $100 he gave his own recruiter). For no one to lose money, the recruiting must go on forever. On a planet with a limited number of people, even if the planet is as large as Earth, one runs out of new recruits rather quickly.

Thus, the result of all these schemes is inevitable: at best, a few people walk away with a lot of money, while most recruits lose whatever money they put into the scheme. In fact, the only way anybody can make money through a pyramid scheme is if other people are defrauded into giving money on a promise of getting something in return when it will be impossible for them to get anything at all in return. That is to say, in plain English, these schemes always constitute fraud. They use deception to get money. That is why they are illegal. They are not illegal because they involve recruiting people to recruit other people to recruit other people. That is perfectly legal and is done to some degree in many legitimate businesses. They are not illegal because they involve giving money to people. It is perfectly legal to give money to people. They are illegal because they involve deceiving people in order to get money from them.

Chain Letters. In the money chain letter, the recruiter sends the new recruits a letter with a list of names on it, including the recruiter's name at the bottom of the list. The recruits are asked to send money to the person whose name is at the top of the list and to add his or her name to the bottom. Money is made solely by getting new recruits to join the chain, adding their names to the list and recruiting others to do the same. In theory, eventually each recruit's name will be at the top of millions of lists and receive millions of dollars. In practice, most people will receive nothing. Anyone can break the chain, thus depriving all those on the list of any possible profit. But even if no one broke the chain, 95% of those who sent out money will get nothing in return.

Ponzi Schemes. A Ponzi scheme, named after Charles Ponzi, who defrauded people in the 1920s using the method, involves getting people to invest in something for a guaranteed rate of return and using the money of later investors to pay off the earlier ones. Who will make money

from such a scheme? Those who start it and those who get in early. Does anyone really make money from these schemes? They must, or else they would have died off long ago. How? If I start the scheme, I just skim off the top and pay off enough people to make it look like it's working, even if that means buying in again at the bottom. I might even be stupid enough to think that I can keep the scheme going when the recruiting has dried up. I can try to get money quickly by some other scheme. For example, I can take a big chunk of money and go to Las Vegas and hope to hit it big. Or I can try to rob a bank.

See also **multi-level marketing**.

Further reading: Bulgatz 1992; Fitzpatrick and Reynolds 1997; Walsh 1998.

pyramidiocy

Having an outlandish, farfetched theory about the origin, nature, or purpose of the Egyptian pyramids. The theories of pyramidiots are barely supported by slender threads of evidence. They serve little purpose except to stand as bad examples of speculative thought and fanciful imagination. For example, Edgar **Cayce** claimed that beings from **Atlantis** helped the Egyptians build the pyramids by showing them how to levitate stones. Charles Berlitz claimed that Atlantis lay beneath the **Bermuda Triangle** and had a pyramid the same size as the Great Pyramid at Giza.

Pyramidiots think Atlantis is the link between the pyramids of Egypt and the pyramids of Mexico. Arguments demonstrating that the ancient Egyptians, Mayans, or Toltecs were intelligent and resourceful enough to build pyramids are to no avail. To give plausibility to theories that are universally rejected in academia one must ignore any refuting evidence. For example, anything that would indicate slow and incremental development, rather than sudden and unprecedented work, is ignored. "Alternative archaeologists" ignore hundreds of years of experimentation with building large pyramidal structures. The failures of the Egyptians in their early attempts at pyramid building indicate that they arrived at the Giza level of construction through trial and error. The thousands of underground tombs with many chambers antedates the patterns used in the pyramids. Ignore their history, and you can make a case that the people who built Giza needed help from Atlantis or from **ancient astronauts** (Feder 2002).

Pyramidiots are not dissuaded by the fact that Egyptian pyramids were primarily funereal, while the pyramids in Mexico were primarily ceremonial. There is no evidence that the Egyptians imitated the Aztecs and the Mayans in performing human sacrifices at their pyramids. Also, the pyramids in Mexico are all step pyramids. Furthermore, intimate knowledge of Egyptian hieroglyphics will not help you read the pictoglyphs on a Mayan temple. Why would the aliens not teach the Egyptians and the Meso-Americans the same form of writing or give them the same plans and purpose for their buildings? (Feder 2002).

Other pyramidiots ascribe super technological or **paranormal** powers to the ancient Egyptians. Traditional explanations in terms of religion, tombs for pharaohs and their families, belief in immortality, slave labor or paid workers, slipways, canals, and so on are rejected by pyramidiots in favor of theories claiming that the pyramids were *power stations* or *water pumps*.

Some pyramidiots claim that the pyramids were built according to some sort of mystical **numerology** to contain coded messages. Some believe that the Great Pyramid at Giza is at the center of the world. Some think the pyramids provide a map of the heavens. Mystical mathematical notions about the pyramids abound. Some believe only God could have designed such a numerical mystery. That almost anything in the universe can be found to have interesting mathematical proportions or be related to several interesting mathematical formulas or astronomical phenomena is of little interest to pyramidiots.

Some pyramidiots think pyramids have healing power and are foci of spiritual **energy**. Still others have believed that razor blades could be kept perpetually sharp by being placed under a pyramid of the same proportions as the Great Pyramid at Giza by focusing cosmic energy and realigning crystals in steel (Hines 1990). That there is no evidence for such beliefs seems to cheer up rather than dishearten pyramidiots.

Further reading: Bullard 1996; De Camp 1977; Edwards 1992; Lehner 1997; Williams: www.skeptics.com.au/journal/paramyth.htm.

Q

quackery

A term that used to be considered derogatory, describing medical charlatanism. As medical charlatanism became more popular and as using pejorative terms became politically incorrect except when used by the formerly oppressed classes, quackery evolved into "**alternative health practices**" and "complementary medicine" by those who practice it, and into "unproven therapies" and "questionable methods" by those who are critical of it. When quackery is mixed with scientific medicine, the latter is called a "mainstream modality" and those who practice it call the result "integrative medicine."

Quackery usually involves integrating **metaphysics** and such things as **spiritualism** or **sympathetic magic** with healing. What quackery lacks in scientific study it makes up for by prescribing overdoses of false hope.

"Quackery" is short for *quacksalver,* and probably has nothing to do with the sound made by ducks. Quacksalver derives from the obsolete Dutch term *Kwakzalver,* meaning to boast of one's salves.

Further reading: Barrett and Butler 1992; Barrett and Jarvis 1993; Gardner 1957; Randi 1989a; Raso 1994, 1995; Stenger 1997b.

R

Raël, Raëlians

A **UFO** cult that follows Claude Vorilhon, a former motor sport journalist and race-car driver. He claims that on December 13, 1973, he was in a volcano near Clermont-Ferrand, France, when he saw a UFO "7 meters in diameter made of a very shiny silver metal and moving in a total silence." He says a radiant being emerged and entrusted him with a message revealing the true origin of mankind. Henceforth he would be known as Raël, which means "messenger" in some language on some planet somewhere.

His followers consider him to be "the prophet of the third millennium." Like all good religious leaders, Raël expects his followers to support him. A 10% tithe is the norm.

He explains his mission in his book *The True Face of God.* According to Taras Grescoe (2000), Vorilhon claims that

> he was taken to the planet of the Elohim in a flying saucer in 1975, where he was introduced to noted earthlings such as Jesus, Buddha, Joseph Smith and Confucius. The Elohim, small human-shaped beings with pale green skin and almond eyes, were apparently the original inspiration for the Judeo-Christian God. They informed Vorilhon that he was the final prophet—sent to relay a message of peace and sensual meditation to humankind under his new name of Raël—before the Elohim would return to Jerusalem in 2025.

Raël claims that the Elohim have taught him that the human race was created from the DNA of aliens some 25,000 years ago. In fact, all life on Earth was created in alien laboratories. Among other things, Raël has also learned that there is no God or **soul** and that cloning is the way to immortality. According to Raël, our alien creators want us to be beautiful and sexy and to enjoy a sensuous life, free from the restrictions of traditional Judeo-Christian morality.

According to Grescoe, "Raël's success seems to derive from providing a structured environment for decadent behavior: He offers a no-guilt playground for hedonism and sexual experimentation" (ibid.). Fortunately, the Raëlians are big on using condoms. They won't spread as much disease that way. However, using condoms won't suffice to deplete their numbers, Raël believes, since he has formed a cloning company called Clonaid that promises to provide assistance to couples willing to have a child cloned from one of them. This service will be offered to any couple, regardless of sexual orientation. Some scientists say that there is no possibility of Clonaid actually working in the near future and dismiss its goals as pure fantasy, yet Advanced Cell Technology (ACT) of Worcester, Massachusetts, has cloned a human embryo. Clonaid should be a reminder of what might happen in the distant future if controls on genetic engineering are not developed to prevent religious fanatics and lunatics from gaining more control of the planet than they already have. In December 2002, Raëlian bishop Dr. Brigitte Boisselier announced that Clonaid had several human clones near birth. Raël admitted that it might be a hoax, but it was worth about "$500 million" in media coverage and "saved me 20 years of work" (January 21, 2003, www .worldnetdaily.com)

The Raëlian headquarters are in Montreal, but the cult is international and claims to have some 50,000 members in 85 countries. They have an "Evidence Page" on their web site, where they offer proof of their prophet's claims, thus relieving us of the burden of having to believe on pure **faith**. Unfortunately, the evidence provided is likely to satisfy only those eager for delusion and **self-deception**. For example, the Raëlians consider UFO sightings as proof of their messenger's claims.

Their attempt at "scientific" evidence will have some appeal to the scientifically illiterate and the logically challenged. The scientific evidence is nothing more than speculation and assumption in juxtaposition to facts. The evidence consists of claiming that we are about to create life in our laboratories and our creations will probably think we are gods. Therefore, it is reasonable to conclude that we were created in laboratories and think of our creators as gods. The rest of the "scientific" evidence consists of a list of scientific accomplishments that, I suppose, are imagined to have occurred elsewhere before the living things on our planet

could have been created in the lab. All of which begs the question as to whether this occurred elsewhere 25,000 years ago.

Apparently, the Raëlians are not bothered by the rather absurd image of a race of superior beings working for thousands of years in a laboratory to create all our insects, fungi, bacteria, viruses, and so on, not to mention all the extinct species. The beings then wait 25,000 years to reveal this secret to a French race-car driver who spots their UFO in a volcano.

The best Raëlian howler is their proof that evolution could not have occurred. They claim scientists have discovered that genes have a DNA repair mechanism (p53) that prevents mutation, an important process in evolution. Species couldn't have diversified if this mechanism was present. p53 was at first thought to be an oncogene but is now thought to be antioncogenic. It is of little interest to the Raëlians, I suppose, that p53 itself mutates (Pergament and Fiddler 2001). And it is pure speculation on their part that the entire genetic code of all species always consists of genes that prevent mutation from occurring. Even if they're right, however, it wouldn't follow that Vorilhon's preposterous UFO tale is true. In the meantime, in a glorious non sequitur, Raël has offered $2,000 to anyone who starts a new religion.

Rama

Frederick P. Lenz, Ph.D. (in English literature) and businessman (Advanced Systems, Inc.) called himself Zen Master Rama. Lenz parlayed his knowledge of Hinduism and Buddhism into a **cult**. In the early 1980s he started calling himself Rama, after the last incarnation of the Hindu deity Vishnu. He started giving seminars in 1982 in Malibu, California. Eventually, thousands of people would pay as much as $5,000 per seminar to be enlightened by this self-proclaimed guru, **psychic**, and miracle worker. Here is what one of his followers said he learned from his master: "Spiritually advanced people work with computers because it makes a lot of money. The more money you make, the better you meditate" (Clark and Gallo 1993: 102).

Rama used a variety of so-called **mind control** techniques to seduce his disciples. He had his subjects stare at him for long hours until they would hallucinate and see Lenz begin to glow or change shapes. Lenz told his followers that having these visions meant they were psychic.

Rama seduced many of his female followers by telling them that he has sex only with women who have a rare sort of **karma**. He also told women that having sex with him would elevate them to a higher plane of consciousness. It is hard for a skeptic to believe that such a line would work with any woman, but apparently it does.

Rama took religious freedom and **Tantric** gullibility to new heights in his book *Surfing the Himalayas: A Spiritual Adventure* (1997). There, he tells us of his adventures "snowboarding through Tantric myetiolem" and offers such bits of wisdom as "Ultimately, thinking is a very inefficient method of processing data" and:

> The relational way of doing things is to move your mind to a fourth condition, a condition of heightened awareness. In a condition of heightened awareness, you elevate your conscious mind above the stream of extraneous data—out of dimensional time and space, so to speak—and you meld your mind instead with the pure intelligent consciousness of the universe.

Bob Frankenberg, Chairman and CEO of Novell, allegedly claims the book "entertains and enlightens" and calls it "a won-

derful contrast of Eastern spirituality and Western pragmatism." Phil Jackson, professional basketball coach, allegedly said the book "[b]rings levity and humor to a subject often relegated to a mundane, boring prospect." The book became a best-seller. Within a year Rama published another cult classic: *Snowboarding to Nirvana*.

Unfortunately, all his Tantric wisdom couldn't save him, and the day before taxes were due in 1998, Rama drowned in Conscience Bay near his residence in the exclusive Old Field section of Setauket on Long Island, New York. Rumor has it that he was stoned when he fell off the dock. An unidentified woman described by police as "incoherent" was found to be in Lenz's house when his body was recovered by police divers. Lenz was 48 at the time of his death. **Cult** expert Joe Szimbart claims Lenz was suffering from liver cancer and committed suicide by overdosing on phenobarbital (*Skeptical Inquirer*, July/August 1998). The Suffolk County Medical Examiner's office says it was Valium. Either way, Rama now snowboards with the fishes.

Ramtha

Ramtha is a 35,000-year-old spirit-warrior who appeared in J. Z. Knight's kitchen in Tacoma, Washington, in 1977. Knight claims that she is Ramtha's **channel**. She also owns the copyright to Ramtha and conducts sessions in which she pretends to go into a trance and speaks Hollywood's version of medieval or Elizabethan English in a guttural, husky voice. She has thousands of followers and has made millions of dollars performing as Ramtha at seminars ($1,000 per appearance) and at her Ramtha School of Enlightenment, and from the sales of tapes, books, and accessories (Clark and Gallo 1993). She must have hypnotic powers. Searching for self-fulfillment, otherwise normal people obey her command to spend hours blindfolded in a cold, muddy, doorless maze. In the dark, they seek what Ramtha calls the "void at the center."

Knight says she used to be "spiritually restless," but not any more. Ramtha, from Atlantis via Lemuria, has enlightened her. He first appeared to her, she says, while she was in business school having extraordinary experiences with **UFOs**. She must have a great rapport with her spirit companion, since he shows up whenever she needs him to put on a performance. It is not clear why Ramtha would choose Knight, but it is very clear why Knight would choose Ramtha: fame and fortune, or simple delusion.

Knight claims to believe that she's lived many lives. If so, one wonders what she needs Ramtha for: She's been there, done that, herself, in past lives. She ought to be able to speak for herself after so many **reincarnations**.

Knight claims that spirit or consciousness can "design thoughts" that can be "absorbed" by the brain and constructed "holographically." These thoughts can affect your life. If this means what I think it means, then Knight has taken the notion of proving the obvious to new heights: She has discovered that one's thoughts can affect one's life.

Knight not only has rewritten the book on neurology, she has also rewritten the book on archaeology and history. The world was not at all like the scholars of the world say it was 35,000 years ago. We were not primitive hunters and gatherers who liked to paint in caves. No; there were very advanced civilizations around then. It doesn't matter that there is no evidence for this, because Knight has rewritten the book of evidence as well. Evidence is what appears to you, even in visions and hallucinations and delusions. Evidence is anything you

feel like making up. So, when you are told that Ramtha came first from Lemuria in the Pacific Ocean, do not seek out scholars to help you understand that ancient civilization. The scholars of the world do not believe Lemuria existed except as a fantasy. When you are told that the Lemurians were a great civilization from the time of the dinosaurs, do not expect to be burdened with evidence. There isn't any evidence. The only mammals around at the time of the dinosaurs were primitive and nonhominid, very much like lemurs. Maybe the Lemurians were really lemurs. No; the Lemurians came from "beyond the North star," according to Knight, which may explain why humans occasionally look to the sky with longing.

But as cool as Lemuria was, it could not compare with its counterpart in the Atlantic Ocean. Knight's story of Ramtha in Atlantis is too bizarre to retell. Let's just say that Ramtha was a warrior who appeared to Edgar **Cayce** and leave it at that. Her story is appealing to those who are not comfortable in today's world. The past *must* have been better. It must have been *safer* then, and people must have been *nobler.* This message is especially appealing to people who feel like misfits.

Ramtha, like Christ, ascended into heaven after his many conquests, including the conquest of himself. He said he'd be back, and he kept his promise by coming to Knight in 1977 while she was in her **pyramidiocy** phase. She put a toy pyramid on her head, and lo and behold if that wasn't a signal for Ramtha to return to the land of the living dead:

> And he looked at me and he said: "Beloved woman, I am Ramtha the Enlightened One, and I have come to help you over the ditch." And, well, what would you do? I didn't understand because I am a simple

person so I looked to see if the floor was still underneath the chair. And he said: "It is called the ditch of limitation," and he said: "And I am here, and we are going to do a grand work together." [www.ramtha.com/html/aboutus/about-jz.stm]

Apparently, the first rule of the wise is: *Beware the ditch of limitation*. Knight's husband-to-be must have fallen into the ditch. He was there at the time Ramtha first invaded his girlfriend's body, but he was so busy lining up pyramids with a compass that he didn't see Ramtha. He did feel The Enlightened One's magnetic charm, however; for, according to Knight, the compass needle was spinning around madly and they saw "ionization" in the kitchen air.

Ramtha then became Knight's personal tutor for two years, teaching her everything from theology to quantum mechanics. He taught her how to have **out-of-body experiences**. The experience was so extraordinary she had to dig very deep for a metaphor to try to convey the bliss she felt: "I felt like . . . like a fish in the ocean."

Her big break came when her son, Brandy, developed "an allergic reaction to life." He had to have a few shots but he was allergic to the allergy shots. Fortunately, "the Ram" (as Knight calls her spirit invader) came to the rescue and taught her **therapeutic touch**. She healed Brandy with **prayer** and her touch "in less than a minute," greatly reducing her medical bills. She had performed a **miracle** and now nothing would stop her from entering the public arena.

Perhaps the reason J. Z. Knight is so successful in getting followers and students is that Ramtha is a feminist. (The fact that Knight is quite attractive herself also might have something to do with her success.) He recognized that if he appeared in his own masculine body, he would perpetuate the myth that God is male and further contribute to the eternal abuse of women.

> That's what he said. So women have been abused by men, and herded by men through religion to perform according to those religious doctrines, and in fact, women were despised by Jehovah. So, he said: "It is important that when the teachings come through, they come through the body of a woman." (ibid.)

This feminization of God must be pleasing to people who are tired of masculine divinities. According to Knight, Ramtha will help people master their humanity and "open our minds to new frontiers of potential."

Further reading: Alcock 1996a; Gardner 1988; Schultz 1989.

Randi Challenge, The

James Randi, a.k.a. The Amazing Randi, magician and author of numerous works skeptical of **paranormal** claims, offers "a one-million-dollar prize to anyone who can show, under proper observing conditions, evidence of any paranormal, supernatural, or occult power." His rules are little more than what any reasonable scientist would require. If you are a mental spoon bender, you can't use your own spoons. If you are going to see **auras**, you will have to do so under controlled conditions. If you are going to do some **remote viewing**, you will not be given credit for coming close in some vague way. If you are going to demonstrate your **dowsing** powers, be prepared to be tested under controlled conditions. If you are going to do **psychic surgery** or experience the **stigmata**, expect to have cameras watching your every move.

For more information on the James Randi Challenge write to: JREF, 201 S.E. 12th St. (E. Davie Blvd.), Fort Lauderdale, FL 33316-1815, USA.

After collecting the million dollars, successful psychics should contact B. Premanand of the Indian Skeptic (www .indian-skeptic.org/html/rules.htm), who will pay 100,000 rupees "to any person or persons who will demonstrate any psychic, supernatural or paranormal ability of any kind under satisfactory observing conditions." Also, the Australian Skeptics (www .skeptics.com.au/features/challenge.htm) will throw in an additional $100,000 (Australian) for the psychic and $20,000 for anyone "who nominates a person who successfully completes the Australian Skeptics Challenge."

reflexology

The massaging of feet to diagnose and cure disease. In the 1930s, Eunice Ingham (1889–1974) applied **Occam's razor** to Dr. William Fitzgerald's teachings in *Zone Therapy* (1917) and dubbed the result *reflexology*. She eliminated all of Fitzgerald's **energy** zones—he said there are 10 such zones in the body—except for the feet. Reflexology is based on the unsubstantiated belief that each part of each foot is a mirror site for a part of the body. The big toe, for example, is considered a reflex area for the head. As **iridology** maps the body with irises, reflexology maps the body with the feet, the right foot corresponding to the right side of the body and the left foot corresponding to the left side of the body. Because the whole body is represented in the feet, reflexologists consider themselves to be **holistic** health practitioners, not foot doctors. Allegedly, the ancient Chinese and Egyptians practiced reflexology, and it is still very popular in Europe (www.reflexeurope.org/ landeninfo.htm).

Practitioners of reflexology claim that they can cure a variety of aches and pains by massaging the correct reflex points on the foot. It is said by those who practice it that reflexology can cure migraine headaches and relieve sinus problems. It can restore harmony to hormonal imbalances and cure breathing disorders and digestive problems. If you have a back problem, a massage on the right spot on the right foot (which might be the left foot in some cases) can alleviate your suffering. If you suffer from circulatory problems or have a lot of tension and stress, reflexology promises relief.

There are many variations of reflexology and many names for these variations, including Zone Therapy, Vacuflex, and Vita Flex. Some chiropodists are also reflexologists, although there is no necessary connection between the two. Some reflexologists deny that they diagnose or treat diseases and claim only to restore "balance" to one's "energy."

Reflexology is often combined with other therapies and practices, such as acupressure, shiatsu, yoga, and tai chi, and it often involves the hands and other body parts or zones, not just the feet. Reflexology seems to be a variation of acupressure, with its notion that there are correspondences between special pressure points and the flow of **chi** to bodily organs. Polarity therapy, a variant of reflexology, replaces the **yin and yang** opposition with the positive/negative energy charges of the sides of the body (the right side is positively charged); massage allegedly restores the proper balance of energy. In polarity therapy, the foot is the site of just one of many key massage points.

One reason foot massage may be so pleasurable and is associated with significant improvement in mood is that the area of the brain that connects to the foot is adjacent to the area that connects to the genitals. There may be some neuronal overlapping. Neuroscientist V. S. Ramachandran writes of a person whose leg was amputated and who experienced orgasms in his phantom foot (1998: 36–37). "The genitals are right next to the foot in the body's brain maps," he notes, and he speculates that this fact may account for foot fetishes.

See also **alternative health practices, massage therapy, placebo effect, pragmatic fallacy,** and **regressive fallacy.**

Further reading: Barrett and Jarvis 1993.

regressive fallacy

The failure to take into account natural and inevitable fluctuations when ascribing causality (Gilovich 1993: 26). Things such as stock market prices, golf scores, and chronic back pain inevitably fluctuate. Periods of low prices, low scores, and little or no pain are eventually followed by periods of higher prices, higher scores, and more pain. To ignore these natural fluctuations and tendencies leads to **post hoc** reasoning regarding their causes.

For example, a professional golfer with chronic back pain or arthritis might try a copper bracelet on his wrist or **magnetic** insoles in his shoes. He is likely to try such gizmos when he is not playing or feeling well. He notices that his scores are improving and his pain is diminishing or gone. He concludes that the copper bracelet or the magnetic insole is the cause. It never dawns on him that the scores and the pain are probably improving due to natural and expected fluctuations. Nor does it occur to him that he could check a record of all his golf scores before he used the gizmo and see whether the same kind of pattern has occurred in the past. If he takes his average score as a base, most likely he would find that after a very low score he tended to shoot not a lower score but a higher score in the direction of his average. Likewise, he would find that after a very high score, he did not tend to shoot a higher score but rather would shoot a lower score in the direction of his average.

This tendency to move toward the *average* away from extremes was called *regression* by Sir Francis Galton in a study of the average heights of sons of very tall and very short parents ("Regression Toward Mediocrity in Hereditary Stature," 1885). He found that sons of very tall or very short parents tend to be tall or short, respectively, but not as tall or as short as their parents.

Many people are led to believe in the causal effectiveness of worthless remedies because of the regressive fallacy. The intensity and duration of pain from arthritis, backache, gout, and other chronic problems fluctuates. A remedy such as **acupuncture**, a **chiropractic** spinal manipulation, or a magnetic belt is likely to be sought when the pain is at its worst. The pain in most cases would begin to lessen after it has peaked. It is easy to deceive ourselves into thinking that the remedy we sought caused our reduction in pain. It is partly because of the ease with which we can deceive ourselves about causality in such matters that scientists do **controlled experiments** to test causal claims.

Further reading: Gilovich 1993.

reiki

Reiki (pronounced "ray-key") is a form of healing through manipulation of *ki,* the Japanese version of **chi**. *Rei* means "spirit" in Japanese, so *reiki* literally means "spirit life force."

Like their counterparts in traditional Chinese medicine who use **acupuncture**, as well as their counterparts in the West who use **therapeutic touch** (TT), the practitioners of reiki believe that health and disease are a matter of the life force being disrupted. Each believes that the universe is full of **energy** that cannot be detected by any scientific instruments but that can be felt and manipulated by people with special training. Reiki healers differ from acupuncturists in that they do not try to *unblock* a person's ki, but rather they try to *channel* the ki of the universe so that the person heals. The channeling is done with the hands, and, like TT, no physical massaging is necessary since ki flows through the body of the healer into the patient via the air. The reiki master claims to be able to draw on the energy of the universe and increase his or her own energy while performing a healing. Reiki healers claim they channel ki into diseased individuals for "rebalancing." Larry Arnold and Sandra Nevins claim in *The Reiki Handbook* (1992) that reiki is useful for treating brain damage, cancer, diabetes, and venereal diseases. If the healing fails, however, it is because the patient is *resisting* the healing energy.

Reiki is very popular among New Age spiritualists, who are very fond of "attunements," "harmonies," and "balances" of energy. Reiki apprentice healers pay up to $10,000 to their masters to become masters themselves. The process involves going through several levels of attunement. One must learn which symbols to use and when to call up the universal life force, how to heal an emotional or spiritual illness, and how to heal someone who isn't present.

Reiki was popularized by Mikao Usui (1865–1926). After fasting and meditating for several weeks, he began hallucinating and hearing voices giving him "the keys to healing."

Further reading: Beyerstein 1997; Jarvis: www.ncahf.org/articles/o-r/reiki.html; Raso 1994, 1995.

reincarnation

The belief that when one dies, one's body decomposes, but one is reborn in another body. The bodies one passes in and out of need not be human. One may have been a Doberman in a past life, and one may be a mite or a beetle in a future life. Some tribes avoid eating certain animals because they believe that the souls of their ancestors dwell in those animals. A man could even become his own daughter by dying before she is born and then entering her body at birth.

The belief in past lives used to be mainly a belief in Eastern religions such as Hinduism and Buddhism, but now is a central tenet of such theories as **Dianetics**, **theosophy**, **past life regression**, and **channeling**. In ancient Eastern religions, reincarnation was considered a bad thing. To achieve the state of ultimate bliss (nirvana) is to escape from the wheel of rebirth. In most, if not all, ancient religions with a belief in reincarnation, the soul entering a body is seen as a metaphysical demotion, a sullying and impure descent. In New Age religions, however, being born again seems to be a kind of perverse goal. Prepare yourself in this life for who or what you want to come back as in the next life.

L. Ron Hubbard, author of *Dianetics* and founder of Scientology, introduced his own version of reincarnation into his new religion. According to Hubbard, past lives need auditing to get at the root of one's "troubles." He also claims that "Dianetics gave impetus to Bridey **Murphy**" and that some Scientologists have been dogs and other animals in previous lives ("A Note on Past Lives" in *The Rediscovery of the Human Soul*). According to Hubbard, "It has only been in Scientology that the mechanics of death have been thoroughly understood." What happens in death is this: The Thetan (spirit) finds itself without a body (which has died) and then it goes looking for a new body. Thetans "will hang around people. They will see a woman who is pregnant and follow her down the street." Then, the Thetan will slip into the newborn "usually . . . two or three minutes after the delivery of a child from the mother. A Thetan usually picks it up about the time the baby takes its first gasp." How Hubbard knows this is never revealed.

From a philosophical point of view, reincarnation poses some interesting problems. What is it that is reincarnated? Presumably, it is the **soul** that is reincarnated, but what is the soul? A disembodied consciousness? A fragment of the universal spirit? In any case, since there is no way to tell the difference between a baby *with* a soul and one *without* a soul, it follows that the idea of a soul is unnecessary to our understanding of a human being.

Reincarnation does seem to offer an explanation for some strange phenomena, such as child prodigies. But explanations in terms of genetics and brain structures account for such prodigies equally well. Reincarnation could explain why bad things happen to good people and why good things happen to bad people: They are being rewarded or punished for actions in past lives (**karma**). But since bad things also happen to bad people and good things also happen to good people, one might well suppose that there is no rhyme or reason why anything happens to anybody. One could explain **déjà vu** experiences by claiming that they are memories of past lives. However, déjà vu experiences may not even involve memories, but if they do, they are best explained as the recalling of events from *this* life, not some past life. **Dreams** could be interpreted as a kind of soul travel and soul memory. But dreams, like child prodigies, are better explained in terms of brain structures and processes.

Further reading: Baker 1990, 1996b; Carroll 1996; Edwards 1996a, 1996b; Spanos 1987–88.

remote viewing

The alleged **psychic** ability to perceive places, persons, and actions that are not within the range of the senses. Remote viewing might well be called *psychic dowsing*. Instead of a twig or other device, one uses psychic power alone to dowse the entire galaxy, if need be, for whatever one wants to visualize: oil, mountains on Jupiter, a lost child, a buried body, a hostage site thousands of miles away, the inside of the Pentagon or the Kremlin, a drawing of a bridge, or a picture of a horse.

Ingo Swann and Harold Sherman claim to have done remote viewing of Mercury and Jupiter. Drs. Russell Targ and Harold Puthoff studied Swann and Sherman and reported that their remote viewing compared favorably to the findings of the Mariner 10 and Pioneer 10 research spacecrafts. Isaac Asimov, however, did a similar comparison and found that 46% of the observation claims of the astral travelers were wrong. Also, only one out of 65 claims made by the remote viewers was a fact that either was not obvious or not obtainable from reference books (Randi 1982a: 68–69).

Targ and Puthoff were not put off by the fact that Swann claimed he saw a 30,000-foot-high mountain range on Jupiter on his astral voyage when there is no such thing. It is hard to imagine why anyone would have faith in such claims. If I told you that I had been to your hometown and had seen a 30,000-foot-high mountain there, and you knew there was no such mountain, would you think I had really visited your town even if I correctly pointed out that there is a river nearby and it sometimes floods? Swann, in an exquisite **ad hoc hypothesis**, now claims that astral travel is so fast that he probably wasn't seeing Jupiter but another planet in another solar system! There really is a big mountain out there on some planet in some solar system in some galaxy.

The CIA and the U.S. Army thought enough of remote viewing to spend millions of taxpayers' dollars on research in a program referred to as "Stargate." The program involved using psychics for such operations as trying to locate Gadhafi of Libya (so our Air Force could drop bombs on him) and a missing airplane in Africa. The mass media, ever watchful of wasteful government programs, did not exhibit much skepticism regarding remote viewing. Typical is the reporting in the Sacramento area. TV news anchors Alan Frio and Beth Ruyak led their nightly Channel 10 program on November 28, 1995, with a story on "exciting new evidence" that remote viewing really works. The same story had appeared that morning in the *Sacramento Bee* in an Associated Press article about Stargate by Richard Cole. "A particularly talented viewer accurately drew windmills when the sender was at a windmill farm at Altamont Pass," Cole wrote. The talented viewer was Joe McMoneagle, a former army psychic spy. Cole based his claim on the testimony of Dr. Jessica Utts, a statistics professor at the University of California, Davis, who was hired by the government to do an assessment of "psychic functioning." Channel 10 interviewed Dr. Utts, who confirmed that there is good reason to believe that Joe McMoneagle does indeed have psychic powers.

McMoneagle was in the army for 16 years, apparently serving some or most of that time as a psychic spy. He claims he helped locate the U.S. hostages taken by Iran during Jimmy Carter's presidency. Now a civilian psychic consultant, McMoneagle has turned his talents to more significant feats, as Dr. Utts demonstrated. She held up a drawing allegedly done by McMoneagle and declared that it was done by remote viewing. Another scientific researcher had gone to the Altamont pass, known for its miles of funny looking windmills on acres of rolling hills. McMoneagle tried to use his psychic powers to see what the researcher

at Altamont was seeing and then draw what he was seeing. The sum total of the evidence for the value of psychic spying consisted of only one drawing and Dr. Utts's word that it looks like the Altamont pass. I will testify that in fact the drawing did have a strong resemblance to the Altamont pass. It also had a strong resemblance to ships on a stormy sea and to debris in a cloudy, stormy sky.

McMoneagle was just one of the psychics studied by Targ and Puthoff at the Stanford Research Institute (a.k.a. SRI International) from 1973 through 1989 and by another outfit with the unassuming name of Science Applications International Corp., which did its research from 1992 through 1994. Utts and Dr. Ray Hyman, a psychologist at the University of Oregon and a skeptic, issued separate reports on these studies. Utts concluded that "psychic functioning has been well established." Hyman disagreed. In his AP article, Cole wrote that Utts and Hyman stated that "the research was faulty in some respects. The government often used only one 'judge' to determine how close the psychics had come to the right answer. That should have been duplicated by other judges." I would assume that Hyman, if not Utts, would have required a bit more of these studies than that they have more judges.

CIA spokesman Mark Mansfield said, "The CIA is reviewing available programs regarding parapsychological phenomena, mostly remote viewing, to determine their usefulness to the intelligence community." He also notes that the Stargate program was found to be "unpromising" in the 1970s and was turned over to the Defense Department. At one time as many as 16 psychics worked for the government and the Defense Intelligence Agency made them available to other government departments. One of the psychics, David Morehouse, was recruited when he took a bullet in the head in Jordan and started having visions and vivid nightmares. He's written a book about it (*Psychic Warrior*) and it is sure to be better received by true believers than Mansfield's disclaimer.

See also **astral projection** and **out-of-body experience**.

Further reading: Cole 1995; Hyman 1995, 1996b; Randi 1982a; Vistica 1995.

repressed memory

The memory of a traumatic event unconsciously retained in the mind, where it is said to adversely affect conscious thought, desire, and action.

It is common to *consciously* repress unpleasant experiences. Many psychologists believe that *unconscious* repression of traumatic experiences such as sexual abuse or rape is a defense mechanism that backfires. The unpleasant experience is forgotten but not forgiven. It lurks beneath consciousness and allegedly causes a myriad of psychological and physical problems from bulimia to insomnia to suicide.

The theory of *unconsciously* repressing the memory of traumatic experiences is controversial. There is little scientific evidence to support either the notion that traumatic experiences are typically unconsciously repressed or that unconscious memories of traumatic events are significant causal factors in physical or mental illness. Most people do not forget traumatic experiences unless they are rendered unconscious at the time of the experience. No one has identified a single case where a specific traumatic experience in childhood was repressed and the repressed memory of the event, rather than the event itself, caused a specific psychiatric or physical disorder in adulthood.

The strength of the scientific evidence for repression depends on exactly how the term is defined. When defined narrowly as intentional suppression of an experience, there is little reason to doubt that it exists. But when we talk about a repression mechanism that operates unconsciously and defensively to block out traumatic experiences, the picture becomes considerably murkier.

Evidence concerning memory for real-life traumas in children and adults indicates that these events—such as the Chowchilla kidnappings, the sniper killing at an elementary school, or the collapse of skywalks at a Kansas City hotel—are generally well remembered. . . . [C]omplete amnesia for these terrifying episodes is virtually nonexistent. (Schacter 1996: 256)

Psychologist Lenore Terr, a defender of **repressed memory therapy** (RMT), argues that repression occurs for *repeated* or *multiple* traumas. Schacter notes that "hundreds of studies have shown that repetition of information leads to improved memory, not loss of memory, for that information." He also notes that people who have experienced repeated traumas in war, even children, generally remember their experiences. A person who suffers a great trauma often finds that she cannot get the event out of her mind or dreams. Terr's theory is that the child becomes practiced at repression to banish the awful events from awareness and forgetting might aid in the child's survival. Her dissociative theory, however, is based on speculation rather than scientific evidence.

Most psychologists accept as fact that *consciously* repressed unpleasant experiences, even sexual abuse, are sometimes *spontaneously* remembered long afterward. However, there is good reason to doubt the validity of memories recovered during RMT. Critics of RMT maintain that many therapists are not helping patients *recover* repressed memories, but are *suggesting* and planting **false memories** of **alien abduction**, **satanic rituals**, and sexual abuse.

See also **hypnosis, multiple personality disorder**, and **unconscious mind**.

Further reading: Ashcraft 1994; Baddeley 1998; Baker 1996a; Hallinan 1997; Loftus 1994; Schacter 1997, 2001.

repressed memory therapy (RMT)

A type of psychotherapy that assumes that problems such as bulimia, depression, sexual inhibition, insomnia, and excessive anxiety are due to unconsciously repressed memories of childhood sexual abuse. RMT assumes that a healthy psychological state can be restored only by recovering and facing these repressed memories of sexual abuse.

Any amount of sexual abuse of children is intolerable. Nevertheless, there is little scientific evidence supporting the notions that childhood sexual abuse almost always causes psychological problems in adults; that memories of childhood sexual abuse are unconsciously repressed; or that recovering repressed memories of abuse leads to significant improvement in one's psychological health and stability.

The Royal College of Psychiatrists in Britain has officially banned its members from using therapies designed to recover **repressed memories** of child abuse. The British Psychological Society, on the other hand, does not ban its members from such therapy, but in a 1995 report urged them "to avoid drawing premature conclusions about memories recovered during therapy." The report noted that a patient's recovered memory may be metaphorical or emanate from **dreams** or fantasies. The report also denied that there is any evidence suggesting that therapists are widely creating **false memories** of abuse in their

patients, a charge levied by members of the False Memory Syndrome Foundation.

In the United States, the American Psychological Association's Working Group on the Investigation of Memories of Childhood Abuse also issued a report in 1995. The report notes that recovered memory is rare. It also states that "there is a consensus among memory researchers and clinicians that most people who were sexually abused as children remember all or part of what happened to them although they may not fully understand or disclose it." According to the APA, "At this point, it is impossible, without other corroborative evidence, to distinguish a true memory from a false one." Thus, says the APA report, a "competent psychotherapist is likely to acknowledge that current knowledge does not allow the definite conclusion that a memory is real or false without other corroborating evidence" (www.apa .org/pubinfo/mem.html).

Many of the more prominent RMT advocates use a checklist approach to diagnose repressed memories of childhood sexual abuse as the cause of a patient's problems, despite the fact that "there is no single set of symptoms which automatically indicates that a person was a victim of childhood abuse" (APA report). Works on child abuse promoting such a notion have been very popular among therapists and talk show hosts, featuring Ellen Bass, E. Sue Blume, Laura Davis, Beverly Engel, Beverly Holman, Wendy Maltz, and Mary Jane Williams. Through **communal reinforcement** many empirically unsupported notions achieve the status of facts. Dr. Carol Tavris writes:

> In what can only be called an incestuous arrangement, the authors of these books all rely on one another's work as supporting evidence for their own; they all endorse and recommend one another's

books to their readers. If one of them comes up with a concocted statistic—such as "more than half of all women are survivors of childhood sexual trauma"—the numbers are traded like baseball cards, reprinted in every book and eventually enshrined as fact. Thus the cycle of misinformation, faulty statistics and invalidated assertions maintains itself. (Tavris 1993)

One significant difference between this group of experts and, say, a group of physicists is that the child abuse experts have achieved their status as authorities not by scientific training but either by experience (they were victims of child abuse or they treat victims of child abuse in their capacity as social workers), or by writing a book on child abuse. The child abuse experts aren't trained in scientific research, which, notes Tavris, "is not a comment on their ability to write or to do therapy, but which does seem to be one reason for their scientific illiteracy."

Before discussing the methods and techniques of RMT, it should be noted that very few recovered memories of childhood sexual abuse first occur spontaneously. When they do, they are usually more likely to be corroborated by evidence than those evoked in RMT therapy. In fact, in some cases corroborative evidence serves as the retrieval cue for the repressed memory. RMT, however, seems to be able to produce recovered memories of sexual abuse in most of its clients. To those practicing RMT, this is proof of its power and effectiveness. To skeptical critics, this is warning sign: the memories are **confabulations** suggested by prodding, suggestive therapy.

Furthermore, even if traumatic memories are repressed sometimes, they are probably done so consciously and deliberately. Many of us choose not to dwell on unpleasant experiences and make a determined effort to wipe them from our memories as far as possible. We hardly desire some hypnotist or therapist to dredge up memories of experiences we've chosen to forget. In short, limited amnesia is best explained neurologically, not metaphysically. We forget things either because we never encoded them strongly enough in the first place or because neural connections have been destroyed or because we choose to forget them.

RMT uses a variety of methods—including **hypnosis**, visualization, group therapy, and **trance writing**—to assist the patient in "remembering" the traumatic event. Hypnosis is risky because it is easy to lead and encourage the patient by suggestive or leading questions. Trance writing has never been proven to have any therapeutic value (Schacter 1996: 271). Group therapy, on the other hand, can lead to communal reinforcement of delusions, if the therapist is not careful. People in the group can encourage others to share bizarre tales without fear of ridicule. The group might not originate the repressed memory, but they might facilitate the birth and nourish the growth of horrendous fantasies.

Using guided imagery or visualization in therapy can also be dangerous. Sherri Hines describes how her therapist used this method to help her retrieve a memory of being abused by her father:

My father would give me a bath and he used to draw on the mirror, draw on the steam, and he would draw cartoon characters. And that was the seed for a memory; we would start with that.

And [my therapist] would tell me, "You're in the bathtub. Your dad is there. He's drawing in the mirror. What is he drawing?" Then he'd say, "OK, now your father's coming over toward you in the bathtub. He's reaching out to touch you. Where is he touching you?" And that's how the memories were created. (Hallinan 1997)

Hines came to believe she was molested by her father and became so depressed she attempted suicide. She is now out of therapy and believes the memories were false and created in therapy.

The case of Diana Halbrook also brings into question the reliability of RMT methods. In a trance writing session, Hallbrook had written that her father had molested her. This was shocking news to her. She went into group therapy and heard bizarre tales of satanic ritual sacrifices. Soon the same kinds of bizarre events appeared in her trance writings, including the recovered memory that she'd killed a baby.

Because Diana Halbrook's ritual abuse memories seem so outlandish, her doubts about the reality of these and her other recovered recollections continued to grow. But these doubts met resistance from the people in her support group and her therapist. "I continually questioned the memories, doubted them, but when I questioned the therapist, he would yell at me, tell me I wasn't giving my 'little girl within' the benefit of the doubt. Tell me that I was in denial. I didn't know what to believe. But I trusted him" (Schacter 1996: 269).

Halbrook got out of the therapy, characterized by Daniel Schacter as "toxic," and no longer believes the outlandish memories. Schacter comments that "the most reasonable interpretation is that the events [recovered in therapy] do not have any basis in reality."

Each of the various methods described above has been very successful in getting patients to remember many things of which they were unaware before therapy. The memories include those not just of being sexually abused as children but of some very bizarre things, such as being abducted by aliens for sexual experimentation or breeding, being forced to participate in satanic rituals, or being traumatized in a past life.

Psychologist Joseph de Rivera claims that in RMT, "rather than help the patient separate truth from fantasy, the therapist encourages the patient to 'remember' more about the alleged trauma. And when the patient has an image—a dream or a feeling that something may have happened— the therapist is encouraged, praises the patient's efforts and assures him or her that it really did happen." This kind of therapy, he says, "confuses the differences between real and fantasized abuse and encourages destruction of families" (de Rivera 1993).

If therapists were planting all kinds of *good* memories in patients' minds, helping them enjoy more satisfying lives and relationships, it is doubtful that there would be much concern about RMT. Some of the memories recovered in RMT are extraordinarily bizarre, so bizarre that one would think that a reasonable person could hardly take them at face value. But RMT therapists are not put off by bizarre recollections. They either take them at face value (as John Mack does of his **alien abduction** patients and others do when interrogating children) or they take them as *artifacts* of the mind, which they then analyze as if they were archaeologists, inferring truth from ruins. Or they claim fantastic memories symbolize real experiences.

Laura Brown, for example, a Seattle psychologist in the forefront of RMT, says that fantastic memories are "perhaps coded or symbolic versions of what really happened." What really happened, she's sure, was sexual abuse in childhood. "Who knows what pedophiles have done that gets reported out later as satanic rituals and cannibalistic orgies?" asks Dr. Brown (Hallinan 1997).

In the past, Brown has criticized the False Memory Syndrome Foundation for being unscientific, but her emphasis on the symbolic nature of fantastic memories has little scientific credibility itself. Where is the scientific evidence that a fantastic memory can be distinguished from a delusion? How do we distinguish memories of real cannibalism from symbolic memories? We usually know what a crucifix or a swastika symbolizes, but what does eating an infant symbolize? Symbols might be ambiguous. How can we be sure that a memory is a symbol of child abuse and not of adult abuse by coworkers, by other children who tormented the patient years ago, or by the therapist him- or herself? How can we be sure it is not a symbol of self-abuse? How can we be sure it is a symbol of any kind of abuse at all? What would distinguish a symbol of abuse from a symbol of *fear* of abuse? For that matter, what would distinguish a symbolic representation of fear of being abused from one representing fear of abusing someone else in the present or a regret of having abused someone else in the past? The dangers and imminent probabilities of misinterpretation of symbolic memories should be obvious, especially when it is not always that clear that a memory really is a symbolic expression at all.

History is replete with examples of what happens when any group of authorities do not have to answer to empirical evidence but are free to define truth as they see fit. None of the examples has a happy ending. Why should it be otherwise with therapy?

See also **hystero-epilepsy**; **multiple personality disorder**; **psychotherapies**, **New Age**; and **unconscious mind**.

Further reading: Baker 1996a; de Rivera 1993; Hallinan 1997; Johnston 1999; Loftus 1994; Ofshe and Watters 1994; Schacter 1997; Singer and Lalich 1996; Wakefield and Underwager 1994.

retroactive clairvoyance

The ability to use hindsight to predict what happened after it has happened. This is the one skill skeptics admit that psychics possess.

Advocates of the prophetic abilities of **Nostradamus** are experts at retroactive clairvoyance, as are those who defend the notion of the **Bible Code**.

See also **shoehorning**.

retrocognition

A type of **clairvoyance** involving **psychic** knowledge of something *after* its occurrence.

My sister related an apparent case of retrocognition to me. She was watching television when a report came on about a woman—Susan Smith of Union, South Carolina—who claimed that her two young children had been kidnapped by a black man who carjacked her in some small town in the South. She claimed the black man drove out near a lake and let her out of the car and drove off with the two children. My sister said she immediately sensed that the children were dead and that they were in the lake. About a week later, the world was told that the woman herself had sent the car into the lake with the children alive and strapped into the back seat. She apparently watched as the car sank into the lake with her sons, drowning them.

It is a sad commentary on our times, but false reports of crimes are not uncommon and mothers killing their children are not uncommon. They are certainly more common than black carjackers kidnapping little white boys. In any case, many people who saw the broadcast probably shared the suspicious feelings that my sister had concerning the mother/murderer. The police in the small southern town were skeptical too, not because they are clairvoyant any more than the rest of us but because they know a little bit about human nature and human behavior. If one was suspicious of the mother's story, the fact that she said she was driven to a lake leaves little to the imagination to fill in the blanks.

We all make judgments about people's stories. Sometimes we're right and sometimes were not. We tend to forget the times we're not. If we didn't, we wouldn't find the occasional correct "feeling" to be so uncanny.

retrospective falsification

A term coined by D. H. Rawcliffe (1959, 1988) to refer to a situation in which an extraordinary story is told, then retold with embellishments and remodeled with favorable points being emphasized while unfavorable ones are dropped; the distorted version becomes part of memory, fixing conviction in a remarkable tale.

The term is also used in psychology to describe the process used by the paranoid as he recalls selective incidents from the past, reshaping them to fit his present needs.

See also **shoehorning**.

Further reading: Cameron and Rychlak 1985; Rawcliffe 1959.

reverse speech

A form of communication arising from the **unconscious mind**, according to David John Oates. He credits himself with the discovery of reverse speech (which really means "a cheap server," since that is what it sounds like in reverse). The unconscious mind, says Oates, is a seat of deep truths, inexorable honesty, and hidden meanings. Furthermore, it sends out backward messages to the conscious mind every 10 to 15 seconds. The conscious mind then reverses the message and directs us to speak in forward speech. To grasp the real meaning of our speech, says Oates, we must tape it and play it backward on a Reversing Machine™, which is conveniently available from Reverse Speech Enterprises™ (www.reversespeech.com/intro.shtml). "The implications are mind-boggling," says Oates of his discovery, "because reverse speech opens up the Truth."

Oates claims that reverse speech is a

form of human communication that is automatically generated by the human brain. It occurs every time we speak and is imbedded backwards into the sounds of our speech. This previously undiscovered function of the mind is the mind's own independent voice speaking from the deepest regions of consciousness. . . . [F]orward speech is from the left brain and Reverse Speech is from the right brain.

Mr. Oates claims to have made many discoveries, including that children learn to speak backward before they learn to speak forward. He believes that what most of the world has taken to be babbling is actually deep thought from the unconscious minds of infants.

Oates is an Australian who started on his road to discovery by accidentally dropping a tape recorder into the toilet while shaving. He "fixed" the recorder, but henceforth the recorder would play only in reverse. His electronic incompetence was not without its rewards. What happened next was pure synchronicity. Not long afterward, some teenagers asked him about **backward satanic messages** in rock music. He was ready to counsel the local youth, since he had the only reverse tape player in the area. He could play tapes backward for them and together they could search for the hidden messages. Oates not only found satanic messages in rock music, he found that if one listens very carefully, one can hear reverse messages in every bit of communication that uses words. For 12 years he labored at uncovering the secret of reverse speech. He has shared his discoveries with the rest of the world in *Beyond Backward Masking: Reverse Speech and The Voice of the Inner Mind* (which, in reverse, sounds something like *Dante Rentifor: Anchovy Tulip Server* by Stone Adjective). This book has an appendix on UFOs, as well.

Oates claims that "reverse speech is the voice of truth" and that "If a lie is spoken forwards, the truth may be spoken backwards. . . . It can be used as a truth detector. It will reveal the truth if a lie is spoken and it will reveal extra facts if they are left out, for example, the name of an accomplice in crime or the location of evidence." Apparently, only criminals who speak in palindromes can avoid Oates's detection.

He also claims that 95% of our thoughts "are below consciousness" and that

> reverse speech can describe unconscious subjects such as personality patterns and behavioral agenda, it can reveal hidden memory and experiences, and it can also describe the state of the physical body. At the deepest levels of human consciousness, Reverse Speech also describes the state of the human **soul** and our relationship with God.

He also claims most reverse speech is metaphorical and is communicated in pictures or parables, "similar to **dreams**."

Furthermore, reverse speech analysis

> can be used as a therapeutic tool for psychoanalysis. Its metaphors will give a detailed map of the mind and often pinpoint precise reasons and causes for problems experienced. Used in conjunction with hypnosis it can be used as an extremely powerful and permanent form of behavioral change.

Employers can use it for employee selection, lawyers for deposition analysis, reporters for politicians' speeches. Its applications are endless.

As with his other claims, Oates provides no support for these notions. However, there is considerable evidence against him.

His notion about reverse speech occurring in the right brain is not supported by empirical study. His claims about infants and speech contradict everything that is known about the development of the human brain and speech in children. The claim that 95% of our thoughts are below the level of consciousness seems to have been pulled out of thin air. The claim that the unconscious mind contains data that reveal hidden truths about a person's behavior and personality, as well as one's physical and spiritual health, while not original with Oates, is highly questionable.

Despite a lack of theoretical grounding, or even a concern with such grounding, Oates provides examples to prove his theory. On the *Geraldo* show (December 26, 1997), Oates said that Patsy Ramsey's forward speech, "We feel that there are at least two people on the face of the earth that know who did this [i.e., murdered her daughter JonBenet] and that is the killer and someone else that person may have confided in" is actually reverse speech for "I'm that person. Seen that rape." When Bill Clinton said, "I try to articulate my position as clearly as possible," he really was saying, "She's a fun girl to kiss." Oates knows this because that is how it sounds to him in reverse, or that is what the metaphors mean. We cannot prove him wrong. If it does not sound like that to us or if we don't grasp the metaphors, it is because the messages "are very quick and fast and are often hidden in the high tones of speech. For this reason speech reversals are very easily missed by most researchers." Furthermore, he says the untutored are not familiar with the language of metaphor. (He has two books that can help us here.) These are **ad hoc hypotheses**. Furthermore, once he has told you that the reverse speech you are about to hear is "She's a fun girl to kiss," you are likely to hear that, regardless of how garbled the message sounds. Call it the power of suggestion or **pareidolia** for the ears. Call it anything but worth investigating much further.

Oates is taken seriously because he is an entertaining novelty feeding the lust for trash gossip that often passes as news and information in our society. He has modeled himself after many others who have found that self-confidence and the ability to understand are so diminished in so many people that the more shocking a person's communication, the more likely it can be trademarked and profitably marketed as a gateway to the secret of life.

rods

Flying insects caught on video. Some hoaxers or very imaginative people have been maintaining that rods are actually an unknown life form of alien origin. But according to Doug Yanega of the Entomology Department at the University of California at Riverside and a member of the Straight Dope Science Advisory Board, rods are

> a videographic artifact based on the frame capture rate of the videocam versus the wingbeat frequency of the insects. Essentially what you see is several wingbeat cycles of the insect on each frame of the video, creating the illusion of a "rod" with bulges along its length. The blurred body of the insect as it moves forward forms the "rod," and the oscillation of the wings up and down form the bulges. Anyone with a video camera can duplicate the effect, if you shoot enough footage of fly-

ing insects from the right distance. [www .straightdope.com/mailbag/mrodhoax .html]

Rods seem to be a favorite topic of **UFO** and **cryptozoology** buffs. One of the more outspoken defenders of alien rods is Jose Escamilla, host of the RoswellRods. com web site. Jose has even brought his story and films to the Learning Channel. Some hilarious photographs of rods have been posted on the Internet at the Escamilla site. A favorite is "the swallow chases a rod," which looks just like a bird going after an insect.

Rolfing

A kind of deep massage and "movement education" (www.rolfusa.com/movement .html) aimed at realigning the body, developed by Ida P. Rolf (1896–1979), a biochemist and therapist. She authored several books on the relationship of form and structure in the human body, including *Rolfing: The Integration of Human Structures* (1977). Her dissertation was on the chemistry of unsaturated phosphatides (1922). According to the Rolf Institute, "Rolfing's foundation is simple: Most humans are significantly out of alignment with gravity, although we function better when we are lined up with the gravitation field."

Dr. Rolf claimed she found a correlation between muscular tension and pent-up emotions. Rolfing is based on the notion that emotional and physical health depends on being properly aligned. To be healthy, one must align the head, shoulders, hips, knees, and ankles. By being properly aligned, gravity enhances personal **energy**, leading to a strong body and healthy emotional state.

Has this claim of the muscular/emotional connection been demonstrated by any **controlled studies**? No, but there are many anecdotes and **testimonials** verifying Rolfing. One practitioner claims that Rolfing is a "scientifically validated system of body restructuring and movement education." She claims that there is scientific proof that each of us has lifelong patterns of tension and that realigning releases this tension, so that "overall personal functioning tends to improve" (www.bnt.com/~rolfer).

It takes 1 to 2 years to complete the Rolfing training from the official Rolf Institute in Boulder, Colorado. The Guild for Structural Integration, which is dedicated to Ida Rolf, seems to be Rolfing in everything but the name. Another school of structural integration is Hellerwork, which will align and integrate not only your body parts but your mind and indivisible **soul** as well.

Further reading: Beyerstein 1997; Barrett and Butler 1992; Barrett and Jarvis 1993.

Rorschach inkblot test

Hermann Rorschach (1884–1922), a psychiatrist with a strong interest in art and inkblots, developed the Rorschach inkblot test, a psychological projective test of personality in which a subject's interpretations of 10 standard abstract designs are analyzed as a measure of emotional and intellectual functioning and integration.

The test is considered "projective" because the patient is supposed to project his or her real personality into the inkblot via the interpretation. The inkblots are purportedly ambiguous, structureless entities that are to be given structure by the interpreter. Those who believe in the efficacy of such tests think that they are a way of getting into the deepest recesses of the patient's psyche or **unconscious mind**. Those who give such tests believe themselves to be experts at interpreting their patients' interpretations.

What evidence is there that an interpretation of an inkblot issues from a part of the self that reveals true feelings or desires, rather than, say, creative expression? What justification is there for assuming that any given interpretation of an ink blot does not issue from a part of the self bent on deceiving others? Or on deceiving oneself, for that matter? In any case, it is a long jump from having desires to committing actions.

Rorschach testing is inherently problematic. For one thing, to be truly projective the inkblots must be considered ambiguous and without structure by the therapist. Hence, the therapist must not make reference to the inkblot in interpreting the patient's responses, or else the therapist's projection would have to be taken into account by an independent party. Then the third person would have to be interpreted by a fourth, ad infinitum. Thus, the therapist must interpret the patient's interpretation without reference to what is being interpreted. Clearly, the inkblot becomes superfluous. You might as well have the patient interpret spots on the wall or stains on the floor. In other words, the interpretation must be examined as if it were a story or dream with no particular reference in reality. Even so, ultimately the therapist must make a judgment about the interpretation, that is, interpret the interpretation. But again, who is to interpret the therapist's interpretation? Another therapist? Then, who will interpret his?

To avoid this logical problem of having a standard for a standard for a standard, some therapists invented standardized interpretations of interpretations. Both form and content are standardized. For example, a patient attending only to a small part of the blot indicates an "obsessive personality"; one seeing figures that are half-human and half-animal indicates alienation, perhaps on the brink of schizophrenic withdrawal from people (Dawes 1994: 148). If there were no standardized interpretations of the interpretations, then the same interpretations by patients could be given equally valid but different interpretations by therapists. No empirical tests have been done to demonstrate that any given interpretation of an inkblot is indicative of any past behavior or predictive of any future behavior. In short, interpreting the inkblot test is about as scientific as interpreting dreams.

To have any hope of making the inkblot test appear to be scientifically valid, it was essential that it be turned into a nonprojective test. The blots can't be considered completely formless, but must be standardized and patient interpretations compared with the standards. This is what John E. Exner did. The Exner System uses inkblots as a standardized test, which at least preserves the illusion of objectivity.

The Rorschach enthusiast should recognize that inkblots, dreams, drawings, or handwriting may be no different in structure than spoken words or gestures. Each is capable of many interpretations, some true, some false, some meaningful, some meaningless. It is an unprovable assumption that dreams or inkblot interpretations issue from a source deep in the subconscious that wants to reveal the "real" self (Dineen 1998). The mind is a labyrinth and it is a pipe dream to think that the inkblot is Ariadne's thread, leading the therapist to the center of the patient.

Roswell

On or around Independence Day, 1947, during a severe thunderstorm near Roswell, New Mexico, an Air Force experiment using high altitude balloons blew apart and fell to Earth. This minor event in the history of reconnaissance turned out to be the Big Bang of UFOlogy. **UFO** enthusiasts have

come to see that Fourth of July as the day an alien spaceship crashed on Earth. Some UFOlogists claim that the U.S. Air Force and other government coconspirators took away the aliens for an interrogation or an autopsy. Some claim that all our modern technology was learned by analyzing and copying the technology of the aliens.

The actual crash site was on the Foster ranch 75 miles north of Roswell, a small town now doing a brisk business feeding the insatiable appetite of UFO enthusiasts. Roswell now houses two UFO museums and hosts an annual alien festival. Shops cater to this curious tourist trade, much as Inverness caters to the Loch Ness crowd. This seems a bit unfair to Corona, New Mexico, which is actually the closest town to the alleged crash site. Roswell is the nearest military base, however, and that is where the remains of the alien craft and its occupants were allegedly taken.

William "Mac" Brazel, foreman of the Foster Ranch, along with a 7-year-old girl, Dee Proctor, found the most famous debris in modern history. They had never seen anything like it before. Millions agree that the stuff was strange. Actually, it was pretty mundane stuff, including a piece of reinforcing tape whose flowerlike design was taken to be alien hieroglyphics. But the Air Force was not consistent in describing the debris and has suggested that ardent UFOlogists have had a little trouble with their source memory. Perhaps what people are recalling as a single event is actually a mixture of several events that occurred in different years (such as weather balloon and nuclear explosion detection balloon tests, airplane crashes with burned bodies, and dumping of featureless dummies from airplanes). The likelihood that the Roswell myth is a reconstruction involving many events over many years is supported by the fact that Roswell was ignored by UFOlogist until Charles Berlitz and William Moore published a book on the subject in 1980, more than 30 years after the event.

UFO buffs trust Berlitz and others with fantastic stories based on 30-year-old memories. That the government made errors and was inconsistent is taken as sufficient evidence that there is a massive conspiracy by the government and mass media. They are trying to conceal the truth from the general public that the aliens have landed. Some even believe that the U.S. government has signed a treaty with the aliens.

Skeptics agree that something crashed near Roswell in 1947, but not an alien craft. Skeptical explanations have varied from weather balloons to secret aircraft to espionage devices. Current conventional wisdom among skeptics is that what was found on the Brazel ranch was part of Project Mogul, a top-secret project testing giant, high-flying balloons to detect Soviet nuclear explosions.

To skeptics, Roswell is a classic example of what D. H. Rawcliffe (1959, 1988) called **retrospective falsification**. An extraordinary story is told, then retold with embellishments and remodeled, with favorable points emphasized while unfavorable ones are dropped. False witnesses put in their two cents. In the case of Roswell, we also have a few unreliable characters who added their delusions, such as Whitley Strieber, Budd Hopkins, and John Mack (see **alien abductions**). There is also Robert Spencer Carr, a high school graduate who liked to be called "Professor Carr." Carr is a hero in the UFO literature, but his stories of flying saucers and alien creatures were all delusions. His son has written: "I am so very sorry that my father's pathological prevarication has turned out to be the foundation on which such a monstrous mountain of falsehoods has been heaped" (Carr 1997). It was that mountain of falsehoods that became part of

the UFO memory, fixating conviction in a remarkable tale. It happened at Fatima (during a time when the only aliens thought to be visiting our planet were messengers from God), and it happened at Roswell. One might think, however, that unlike the belief in our Lady of Fatima and other apparitions from the supernatural world, Roswell might be settled some day since it involves testable hypotheses and refutable claims. Don't count on it. UFO enthusiasts are every bit as devoted to their belief system as religious devotees are to theirs. Evidence and rational argument are of little concern to those who consider science fiction to be a wiser guide than science, logic, and reasonable probability.

See also **Area 51** and the **Aztec UFO hoax**.

Further reading: Frazier 1997; Klass 1997b; Korff 1997; Peebles 1994; Sagan 1995; Saler et al. 1997.

runes

The characters of ancient alphabets: Teutonic (24 letters), Anglo-Saxon (32 letters), and Scandinavian (16 letters). Runic characters are similar to Latin letters, except that they tend to have few curves and consist mostly of straight lines, suitable for carving with knives. Runic letters were used for over 1,000 years. For most people, the runic alphabet died out sometime between the 13th and 16th centuries. But for those special New Age people with one foot in the world of secrets and the other in the world of mysteries, runes are used as a form of **divination**. Runes may have gotten their reputation for being tools of divination in 1639 when the Catholic Church banned runes because they were said to be used to cast magic spells or communicate with the devil (www.tarahill.com/runes/runehist.html). Many New Agers seem to like the writer J. R. R. Tolkien, so the fact that his fictional Hobbits used a kind of runes in their writing may have enhanced the association of runic letters with magic and mystery.

The Norse used Runic characters mostly for practical purposes, such as marking graves, identifying property, or defacing other's graves and property with graffiti, such as at Maes Howe in Orkney, Scotland. New Agers ignore these uses and prefer to side with superstitious 12th-century Norsemen and -women who thought they could see the future in alphabetic characters on wood or stone.

The word "rune" derives from the Old Norse and Old English *run,* which means "mystery." The real mystery is why anyone would think that writing the letters of an alphabet on little pieces of wood or stone, putting them in a bag, and then drawing them out and throwing them or laying them down in certain ways would answer their questions, give them direction or guidance, or help them make good decisions.

How is it that random alphabetic stone selection can be so useful? Easy. *Anything* can be a source of transformation and breakthrough if you decide to let it be. Runes, **tarot** cards, the *I Ching*, **enneagrams**, **Myers-Briggs**—anything can be used to stimulate self-reflection and self-analysis. Anything can be used to justify coming to a decision about an unresolved matter. Coming to a decision brings relief, reduces anxiety, and may well seem like a breakthrough and transformation. Using something such as rune stones to help make your decision seems to relieve you of responsibility for it. The stones and your subconscious mind made the choice for you, so you are off the hook if anything goes wrong. Furthermore, since there are no standard interpretations, you can always change your initial interpretation to fit new facts or desires.

Further reading: Davidson 1989; Hutton 1993.

S

Sai Baba

An Indian guru born in 1926 named Satyanarayana Raju, who now goes by the name Bhagavan Sri Sathya Sai Baba, or Sai Baba for short. His followers, who allegedly number in the several millions, believe he is divine. He claims to have **paranormal** powers and to be able to work **miracles**. Two of his favorite miracles are to make ashes materialize for poor people and to make jewelry materialize for rich people.

The film *Guru Busters* (Equinox) demonstrated that these alleged miracles are magician's **conjuring**. Yet it took some very good camera work to expose Sai Baba. He fools even the smartest. For example, in the film an Indian physicist who had been given a ring allegedly materialized by Sai Baba declared that a magician could not possibly fool him. The physicist said that he had a doctorate from Harvard. He should have recalled the words of Richard Feynman: I'm wise enough to know I can be fooled.

Guru Busters depicts Indian Science and Rationalists' Association (ISRA) members as they travel throughout India debunking and exposing as frauds local **fakirs**, godmen, and godwomen. ISRA debunkers are to India what James **Randi** and the Committee for the Scientific Investigation of the Paranormal are to America and Canada. They utilize scientific and rational principles to expose the magical art of illusion used by Hindu mystics in performing feats of **levitation** and other alleged miracles.

Being a godman is not without its problems, however. According to Michelle Goldberg of Salon.com, the master has been accused of several human sins, including child molestation. Goldberg says that there is a "growing number of ex-devotees who decry their former master as a sexual harasser, a fraud and even a pedophile." She notes, however, that this trouble "has hardly put a dent in his following" (http://www.salon.com/people/feature/2001/07/25/baba/index.html).

See also **Indian Rope Trick**.

saint

A former human being, now dead, whose spirit is said to dwell in heaven with God. Such spirits are identified by their having belonged to heroically virtuous or holy people when attached to their bodies on earth. (The word "saint" derives from *sanctus,* the Latin word for holy.)

Some spirits are officially recognized as saints by Christian ecclesiastical authority in a process known as canonization. Different ecclesiastical authorities used different criteria and hence have different canons, or catalogues, of saints.

Keeping a canon of saints assists in recruitment of new church members, mythologizes the faith, and allows for currying favor from subordinates close to the boss. Saints are *venerated,* not worshipped; that is, they are admired and sought as intercessors because of their special place in the hierarchy. Saints are in the inner circle, so to speak, and because of their status a word from them to the boss might be sufficient to get a wish granted.

Why saints would intercede for the living seems inexplicable to the logical mind. First, they have nothing to gain by acting as anybody's intercessor. They are already in glory, and their glory does not depend on others reaching glory, and there is no reason they should prefer the glory of one person over another. Earthly beings might grant favors only to those who ask, but supernatural beings would have no reason for favoring only those who curry favor from them. Second, there is no reason why God would be more accessible to the prayers of a saint than to those of a holy person on Earth. Why use a middleman when you can go directly to the source? Third, if God would not listen to an unholy and unworthy person who wants a favor, why would God listen to a saint's plea for such a person? The unworthy shouldn't get a hearing from the boss *or* his underlings. Were it not for their supposed utility here on Earth, saints would be superfluous to humans.

That sainthood is valued for its intercessory value is clearly indicated by the fact that the primary method of identifying who will be canonized is by the performing of **miracles**. To even be considered for canonization, you not only must have led an exemplary holy life, but must have performed several miracles that show you are answering the **prayers** of those who pray exclusively to you. Such miracles are identified by a theological board and require some sort of connection to an allegedly miraculous cure. For example, Katherine Drexel, an heiress from Philadelphia who became a nun, was canonized because several cures have been attributed to her intercession. For example, Drexel's spirit is being credited with being instrumental in the "cure" of the temporary deafness of a young girl. Edith Stein, who was recently canonized, allegedly interceded to save the life of a young girl who had swallowed an obviously nonlethal dose of Tylenol. Pope John Paul II, who canonized the **pious fraud** Padre Pio, believes the controversial cult figure and alleged **stigmatic** cured a woman of cancer.

In his effort to provide role models for the faithful, Pope John Paul II has added more than 450 names to the canon of saints since he took over as the head of the Roman Catholic Church in 1978. That's about 150 more than have been sainted in the past *400 years*.

St. Malachy

See Bishop **Malachy O'Morgair**.

Santa Claus

Santa Claus is one of the most famous IFPs (identified flying persons) in history. The hirsute gift giver has had more sightings than **Bigfoot**, **UFOs**, and the Virgin Mary combined. The innocent and pure witnesses to the jolly one decked out in sartorial crimson flailing away at his flying reindeer are legion. Who can mistrust a child, much less billions of children? Surely these witnesses are reliable. There is no proof that they are suffering from any mental derangement. They have no motive for lying. The only plausible explanation for these sightings is that they are genuine. There is no reason to think that all these witnesses are **confabulating**. If there is nothing to this belief, then why do so many people believe it? There is no way this could be an example of **communal reinforcement** of a false idea or delusion. This must be a genuine vision.

Cynical skeptics are wont to note that the belief in the Christmas gift giver requires acceptance of the hypothesis that in a single evening the infrequent flyer visits all the homes in America and the homes of Americans everywhere else on Earth. Even if the speedy one spent a single second at each home and took no time to travel between homes, it would take him several years to complete his rounds. Obviously, a **miracle** happens every Christmas! That is the only logical explanation for flying reindeer traveling at tachyonic speed carrying hundreds of thousands of pounds of weightless presents. What else could it be?

Further reading: Sagan 1972.

Sasquatch

See **Bigfoot**.

Satan

The adversary of God, who is Good personified. Thus, Satan is Evil personified.

Many followers of the Bible consider Satan to be a real being, a spirit created by God. Satan and the other spirits who followed him rebelled against God. The Creator cast them out from Heaven. Theologians might speculate as to why the Almighty did not annihilate the fallen angels, as He is said to have done to his other creations when they failed to be righteous (save Noah and his family, of course). Satan was allowed to set up his own kingdom in Hell and to send out devils to prowl the earth for converts. The demonic world seems to have been allowed to exist for one purpose only: to tempt humans to turn away from God. Why God would allow Satan to do this is explained in the Book of Job. When Job asks why God let Satan torment him, the answer is blunt and final: *Hath thou an arm like the Lord?* Theologians interpret the story of Job in many different ways, but my interpretation is that the moral of the story is that nobody knows why God lets Satan live and torment us. God is God and can do whatever He wants. Ours is not to question why, ours is but to do and die.

Satan, being a spirit, is neither male nor female. However, like his Creator, Satan is usually referred to as a masculine being. Many believe that Satan, or the Devil as he is often called, can possess human beings. Possession is bodily invasion by the Devil. Many religions still perform **exor-**

cisms on those considered to be possessed. Throughout the centuries, many pious religious people have erroneously considered those with certain mental or physical illnesses to be possessed by Satan.

More frequent than outright possession, however, has been the accusation of being in consort with the Devil. Satan is believed to have many powers, among them the power to manifest himself in human or animal form. For most of the history of Christianity there are reports of Satan having sex with humans, either as an incubus (male devil) or succubus (female devil). **Witches** and **sorcerers** were thought by many to be the offspring of such unions. They are considered especially pernicious because they inherit some of the Devil's powers. Pope Innocent VIII proclaimed in a Bull that "evil angels," that is, devils, were having sex with many human men and women. He was not the first to have made this claim. Others before him, such as Thomas Aquinas, had explored this territory in great detail. Thomas reminds us that since Satan is not human, he can't produce human seed. So he must transform himself into a woman, seduce a man, keep the seed, transform into a male, seduce a woman, and transfer the seed. Something of the Devil is captured by the seed along the way, so the offspring are not normal. Apparently it took Satan a long time to figure out that if he wanted to control the world, the best way to do it would be to breed with humans. Invading our bodies would be more efficient and effective than trying to invade our minds. But the Pope and many other pious men had a plan to exterminate the diabolical offspring: They would torture and burn them all!

According to Carl Sagan, accounts of diabolical intercourse are common cultural phenomena: "Parallels to incubi include Arabian djinn [**jinn**], Greek satyrs, Hindu bhuts, Samoan hotua poro, Celtic dusii . . ."

(Sagan 1995: 124). However, as a child being instructed in the ways of Satan by Dominican Sisters, I was assuredly not told stories of nuns being raped by incubi in priests' clothing. The Devil was there to tempt us to sin, pure and simple. He was not there to have sex with us or engage in reproductive experimentation or breed a race of witches and magi. His main temptations would be sexual, however. Surely, it has occurred to many observers that the fear of Satan seems very much like fear of our own sexuality.

One of the more interesting aspects of Satanology is the recurring theme of humans making a pact with the Devil. The Faust legend is the best-known of these: In exchange for one's soul, Satan will bestow one with wealth or power for a specified time. In most versions of the story, Faust tricks the Devil and avoids payment. In the original, the Devil mutilates and kills Faust at the end of the contract. His brains are splattered on the walls of his room, his eyes and teeth lay on the floor, and his corpse rests outside on a dunghill (Smith 1952: 269).

The closest thing we have to such stories today are **alien abduction** and **star child** stories. Aliens from outer space have replaced Satan in modern mythology. But instead of being burnt at the stake, today's victims get therapy and are guests on talk shows.

Finally, there are the modern-day Satanists who find solace and power in occult **magick** but especially in anything anti-Christian. They draw their inspiration from the great works of imagination in art, literature, and policy created primarily by pious Christians in their zealous wars against their enemies but also those created by pre-Christian **cults** such as the Egyptian cult of Set or by non-Christian occultists such as Aleister **Crowley** and Anton LaVey. Today's Satanists have been blamed by pious Christians for ritual murders of children, mutilation and sacrificial killings of animals, writing **backward satanic messages** on musical recordings instructing people to kill, sending **subliminal** or secret messages through diabolical symbols on pizza boxes or soap wrappers, and causing the decay of morals and civilization as we know it. This is just in America and Europe. Satan has many brothers and sisters in other places around the world that go by many names and oppose many gods.

Further reading: Carus 1996; de Givry 1971; Hicks 1991; Hill 1995.

satanic ritual abuse (SRA)

The name given to the allegedly systematic abuse of children by Satanists.

In the mid-1970s, there were widespread allegations of the existence of a well-organized intergenerational satanic cult whose members sexually molest, torture, and murder children across the United States. In the 1980s there was a panic regarding SRA, which was largely triggered by a fictional book called *Michelle Remembers* (1980). The book was published as fact but has subsequently been shown to be a hoax by at least three independent investigators (www.religioustolerance.org/jud_blib.htm). Nevertheless, the allegations were widely publicized on radio and television talk shows, especially on Geraldo Rivera's show (www.religioustolerance.org/geraldo.htm).

A 4-year study in the early 1990s found the allegations of SRA to be without merit. The study was conducted by University of California at Davis psychology professors Gail S. Goodman and Phillip R. Shaver, in conjunction with Jianjian Qin of UC Davis and Bette I. Bottoms of the University of Illinois at Chicago. Their study was supported by the National Center on Child Abuse and Neglect. The researchers

investigated more than 12,000 accusations and surveyed more than 11,000 psychiatric, social service, and law enforcement personnel. The researchers could find no unequivocal evidence for a single case of satanic cult ritual abuse.

Another study, published in 1992 by Kenneth V. Lanning, a Supervisory Special Agent at the FBI Academy, came to the same conclusion: There is no good evidence of a single case of SRA. Lanning has investigated SRA since 1981 (www.religioustolerance.org/ra_rep03.htm).

If there are thousands of baseless accusations, how do they originate? Most of them are said to originate with children. Since there is a widespread belief that children wouldn't make up stories of eating other children or being forced to have sex with giraffes after flying in an airplane while they were supposed to be in day care, the stories are often taken at face value by naive prosecutors, therapists, police officers, and parents. Yet children are unlikely to invent stories of satanic ritual abuse on their own. Therapists, district attorneys, police, and parents themselves stimulate the stories. There is ample evidence that therapists and law enforcement personnel encourage and reward children for making and accepting the suggestions of bizarre abusive behavior. The investigators also discourage truth by refusing to accept no for an answer, forcing children to undergo interrogations until the interrogator gets what he or she is after.

See also **repressed memory therapy**.

Further reading: Eberle 1993; Loftus 1994; Nathan and Snedeker 1995; Ofshe 1994; Victor 1993; Wright 1995.

Scallion, Gordon-Michael

A man who claims to have the gift of prophecy like **Nostradamus** and Edgar **Cayce**. Like Cayce, his visions have taken him to **Atlantis**, and like Nostradamus and Cayce, his head is filled with visions of disasters and apocalypses. In fact, Scallion's head is filled with many of the same visions Cayce claimed to have had. Coincidence? Not likely. He has predicted earthquakes in California and hurricanes in Florida. His predictions are so wild that his followers seem not to care that his accuracy is on par with Jeane **Dixon's**. Skeptics might think Scallion is a plagiarist; true believers might think he, Cayce, Dixon, and Nostradamus tapped into the **Akashic record**.

Scallion is actually a prophetic industry that he calls the Matrix Institute. One of his more popular items for sale is a map of a future Earth as seen by Scallion. On this map—which he claims will be the true map of the world by at least the year 2002—California is nothing but a few islands in the Pacific and Denver is oceanfront property. (This is a revision of an earlier prophecy that also proved false.) Eventually, says Scallion, the United States will restructure itself as 13 colonies, proving that what goes around comes around.

Scallion claims to have first noticed his gift in 1979 while hospitalized. It was then that he started hearing voices and having cataclysmic visions. Soon after he was healed, he started to believe that he also was given the gift of healing along with the gift of hallucination. He became another Edgar Cayce, doing readings and healings and giving lectures to all who would listen. Soon he founded his own newsletter, the *Earth Changes Report,* to keep track of all his apocalyptic dreams and hallucinations, which began arriving at a furious rate. The visions now began appearing on his computer screen. Scallion's visions can appear on your computer screen should you so desire, since he provides an online version of *Earth Changes*.

Those living in Palm Springs should know that according to Scallion, you were hit with a 9.0 earthquake sometime between 1995 and 1997. Sonoma County (north of San Francisco) was hit by an 8.5 quake during that period, as well. If you didn't feel it, that may be because they didn't occur as predicted by Scallion. Nor did volcanic ash cover the whole planet.

His recent poor track record was predictable, however, since he predicted that California would be in the Pacific Ocean by May 1993 and Denver would be on the Pacific coast by 1998. Also, one-fourth of Alaska was supposed to be gone by now.

No wonder scientists are not rushing to verify his latest vision: underground volcanoes fuel El Niño.

science

Science (at least natural science, which is the focus of this entry) is first and foremost a set of logical and empirical methods that provide for the systematic observation of empirical phenomena in order to understand them. We think we understand empirical phenomena when we have a satisfactory theory that explains how the phenomena work, what regular patterns they follow, or why they appear to us as they do. Scientific explanations are of natural phenomena in naturalistic terms rather than of supernatural phenomena, although science itself requires neither the acceptance nor the rejection of the supernatural.

Science is also the organized body of knowledge about the empirical world that issues from the application of logical and empirical methods.

Science consists of several specific sciences, such as biology, physics, chemistry, geology, and astronomy, which are defined by the type and range of empirical phenomena they investigate.

Finally, science is also the application

of scientific knowledge, as in the altering of rice with daffodil and bacteria genes to boost the vitamin A content of rice.

There is no single scientific method. Some of the methods of science involve *logic,* for example, drawing inferences or deductions from hypotheses or thinking out the logical implications of causal relationships in terms of necessary or sufficient conditions. Some of the methods are *empirical,* such as making observations, designing **controlled experiments**, or designing instruments to use in collecting data.

Scientific methods are impersonal. Whatever one scientist is able to do, other scientists should be able to duplicate. When a person claims to measure or observe something by some purely *subjective* method, which others cannot duplicate, that person is not doing science. When scientists cannot duplicate the work of another scientist, that is a clear sign that the scientist has erred in design, methodology, observation, calculation, or calibration.

Science does not assume it knows the truth about the empirical world a priori. Science assumes it must *discover* its knowledge. Those who claim to know empirical truth a priori (as in **creationism** or **intelligent design**) cannot be talking about scientific knowledge. Science presupposes a regular order to nature and assumes there are underlying principles according to which natural phenomena work. It assumes that these principles or laws are relatively constant. But it does not assume that it can know a priori either what these principles are or what the actual order of any set of empirical phenomena is.

A scientific theory is a unified set of principles, knowledge, and methods for explaining the behavior of some specified range of empirical phenomena. Scientific theories attempt to understand the world of observation and sense experience. They

attempt to explain how the natural world works.

A scientific theory must have some logical consequences we can test against empirical facts by making predictions based on the theory. The exact nature of the relationship of scientific theory, facts, and predictions is something about which philosophers widely disagree (Kourany 1997).

It is true that some scientific theories, when they are first developed and proposed, are often little more than guesses based on limited information. On the other hand, mature and well-developed scientific theories systematically organize knowledge and allow us to explain and predict wide ranges of empirical events. In either case, however, one characteristic must be present for the theory to be scientific. The distinguishing feature of scientific theories is that they are "capable of being tested by experience" (Popper 1959: 40).

To be able to test a theory by experience means to be able to predict certain observable or measurable consequences from the theory. For example, from a theory about how physical bodies move in relation to one another, one predicts that a pendulum ought to follow a certain pattern of behavior. One then sets up a pendulum and tests the hypothesis that pendulums behave in the way predicted by the theory. If they do, then the theory is *confirmed*. If pendulums do not behave in the way predicted by the theory, then the theory is *falsified*. (This assumes that the predicted behavior for the pendulum was correctly deduced from your theory and that your experiment was conducted properly.)

The fact that a theory passed an empirical test does not *prove* the theory, however. The greater the number of severe tests a theory has passed, the greater its degree of *confirmation* and the more reasonable it is to accept it. However, to confirm is not the same as to prove logically or mathematically. No scientific theory can be proved with absolute certainty.

Furthermore, the more tests that can be made of the theory, the greater its empirical content (Popper 1959: 112 and 267). A theory from which very few empirical predictions can be made will be difficult to test and generally will not be very useful. A useful theory is rich or fecund, that is, many empirical predictions can be generated from it, each one serving as another test of the theory. Useful scientific theories lead to new lines of investigation and new models of understanding phenomena that heretofore have seemed unrelated (Kitcher 1983).

However, even if a theory is very rich and even if it passes many severe tests, it is always possible that it will fail the next test or that some other theory will be proposed that explains things even better. Logically speaking, a currently accepted scientific theory could even fail the same tests it has passed many times in the past. Karl Popper calls this characteristic of scientific theories *falsifiability.*

A necessary consequence of falsifiability is *fallibility.* Scientific facts, like scientific theories, are not infallible certainties. Facts involve not only easily testable perceptual elements; they also involve interpretation. We should remember that science, as Jacob Bronowski put it, "is a very human form of knowledge. . . . Every judgment in science stands on the edge of error. . . . Science is a tribute to what we can know although we are fallible" (Bronowski 1973: 374). "One aim of the physical sciences," he said, "has been to give an exact picture of the material world. One achievement of physics in the twentieth century has been to prove that aim is unattainable" (p. 353).

Scientific knowledge is human knowledge and scientists are human beings. They are not gods, and science is not infallible. Yet the general public often thinks of scientific claims as absolutely certain truths. They think that if something is not certain, it is not scientific, and if it is not scientific, then any other nonscientific view is its equal. This misconception seems to be, at least in part, behind the general lack of understanding about the nature of scientific theories.

Another common misconception is that since scientific theories are based on human perception, they are necessarily relative and therefore do not tell us anything about the real world. Science, according to certain postmodernists, cannot claim to give us a true picture of what the empirical world is really like; it can tell us only how it appears to scientists. There is no such thing as scientific truth. All scientific theories are mere fictions. However, just because there is no one, true, final, godlike way to view reality does not mean that every viewpoint is as good as every other. Just because science can give us only a *human* perspective does not mean that there is no such thing as scientific truth. When the first atomic bomb went off as some scientists had predicted it would, another bit of truth about the empirical world was revealed. Bit by bit we are discovering what is true and what is false by empirically testing scientific theories. To claim that those theories that make it possible to explore outer space and the inner atom are "just relative" and "represent just one perspective" of reality is to profoundly misunderstand the nature of science and scientific knowledge.

The late paleoanthropologist and science writer Stephen Jay Gould reminds us that in science "fact" can mean only "confirmed to such a degree that it would be perverse to withhold provisional assent"

(Gould 1983: 254). However, facts and theories are different things, notes Gould, "not rungs in a hierarchy of increasing certainty. Facts are the world's data. Theories are structures of ideas that explain and interpret facts." In Popper's words: "Theories are nets cast to catch what we call 'the world': to rationalize, to explain, and to master it. We endeavor to make the mesh ever finer and finer."

To the uninformed public, facts contrast with theories. Nonscientists commonly use the term "theory" to refer to a speculation or guess based on limited information or knowledge. However, when we refer to a scientific theory, we are referring not to a speculation or guess but to a systematic explanation of some range of empirical phenomena. Nevertheless, scientific theories vary in degree of certainty from the highly improbable to the highly probable. That is, there are varying degrees of evidence and support for different theories; some are more reasonable to accept than others.

There are, of course, many more facts than theories, and once something has been established as a scientific fact (e.g., that the earth goes around the sun), it is not likely to be replaced by a better fact in the future. But the history of science clearly shows that scientific theories do not remain forever unchanged. The history of science is, among other things, the history of theorizing, testing, arguing, refining, rejecting, replacing, more theorizing, more testing, and so on. It is the history of theories working well for a while, **anomalies** occurring (new facts being discovered that do not fit with established theories), and new theories being proposed and eventually replacing the old ones partially or completely (Kuhn 1996). It is also the history of rare geniuses—such as Newton, Wallace, Darwin, and Einstein—finding new and better ways of explaining natural phenomena.

Science is, as Carl Sagan put it, a candle in the dark. It shines a light on the world around us and allows us to see beyond our superstitions and fears, beyond our ignorance and delusions, and beyond the magical thinking of our ancestors, who rightfully fought for their survival by fearing and trying to master occult and supernatural powers.

Jacob Bronowski put it all in perspective in one scene from his televised version of the *Ascent of Man*. In the episode on "Knowledge and Certainty," he went to Auschwitz, walked into a pond where the ashes were dumped, bent down and scooped up a handful of muck, and said:

> It is said that science will dehumanize people and turn them into numbers. That is false, tragically false. Look for yourself. This is the concentration camp and crematorium at Auschwitz. This is where people were turned into numbers. Into this pond were flushed the ashes of some four million people. And that was not done by gas. It was done by ignorance. When people believe that they have absolute knowledge, with no test in reality, this is how they behave. This is what men do when they aspire to the knowledge of gods. (Bronowski 1973: 374)

See also **naturalism** and **pseudoscience**.

Further reading: Beveridge 1957; Copi 1998; Friedlander 1972; Gardner 1981; Giere 1998; Gould 1979, 1982; Koertge 1998; Sagan 1979, 1995; Sokal and Bricmont 1998.

scientism

In the strong sense, scientism is the self-annihilating view that only scientific claims are meaningful, which is not a scientific claim and hence, if true, not meaningful. Scientism is either false or meaningless.

In the weak sense, scientism is the view that the methods of the natural **sciences** should be applied to any subject matter.

See also **naturalism**.

Scientology

See **Dianetics**.

scrying

A type of **divination**. To scry or descry is to spy out or discover by the eye objects at a distance. In **occult** literature, the term is used to describe the act of gazing at a shiny stone or mirror or into a **crystal** ball to see things past and future. Occultists claim that if one concentrates hard enough while scrying, one can conjure up the dead by clearing out the consciousness, which opens a direct line to the other world.

See also Raymond **Moody**.

Further reading: de Givry 1971; Randi 1995.

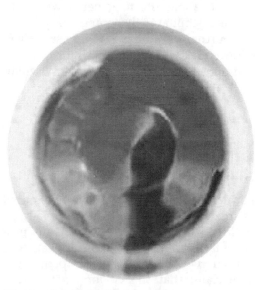

"Scrying Spinoza's God."

séance

A **spiritualist** meeting, allegedly to receive communications from the dead. Usually, a group is led by a **medium** in a very dark room (to make deception easier) who, often with an assistant, produces noises and voices or objects, and moves things about the room, insisting these are caused by spirits of the dead.

See also **apport** and **automatic writing**.

Further reading: Keene 1997; Rawcliffe 1988.

second sight

See **clairvoyance**.

selective thinking

The process whereby one selects favorable evidence for remembrance and focus while ignoring unfavorable evidence for a belief. The success of **cold readings** by **mentalists** and **psychics** depends on the selective thinking of their subjects.

See also **confirmation bias**, **self-deception**, and **wishful thinking**.

Further reading: Gilovich 1993; Martin 1998; Randi 1982a.

self-deception

The process or fact of misleading ourselves to accept as true or valid what is false or invalid. Self-deception, in short, is a way we justify false beliefs to ourselves. Self-deception is quite common. For example, 94% of university professors think they are better at their jobs than their colleagues; 25% of college students believe they are in the top 1% in terms of their ability to get along with others; and 70% of college students think they are above average in leadership ability. Only 2% think they are below average (Gilovich 1993: 77). Some 85% of medical students think it is improper for politicians to accept gifts from lobbyists. Only 46% think it's improper for physicians to accept gifts from drug companies (Dr. Ashley Wazana, *Journal of the American Medical Association,* 283, no. 3, January 19, 2000, p. 273).

When philosophers and psychologists discuss self-deception, they usually focus on unconscious motivations and intentions. They also usually consider self-deception a bad thing, something to guard against. To explain how self-deception works, they focus on **wishful thinking**, self-interest, prejudice, desire, insecurity, and other psychological factors unconsciously affecting in a negative way the will to believe. A belief so motivated is usually considered more flawed than one due to lack of ability to evaluate evidence properly. The former is considered to be a kind of *moral* flaw, dishonest and irrational. The latter is considered to be a matter of fate: Some people are just not gifted enough to make proper inferences from the data of perception and experience.

Fortunately, it is not necessary to know whether self-deception is due to unconscious motivations or cognitive incompetence in order to know that there are certain situations where self-deception is so common that we must systematically take steps to avoid it. Such is the case with all extraordinary beliefs. We must be on guard against the tendencies to

- misperceive random data and see patterns where there are none (**apophenia**);
- misinterpret incomplete or unrepresentative data and give extra attention to confirmatory data while drawing conclusions without attending to or seeking out disconfirmatory data (**confirmation bias**);

- make biased evaluations of ambiguous or inconsistent data, tending to be uncritical of supportive data and very critical of unsupportive data (**selective thinking**). (Gilovich 1993)

It is because of these tendencies that scientists require clearly defined, double-blind, randomized, repeatable, publicly presented **controlled studies**.

Many people think that as long as they use their intelligence, they are unlikely to deceive themselves. Yet many intelligent people have invested in numerous fraudulent products that promised to save money, the environment, or the world. Likewise, many intelligent people believe in **astrology**, **biorhythms**, **crystal power**, **dowsing**, **ESP**, **facilitated communication**, the **miracles** of **Sai Baba**, **therapeutic touch**, and a host of other notions that seem to have been clearly refuted by the scientific evidence.

On the other hand, self-deception may not always be a flaw and may even be beneficial at times. If we were too brutally honest and objective about our own abilities and about life in general, we might become debilitated by depression.

Further reading: Fingarette 2000; Gilovich 1993; Kahane 1997; Kruger and Dunning 1999; Taylor 1989; Wiseman 1997.

sensitive

See **intuitives**.

shark cartilage

Powdered shark cartilage has been touted as an alternative cancer cure, especially by William Lane, Ph.D., whose company produces it under the name BeneFin. Lane has written two books, both trumpeting the false claim that sharks don't get cancer.

Sharks do get cancer, even cartilage cancer. On June 29, 2000, Lane was prohibited by the Federal Trade Commission from claiming that "BeneFin or any other shark cartilage product prevents, treats or cures cancer" until he has substantial evidence to support his claims.

Dr. Lane is an example of why alternative medicine is usually either useless or harmful. He took a little bit of knowledge, generalized from it, started a company to produce the miracle cure, wrote books and misleading promotional pieces supporting his company's research and product, and got a major news show to do a shoddy, uncritical story that suggested that maybe there was something to the miracle cure. Finally, he responded to criticism with the claim that his critics were conspiring to stifle him because his research was somehow a threat to traditional medical practitioners.

There are no scientific studies done by independent researchers with proper controls that have substantiated the claim that shark cartilage is a useful treatment for cancer or anything else. In at least one such study, the treatment was found ineffective. Other studies are under way.

Lane got his inspiration from the work of real scientists who injected bovine and shark cartilage into the bloodstreams of rabbits and mice with cancer. The injections greatly inhibited angiogenesis, the growth of blood vessels that supply nutrients to the cancerous cells. However, not all cancers rely on angiogenesis. Most researchers doubt that cartilage taken orally will result in significant quantities making it to the site of a tumor. They believe that there is a protein in cartilage that affects angiogenesis and that the protein would be digested rather than absorbed into the bloodstream, where it might find its way to a tumor. However, injecting shark cartilage directly into the human bloodstream might result in an unfavorable immune system response.

See also **alternative health practices**.

Further reading: Kava 1995; McCutcheon 1997.

shoehorning

The process of force-fitting some current event into one's personal, political, or religious agenda. So-called **psychics** frequently shoehorn events to fit vague statements they made in the past. This is an extremely safe procedure, since they can't be proven wrong and many people aren't aware of how easy it is to make something look like confirmation of a claim *after* the fact, especially if you give them wide latitude in making the "shoe" fit.

After the terrorist attacks on the World Trade Center and the Pentagon on September 11, 2001, fundamentalist Christian evangelists Jerry Falwell and Pat Robertson shoehorned the events to fit their agenda. They claimed that "liberal civil liberties groups, feminists, homosexuals and abortion rights supporters bear partial responsibility . . . because their actions have turned God's anger against America." According to Falwell, God allowed "the enemies of America . . . to give us probably what we deserve." Robertson agreed. The American Civil Liberties Union has "got to take a lot of blame for this," said Falwell, and Robertson agreed. Federal courts bear part of the blame, too, said Falwell, because they've been "throwing God out of the public square." Also "abortionists have got to bear some burden for this because God will not be mocked," said Falwell, and Robertson agreed.

Neither Falwell nor Robertson has any way of proving any of their claims, as they are by their nature unprovable. But such claims can't be *dis*proved, either. Their main purpose is to call attention to their agenda and get free publicity in the news media.

See also **confirmation bias**, **retroactive clairvoyance**, and **self-deception**.

shotgunning

A **cold reading** trick used by **psychics** whereby one rapidly asks questions and seems to provide a large quantity of information to a subject. The subject actually supplies most of the information. Then, due to **selective thinking**, the subject often becomes convinced that the psychic is truly in touch with the other world.

For example, a dialogue between a psychic and a subject might go like this:

P. I'm getting a warm feeling in my stomach.
S. Uh huh.
P. Does that ring a bell?
S. No.
P. Are you sure? It could be a sign. Maybe something in the heart. . . . a definite pain in the chest . . .
S. My father had a heart attack before he died.
P. Right. He says he thought it was just gas. It was a sudden death, right?
S. Right.
P. No warning. Right?
S. Right. Well, actually he'd had two heart attacks before the final one.
P. Right. The feeling I'm getting is very strong in the chest. Like he had been suffering there a long time . . . years maybe . . .
S. Right.
P. I'm sensing something shiny now. . . . jewelry perhaps . . . maybe a watch . . . something . . .
S. We buried him with his Club Med pin. He wanted to take it with him.
P. Right. I see that. He wanted to be buried with a special pin.

S. Yes, his Club Med pin.
P. He loved Club Med.
S. No, but always wanted to take a vacation.
P. Well, he says he's enjoying his vacation now.
(Laughter and fade.)

shroud of Turin

A woven cloth about 14 feet long and 3.5 feet wide with an image of a man on it. Actually, it has two images, one of the back of the body and the other of the front, so that when the shroud is laid out flat one sees the tops of the heads meeting in the middle. If the shroud were really wrapped over a body there should be a space where the two heads meet. Nevertheless, the image is believed by many to be a negative image of the crucified Christ, and the

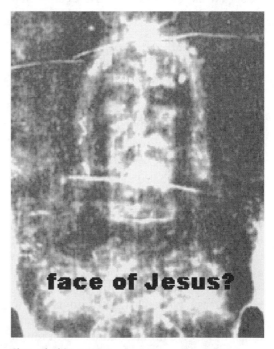

Shrouded in mystery.

shroud is believed to be his burial shroud. Most skeptics think the image is a painting and a **pious fraud**, and agree with the following assessment:

> All empirical evidence and logical reasoning concerning the Shroud of Turin will lead any objective, rational person to the firm conclusion that the Shroud is an artifact created by an artist in the fourteenth-century. (Schafersman 1998)

The shroud is kept in the cathedral of St. John the Baptist in Turin, Italy. Apparently, the first historical mention of the "shroud of Turin" is in the late 16th century, when it was brought to the cathedral in that city, though it allegedly was discovered in Turkey during one of the "Holy" Crusades. In 1988, the Vatican allowed the shroud to be dated by three independent sources: Oxford University, the University of Arizona, and the Swiss Federal Institute of Technology. Each dated the cloth around 1350. The shroud allegedly was in a fire during the early part of the 16th century, and according to believers in the shroud's authenticity, that is what accounts for the carbon dating of the shroud as being no more than 650 years old. Dr. Walter McCrone calls this "ludicrous" because "the linen cloth samples were very carefully cleaned before analysis at each of the C-dating laboratories" and the samples "are routinely and completely burned to CO_2 as part of a well-tested purification procedure" (www.mcri.org/Shroud.html).

Skeptics believe that the shroud of Turin is just another religious relic invented to beef up the pilgrimage business or impress infidels. The case for the forged shroud is made most forcefully by Joe Nickell in his *Inquest on the Shroud of Turin* (1987), which was written in collaboration with a panel of scientific and technical experts. The author claims that historical, iconographic, pathological, physical, and chemical evidence points to inauthenticity. The shroud is a 14th-century painting, not a 2,000-year-old cloth with Christ's image.

Author and microchemist Walter McCrone (1999) has also declared the shroud is a fake. His thesis is that "a male model was daubed with paint and wrapped in the sheet to create the shadowy figure of Christ." The model, covered in red ochre, "a pigment found in earth and widely used in Italy during the Middle Ages . . . pressed his forehead, cheekbones and other parts of his head and body on to the linen to create the image that exists today. Vermilion paint, made from mercuric sulfide, was then splashed onto the image's wrists, feet and body to represent blood." For his work, McCrone was awarded the American Chemical Society's Award in Analytical Chemistry.

The shroud has many defenders, however, who believe they have demonstrated that the cloth is not a forgery, dates from the time of Christ, is of miraculous origin, and has Christ's blood on it. Defenders of the shroud's authenticity claim that the shroud contains a negative image of a crucifixion victim, of a man brutally beaten in a way that corresponds to the way Jesus is thought to have been treated. It is claimed that there is type AB blood on the shroud. Skeptics deny it. Blood has not been identified on the shroud directly, but it has been identified on sticky tape that was used to lift fibrils from the shroud (Nickell 1993). Dried, aged blood is black. The stains on the shroud are red. Forensic tests have identified them as red ocher and vermilion tempera paint. However, tests by John Adler and Alan Heller, two scientists who have examined the shroud and attest to its being 2,000 years old, have identified blood on the shroud. If it is blood, it could be the blood of anyone who has handled or come near the shroud since its origin and would have no bearing on the age of the shroud or on its authenticity.

It is claimed that the cloth has some pollen and images on it that are of plants found only in the Dead Sea region of Israel. Avinoam Danin, a botanist from Hebrew University of Jerusalem, claims he has identified pollen from the tumbleweed *Gundelia tournefortii* and a bean caper on the shroud (www.shroud.com/danin.htm). He claims this combination is found only around Jerusalem. Some believers think the crown of thorns was made of this type of tumbleweed. However, Danin did not examine the shroud itself. His sample of pollens originated with Max Frei, who tape-lifted pollen samples from the shroud. Frei's pollens have been controversial from the beginning. Frei, who once pronounced the forged "Hitler Diaries" genuine, probably introduced the pollens himself or was duped and innocently picked up pollens another pious fraud had introduced (Nickell 1987; Schafersman 1998).

Danin and his colleague Uri Baruch also claim that they found impressions of flowers on the shroud and that those flowers could come only from Israel. However, the floral images they see are hidden in mottled stains much the way the image of Jesus is hidden in a tortilla or the image of Mary is hidden in the bark of a tree. The first to see flowers in the stains was a psychiatrist, who was probably an expert at seeing personality traits in inkblots (Nickell 1994a).

Danin notes that another relic believed to be the burial face cloth of Jesus (the Sudarium of Oviedo in Spain) contains the same two types of pollen grains as the shroud of Turin and also is stained with type AB blood. Since the Sudarium is believed to have existed before the 8th century, according to Danin, there is "clear evidence that the shroud originated before the eighth century." The cloth is believed to have been in a chest of relics from at least the time of the Moorish invasion of Spain. It is said to have been in the chest when it was opened in 1075. But since there is no blood on the shroud of Turin and there is no good reason to accept Danin's assumption that the pollens were on the shroud from its origin, this argument is spurious.

In any case, the fact that pollens found near the Dead Sea or Jerusalem were on the shroud means little. Even if the pollens weren't introduced by some pious fraud, they could have been carried to the shroud by anyone who handled it. In short, the pollens could have originated in Jerusalem at any time before or after the appearance of the shroud in Italy.

Moreover, that there are two cloths believed to have been wrapped around the dead body of Jesus does not strengthen the claim that the shroud is authentic but weakens it. How many more cloths are there that we don't know about? Were they mass-produced like pieces of the true cross, straw from Christ's manger, and chunks of Noah's ark? That cloths in Spain and Italy have identical pollens and bloodstains is a bit less than "clear evidence" that they originated at the same time, especially since there is clear evidence that the claim that they have identical pollens and bloodstains is not true. But even if it were true, it would be of little value in establishing that either of these cloths touched the body of Jesus.

To the believer, however, it is not the scientific proof of the shroud's authenticity that gives the shroud its special significance. It is the **faith** in the miraculous origin of the image that defines their belief. The **miracle** is taken as a sign that the resurrection really happened and that Jesus was divine. Even if it is established beyond any reasonable doubt that the shroud originated in Jerusalem and was used to wrap up the body of Jesus, would that prove Jesus rose from the dead? Only religious faith can sustain such a belief. No amount of physical evidence could ever demonstrate that a man was God, was also his own Father, and was conceived without his mother ever having had sex. Thus, no matter how many brilliant scientists marshal forth their brilliant papers with evidence for images of Biblical ropes, sponges, thorns, spears, flowers, tumbleweeds, blood, and so on, none of it has the slightest relevance for proving these matters of faith.

Further reading: Nickell 1993.

Silva method (Silva mind control)

A self-help program for increasing one's IQ, developing **clairvoyance**, and using the mind to heal the body and find God, among other things. The program promises to teach you to "use the untapped power of your mind to accomplish whatever you desire" (www.silvamethod.com/about.htm). The program is a hodgepodge put together by trial and error by José Silva (1914–1999), an electronics repairman who had a voracious appetite for literature in psychology, **parapsychology**, and religion. He studied **hypnosis**, hoping to use it to increase the IQ of his children, but became interested in developing **psychic** abilities after he became convinced that one of his daughters was clairvoyant.

According to Silva, he began using his method in 1944 on family and friends, but the program that now goes by his name started in the 1960s. He called his program "subjective education" and "psychorientology," which he defined as "Educating the mind to function consciously within its own psychic dimension . . . becoming aware of the enormity of human potential and learning how to actualize this potential for the betterment of humanity."

The instruction emphasizes positive thinking, visualization, meditation, and self-hypnosis. One key element of the course "consists of 'visiting' absent persons imagined by students and performing diagnoses on them" (Randi 1995: 218). Silva became convinced that most personal and world problems are due to "using only logical, intellectual, objective means to correct problems." He claimed that "only 10 percent of humanity think with the right brain hemisphere," and that these 10% are geniuses. The other 90% "use only the left brain hemisphere to think with. They do both their thinking and acting with only the left brain hemisphere."

Silva seems to have based his notions about the split-brain on the work of Roger Sperry and his colleagues. Silva, like many others who latched onto this split-brain model, seems to have modified it to suit his purposes and beliefs. Much work has been done on the brain since Sperry. Without putting too fine a point on it, nearly everything Silva said about the brain is wrong. For a more accurate picture of what scientists think about this split-brain distinction, see "Left Brain Right Brain" by John McCrone (*New Scientist,* July 3, 1999). The truth is that nobody thinks or acts only with their left hemisphere unless the right one is damaged or gone. In any case, it seems that Silva latched onto the split-brain theory after he had developed his subjective education program. I have no idea where he got the notion that geniuses don't use their left hemisphere.

Silva thought he had found not only the method to make people smarter, healthier, and happier, but also *why* his method works. It is because he trained his subjects to think with their right brain.

Those who think and act with only the left-brain hemisphere get sick more often with psychosomatic health problems. They are more accident-prone. They make more mistakes. They are less successful in life. When people think with the right brain hemisphere and act with the left, the results are just the opposite: They are healthier, less accident prone, make fewer mistakes, and are more successful in life. [www.silvacourse.com/josespeaks.html]

How he knew these things is not clear. He claimed that left-brainers are functioning at the beta wave level, while right-brainers are at the **alpha wave** level. Silva believed that alpha waves are significantly better than beta waves. Actually, alpha waves increase in meditation and under hypnosis, indicating lack of focus or visual stimulation, not some higher brain activity. Both sides of the brain produce alpha waves. The evidence is very speculative and tenuous that geniuses and creative artists are primarily functioning at the alpha wave level. Beta waves are evidence of concentration and heightened mental activity. If they are too extreme, they can be indicative of stress. But no one in his or her right mind should want to *eliminate* beta waves, unless he or she is trying to relax or go to sleep, in which case one would want fewer beta waves (and more alpha waves) if relaxation is desired, and more delta waves if sleep is desired.

Silva claimed that right-brain thinking allows people to use information stored in the **unconscious mind**. He thought that right-brain thinking psychically connects us to all information on the planet and connects us with higher intelligence, which he also refers to as "Christ consciousness." This allows us to get the guidance and help that we need in solving our problems. It is also what led some critics to think that the method is actually a religious movement.

The Silva method has many satisfied customers. Some claim their backhand in tennis has improved, others claim they were able to quit smoking or lose weight because of it. Still others say that they are happier and healthier than they've ever been, thanks to the Silva method. It's been around for over 40 years and claims to have instructors in 107 countries.

See also **pragmatic fallacy**.

Sitchin, Zecharia

Along with Erich von **Däniken** and Immanuel **Velikovsky**, Sitchin is one of the holy trinity of **pseudohistorians**. Each begins with the assumption that ancient myths are not myths but historical and scientific texts. Sitchin's claim to fame is announcing that he alone correctly reads ancient Sumerian clay tablets. All other scholars have misread these tablets, which, according to Sitchin, reveal that gods from another planet (Niburu, which orbits our Sun every 3,600 years) arrived on Earth some 450,000 years ago and created humans by genetic engineering of female apes. Niburu orbits beyond Pluto and is heated from within by radioactive decay, according to Sitchin. No other scientist has discovered that these descendents of gods blew themselves up with nuclear weapons some 4,000 years ago. Sitchin alone can look at a Sumerian tablet and see that it depicts a man being subjected to radiation. He alone knows how to correctly translate ancient terms allowing him to discover such things as that the ancients made rockets. Yet he doesn't seem to know that the seasons are caused by the earth's tilt, not by its distance from the sun.

Sitchin was born in Russia, was raised in Palestine, and graduated from the University of London with a degree in economic history. He worked for years as a journalist and editor in Israel before settling in New York.

Sitchin, like Velikovsky, presents himself as erudite and scholarly in a number of books, including *The Twelfth Planet* (1976) and *The Cosmic Code* (1998). Both Sitchin and Velikovsky write knowledgeably of ancient myths and both are nearly scientifically illiterate. Like von Däniken and Velikovsky, Sitchin weaves a compelling and entertaining story out of facts, misrepresentations, fictions, speculations, misquotes, and mistranslations. Each begins with their beliefs about ancient visitors from other worlds and then proceeds to fit facts and fictions to their basic hypotheses. Each is a master at ignoring inconvenient facts, making mysteries where there were none before and offering their alien hypotheses to solve the mysteries. Their works are very attractive to those who love a good mystery and are ignorant of or indifferent both to scientific knowledge and to the nature and limitations of scientific research.

Sitchin promoted himself as a Biblical scholar and master of ancient languages, but his real mastery was in making up his own translations of Biblical texts to support his readings of Sumerian and Akkadian writings:

> He's let us know he's going to twist the translations around to support his thesis. Indeed, a reader of Sitchin's book would do well to keep a couple of Bibles handy to check up on the verses Sitchin quotes. Many of them will sound odd or unrecognizable because they have been translated from their familiar form (this is made harder by the fact that Sitchin rarely tells you just which verse he is quoting). This would be much more acceptable if he wasn't using the twisted translations to support the thesis that led to the twisted translations. [Hafernik: www.geocities.com/Area51/Corridor/ 8148/hafernik.html]

Most of Sitchin's sources are obsolete. He received nothing but ridicule from scientific archaeologists and scholars familiar with ancient languages. His most charming qualities seem to be his vivid imagination and his complete disregard for established facts and methods of inquiry, traits that are apparently very attractive to some people.

See also **Raël**.

Further reading: Heiser: www .facadenovel.com/sitchinerrors.htm; Paynter: www.geocities.com/Area51/Corridor/ 8148/zindex.html.

sixth sense

A term sometimes used to refer to **psychic** abilities such as **channeling**, **clairaudience**, **ESP** (**telepathy**, **clairvoyance**, or **precognition**), or **telekinesis**.

Intuitives think they possess the sixth sense.

sleep paralysis

A condition that occurs just before a person drops off to sleep (**hypnagogic state**) or just before they fully awaken from sleep (**hypnopompic state**). The condition is characterized by being unable to move or speak. It is often associated with a feeling that there is some sort of presence, a feeling that often arouses fear but is also accompanied by an inability to cry out. The paralysis may last only a few seconds. The description of the symptoms of sleep paralysis is similar to the description many alien abductees give in recounting their abduction experiences. Sleep paralysis is thought by some to account for not only many **alien abduction** delusions, but also other delusions involving **paranormal** or supernatural experiences.

Sleep paralysis is something many people experience once or twice in a lifetime, but it is a frequent occurrence of those suffering from narcolepsy.

Further reading: Blackmore 1998; Hufford 1989.

Sokal hoax

In its 1996 spring/summer issue, the journal *Social Text* published an article by Allan Sokal, a professor of physics at New York University, entitled "Transgressing the Boundaries: Towards a Transformative Hermeneutics of Quantum Gravity" (pp. 217–252). According to Sokal, the article was a hoax submitted to see whether "a leading journal of cultural studies" would "publish an article liberally salted with nonsense if (a) it sounded good and (b) it flattered the editors' ideological preconceptions."

Sokal claims that the editors, had they been scrupulous and intellectually competent, would have recognized from the first paragraph of his essay that it was a parody. The physicist says he was "troubled by an apparent decline in the standards of intellectual rigor in certain precincts of the American academic humanities." The hoax was his way of calling attention to this decline.

In his article, Sokal attacks "the dogma imposed by the long post-Enlightenment hegemony over the Western intellectual outlook" that there is an external world governed by laws of nature which we can understand imperfectly using the scientific method. He also claims that "physical 'reality' . . . is at bottom a social and linguistic construct." Furthermore, he says,

> Throughout the article, I employ scientific and mathematical concepts in ways that few scientists or mathematicians could possibly take seriously.
>
> For example, I suggest that the "morphogenetic [*sic*] field"—a bizarre New Age idea due to Rupert Sheldrake—constitutes a cutting-edge theory of quantum gravity. This connection is pure invention; even Sheldrake makes no such claim. I assert that Lacan's psychoanalytic speculations

have been confirmed by recent work in quantum field theory. Even nonscientist readers might well wonder what in heavens' name quantum field theory has to do with psychoanalysis; certainly my article gives no reasoned argument to support such a link.

In sum, I intentionally wrote the article so that any competent physicist or mathematician (or undergraduate physics or math major) would realize that it is a spoof. Evidently the editors of *Social Text* felt comfortable publishing an article on quantum physics without bothering to consult anyone knowledgeable in the subject. (Sokal 1996)

Such lax editing might be expected in a New Age magazine, where preposterous and unfounded claims about **paranormal** or spiritual **energy** being validated by quantum mechanics are commonplace. But Sokal thinks we should expect more of a prestigious journal edited by distinguished scholars in the humanities. But why did he pick on this particular journal?

Sokal hoaxed *Social Text* for political reasons. Both he and the journal are leftist politically, but Sokal considers the New Left to be guilty of "epistemic relativism." (Is this another hoax?) He seems particularly annoyed that the New Left promotes the notion that reality is a social construction. Furthermore, the New Left has created "a self-perpetuating academic subculture that typically ignores (or disdains) reasoned criticism from the outside." Apparently, Sokal wanted to criticize the "epistemic relativism" and "social constructivism" of the New Left in a New Left journal but felt the only way they would let him do so would be if he pretended to share their ideology.

Many have pointed out the profound implications of this hoax. At the very least, articles should be reviewed by experts in the field covered by the article. The editors should check sources and references named in the article.

Above all, however, the Sokal hoax demonstrates how willing we are to be deceived about matters in which we strongly believe. We are likely to be more critical of articles that attack our position than we are of those that we think support it (Gilovich 1993). We should not forget, however, that this tendency to **confirmation bias** affects physicists as well as professors in the social sciences and the humanities.

See also **communal reinforcement**, **self-deception**, **subjective validation**, and **wishful thinking**.

Further reading: Gross and Levitt 1997; Koertge 1998; *Lingua Franca* 2000.

sorcery

Literally, "**divination** by casting lots" (from the Latin *sortiarius,* one who casts lots). Sorcery is also often identified with **witches** and black magic, both of which involve getting power from association with evil spirits or **Satan**. Sorcery is often associated with using **magick** potions and casting **spells**.

soul (spirit)

A nonphysical entity capable of perception and self-awareness. Souls are often believed to be immortal.

If ever there were an entity invented for human wish fulfillment, the soul is it. Thomas Hobbes claimed that the concept of a nonsubstantial substance is a contradiction. He could not imagine a nonphysical entity having any properties, much less thoughts and perceptions. Yet billions of people have believed in a nonspatial perceiver that can perceive without any sense organs. Descartes' *Metaphysical Meditations* tries to convince us that the mind, not the body, is what has sense perception. Leibniz

conceived of an infinite variety of bodiless perceivers. Most skeptics, however, think that belief in a nonphysical perceiver requires an act of **faith**.

See also **astral projection** and **dualism**.

Further reading: Freud 1927; Hobbes 1660; O'Connor 1993; Ryle 1949; Sacks 1984, 1985.

speed reading

The purported ability to read thousands of words a minute. For example, Howard Berg claims to be able to read 25,000 words a minute by reading "15 lines at a time backwards and forwards." That's about 80 to 90 pages a minute. Tolstoy's *War and Peace* should take Berg about 15 minutes to read.

George Stancliffe claims he has taught a woman with a reading disability to read 18,000 words a minute. Such a feat, he says, is common in children, but rare in adults (www.selfgrowth.com/articles/stancliffe.html). One wonders what concern Berg or Stancliffe have for comprehension.

There seems to be only one person who can read at such speeds with near-perfect comprehension. His name is Kim Peek, and he has the ability to read two pages simultaneously, one with each eye, with 98% retention. Nobody knows how he does it, but he was born without a corpus callosum, that bundle of nerves that connects the right and left hemispheres of the brain. However, others have also been born with no corpus callosum, or have had it surgically disconnected, without resulting in an increase in reading or retention abilities. Kim Peek can recall most of the contents of some 7,600 books. But since nobody knows how Kim Peek does it, nobody can teach this skill to others. The rest of us must focus both eyes on one page and read at a much slower rate than Peek. (Peek was partly the model for Raymond, the *savant* played by Dustin Hoffman in the movie *Rain Man*.)

Anne Cunningham, a University of California at Berkeley education professor and an expert on reading, reports that tests measuring saccades (small rapid jerky movements of the eye as it jumps from fixation on one point to another) while reading have determined that the maximum number of words a person can accurately read is about 300 a minute. "People who purport to read 10,000 words a minute are doing what we call skimming," she said. Speed in reading is mainly determined by how fast a reader can understand the words and expressions one is reading. The fastest readers are those with excellent "recognition vocabularies." Faster readers can see words and understand them faster than slower readers. To improve one's speed at reading, she says, one should work on comprehension and study strategies (Robertson 1999).

Others claim that "the average college student reads between 250 and 350 words per minute on fiction and non-technical materials" and that a good reading speed is 500 to 700 words per minute (www.ucc .vt.edu/stdysk/suggest.html). It does seem intuitively true that one could speed up one's reading by spending less time between eye movements, taking in more words with each fixation, and always moving forward, rather than skipping back to reread something. Having a good recognition vocabulary would certainly speed these processes up. Conscious practice at improving one's speed should also help.

Berg has repackaged the Evelyn Woods Reading Dynamics course, one popular several decades ago with people such as John F. Kennedy. A reporter who attended one of Berg's classes noted that in his 5-hour course, Berg hadn't said much about comprehension, except to suggest that it would come with practice. This did not deter several of the 35 students, who had paid $51 each for the class, from purchasing audiotapes for $65 (Robertson 1999).

Those desiring to increase the speed of their reading would do better to enroll in a community college course devoted to building study skills, vocabulary, and reading comprehension. It would cost them less, and they would not end up wasting their time trying to read 10 lines at a time, backward and forward. They would also avoid the frustration that will be inevitable when they find that while they can skim through material at a greater rate than they can read it, the utility of such a skill is limited—good for most of what's likely to be in the daily newspaper, for example, but not for studying physics or reading a good novel. Skimming makes both comprehension and taking pleasure in words or ideas next to impossible.

spell

A word, formula, or **incantation** believed to have magical powers for good or ill. Those who believe they can access occult powers or communicate with helpful spirits cast spells. Spells can be broken by counterspells or **exorcisms**.

See also **sorcery**.

spirit

See **soul**.

spiritualism (spiritism)

The belief that the human personality survives death and can communicate with the living through a sensitive **medium**. The spiritualist movement began in 1848 in upstate New York with the Fox sisters, who claimed that spirits communicated with them by rapping on tables. (The raps were actually made by cracking their toe joints.) By the time the sisters admitted their fraud some 30 years later, there were tens of thousands of mediums holding **séances** where spirits entertained with numerous magical tricks such as making sounds, materializing objects, making lights glow, levitating tables, and moving objects across the room. The mediums demonstrated every variety of **psychic** power, from **clairaudience** and **clairvoyance** to **telekinesis** and **telepathy**. Repeated charges of fraud did little to stop the spiritualist movement until the 1920s, when magicians such as Houdini exposed the techniques and methods of deceit used by mediums to fool even the wisest and holiest of men and women.

For many, spiritualism was scientific proof of life after death, and didn't involve any of the superstitious nonsense of religion. By the late 20th century, **channelers** and **clairaudients** had largely replaced spiritualism.

Further reading: Houdini 1924, 1981; Keene 1997; Randi 1982a, 1995; Rawcliffe 1988; Stein 1996b; Tanner 1994.

spontaneous human combustion (SHC)

The alleged process of a human body catching fire as a result of heat generated by internal chemical action. While no one has ever witnessed SHC, several deaths involving fire have been attributed to SHC by investigators and storytellers.

In the literature, spontaneous human combustion is almost exclusively reserved for corpses. One 17th-century tale, however, claims that a German man self-ignited due to his having drunk an excessive amount of brandy. If drinking a great quantity of spirits caused self-combustion, there should be many more cases to study than this isolated report about one man.

Many of the SHC stories have originated with police investigators who have been perplexed by partially ignited corpses near unburnt rugs or furniture. "What else could it be?" they ask. Many of the allegedly spontaneously combusted corpses are of elderly people who may have been murdered or who may have ignited themselves accidentally. Yet self-ignition due to dropping a lit cigarette or ignition due to another person is ruled out by the investigators as unlikely. Even when candles or fireplaces present a plausible explanation for the cause of a fire, investigators sometimes favor an explanation that requires belief in an event that has never been witnessed in all of human history and whose likelihood is extremely implausible.

The physical possibilities of SHC are remote. Not only is the body mostly water, but aside from fat tissue and methane gas, there isn't much that burns readily in a human body. To cremate a human body requires enormous amounts of heat over a long period of time. To get a chemical reaction in a human body that would lead to ignition would require some doing. If the deceased had recently eaten an enormous amount of hay that was contaminated with bacteria, enough heat might be generated to ignite the hay, but not much besides the gut and intestines would probably burn. Or if the deceased had been eating the newspaper and drunk some oil and was left to rot for a couple of weeks in a well-heated room, his gut might ignite.

It is true that the ignition point of human fat is low, but to get the fire going would probably require an external source. Once ignited, however, some researchers think that a "wick effect" from the body's fat would burn hot enough in certain places to destroy even bones. To prove that a human being might burn like a candle, Dr. John de Haan of the California Criminalistic Institute wrapped a dead pig in a blanket, poured a small amount of gasoline on the blanket, and ignited it. Even the bones were destroyed after 5 hours of continuous burning. The fat content of a pig is very similar to the fat content of a human being. The damage to the pig, according to Dr. De Haan, "is exactly the same as that from supposed spontaneous human combustion."

In their investigation of a number of SHC cases, Dr. Joe Nickell and Dr. John Fischer found that when the destruction of the body was minimal, the only significant fuel source was the individual's clothes, but where the destruction was considerable, additional fuel sources increased the combustion (Nickell and Fischer 1991). Materials under the body help retain melted fat that flows from the body and serves to keep it burning.

Further reading: Nickell 1996b; Williams 1998.

star child

A star child is the offspring of a human and an alien.

See also **Raël** and Zecharia **Sitchin**.

stichomancy

Literally, "**divination** from lines." It is the practice of seeking answers to the great metaphysical questions, as well as trying to gain insight into the meaning of existence and reality, by reading random passages from a book such as the Bible or the *I Ching*.

stigmata

Wounds believed to duplicate the wounds of Christ's crucifixion that appear on the hands and feet, and sometimes on the side and head, of a person. The fact that the stigmata appear differently on victims is strong evidence that the wounds are not genuinely miraculous (Wilson 1988).

St. Francis of Assisi (1182–1226), de-voted to imitate Christ in all ways, apparently inflicted himself with wounds and perpetrated the first stigmatic fraud. There have been many others since, including Magdalena de la Cruz (1487–1560) of Spain (who admitted her fraud when she became seriously ill) and Therese Neumann of Bavaria (1898–1962). The latter reportedly survived for 35 years eating only the "bread" of the Holy Eucharist at Mass each morning. She was also said to be **clairvoyant** and capable of **astral projection**. One of the more recent stigmatics, Fr. James Bruce, claimed that religious statues wept in his presence. This was in 1992 in a suburb of Washington, D.C., where strange things are common. Needless to say, he packed the pews. He now runs a parish in rural Virginia where the **miracles** have ceased.

Self-inflicted wounds are common among people with certain kinds of brain disorders. Claiming that the wounds are miraculous is rare and is more likely due to excessive religiosity than to a diseased brain, though both could be at work in some cases.

The likelihood that the wounds are psychosomatic (psychogenic purpuras), manifested by tortured souls, seems less likely than hoaxing in most cases. There are two main reasons for believing the stigmata are usually self-inflicted, rather than psychosomatic or miraculous. The first is that no stigmatic ever manifests these wounds from start to finish in the presence of others. Only when they are unwatched do they start to bleed. And the second is that Hume's rule in "Of Miracles" is that when an alleged miracle occurs, we ask ourselves which would be more miraculous, the alleged miracle or that we are being hoaxed? Reasonableness requires us to go with the lesser of two miracles, the less improbable, and conclude that we are witnessing not miracles but **pious frauds**. All 32 or so recorded cases of those with stigmata have been Roman Catholics, and all but four of those cases were women. No case of stigmata is known to have occurred before the 13th century, when the crucified Jesus became a standard icon of Christianity in the West. Reasonableness seems to require the nonmiraculous explanation.

One of the latest to be added to the list of alleged stigmatics is Audrey Santo, a child who has been in a coma since 1987, when she was 3 years old. Her mother calls her a **victim soul**. One might wonder who would be inspired by the concept of a God who renders a child comatose and then inflicts wounds on her? Joe Nickell thinks he has the answer:

> People seem to hunger for some tangible religious experience, and wherever there is such profound want there is the opportunity for what may be called "pious fraud." Money is rarely the primary motive, the usual impetus being to seemingly triumph over adversity, renew the faith of believers, and confound the doubters. (Nickell 1999)

People also don't want to think God would allow purposeless and gratuitous pain. They also like to feel special. What could be more special than being chosen to suffer the Savior's wounds and torments? What could please God more than being a living proof of God's existence?

Being truthful might be a good start.

Further reading: Nickell 1993, 1996a, 2000b.

subconscious

See **unconscious mind**.

subjective validation

See **Forer effect**.

subliminal

Literally, "below the liminal" (the smallest detectable sensation), that is, below the threshold of *conscious* perception.

There is a widespread belief, not strongly supported by empirical research, that a person's behavior can be significantly affected by subliminal messages. It is believed that one can influence behavior by surreptitiously appealing to the **unconscious mind** with words and images. If this were true, then advertisers could manipulate consumer behavior by hiding subliminal messages in their ads. The government, or Aunt Hildy for that matter, could control our minds and bodies by secretly communicating to us subliminally. Learners could learn while listening to music embedded with subliminal messages. Unfortunately, "years of research has resulted in the demonstration of some very limited effects of subliminal stimulation" and very little support for its efficaciousness in behavior modification (Hines 1990: 312). One study in "terror management," for example, found that subliminal reminders of death can significantly affect behavior, increasing defensive behaviors, including scapegoating (J. Arndt et al., "Subliminal Presentation of Death Reminders Leads to Increased Defense of the Cultural Worldview," *Psychological Science* 8 [1997]: 379–385).

The fact that there is little empirical support for the usefulness of subliminal messaging has not prevented numerous industries from producing and marketing tapes that allegedly communicate directly with the unconscious mind, encouraging the "listener" not to steal, to have courage, or to believe in his or her power to accomplish great things. Consumers spend more than $50 million each year on subliminal self-help products (*Journal of Advertising Research,* reported by Dennis Love, *Sacra-*

mento Bee, September 14, 2000). Yet a study of 237 motivated students who listened to audiotapes that were supposed to improve memory or enhance self-esteem through subliminal suggestions found "neither the memory nor the self-esteem tapes produced their claimed effects" (Greenwald et al., "Double-Blind Tests of Subliminal Self-Help Audio-Tapes," *Psychological Science* 2 [1991]: 119). One interesting finding of this study was that more than a third of the subjects were convinced the tapes worked as advertised. So, one might justifiably conclude that the tapes are not a total waste of money.

It is true that we can perceive things even though we are not conscious of perceiving them (see **memory** for a discussion of implicit memory and see **unconscious mind** for discussion of blindsight). However, for those who put messages in tapes and then record music over the messages so that the messages are drowned out by the music or other sounds, it might be useful to remember that if the messages are drowned out by other sounds, the only perceptions one can have are of the sounds drowning out the messages. There is no evidence of anyone hearing a message that is buried beneath layers of other sounds to the point where the message does not distinctly stand out. Of course, if the message distinctly stood out, it would not be subliminal.

The belief in the power of subliminal messaging to manipulate behavior seems to have originated in 1957 with James Vicary, an advertising promoter who claimed to increase popcorn sales by 58% and Coca-Cola sales by 18% in a New Jersey movie theater simply by flashing very briefly the messages "Drink Coca-Cola" and "Hungry—Eat Popcorn." Even though the claim has been shown to be a hoax, and even though no one has been able to duplicate the event, belief in the legend lingers. This story and several others were retold by

Vance Packard in *The Hidden Persuaders* (1957), a book that became required reading for a generation of college students.

Belief in subliminal messaging reached a surreal apex in 1980 with the publication of Wilson Bryan Key's *The Clam-Plate Orgy and Other Subliminals the Media Use to Manipulate Your Behavior.* The book has been reissued under the sexier title *Subliminal Adventures in Erotic Art.* Key claims that advertisers use subliminal messaging of a very serious sexual nature in order to manipulate behavior, including imbedding sexy figures and the word "sex" in images of things such as ice cubes and food. While carefully examining a Howard Johnson's menu, Key saw that the plate of clams pictured on the menu actually portrayed a sexual orgy. Among Key's many unfounded claims is that the unconscious mind processes subliminal messages at the speed of light. Actually, the fastest brain process chugs along at some 40 mph (Hines 1990).

See also **hypnosis**, **mind control**, and **pareidolia**.

Further reading: Volkey and Read 1985.

sunk-cost fallacy

When one makes a hopeless investment, one sometimes reasons: "I can't stop now, otherwise what I've invested so far will be lost." This is true, of course, but irrelevant to whether one should continue to invest in the project. Everything one has invested is lost regardless. If there is no hope for success in the future from the investment, then the fact that one has already lost a bundle should lead one to the conclusion that the rational thing to do is to withdraw from the project.

To continue to invest in a hopeless project is irrational. Such behavior may be a pathetic attempt to delay having to face the consequences of one's poor judgment. The irrationality is a way to save face, to appear to be knowledgeable, when in fact one is acting like an idiot. For example, it is now known that Lyndon Johnson kept committing thousands and thousands of U.S. soldiers to Vietnam after he had determined that the cause was hopeless and that the U.S. could never defeat the Viet Cong.

This fallacy is also sometimes referred to as the Concorde fallacy, after the method of funding the supersonic transport jet jointly created by the governments of France and Britain. Despite the fact that the Concorde is beautiful and as safe as any other jet transport, it was very costly to produce and suffered some major marketing problems. There weren't many orders for the plane. Even though it was apparent there was no way this machine would make anybody any money, France and England kept investing deeper and deeper, much to the dismay of taxpayers in both countries.

Further reading: Belsky and Gilovich 2000; Dawes 2001.

sympathetic magic

Sympathetic magic is the **metaphysical** belief that like affects like and is the basis for most forms of **divination**. The lines, shapes, and patterns in entrails, stars, thrown dirt, folded paper, or the palm of the hand are believed to be magically connected to the empirical world—past, present, and future. Sympathetic magic is also the basis for such practices as sticking needles into figurines representing enemies, as is done in voodoo. The pins and needles stuck in a doll are supposed to magically cause pain and suffering in the person the doll symbolizes.

Sympathetic magic is the basis of **psychometry**, the claim of **psychic detectives** that touching an item belonging to a victim gives them magical contact with the victim.

Psychologist Barry Beyerstein believes that sympathetic magic is the basis for many New Age notions such as "resonance," the idea that if things can be mentally associated, they can magically influence each other. Beyerstein also explains many notions of **graphologists** as little more than sympathetic magic (Beyerstein 1996b).

Sympathetic magic is probably the basis for such notions as **karma**, **synchronicity**, many theories on the interpretation of **dreams**, eating the heart of a brave but defeated warrior foe, throwing spears at painted animals on cave walls, wearing the reindeer's antlers before the hunt, having rape rituals to increase the fertility of the crops, or taking Holy Communion to infuse the participant with divinity. Sympathetic magic is surely the basis for **homeopathy**.

Anthropologists consider magical thinking a precursor to scientific thinking. Magical thinking shows a concern with control over nature through understanding cause and effect. Nevertheless, the methods of magic, however empirical, are not scientific. Such thinking may seem charming when done by our ancestors living thousands of years ago, but today such thinking may indicate a profound ignorance or indifference toward a scientific understanding of the world. Most of us, from time to time, undoubtedly slip into this primitive mode of thinking, but a bit of reflection should wake us up to the fact that oysters are not an aphrodisiac, having a bit of good luck is not likely to influence our chances of winning the lottery that day, and a crab tattoo is not going to cure cancer.

Further reading: Beyerstein and Beyerstein 1991.

synchronicity

See Carl **Jung**.

T

tachyon, takionics

A tachyon is a theoretical particle or wave that travels faster than the speed of light. Tachyons exist in a theoretical world where objects have negative mass and time goes backward. Tachyon energy is used to scan "subspace," among other things, on the sci-fi fantasy program, *Star Trek Voyager.* So far, there is no empirical evidence for the existence of tachyons. "If they do exist, tachyons would be extremely difficult to utilize under our current understanding of physics," says NASA scientist Tom Bridgman (imagine .gsfc.nasa.gov/docs/ask_astro/answers/ 970612b.html). Despite being theoretical and, if real, difficult to utilize, and if utilized, of unknown value, tachyons are the main ingredient in a line of New Age products that range from beads, belts, and shoe inserts to sweatbands, power pillows, massage oils, and vials of tachyon water. And of course, there are books, such as *Tachyon Energy: A New Paradigm in Holistic Healing* by Gabriel Cousens and David Wagner (2000).

A few enterprising New Agers claim that they know tachyons exist and they have harnessed their power. For example, Fred Pulver of the Carbondale Center for Macrobiotic Studies (formerly known as Biotech Industries of Carbondale Colorado) claims to know that "the Tachyon Field supplies the **energy** needs of all living organisms until balance is achieved, then it eases until called upon again. As it is needed, and a depletion occurs, it rushes in until balance is achieved once again" (www.macrobiotic.org/health11.html). But just in case nature fails to keep you in tachyon balance, you can get all the tachyon power you need from one of Pulver's takionic products. (The reason for the spelling difference has to do with the fact that common words such as "tachyon" usually cannot be trademarked, and "tachyonized" was already trademarked by Advanced Tachyon Technologies.)

Pulver makes some incredible claims. For example, "Motors have been built which draw upon the Tachyon Field for energy. They exhibit strange behavior, such as increasing in speed the longer they run, even though they are connected to no visible power source" (ibid.). Where are these motors? Perhaps they are in the **UFOs** hovering above Earth in search of cattle to mutilate, crops to carve, or people to experiment on.

According to Pulver, his takionic products,

> with their aligned atomic polarities, enhance the body's natural ability to draw from the Tachyon Field for its energy needs. Athletes have discovered that Takionic products allow them to perform faster and longer, and shorten recovery time. As conduits for input from the Tachyon Field, Takionic products are proving themselves in the sports performance arena. (ibid.)

Who are these athletes? The ones wearing magnetic shoes and crystals to ward off bad energy?

Pulver has developed an "alternative" physics to explain his products' magical qualities:

> The Tachyon Field is extremely dense. This density cannot be measured because it is a negative state, mirroring the universe of positive density which we inhabit. The theory of negative density is supported by an observable phenomenon: a perpetually expanding physical universe which is brought into being

through pressure exerted by expansion of the invisible one. Pressure exerted by the Tachyon Field upon our physical universe indicates the existence of an invisible, highly dense universe, the Tachyon Field. (ibid.)

The tachyon field may be dense, but it would have to go a long way to match the density of any "alternative" physicist who finds this gibberish compelling. There seems to be something missing here, such as a fundamental grasp of reality. We are asked to believe in an invisible universe causing pressure on ours, thereby forming energy that "cannot be proven by instrumentation currently available" but that we can use to explain just about anything we want, including divine omnipresence:

> Tachyon theory is holistic because it accepts the notion of two interdependent universes which are actually indivisible: the visible, sub-light speed universe and an invisible, faster-than-light one. Tachyon theory also substantiates omnipresence, a purely metaphysical concept. God is omnipresent (simultaneously existing everywhere). Omnipresent existence can only occur at faster-than-light speeds, since slower-than-light travel takes time to cross space. Therefore, omnipresence can only be an attribute of a Tachyon Universe where time and space are uniform. (ibid.)

Pulver is also an alternative metaphysician! The speculation seems infinite. "The nervous system and brain are nothing but an extremely sophisticated antenna and receiver to absorb, process and transform the resources of the Tachyon Field." "Healers have learned to access the Tachyon Field's resources for its healing powers more successfully than the average person has."

All the above claims are by way of introduction to Pulver's line of takionic products: There are takionic beads, 10 for $118.95, which are said to have antennae that "focus the beneficial tachyon energy." There is a takionic belt for $268.95, which is said to help improve circulation and increase strength. Takionic water comes in a small vial for $27.95 and is "pure and cluster-free."

Another New Age business, Advanced Tachyon Technologies (ATT) of Santa Rosa, California, offers a much broader array of products, including some for cats and dogs. ATT has products that can enhance your love life (including one called "panther juice") and your athletic skills, not that the two are mutually exclusive. They have products to ease your pain and improve your brain. They have **chakra** balancing kits and a Tachyonized Silk Meditation Wrap to help you "meditate with your lover before making love."

Further reading: Park 1997; Pickover 1998; Wall 1995.

talismans

Cut figures or engravings, such as on a coin, which have magical powers to avert evil or bring about good.

See also **amulets**, **charms**, and **fetishes**.

Tantra

A type of Hindu or Buddhist scripture, or the rituals and practices described therein. They deal especially with meditative techniques and rituals involving sexual practices.

tarot cards

Cards used in fortune-telling. A few years ago, tarot cards would have conjured up images of Gypsies, who didn't begin using tarot cards until the 20th century. Today, the cards are popular among occultists and

New Agers in all walks of life. According to Grillot de Givry (1971),

> The tarot is one of the most wonderful of human inventions. Despite all the outcries of philosophers, this pack of pictures, in which destiny is reflected as in a mirror with multiple facets, remains so vital and exercises so irresistible an attraction on imaginative minds that it is hardly possible that austere critics who speak in the name of an exact but uninteresting logic should ever succeed in abolishing its employment.

The modern tarot deck has been traced back to 15th-century Italy and a trick-taking game called "triumphs" (*tarots* in French; Decker 1996). The traditional tarot deck consists of two sets of cards, one having 22 pictures (the major arcana), such as the Fool, the Devil, Temperance, the Hermit, the Sun, the Lovers, the Hanged Man, and Death. The other set (the minor arcana) has 56 cards with kings (or lords), queens (or ladies), knights, and knaves (pages or servants) of sticks (or wands, cudgels, or batons), swords, cups, and coins.

There are many different tarot decks used in **cartomancy**. The meanings of the figures and numbers on tarot cards vary greatly among tarot readers and advocates, many of whom find connections between tarot and **astrology**, the *I Ching*, ancient Egypt, and various occult and mystical notions.

The oldest playing cards date back to 10th-century China, but the four suits of tarot and modern playing cards probably originated with a 14th-century Muslim deck (ibid.). According to de Givry, in the modern 52-card deck of ordinary playing cards, sticks or wands = clubs (and announce news); swords = spades (and presage unhappiness and death); cups = hearts (and presage happiness); coins = diamonds (and presage money). According to Ronald Decker, the Muslim sticks represented polo sticks. As Europeans were not yet familiar with polo, they changed the suit of sticks to that of wands, cudgels, or batons.

Tarot cards are usually read by a fortune-teller, though in these days of New Age enterprise, anyone can buy a deck with instructions on how to discover your real self and actualize your true potential. Why anyone's fate would be mysteriously contained in playing cards is a mystery; although **sympathetic magic** seems to play a role.

There is a romantic irresistibility to the notion of shuffling the cards and casting one's fate, to putting one's cards on the table for all to see, to drawing into the unknown, to having one's life laid out and explained by strangers who have the gift of **clairvoyance**, to gamble on the future, and so on. The idea of staring at a picture card and letting it reveal the future or mirror the soul is not one that austere critics are likely to find tantalizing, but the thought of such visionary mysticism obviously has its attraction. Centuries of scientific advancement and learning have not diminished the popularity of occult guidance systems such as the tarot, astrology, **crystal balls**, **enneagrams**, **graphology**, the *I Ching*, the **Ouija board**, **palmistry**, and the like.

Further reading: Decker 1996; Decker, Dummett and Depaulis 1996; de Givry 1971.

Tart, Charles (1937–)

A **parapsychologist** with a Ph.D. in psychology (University of North Carolina, 1963), known for his work on **astral projection**, **ESP**, LSD, **lucid dreams**, and marijuana. After retiring from the University of California at Davis psychology department, he joined the Institute of Transpersonal Psychology in Palo Alto and spent a year developing a curriculum for Robert Bigelow's now-defunct endowed Chair of Consciousness Studies at the University of Nevada at Las Vegas (UNLV). Bigelow, a wealthy Las Vegas businessman with a penchant for funding **paranormal** research, gave nearly $4 million to UNLV to teach courses on such subjects as **dreams**, **hypnosis**, meditation, **out-of-body experiences**, **telepathy**, and that ever-popular subject among college students, drug-induced **altered states of consciousness** (ASCs). (In 1971, Tart authored *On Being Stoned: A Psychological Study of Marijuana Intoxication.*)

Early in his career, Tart edited a psychology text, *Altered States of Consciousness* (1969) and wrote several of the articles in his anthology. He defined an ASC as one in which an individual "clearly feels a qualitative shift in his pattern of mental functioning." For those who prefer a behaviorist definition, he offered the following: "an ASC is a hypothetical construct invoked when an S's behavior (including the behavior of verbal report) is radically different from his ordinary behavior." Tart believes that Yoga and Zen had long been tapping into ASCs and that there was something mystical or spiritual, something superior or higher, about these ASCs. For Tart, ASCs are a gateway to a higher consciousness, to the realm of the paranormal and the **spiritual**.

Tart considers a hypnotized person to be in an altered state, and one of the more unusual uses of hypnosis is described in his article "Psychedelic Experiences Associated With a Novel Hypnotic Procedure, Mutual Hypnosis." Tart's scientific experiment involved two subjects, or Ss, called A and B. Tart had A hypnotize B. Then, while under hypnosis, B hypnotized A. Then A would deepen B's hypnotic state; then B would deepen A's hypnotic state, and so on. He wanted to see whether he could increase the depth of hypnosis a given S could reach by having S "en rapport," defined as "the special relationship supposed to exist between hypnotist and S." Says Tart: "I reasoned that if rapport was greatest in deep hypnotic states, a technique which markedly increased rapport would likely increase the depth of hypnosis" (1969: 292). Tart concluded: "Although this report is based on only two Ss, the results with them were dramatic enough to warrant considerable research on mutual hypnosis" (ibid.: 307). He notes that mutual hypnosis "might offer a way to produce psychedelic experiences in the laboratory without the use of drugs and with more flexibility and control than is possible with drugs" (ibid.: 308).

Tart explains how he first got interested in the paranormal in the following story told at a talk he gave in Casper, Wyoming:

There was a time, years ago, when I was highly skeptical of any paranormal claims of any kind. One of the things that convinced me that there must be something to this is a strange experience that I personally went through. It was wartime. I was at Berkeley, California, and everybody was working overtime. . . . [T]he young lady who was my assistant at the time worked with me until very late this one night. She finally went home; I went home. Then the very next day she came in, all excited. . . . She reported that during this night she had suddenly sat bolt upright in her bed, convinced that something terrible had happened. "I had a terrible sense of foreboding," she said, but she did not know what had happened. "I immediately swung out of bed and went over to the window and looked outside to see if I could see anything that might have happened like an accident. I was just turning away from the window and suddenly the window shook violently. I couldn't understand that. I went back to bed, woke up the next morning and listened to the radio." A munitions ship at Port Chicago had exploded. It literally took Port Chicago off the map. It leveled the entire town and over 300 people were killed. . . . She said she had sensed the moment when all these people were snuffed out in this mighty explosion. How would she have suddenly become terrified, jumped out of bed, gone to the window, and then—from 35 miles away, the shock wave had reached Berkeley and shook the window? (Randi 1992)

There is no need to perceive this event as paranormal, according to James **Randi**, who tape-recorded the story. A shock wave travels at different speeds through the ground and through the air. The difference over 35 miles would be about 8 seconds. Most likely the shaking earth woke up the young lady in a fright, and 8 seconds later the window shook. She and Tart assumed that the explosion took place when the window shook, making her experience inexplicable by the known laws of physics. This explanation only makes sense, however, if one ignores the known laws of physics.

Tart once wrote, "The implications of ESP for understanding human nature are enormous, and call for extensive, high quality scientific research" (letter to the *New York Review,* February 19, 1981). Yet Tart and other parapsychologists seem to have made little headway in justifying the first claim or in living up to the second (Randi 1982a: 153; Gardner 1981: 211).

See also Raymond **Moody**.

Further reading: Gordon 1987; Hansel 1989; Marks 1989.

telekinesis

The movement of objects by scientifically inexplicable means, as by the exercise of an **occult** power.

See also **psychokinesis**.

Further reading: Gardner 1957; Houdini 1924, 1981; Randi 1982a, 1982b.

telepathy

Literally, "distance feeling." The term is often a shortened version of mental telepathy and refers to mind-reading, discerning another's thoughts through **ESP**.

teleportation

The act or process of moving an object or person by **psychokinesis**. The term originated with Charles **Fort**, although he used it to describe magical transport between Earth and the heavens.

tensegrity

See Carlos **Castaneda**.

testimonial evidence (anecdotes)

Testimonials and vivid anecdotes are one of the most popular and convincing forms of evidence presented for beliefs in the transcendent, **paranormal**, and **pseudoscientific**. Nevertheless, testimonials and anecdotes in such matters are of little value in establishing the probability of the claims they are put forth to support. Sincere and vivid accounts of one's encounter with an **angel**, an alien, a **ghost**, **Bigfoot**, a child claiming to have lived before, purple **auras** around dying patients, a miraculous **dowser**, a **levitating** guru, or a **psychic surgeon** are of little value in establishing the reasonableness of believing in such matters. Such accounts are inherently subjective, inaccurate, unreliable, and biased. They are on par with televised accounts of satisfied customers of the latest weight loss program or the tastiness of margarine.

The testimonial of personal experience in paranormal or supernatural matters has no scientific value. If others cannot experience the same thing under the same conditions, then there will be no way to verify the experience. If there is no way to test the claims made, then there will be no way to tell whether the experience was a delusion or was interpreted correctly. If others can experience the same thing, then it is possible to make a test of the testimonial and determine whether the claim based on it is worthy of belief.

Testimonials regarding paranormal experiences are scientifically worthless because **selective thinking** and **self-deception** are too powerful and must be controlled for. Most **psychics** do not even realize that they need to do a **controlled test** of their powers to rule out the possibility that they are deceiving themselves. They are satisfied with their experiences as psychics. Controlled tests of psychics will prove once and for all that they are not being selective in their evidence gathering, that is, that they are counting only the apparent successes and conveniently ignoring or underplaying the misses. Controlled tests can also determine whether other factors, such as cheating, might be involved.

If such testimonials are scientifically worthless, why are they so popular and why are they so convincing? There are several reasons. Testimonials are often very vivid and detailed, making them appear very believable. They are often made by enthusiastic people who seem trustworthy and honest and who lack any reason to deceive us. They are often made by people with some semblance of authority, such as those who hold a Ph.D. in psychology or physics. To some extent, testimonials are believable because people *want* to believe them. Often, one anticipates with hope some new treatment or instruction. One's testimonial is given soon after the experience while one's mood is still elevated from the desire for a positive outcome. The experience and the testimonial it elicits are given more significance than they deserve.

Finally, it should be noted that testimonials are often used in many areas of life, including medical science, and that giving due consideration to such testimonials is considered wise, not foolish. A physician will use the testimonies of his or her patients to draw conclusions about certain medications or procedures. For example, a physician will take anecdotal evidence from a patient about a reaction to a new medication and use that information in deciding to adjust the prescribed dosage or to change the medication. This is quite reasonable. But the physician cannot be selective in listening to testimony, listening only to those claims that fit his or her own prejudices. To do so is to risk harming one's patients. Nor should the average person be selective when listening to testimonials regarding some paranormal or occult experience.

Further reading: Stanovich 1992.

Texas sharpshooter fallacy

The name epidemiologists give to the **clustering illusion**. The term refers to the story of the Texan who shoots holes in the side of a barn and then draws a bull's-eye around the bullet holes. Individual cases of disease are noted and then boundaries are drawn, giving the illusion of a large cluster of cases in a small area and leading to the search for a causal connection between some local environmental factor and the disease (Gawande 1999). Correlations that are expected by the laws of chance thus appear statistically significant (i.e., not due to chance).

Of the thousands of studies of cancer clusters investigated by scientists in the United States, "not one has convincingly identified an underlying environmental cause" (ibid.).

Further reading: Gilovich 1993; Gilovich and Tversky 1985.

theist (theism)

A theist is one who denies that God does not exist. Theism is the denial of atheism.

theosophy

"Theosophy," literally, "divine wisdom," refers either to the mysticism of philosophers who believe that they can understand the nature of God by direct apprehension, without revelation, or to the esotericism of eclectic collectors of mystical and occult philosophies who claim to be handing down the great secrets of some ancient wisdom. Here we treat only the latter.

Theosophic esotericism begins with Helena Petrovna Blavatsky (1831–1891), usually known as Madame Blavatsky, one of the cofounders of the Theosophical Society in New York in 1875. The esoteric theosophical tradition of Blavatsky is indebted to several philosophical and religious traditions, including Zoroastrianism, Hinduism, Gnosticism, Manichaeism, and the Cabala.

Blavatsky claims she spent several years in Tibet and India being initiated into occult mysteries by various "masters" (mahatmas or adepts), especially the Masters Morya and Koot Hoomi, who had **astral bodies** and **psychic** powers and were the sacred keepers of "Ancient Wisdom." She claimed these masters want to unite all humanity in a brotherhood, despite the fact that they dwell in the remotest regions of the world and apparently have as little contact with the rest of us as possible.

Blavatsky had an overpowering personality and was knowledgeable of the tricks of spiritualists, having worked for one in Egypt. In the early days of the Theosophical Society, she used trickery to deceive others into thinking she had **paranormal** powers. She faked the materialization of a tea cup and saucer and written messages from her masters, presumably to enhance her credibility.

In 1875 she founded the Theosophical Society in New York City in collaboration with Henry Steele Olcott, a lawyer and writer, and W. Q. Judge. She met Olcott in 1874 while he was investigating the spiritualism of the Eddy brothers in Vermont. They continued to meet with other like-minded seekers and together founded their society. A few years later, she and Olcott went to India together and established Theosophical headquarters there. She left under a cloud of suspicion in 1885 after the faked materializations.

In 1888 she published her major work, *The Secret Doctrine.* The book "is an attempt . . . to reconcile science, the Ancient Wisdom, and human culture through . . . cosmology, history, religion, and symbolism" (Ellwood 1996). According to Blavatsky, "The chief aim of the . . . Theosophical Society [was] to reconcile all religions, sects and nations under a common system of ethics, based on eternal verities."

One might wonder why, if Theosophy is so ancient and universal, it was so unknown until 1875. Madame had an answer. This was due to "willing ignorance." We humans have lost "real spiritual insight" because we are too devoted to "things of sense" and have for too long been slaves "to the dead letter of dogma and ritualism. . . . But the strongest reason for it," she said, "lies in the fact that real Theosophy has ever been kept secret." There were several reasons why this was so:

> Firstly, the perversity of average human nature and its selfishness, always tending to the gratification of personal desires to

the detriment of neighbours and next of kin. Such people could never be entrusted with divine secrets. Secondly, their unreliability to keep the sacred and divine knowledge from desecration. It is the latter that led to the perversion of the most sublime truths and symbols, and to the gradual transformation of things spiritual into anthropomorphic, concrete, and gross imagery—in other words, to the dwarfing of the god-idea and to idolatry. *(The Key to Theosophy)*

What was any different in the late 19th century? If at that time humans were any less perverse, selfish, materialistic, or profane, this would come as a great shock to social historians.

The so-called Ancient Wisdom is an eclectic compilation of Hindu, Egyptian, Gnostic, and other exotic scriptures and teachings, mixed with neo-Platonism, **occultism**, and stories such as the **Atlantis** myth.

When ignorant of the true meaning of the esoteric divine symbols of nature, man is apt to miscalculate the powers of his soul, and, instead of communing spiritually and mentally with the higher, celestial beings, the good spirits (the gods of the theurgiests of the Platonic school), he will unconsciously call forth the evil, dark powers which lurk around humanity—the undying, grim creations of human crimes and vices—and thus fall from theurgia (white magic) into goetia (or black magic, sorcery). *(What Is Theosophy?)*

According to Madame, "no one can be a true Occultist without being a real Theosophist; otherwise he is simply a black magician, whether conscious or unconscious."

The reader may wonder why Theosophy isn't universally recognized as the salvation of mankind. For some it may have been the messenger that kept them away. Many people are not likely to take seriously a Russian noblewoman who claimed to have had childhood visions of a tall Hindu who eventually materialized in Hyde Park and became her guru and advisor. Many skeptics scoff at her noble origins and subsequent employment as a circus performer and séance assistant, and do not gloss over the charges of deception. For others, it may be the doctrines that keep them away. Despite the stated moral goals, and the desire for peace on earth and good will toward men and women, there is the small problem of Aryans, **astral bodies**, **Atlantis**, evolution of spiritual races, paranormal powers, and so on. Finally, others may be repelled by the self-discipline required by Theosophy:

the foremost rule of all is the entire renunciation of one's personality—i.e., a pledged member has to become a thorough altruist, never to think of himself, and to forget his own vanity and pride in the thought of the good of his fellow-creatures, besides that of his fellow-brothers in the esoteric circle. He has to live, if the esoteric instructions shall profit him, a life of abstinence in everything, of self-denial and strict morality, doing his duty by all men.

[E]very member must be either a philanthropist, or a scholar, a searcher into Aryan and other old literature, or a psychic student. *(The Key to Theosophy)*

It is not an easy life, pursuing the path of the mahatmas and the Ancient Wisdom, striving to unite all humankind into a Great Brotherhood of spiritually evolved beings with secret knowledge of such great vacation spots for astrals as Atlantis. The dream of a Brotherhood of Man remains a dream, but Madame has made her mark, as there are Theosophical societies all over the world.

Further reading: Ellwood 1996; Randi 1995; Washington 1996.

therapeutic touch (TT)

A type of **energy** medicine whereby a therapist moves his or her hands over the patient's **aura** or energy field, allegedly directing the flow of **chi** or **prana** so the patient can heal. TT is based on the belief that each living thing has a life energy field that extends beyond the surface of the body and generates an aura. This energy field can become unbalanced, misaligned, obstructed, or out of tune. Energy healers think they can feel and manipulate this energy field by making movements that resemble massaging the air a few inches above the surface of the patient's body. Energy healers also think that they can transfer some of their own life energy to the patient. These airy manipulations allegedly restore the energy field to a state of balance or harmony, to a proper alignment, or otherwise unblock a clog in the field. This restoration of integrity to the field is thought to make it possible for the body to heal itself. Yet something seems amiss. We have scientific devices that can measure extremely minute energy levels emitted from an object, as well as the wavelengths of light reflected from the object. Human tissue is about a million times less sensitive than something like a PET scanner, yet therapeutic touch healers want us to believe they can *feel* auras, that their human senses can perceive a part of objective reality that our most sophisticated energy detection devices cannot measure.

A nurse and a theosophist created TT. Dolores Krieger, Ph.D., R.N., and a faculty member at New York University's Division of Nursing, began TT in the early 1970s. She was convinced that the palms of the hands are **chakras** and can channel healing energy. She is the author of *Therapeutic Touch: How to Use Your Hands to Help and to Heal* (1979) and several other books on TT. Dora Kunz, president of the Theosophical Society of America, was Krieger's mentor and an **intuitive** healer. TT is practiced primarily by nurses, although it is apparently being practiced worldwide by all kinds of alternative healers and laypersons.

Practitioners admit that there has never been any scientific detection of a human energy field. This, they say, is because of the inadequacies of our present technology. One with a trained sense, however, is allegedly able to detect the human energy field and assess its integrity. Despite the obvious **metaphysical** basis for this **quackery**, defenders of TT claim it is scientific because it is based on quantum physics. A grant proposal to study therapeutic touch on burn victims asserts: "Quantum theory states that all of reality is made up of energy fields and that over 99% of the universe is simply space." Another defender claims:

> The underlying principles upon which this technique is based include acceptance of the Einstein paradigm of a complex, energetic field-like universe (i.e., the existence of a Life energy flowing through and around all of us). Further, if life is characterized by an interchange of various qualities of energy, it can be assumed that any form of obstruction—either within the organism or between the organism and the environment—is contrary to Nature's tendencies and therefore unhealthy. In practicing Therapeutic Touch, one attempts to influence this energy imbalance towards health to restore the integrity of this field. In this way the TT practitioner does not so much "heal" the patient as facilitate the patient's own healing processes, by gently manipulating the body's energy flow and adjusting it as a whole. With the achievement of balance in mind, body and spirit, we have a truly holistic approach. (Rebecca

Witmer, "Hands That Heal: The Art of Therapeutic Touch," *Healing Arts,* 1995)

It is not true that Einstein had a paradigm that included the notion of "a Life energy flowing through and around all of us." He may have written of interchanges of *quantities* of energy. Many physicists have written of such things as transforming mechanical energy into electrical energy, for example, but would the typical physicist understand the expression "life is an interchange of *qualities* of energy"? From this notion Ms. Witmer infers that any form of obstruction within the organism or between the organism and the environment is contrary to Nature's tendencies and therefore unhealthy. This seems like a non sequitur, but she goes on: "if life is characterized by an interchange of various qualities of energy, it can be assumed that any form of obstruction—either within the organism or between the organism and the environment—is contrary to Nature's tendencies and therefore unhealthy." This is an alternative logic using an alternative science to support an alternative therapy.

It might be true that an obstruction within an organism is contrary to nature's tendencies, if by that we mean such things as: blockage of an air passage is unhealthy or blocked arteries are unhealthy. Yet most rational patients with such blockages would probably want someone to *physically* unblock the passageway. A rational person would not think that a mystic waving her hands over one's energy field would ever remove such blockage. On the other hand, for most organisms the environment is mostly obstructions. This may not be healthy, but it is certainly **natural**. In any case, what does it mean to say that "it is unhealthy to go contrary to Nature's tendencies"? Are hurricanes, tornados, volcanos, floods, lightning bolts, and earthquakes contrary to nature's tendencies?

How could they be, since they are part of nature? Is the lion eating the gazelle contrary to nature's tendencies?

One might wonder why a group of otherwise intelligent, highly trained professionals such as nurses would be attracted to something like TT. Ms. Witmer might have the answer. She writes: "Those who practice Therapeutic Touch often report reaping benefits for themselves. For example, the ability of TT to reduce burnout in health care professionals has been well-documented." The TT therapist has powers physicians don't have: secret, mystical powers that only the practitioner can measure. You get a lot of positive feedback. You can't hurt anyone because you're not even touching them, much less invading their body with drugs or surgical instruments. You network and those in your network feed off of each other's enthusiasm. There is a great deal of **communal reinforcement**. Many patients swear they can feel your good work. You feel revitalized, empowered.

Why do so many *patients* testify to the benefits of therapeutic touch or other bogus therapies? Some commit the **regressive fallacy**. Most **testimonials** are not followed up. They are based on immediate or early impressions. Both therapist and patient are deceived into thinking a temporary lift, which may be due to expectation or mood change, is significant and will last. Or credit is given to TT when the real causative agent was a concurrent treatment (e.g., drugs or surgery). Also, the feelings associated with illness or injury can be quite complex, involving not just pain but various emotions and desires. The patient may be anxious and fearful or hopeful and optimistic. The intervention of any caring therapist can profoundly affect these feelings. The patient may feel better, but the feeling may have nothing to do with being cured or healed. There is scientific evidence that supportive therapy of breast cancer patients improves mood and pain control, but not longevity (Goodwin 2001). It may be that therapies such as TT have a similar effect on mood, though they do nothing to curtail the illness or disease itself. Elevated mood may be misinterpreted as improved health. Watching a Buster Keaton movie might have induced the same improvement.

See also **alternative health practices**, **aura therapy**, **Ayurvedic medicine**, and **reiki**.

Further reading: Barrett 2001b; Clark and Clark 1984; Fienberg 2000; Gilovich 1993; Hover-Kramer 1996; Montagu 1986; Rosa et al. 1998; Seebach 2000; Selby and Scheiber 1996, 2000; Stenger 1997b; Williams 1989.

thought field therapy (TFT)

A type of cognitive therapy disguised as traditional Chinese medicine. Dr. Roger Callahan, a cognitive psychologist, developed TFT in 1981. While treating a patient for water phobia, he asked her to think about water as he tapped her stomach. He says that the patient claimed she suddenly overcame her lifetime fear of water. He attributes the cure to his tapping, which he thinks unblocked **energy** in her stomach meridian.

TFT allegedly "gives immediate relief for post traumatic stress disorder (PTSD), addictions, phobias, fears, and anxieties by directly treating the blockage in the energy flow created by a disturbing thought pattern. It virtually eliminates any negative feeling previously associated with a thought" (www.thoughtfield.com/about .htm).

The theory behind TFT is that negative emotions cause energy blockage, and if the energy is unblocked, then the fears will disappear. Tapping acupressure points is thought to be the means of unblocking the energy. Allegedly, it takes only 5 to 6 minutes to elicit a cure. Dr. Callahan claims an 85% success rate. He even does cures over the phone using "Voice Technology" on infants and animals. By analyzing the voice, he claims he can determine what points on the body of the patient should be tapped.

Dr. Callahan has a theory that thoughts have fields and that these fields have an effect on the body. He also claims that there is a one-to-one correspondence (isomorphism) between perturbations caused by negative emotions and specific energy meridian points on the body. He claims to know the exact algorithm (where to tap) for each kind of perturbation. How he knows any of this is not clear, though it appears he made up the theory to fit with ancient Chinese beliefs in **chi** and meridians, and he seems to have figured out the algorithms by trial and error or by studying traditional acupressure points. He seems not to have done any **controlled studies** to rule out **confirmation bias** or **self-deception**. He relies on anecdotes to support his beliefs. Hence, he cannot be sure that the effects he observes are not due to standard cognitive therapy techniques (including having the patient think about what frightens him or her) rather than to the tapping on particular pressure points.

See also **Eye Movement Desensitization and Reprocessing** and **yin and yang**.

Further reading: Gaudiano and Herbert 2000.

trance writing

See **automatic writing**.

Transcendental Meditation (TM)

A set of Hindu meditation techniques introduced to the Western world by Maharishi Mahesh Yogi, dubbed the "giggling guru" because of his habit of constantly giggling during television interviews (Gardner 1995a). TM allegedly brings the practitioner to a special state of consciousness often characterized as enlightenment or bliss. The method involves repeating a mantra, an allegedly special expression that is often nothing more than the name of a Hindu god. Disciples pay hundreds of dollars for their mantras. They are led to believe that theirs is special and chosen just for them. The claim of uniqueness for the mantra is just one of many questionable claims made by TM leaders.

The TM web site claims it is a "program" and that it is "scientifically validated":

> Over 500 scientific studies conducted at more than 200 universities and research institutions in 33 countries have documented the benefits of Transcendental Meditation (TM) for mind, body, behavior, and environment. [www.mum.edu/tm_program/welcome.html]

However, TM is actually a religious business or **cult**. The claim of scientific validation is extremely misleading. One must take with a grain of salt claims such as the following:

> The Transcendental Meditation (TM) program of Maharishi Mahesh Yogi is the single most effective technique available for gaining deep relaxation, eliminating stress, promoting health, increasing creativity and intelligence, and attaining inner happiness and fulfillment. [www.tm.org]

These exaggerated claims are based mainly on the attempt to deceive people into thinking that any study done anywhere on the benefits of relaxation techniques validates TM.

The TM movement began in 1956 in India and is now worldwide, claiming more than 5 million followers, though the actual number of TM advocates is probably much smaller. Many know of TM because of the Beatles and other celebrities such as Mia Farrow and Donovan, who hung around at the Maharishi's ashram in the late '60s. It may be that the Beatles found that money and fame weren't all they're made out to be, and like many others they turned to the East for help in finding the happiness and fulfillment they couldn't get from fame, fortune, and drugs. Many think meditation offers a way to a high higher than any drug and a power stronger than all others: the power of self-control. It also has the pleasant side effect of leaving one feeling relaxed and content, as long as one's guru isn't charging too much for the lessons, financially or psychologically, and isn't constantly harassing you to recruit others into the happy, happy cult.

One of the main appeals of TM has been its claim to be a scientific means to overcome stress. TM claims to be based on the "Science of Creative Intelligence," in which one may get a graduate degree at the Maharishi University of Management (MUM; formerly Maharishi International University) in Fairfield, Iowa. MUM offers "a full range of academic disciplines for successful management of all fields of life." It is also the source for a number of health and beauty products for sale to those who want a perfect body to go with the perfect mind (www.theraj.com).

TM recruiting literature is full of charts and graphs allegedly demonstrating scientifically the wonders of TM. Things such as metabolic rate, oxygen consumption rate, bodily production of carbon dioxide, hormone production, and brain waves are measured and charted and graphically presented to suggest that TM really takes a person to a new state of consciousness. The truth is that most TM scientists do not do control group studies and, in fact, are on par with most **parapsychologists** when it comes to experimental design and controls. That is, their work is incompetent, if not fraudulent. Some of the studies are simply trivial: You can get some of the same physiological results by relaxing completely. Nevertheless, according to TM advocates, tests have shown that TM produces neurophysiological signatures that are distinctly different from relaxation and rest. Critics disagree, however, and cite studies suggesting that TM may be hazardous to your health. For example, a German study done in 1980 found that three-fourths of 67 long-term transcendental meditators experienced adverse health effects (www .trancenet.org/research/chap2.shtml). However, one should be cautious in drawing any strong conclusions from this small study. TM may attract people who are stressed out and are seeking relief; many of those who have physical or psychological problems after meditating may have had them *before* they started meditating. Thus, for many in the study, meditation may not have caused their problems, but it didn't relieve them, either.

Probably the least believable claim of TMers is that they can fly or hover. TM loudly promoted **levitation** in its early days. Television news programs featured clips of TMers hopping around in the lotus position, claiming to be flying. Apparently, this claim was too easily disproved, and now TMers do not claim to be able to fly or hover. Some advocates, however, claim they can achieve a range of supernatural or paranormal powers through TM, including invisibility. Apparently, since television is a visual medium, this skill has gone largely unnoticed.

One of the demonstrable powers claimed by TM is the "Maharishi effect." This is another so-called scientifically demonstrated fact: "collective meditation causes changes in a fundamental, unified physical field, and . . . those changes radiate into society and affect all aspects of society for the better" (minet.org/markovsky-critique.1). One TM study by a MUM physics professor, Dr. Robert Rabinoff, claimed that the Maharishi effect was responsible for reducing crime and accidents while simultaneously increasing crop production in the vicinity of Maharishi University in Fairfield, Iowa. James **Randi** checked with the Fairfield Police Department the Iowa Department of Agriculture, and the Department of Motor Vehicles and found that the Rabinoff's data was invented (Randi 1982a: 99–108). The Maharishi effect is as real as the **hundredth monkey effect**.

Apparently, MUM's accounting practices were on par with their scientific research. MUM hired attorney Anthony D. DeNaro in 1975 as director of grants administration and legal counsel. In an affidavit signed in 1986 and presented to the U.S. District Court for the District of Columbia, DeNaro stated that

it was obvious to me that [the] organization was so deeply immersed in a systematic, wilful pattern of fraud including tax fraud, lobbying problems and other deceptions, that it was ethically impossible for me to become involved further as legal counsel.

I discussed this with Steve Druker [the university's executive vice president], but agreed to remain as Director of Grants provided certain conditions and restrictions were met. In practice, however, because I recognized a very serious and deliberate pattern of fraud, designed, in part, to misrepresent the TM movement as a science (not as a cult), and fraudulently claim and obtain tax-exempt status with the IRS, I was a lame duck Director of Grants Administration. [www.trancenet .org/law/denarot.shTMl]

According to DeNaro, "there is no difference at all between other meditation techniques, and TM except the much-publicized propaganda and advertising claims." He also claims in his affidavit that MUM was characterized by a

> disturbing denial or avoidance syndrome . . . even outright lies and deception are used to cover-up or sanitize the dangerous reality on campus of very serious nervous breakdowns, episodes of dangerous and bizarre behavior, suicidal and homicidal ideation, threats and attempts, psychotic episodes, crime, depression and manic behavior that often accompanied roundings (intensive group meditations with brainwashing techniques).

Defenders of TM claim that DeNaro is just a disgruntled former member whose comments are sour grapes. Maybe, but he is not alone.

Patrick Ryan is a graduate of MUM and practiced TM for 10 years. He founded a support group for former members (TM-Ex). Ryan also claims TM is not simply a "harmless way to relax through meditation." He agrees with DeNaro that TM uses a good deal of deception:

> In its advertising, TM emphasizes the practical benefits of meditation—particularly the reduction of stress. TM promoters show videos of members from all walks of life testifying to its benefits. TM sales pitches are full of blood pressure charts, heart-rate graphs, and other clinical evidence of TM's effectiveness. Not mentioned is the fact that scientific tests show similar benefits can be obtained by listening to soothing music, or by performing basic relaxation exercises available in

books costing a couple of dollars. After a TM student pays up to $400 and receives his own personal mantra to chant, he is told never to reveal it to another. Why? Because the same "unique" mantra has been given—on the basis of age—to thousands of people. [www.freedomofmind .com/groups/meditation/ryanexcerpt.asp]

What other relaxation program has a support group for ex-relaxers?

TM, like other religious groups these days, is heavily involved in politics. The Natural Law Party is TM's attempt to introduce its **metaphysical** teachings and practices into every aspect of American life: education, health, economics, prison reform, energy, the environment—they even have a policy on healthy foods.

There have also been attempts to introduce TM into public schools. For example, the March 1, 1995, edition of the *Sacramento Bee* (p. B4) reports that John Black, director of a TM program in Palo Alto, California, tried to convince officials in San Jose to let him teach TM in the schools. Meditation in the classroom, he claims, will increase test scores, reduce teenage pregnancies, rid campuses of violence and drugs, and diminish teacher burnout. This powerful message was delivered at a free forum for teachers and meditators called "Solving the Crisis in Our Schools."

It may be true that people such as John Black really believe that TM can do all these things, but they have no proof that TM in the schools will accomplish any of these noble goals. John Black says that "the crisis in the schools is that people are stressed out." He may be right, but it is doubtful that his claim is even intelligible. Wisely, school officials have remained unconvinced. Even a newspaper ad in which Maharishi Mahesh Yogi himself offered "a proven program to eliminate crime in San Jose" for a mere $55.8 million a year couldn't convince city hall. Similar ads were placed in several major newspapers around the country. There were no takers.

Who said you can't trust city hall?

See also **Ayurvedic medicine**.

Further reading: Gardner 1995a; Randi 1982a.

trepanation

The process of cutting a hole in the skull. According to John Verano, a professor of anthropology at Tulane University, trepanation is the oldest surgical practice and is still performed ceremonially by some African tribes (Colton 1998). A trepanned skull found in France was dated at about 5000 B.C.E. About 1,000 trepanned skulls from Peru and Bolivia date from 500 B.C.E. to the 16th century (Lawson 2000).

Bart Huges (b. 1934), a medical school graduate who has never practiced medicine except for a bit of self-surgery, believes that trepanation is the way to higher consciousness. He wanted to be a psychiatrist but failed the obstetrics exam and so never went into practice. So he says (Mellen 1966–67). In 1965, after years of experimentation with LSD, cannabis, and other drugs, Dr. Huges realized that the way to enlightenment was by boring a hole in his skull. He used an electric drill, a scalpel, and a hypodermic needle (to administer a local anesthetic). The operation took 45 minutes. How does it feel to be enlightened? "I feel like I did when I was 14," says Huges. He also claims that "a genius is one to whom the knowledge of the difference between yes and no is innate."

What led Dr. Huges to believe that trepanation would lead to enlightenment? His first insight came when he was taught that he could get high by standing on his head. He came to believe that by perma-

nently relieving pressure he could increase the flow of blood to the brain and achieve his goal. After he took a little mescaline he soon understood what was going on. "I recognized that the expanded consciousness was attributed to an increase in the volume of blood to the brain."

In the past, trepanation was used either to relieve pressure on the brain caused by disease or trauma or to release evil spirits. The former is still an accepted medical procedure. The latter has died out in those parts of the world where scientific understanding has replaced belief in invading demons. Huges has yet to command a large following of trepanners, but he has managed to attract a few supporters with holes in their heads. One of his most illustrious pupils is Amanda Fielding from Oxford, England, who not only lived through the filming of her self-surgery but also became a candidate for Parliament. She received 40 votes from the people of Chelsea in 1978 where she ran on the promise of free trepanation from the National Health Service.

Fielding maintains that having a hole in her head allows more oxygen to reach her brain and helps expand her consciousness. It's safer than LSD, she says, apparently convinced that those are her only two options to expand her consciousness. She claims she now has more energy and inspiration, and is on a "permanent natural 'high.'" She claims the trepanned are "better prepared to fight neurosis and depression and less likely to become prone to alcoholism and drug addiction" (Pain 2000). One could say that Fielding is very open-minded for a politician.

Further reading: Colton 1998; Lawson 2000; Michell 1987.

true-believer syndrome

An expression coined by M. Lamar Keene to describe an apparent cognitive disorder characterized by believing in the reality of **paranormal** or supernatural events after one has been presented overwhelming evidence that the event was fraudulently staged.

> The true-believer syndrome merits study by science. What is it that compels a person, past all reason, to believe the unbelievable. How can an otherwise sane individual become so enamored of a fantasy, an imposture, that even after it's exposed in the bright light of day he still clings to it—indeed, clings to it all the harder? (Keene 1997)

Keene is a reformed phony **psychic** who exposed religious racketeering—to little effect, apparently. Phony **channelers**, faith healers, psychics, **tarot card** readers, and televangelist miracle workers are as abundant as ever.

Keene believes that "the true-believer syndrome is the greatest thing phony **mediums** have going for them" because "no amount of logic can shatter a **faith** consciously based on a lie." That those suffering from true-believer syndrome are consciously lying to themselves hardly seems likely, however. Perhaps from the viewpoint of a fraud and hoaxer, the mark who is told the truth but who continues to have faith in you must seem to believe what he knows is a lie. Yet this type of **self-deception** need not involve lying to oneself. To lie to oneself would require admission that one believes what one knows is false. This does not seem logically possible. One can't believe or disbelieve what one *knows*. (*Belief* is distinct from *belief in,* which is mainly a matter of trust.) Belief and disbelief entail the possibility of error; knowledge implies that error is beyond reasonable probability. I may have overwhelming evidence that a psychic is a phony, yet still believe that paranormal events occur. I may be deceiving myself in such a case, but I don't think it is correct to say I am lying to myself.

It is possible that those suffering from true-believer syndrome simply do not believe that the weight of the evidence before them revealing fraud is sufficient to overpower the weight of all those many cases of supportive evidence from the past. The fact that the supportive evidence may even have been supplied by the same person exposed as a fraud is suppressed. There is always the hope that no matter how many frauds are exposed, at least one of the experiences might have been genuine. No one can prove that *all* psychic miracles have been frauds; therefore, the true believer may well reason that he or she is justified in keeping hope alive. Such thinking is not completely illogical, though it may seem pathological to the one admitting the fraud.

It does not seem as easy to explain why the true-believer continues to believe in—that is, *trust*—the psychic once he has admitted his deception. Trusting someone who reveals he is a liar and a fraud is irrational, and such a person must appear crazy to the hoaxer. Some true believers may well be mad, but some may be deceiving themselves by assuming that it is possible that a person can have psychic powers without knowing it. Thus, one could disbelieve in one's psychic ability, yet still actually possess paranormal powers. Just as there are people who believe they have psychic powers but really don't have them, perhaps there are some people who really have psychic powers but do not believe they do.

In any case, there are two types of true believers, though they are clearly related. One is the kind Keene was referring to, namely, the type of person who believes in paranormal or supernatural things contrary to the evidence. Their faith is unshakable even in the face of overwhelming evidence against them, for example, those who refused to disbelieve in the **Carlos hoax** once the scam was revealed. Keene's examples are mostly of people who are so desperate to communicate with the dead that no exposé of fraudulent mediums (or channelers) can shake their faith in **spiritualism** (or channeling). The other is the type described by Eric Hoffer in his book *The True Believer* (1951). This type of person is irrationally committed to a *cause,* such as murdering doctors who perform abortions, or to a *guru,* such as Jim Jones, whose followers committed mass suicide.

True-believer syndrome may account for the popularity of Deepak **Chopra**, Uri **Geller**, **Sai Baba**, and James Van Praagh, but the term does not help us understand why people believe in the psychic or supernatural abilities of such characters, despite the underwhelming evidence that they are genuine. Since by definition those suffering from true-believer syndrome are irrationally committed to their beliefs, there is no point in arguing with them. Evidence and logical argument mean nothing to them. Such people are by definition deluded in the psychiatric sense of the term: They believe what is false and are incapable of being persuaded by evidence and argument that their notions are in error.

Clearly, if there is any explanation for true-believer syndrome, it must be in terms of the satisfaction of emotional needs. But why some people have such a strong emotional need to believe in immortality or racial or moral superiority, or even that the latest fad in management must be pursued with evangelical zeal, is perhaps unanswerable. It may have to do with insecurity. Eric Hoffer (ibid.) seemed to think so:

> The less justified a man is in claiming excellence for his own self, the more ready he is to claim all excellence for his nation, his religion, his race or his holy cause. . . .

> A man is likely to mind his own business when it is worth minding. When it is not, he takes his mind off his own meaningless affairs by minding other people's business. . . .

> The fanatic is perpetually incomplete and insecure. He cannot generate self-assurance out of his individual resources— out of his rejected self—but finds it only by clinging passionately to whatever support he happens to embrace. This passionate attachment is the essence of his blind devotion and religiosity, and he sees in it the source of all virtue and strength. . . . He easily sees himself as the supporter and defender of the holy cause to which he clings. And he is ready to sacrifice his life.

Hoffer also seemed to think that true-believer syndrome has something to do with the desire to give up all personal responsibility for one's beliefs and actions: to be free of the burden of freedom. Perhaps Hoffer is right for many of the more severe cases, but many of the lesser ones may have to do with little more than **wishful thinking**. For many people, the will to believe at times overrides the ability to think critically about the evidence for and against a belief.

Further reading: Haught 1995b, 1996; Randi 1982b, 1989a; Raymo 1998.

U

UFO

Acronym for "unidentified flying object." However, "UFO" is used by many people to refer to an object that they believe is a possible or actual alien spacecraft. Properly used, however, "UFO" should be restricted to flying or apparently flying objects that cannot be identified with reasonable probability as being meteors, disintegrating satellites, flocks of birds, aircraft, lights, weather balloons, reflections, or other natural or earthly objects. So far, nothing has been positively identified as an alien spacecraft in a way required by commonsense and science. That is, there is no physical evidence in support of either a UFO flyby or landing except for photographs and video recordings.

If any photo or video provided clear evidence of alien presence, there would be no disputing the claim that some UFOs are alien spacecraft. Other physical evidence, such as alleged debris from alien crashes, burn marks on the ground from alien landings, or implants in noses or brains of **alien abductees**, has turned out to be quite terrestrial, including forgeries.

The main reasons for believing in UFOs are the testimony of many eyewitnesses, the willingness to trust people telling fantastic stories, the tendency to distrust contrary sources as being part of a conspiracy to cover up the truth, and a desire for contact with the world above. According to Paul Kurtz, belief in aliens in UFOs is akin to belief in supernatural beings:

> UFOlogy is the mythology of the space age. Rather than angels . . . we now have . . . extraterrestrials. It is the product of the creative imagination. It serves a poetic and existential function. It seeks to give man deeper roots and bearings in the universe. It is an expression of our hunger for mystery . . . our hope for transcendental meaning. The gods of Mt. Olympus have been transformed into space voyagers, transporting us by our dreams to other realms.

Dr. J. Allen Hynek, astronomer, foremost proponent of UFOs, and the one who came up with the expression "close encounters of the third kind," defines a UFO as:

> the reported perception of an object or light seen in the sky or upon land the appearance, trajectory, and general dynamic and luminescent behavior of which do not suggest a logical, conventional explanation and which is not only mystifying to the original percipients but remains unidentified after close scrutiny of all available evidence by persons who are technically capable of making a common sense identification, if one is possible.

That is, if intelligent people cannot devise a rational explanation for an observation of a UFO, then it is reasonable to conclude that what was observed was an alien craft. A skeptic might think that "all available evidence" is most likely not all that should be considered before concluding that aliens have landed. Just because some scientist, pilot, Air Force Colonel, or doctor of philosophy cannot think of a logical explanation for an observation does not mean that there isn't one. It seems more reasonable to believe that the only reason we cannot explain these sightings by conventional means is because we do not have all the evidence. Most unidentified flying objects are eventually identified as hoaxes, aircraft, satellites, weather balloons, astronomical events or other natural phenomena. In studies done by the Air Force, less than 2% of UFO sightings remain unidentifiable. It is more probable that with more information those 2% would be identified as meteors or aircraft, for example, than as alien spacecraft.

Many UFOlogists think that if eyewitnesses such as Whitley Strieber, Betty and Barney Hill, or other alleged alien abductees are not insane or evil, then they cannot be deluded and are to be trusted with giving accurate accounts of alien abduction. Yet it seems obvious that most sane, good, normal people are deluded about many things and not to be trusted about certain things. While it is generally reasonable to believe the testimony of sane, good, normal people with no ulterior motive, it does not follow that unless you can prove a person is deranged, evil, or a fraud that you should trust his or her testimony about any claim whatsoever. When the type of claim being made involves the incredible, then additional evidence besides eyewitness testimony is required.

UFOlogists reject the conclusions of *Project Blue Book,* the U.S. Air Force report that states, "After twenty-two years of investigation . . . none of the unidentified objects reported and evaluated posed a threat to our national security." (It was in this Blue Book that Edward Ruppelt coined the term "UFO.") UFOlogists are unimpressed with the Condon Report, as well. Edward U. Condon was the head of a scientific research team that was contracted to the University of Colorado to examine the UFO issue. His report concluded that "nothing has come from the study of UFOs in the past 21 years that has added to scientific knowledge . . . further extensive study of UFOs probably cannot be justified in the expectation that science will be advanced thereby."

It is argued by UFOlogists that the government is lying and covering up alien landings and communication. However, there is no evidence for this other than a general distrust of the government and the fact that many government officials have lied, distorted the truth, and been mistaken when reporting to the general public. The fact is that the CIA has shown little interest in UFOs since about 1950, except to encourage UFOlogists to believe that reconnaissance flights might be alien craft. UFOlogists do not show the same kind of skepticism toward the mass media as they

do toward the government. For example, they didn't criticize the work of NBC, which produced two dozen programs called *Project UFO,* said to be based on *Project Blue Book.* NBC suggested that there were documented cases of alien spacecraft sightings. The programs, produced by Jack Webb of *Dragnet* fame, distorted and falsified information to make the presentation look more believable. No UFOlogist took NBC to task. To the skeptic, NBC was pandering to the taste of the viewing audience.

Finally, it should be noted that UFOs are usually observed by untrained skywatchers and almost never by professional or amateur astronomers, people who spend inordinate amounts of time observing the heavens above. One would think that astronomers would have spotted some of these alien craft. Perhaps the crafty aliens know that good scientists are skeptical and inquisitive. Such beings might pose a threat to the security of a story well told.

See also **alien abductions**, **Area 51**, **cattle mutilations**, **crop circles**, **flying saucers**, **Men in Black**, **Roswell**, and **Santa Claus**.

Further reading: Condon 1969; Dudley 1999; Frazier 1997; Klass 1988; Kurtz 1986; Randi 1982a; Sagan 1972, 1979, 1995; Sheaffer 1989.

unconscious mind (subconscious)

According to classical Freudian psychoanalysis, the unconscious is a part of the mind that stores repressed memories of traumatic experiences. The theory of repression maintains that some experiences are too painful to be reminded of, so the mind stuffs them in the cellar. These painful **memories** manifest themselves in neurotic or psychotic behavior and in dreams. However, there is little scientific evidence either for the unconscious repression of traumatic experiences or for repressed memories being significant causal factors in neurotic or psychotic behavior.

Some, such as Carl **Jung** and Charles **Tart**, think the unconscious mind is a reservoir of transcendent truths, although it might be more accurately said that the unconscious is an ocean of mundane feelings and thoughts.

It would be absurd to reject the notion of the unconscious mind simply because we reject the Freudian or Jungian notions of the unconscious. We should recognize that it was Freud more than anyone else who forced us to recognize unconscious factors as significant determinants of human behavior. Furthermore, it seems obvious that much, if not most, of one's brain's activity occurs without our awareness or consciousness. Consciousness or self-awareness is obviously the proverbial tip of the iceberg.

Yet it does not seem appropriate to speak of any of the following cases as involving the unconscious mind, even though the perceivers are not aware of what they are perceiving.

1. *Blindness denial.* There are cases of brain-damaged people who are blind but who are unaware of it.
2. *Jargon aphasia.* There are cases of brain-damaged people who speak unintelligibly but are not aware of it.
3. *Blindsight.* There are cases of brain-damaged people who see things but are unaware of it.
4. *Oral/verbal dissociation.* There are cases of brain-damaged people who cannot orally tell you what you just said, but they can write it down correctly. Furthermore, they can't remember what they wrote down or what it refers to. (Sacks 1984, 1985, 1995; Schacter 1996)

It might be less confusing to abandon talk of the unconscious mind and refer instead

to lost memory, fragmented memory, or implicit memory (the last term coined by Daniel Schacter and Endel Tulving).

In any case, it is not "unconscious repression" of traumatic experiences that causes memories to be lost. The evidence shows that the more traumatic an experience, the more likely one is to remember it, unless one is either very young or later experiences brain damage (Schacter 1996). On the other hand, memories of traumatic experiences are often *consciously* repressed.

Implicit memories or fragments of memories may be locked away because of *inattentiveness* in the original experience or because the original experience occurred at an age when the brain was not fully developed. There are numerous situations—such as **cryptomnesia**—where memory can be manifested without awareness of remembering. Many fragments of pleasant experiences, such as the name of a place or a product, may be influencing present choices without one's being aware of it. But such unconscious memories, even though pervasive, are not quite what Freud or Jung meant by the unconscious. "Implicit memory" may be a far more mundane concept than Freud's "dynamic unconscious mind" or Jung's "collective unconscious," but it is more significant since it reaches into every aspect of our lives and can be subjected to scientific testing.

See also **repressed memory** and **repressed memory therapy**.

Further reading: Churchland 1986; Kandel and Schwartz 2000; Nrretranders 1999; Schacter 1997, 2001; Watters and Ofshe 1999.

unicorn

A creature from fables, usually depicted as a white horse with a spiral horn protruding from its forehead. The unicorn is also a symbol of virginity and in Christian iconography is sometimes used to represent the Virgin Mary. Medieval and Renaissance tapestries often feature the unicorn.

Urantia Book, The

According to the Urantia Book Fellowship (UBF), *The Urantia Book* (UB) is

> an anthology of 196 "papers" indited [i.e., dictated] between 1928 and 1935 by superhuman personalities. . . . The humans into whose hands the papers were delivered are now deceased. The means by which the papers were materialized was unique and is unknown to any living person.

The UB Fellowship was founded in 1955 as the Urantia Brotherhood and is an association of people who say they have been inspired by the "transformative teachings" of the UB. According to the UBF, these "superhuman personalities" are from another world. They synthesized the work of more than 1,000 human authors in a variety of fields, including an "astronomical-cosmological organization of the universe" unknown to modern science and an elaborate extension on the life of Jesus (700 pages). The UB also reveals that the "Universe is literally teeming with inhabited planets, evolving life, civilizations in various states of development, celestial spheres, and spirit personalities." In short, the UB is over 2,000 pages of "revelations" from superhuman beings that "correct" the errors and omissions of the Bible. "Urantia" is the name these alleged superhumans gave to our planet. According to these supermortal beings, Earth is the 606th planet in Satania, which is in Norlatiadek, which is in Nebadon, which is in Orvonton, which revolves around Havona, all of which revolves around the center of infinity where God dwells.

Author Martin Gardner is skeptical of the UBF's claims. He believes the UB has

very real human authors. Originally, he says, the UB was the Bible of a cult of separatist Seventh Day Adventists, allegedly channeled by Wilfred Kellogg and edited by founder William Sadler, a Chicago psychiatrist. According to Gardner, in addition to an array of bizarre claims about planets, names of angels, and so on, *The Urantia Book* contains many Adventist doctrines (Gardner 1995b). Sadler died in 1969 at the age of 94, but his spiritual group lives on. He got his start working for Dr. John Harvey Kellogg, Adventist surgeon, health and diet author, and brother of cornflake king William Keith Kellogg. These are the same Kellogg brothers who were featured and lampooned in the movie *The Road to Wellville*.

One can easily understand why Gardner suspects that the UB has human rather than superhuman origins. The book has all the traits of humanity. For example, our human philosophers and theologians are mimicked perfectly in passages such as the following:

> The philosophers of the universes postulate a Trinity of Trinities, an existential-experiential Trinity Infinite, but they are not able to envisage its personalization; possibly it would equivalate to the person of the Universal Father on the conceptual level of the I AM. But irrespective of all this, the original Paradise Trinity is potentially infinite since the Universal Father actually is infinite. [Foreword XII, The Trinities: www.urantiabook.org]

Any medieval casuist would be proud of such writing and thinking.

> Primary supernaphim are the supernal servants of the Deities on the eternal Isle of Paradise. Never have they been known to depart from the paths of light and righteousness. The roll calls are complete; from eternity not one of this magnificent

host has been lost. These high super-naphim are perfect beings, supreme in perfection, but they are not absonite, neither are they absolute. (paper 27)

Some UBFers are attracted not so much to the theology but to its great insights. Here are a few of those insights culled from paper 100, "Religion in Human Experience." Ask yourself whether a superhuman being was necessary to reveal these gems:

- Give every developing child a chance to grow his own religious experience.
- Religious experience is markedly influenced by physical health, inherited temperament, and social environment.
- Spiritual development depends, first, on the maintenance of a living spiritual connection with true spiritual forces.
- The goal of human self-realization should be spiritual, not material.
- Jesus was an unusually cheerful person, but he was not a blind and unreasoning optimist.

If these philosophical, theological, or spiritual insights do not impress you, then you might want to consider the *scientific* insights of the UB, such as the resurrection of the pre-Adamite thesis of French ecclesiastic Isaac de la Peyrère (1596–1676), who felt compelled to believe that the Bible is the history of the Jews, not of all people, and that in order to explain things such as racial differences, the most reasonable hypothesis is that races of people existed before Adam and Eve (Popkin 1987).

   Not everyone agrees with Gardner's claim that Wilfred Kellogg channeled *The Urantia Book*. Ernest Moyer, for example, believes that the UB is a revelation from God that appeared "out of thin air" in fully developed form, exactly as we know it today (www.world-destiny.org/brief.htm). Moyer claims that Sadler was put through a lengthy process by our "planetary supervi-

sors" in order to prepare him to accept the UB as true revelations. The process began by introducing Sadler to the Sleeping Subject (SS), whose nocturnal ramblings would later be understood to be preparatory messages from extraterrestrial "midwayers." According to Moyer, "SS was a member of the Chicago Board of Trade, a highly pragmatic, hard-nosed businessman who did not believe in 'psychic' phenomena or any such nonsense." Why SS was selected for this task is unknown, but Moyer assures us that the midwayers never took over SS's mind and came only at night when SS was unconscious so as not to disrupt his life too much. Moyer contrasts this with the evil spirit who invaded Edgar **Cayce** during the daytime, a sure sign that Cayce was a false prophet. Sadler was selected, according to Moyer, because of his personality and training.

   Moyer is convinced that we are on the verge of a nuclear holocaust and that the UB offers advice on how to save oneself from destruction and what to do afterward. This is all part of God's plan, as revealed to Sadler. According to Moyer, "God is using this technique to screen the human race."

   God allegedly tried this once before, with water instead of nuclear bombs. If at first you don't succeed . . .

   Further reading: Gardner 1992; Popkin 1987.

urine therapy

Any of several uses of urine to prevent or cure sickness, to enhance beauty, or to cleanse the bowels. Most devotees drink the midstream of their morning urine. Some prefer it straight and steaming hot; others mix it with juice or serve it over fruit. Some prefer a couple of urine drops mixed with a tablespoon of water applied sublingually several times a day. Some wash themselves in their own golden fluid to improve their skin quality. Many modern Japanese women are said to engage in urine bathing. The truly daring use their own urine as an enema. Urine is not quite the breakfast of champions, but it is the elixir of choice of a number of holy men in India, where drinking urine has been practiced for thousands of years. The drink is also the preferred pick-me-up for a growing number of **naturopaths** and other advocates of nature cures. The main attractions of this ultimate home brew are its cost, availability, and portability.

Many advocates claim that urine is a panacea. There is practically nothing it won't cure. Urine is said to be effective against the flu, the common cold, broken bones, toothache, dry skin, psoriasis, and all other skin problems. It is said to deter aging and is helpful with AIDS, allergies, animal and snake bites, asthma, heart disease, hypertension, burns, cancer, chemical intoxication, chicken pox, enteritis, constipation, and pneumonia. Urine is said to be effective against dysentery, edema, eczema, eye irritation, fatigue, fever, gonorrhea, gout, bloody urine, small pox, immunological disorders, infections, infertility, baldness, insomnia, jaundice, hepatitis, Kaposi's sarcoma, leprosy, lymphatic disorder, morning sickness, hangover, obesity, papilloma virus, parasitoses, gastric ulcer, rheumatism, birth marks, stroke, congestion, lumbago, typhus, gastritis, depression, cold sore, tuberculosis, tetanus, Parkinson's disease, foot fungus, diabetes, and other endocrine related diseases. Some enthusiasts see urine therapy as a divine manifestation of cosmic intelligence. They use urine to unleash their *kundalini,* sending it straight into the third eye, bringing instant enlightenment (www .aznewage.com/urine.htm).

With such wondrous properties, it is amazing that science bothered developing medicine when it had the key to good health already in the bottle, so to speak. Each of us is a walking pharmacopoeia.

According to urinophiles, the medical establishment has conspired to keep us ignorant of the wonder drug we all carry in our bladders. One self-proclaimed expert on the subject claims that

> the medical community has already been aware of [urine's] astounding efficacy for decades, and yet none of us has ever been told about it. Why? Maybe they think it's too controversial. Or maybe, more accurately, there wasn't any monetary reward for telling people what scientists know about one of the most extraordinary natural healing elements in the world. [www .all-natural.com/urine.html]

This is a common argument from defenders of alternative therapies: The greed of medical doctors leads them to conspire against **chiropractors**, **chelation** therapists, and other alternative practitioners. The evidence for this conspiracy wouldn't fill a specimen beaker. Part of the alleged conspiracy to keep us ignorant of the wonders of our own wastewater is the fact that many people think urine is poisonous. Urine is generally not toxic and you will not die of uremic poisoning if you start your day off with a cup of it. However, it hardly seems fair to blame the medical establishment for the general public's ignorance on this matter. In any case, just because something is not toxic does not mean it is good for you. Hair is not toxic, either, and even though it might be a good source of roughage, it is generally not desirable to put hair in food.

Furthermore, while it is true that some of the constituents of urine are being used and tested for their potential or actual therapeutic value, it does not follow that drinking one's urine is therapeutic. It may be discovered that one of the chemicals in human urine is effective for fighting cancer. However, drinking one's own urine is unlikely to supply enough of any cancer-fighting substance to do any good. It is also true that some of the substances in urine are good for you. For example, if you are ingesting more vitamin C (a water-soluble vitamin) than your body needs or can process, you will excrete it in your urine. It doesn't follow that drinking your urine is a good way to get vitamin C into your body. An orange or a tablet might be preferable. However, if you are urinating excess vitamin C, what do you think your body will do with the vitamin C you ingest with your urine? If you guessed that it would get rid of it, you guessed right. The reason your urine contains vitamins and minerals is because your body didn't need them or couldn't use them. You might as well pour water into a full glass as reuptake your excess vitamins and minerals. Even urea, which can be toxic in very high doses, occurs in such minute quantities in the average person's urine that there is very little chance of poisoning from drinking one's own urine.

Unfortunately, however, not everybody can just jump right in and start drinking his or her own urine without negative side effects. The Chinese Association of Urine Therapy warns:

Common symptoms include diarrhea, itch, pain, fatigue, soreness of the shoulder, fever, etc. These symptoms appear more frequently in patients suffering long term or more serious illnesses, and symptoms may repeat several times. Each episode may last 3–7 days, but sometimes it may last one month, or even worse over 6 months. It is a pity that many give up urine therapy because of such bad episode [sic]. Recovery reaction is just like the darkness before sunrise. If one persists and overcomes the difficulty, one can enjoy the eventual happiness of healthy life. [www.auto-urine.com/english.htm]

These same people advise, "All kinds of throat inflammation can be helped by gargling with urine to which a bit of saffron has been added" and "drinking one ounce of urine . . . is more beneficial to the average person than a fully staffed multi-billion dollar medical center." I was unable to find their evidence for these claims. Perhaps the evidence was produced at the First World Conference on Urine Therapy, which took place in India in February 1996. Or maybe it came up in 1998 during the Second World Conference on Urine Therapy, held in Germany.

The origin of this practice seems to be certain religious rites among Hindus, where it is called "amaroli" in Tantric religious traditions. The Tantric tradition is known for flouting conventional behavior as a means of establishing the moral superiority of its practitioners. It is also possible that this practice is related to superstitions based on sympathetic magic. Since urine is emitted from the same bodily organ used in sex, perhaps it was thought that by drinking one's urine one was swallowing some sort of sexual energizer. In any case, it is unlikely that Indians some 4,000 years ago had scientific reasons for drinking their own urine.

Another rather unscientific notion that seems to be accepted by urinophiles is that urine is really blood, since it is the byproduct of blood filtering by the kidneys. Yet if you need a blood transfusion, it is unlikely that urine will work just as well as blood.

Another misleading claim being made by urinophiles is that amniotic fluid is nothing but urine—fetal urine. If it is good for the fetus, it should be good for all of us. Here is what urine expert Martha Christy has to say on the subject:

the amniotic fluid that surrounds human infants in the womb is primarily urine. Actually, the infant "breathes in" urine-filled amniotic fluid continually, and without this fluid, the lungs don't develop. Doctors also believe that the softness of baby skin and the ability of in-utero infants to heal quickly without scarring after pre-birth surgery is due to the therapeutic properties of the urine-filled amniotic fluid. [www.all-natural.com/urine.html]

Some of the chemicals found in amniotic fluid are not going to be found in most urine samples. It is misleading to claim that amniotic fluid is primarily urine. It would be more accurate to say that they are both primarily water. I don't know what doctors she is talking about, but most parents will tell you that when their babies came out of the womb their skin was anything but beautiful. Comparisons to wrinkly prunes are quite common. So is comparison to one's skin after being in the swimming pool for a long time. The baby's skin becomes soft only after it has been out of its liquid environment for some time. There is a reason for that, according to Kim Kelly, a naturopathic doctor and nurse from Seattle. Newborns don't produce oil from their sebaceous glands until several weeks after their birth, which is why they often appear to have dry, flaky skin. Rather than amniotic fluid contributing to soft skin, according to Kelly, babies in the womb are protected by vernix, a creamy substance that serves as a barrier between the baby and the amniotic fluid. So, unless your urine is full of vernix, using it as a skin lotion is unlikely to work as a moisturizer.

What is urine? Urine is usually yellow or clear, depending on a person's health and diet. It usually has an ammonia-like odor due to the nitrogenous wastes that make up about 5% of the fluid (the remaining 95% is water). Certain foods can affect the odor, however. For example, asparagus breaks down into several sulfur-containing compounds and imparts a putrid odor on excretion.

Urine is a slightly acidic fluid that carries waste from the kidneys to the outside world. The kidneys have millions of nephrons, which filter toxins, waste, ingested water, and mineral salts out of the bloodstream. The kidneys regulate blood acidity by excreting excessive alkaline salts when necessary. The chief constituent of the nitrogenous wastes in urine is urea, a product of protein decomposition. Urea is, among other things, a diuretic. Average adult urine production is from 1 to 2 quarts a day. The bladder, where urine is stored for discharge, holds on average about 16–20 ounces of fluid, though the average discharge is about half that amount. In addition to uric acid, ammonia, and creatine, urine consists of many other waste products in minute quantities.

Being a waste product does not mean that a substance is toxic or harmful. It means that the body cannot absorb the substance at the present time. However, it is not likely to be healthful or useful except for those rare occasions when one is buried beneath a building or lost at sea for a week or two. In such situations drinking one's own urine might be the difference between life and death. As a daily tonic, however, there are much tastier ways to introduce healthful products into one's blood stream.

Further reading: Beyerstein 1997.

V

vampire

A mythical creature that overcomes death by sucking the blood from living humans. The most common variation of the myth portrays the vampire as a (un)dead person who rises from the grave at night to seek his victim from the realm of the living. The vampire is a popular theme of film makers who have started with Bram Stokers's novel (*Dracula*) and added a number of variations to the theme, for example, the ability to fly (like the vampire bat), a lust for beautiful women as victims who then become vampires on being bitten, fear of the symbol of the Christian cross, the repelling power of garlic or garlic flowers, and death by sunlight or by a special stake driven through the heart (a fitting death for a character based on the 15th-century warrior, Vlad the Impaler).

Legends of blood-sucking creatures are found in many cultures throughout history. One of the more popular bloodsuckers of our age is the **chupacabra**. Another is the alien **cattle mutilator**. The vampire is also a popular literary subject. Hence there are numerous descriptions of the origin, nature, and powers of vampires. What seems to be universal about vampire myths is their connection with the fear of death and the desire for immortality. Many peoples have practiced the ritual drinking of blood to overcome death. The Aztecs and Toltecs, for example, ate the hearts and drank the blood of captives in ritual ceremonies, most likely to satisfy the appetites of their gods and perhaps to gain for themselves power, fertility, and immortality. Also typical were the rites of Dionysus and Mithras, where the drinking of animal blood was required in the quest for immortality. Even today, some Christians believe that their priests perform a magical transubstantiation of bread and wine into the body and blood of Christ to be eaten and drunk in the quest to join God in eternal life.

We might say we've made progress in our ritualistic quest to overcome death. First, we sacrificed humans and drank their blood to keep the gods alive and happy, or to join them in overcoming death. We later came to substitute bulls or other animals for humans to achieve our goal. Finally, we progressed to a vegetarian menu of bread and wine. Even so, the basic truth is depressing: For anything to live, something or someone else must die. Whether this truth sets you free depends, I suppose, on your place at or on the dinner table.

This cultural link between vampirism and the quest for immortality seems to have been subordinated in literature and film, where other themes, such as blood for blood's sake, fear for fear's sake, sex for rating's sake, or entrance into the realm of the **occult**, seem to dominate. One sign of the cultural deterioration of our ancestor's noble quest for immortality can be seen in the modern secondary meaning of "vampire": a woman who exploits and ruins her lover. Another example of deterioration can be seen in the numerous web sites on vampires that appeal to occult or New Age interests, such as entering the so-called dark side of reality, gaining power, establishing a unique identity as a special person, or selling commercial products and games.

Apparently, role playing and masquerading as vampires is not enough to satisfy the bloodlust of some people, and covens or cults of "vampires" have emerged among some occultists. They seek blood to give them power or a sexual rush or to establish a unique and special fictional persona based on creating fear and mystery in others. Unlike our ancient ancestors, their power is sought not because of fear based on ignorance and misunderstanding of nature but because of ignorance and misunderstanding of themselves. Like other **cults**, these vampire covens are attractive to the young and the weak. Just a few years ago, such "vampyres" would have been considered ill or evil. Today, they are said to have an "alternative lifestyle."

Further reading: Anscombe 1994; Barber 1988; Clark 1995; Gelder 1994.

vastu

Vastu is India's version of China's **feng shui**. Vastu goes by many names (Vedic architecture, Sthapatya Ved, vastu vidya, and vastu shastra) but has one goal: to create buildings in harmony with nature and thereby increase happiness. However, this can only be done with the help of **astrology** and **numerology**.

One must understand all the planetary influences and numerological connections to the cosmos in order to have harmony in rooms such as the kitchen. If things are done properly, "the meals get cooked better, assimilation improves and frittering of energy gets checked" (www.indianest.com/vastu/v01.htm). Different qualities of energy are dispersed by different things at different times, and if one is not attuned to these changes, one will be out of tune. If your house is not aligned properly, you could get sick. I am fortunate I live in a house the entrance of which faces east. Many people in my neighborhood whose houses face west or south are more likely to suffer such things as poverty, negativity, lack of success, disease, and, of course, *anger* at being so poor, sick, and unsuccessful.

One cannot deny that a poorly designed workspace, kitchen, bathroom, or bedroom can cause a lot of stress, but one will search in vain for the scientific evidence that one's kitchen must be in harmony with moon energies or that sickness will befall you if your entrance faces west.

The main promoters of vastu in America are the advocates of **Transcendental Meditation**. They are fond of making such unsubstantiated claims as that brain physiology is significantly different when one faces east because neurons fire differently in the thalamus when facing east (www.mgc-vastu.com/html/book/abook1_4.html). They claim that if we could only build 1 billion new houses in the proper fashion, we could all live invincibly in peace and harmony with each other and the universe. They make many claims about the "natural law" (no relation to scientific laws of nature), "cosmic consciousness," the Maharishi effect, yogic flying, mantra chanting, and direct links of planets to brain parts. None of these claims has any scientific basis. In any case, one doesn't need a Vedic astrologer to tell you where the sun rises and sets, or what clutter looks like.

Velikovsky, Immanuel

In 1950, Macmillan published Immanuel Velikovsky's *Worlds in Collision,* a book that asserts, among many other things, that the planet Venus did not exist until recently. Some 3,500 years ago it was ejected from the planet Jupiter and became a comet that grazed Earth a couple of times before settling into its current orbit. Velikovsky (1895–1979), a psychiatrist by training, does not base his claims on astronomical evidence and scientific inference or argument. Instead, he argues on the basis of ancient cosmological myths from places as disparate as India and China, Greece and Rome, Assyria and Sumer. For example, ancient Greek mythology asserts that the goddess Athena sprang from the head of Zeus. Velikovsky identifies Athena with the planet Venus, though the Greeks didn't. The Greek counterpart of the Roman Venus was Aphrodite. Velikovsky identifies Zeus (whose Roman counterpart was the god Jupiter) with the planet Jupiter. This myth, along with others from ancient Egypt, Israel, and Mexico, is used to support the claim that "Venus was expelled as a comet and then changed to a planet after contact with a number of members of our solar system" (Velikovsky, *Worlds in Collision* [1972], p. 182; all page cites are to this edition).

Furthermore, Velikovsky then uses his Venus-the-comet claim to explain several events reported in the Old Testament as well as to tie together a number of ancient stories about flies. For example:

Under the weight of many arguments, I came to the conclusion—about which I no longer have any doubt—that it was the planet Venus, at the time still a comet, that caused the catastrophe of the days of Exodus. (181)

When Venus sprang out of Jupiter as a comet and flew very close to the earth, it became entangled in the embrace of the earth. The internal heat developed by the earth and the scorching gases of the comet were in themselves sufficient to make the vermin of the earth propagate at a very feverish rate. Some of the plagues [mentioned in Exodus] like the plague of the frogs . . . or of the locusts must be ascribed to such causes. (192)

The question arises here whether or not the comet Venus infested the earth with vermin which it may have carried in its trailing atmosphere in the form of larvae together with stones and gases. It is significant that all around the world people have associated the planet Venus with flies. (193)

The ability of many small insects and their larvae to endure great cold and heat and to live in an atmosphere devoid of oxygen renders not entirely improbable the hypothesis that Venus (and also Jupiter, from which Venus sprang) may be populated by vermin. (195)

Who can deny that vermin have extraordinary survival skills? But the cosmic hitchhikers Velikovsky speaks of are in a class all of their own. How much energy would have been needed to expel a "comet" the size of Venus, and how hot must Venus have been to have cooled down only to its current surface temperature of 750°K during the last 3,500 years? What evidence is there that any locust larvae could survive such temperatures? To ask such questions would be to engage in scientific discussion, but one will find very little of that in *Worlds in Collision*. What one finds instead are exercises in comparative mythology, philology, and theology, which together make up Velikovsky's planetology. That is not to say that his work is not an impressive exercise and demonstration of ingenuity and erudition. It is very impressive, but it isn't science. It isn't even history.

What Velikovsky does isn't science because he does not start with what is known and then use ancient myths to illustrate or illuminate what has been discovered. Instead, he is indifferent to the established beliefs of astronomers and physicists and seems to assume that someday they will find the evidence to support his ideas. He seems to take it for granted that the claims of ancient myths should be used to support or challenge the claims of modern astronomy and cosmology. In short, like the **creationists** in their arguments against evolution, he starts with the assumption that the Bible is a foundation and guide for scientific truth. Where the views of modern astrophysicists or astronomers conflict with certain passages of the Old Testament, the moderns are assumed to be wrong. Velikovsky, however, goes much further than the creationists in his reliance on ancient stories, for Velikovsky welcomes *all* ancient myths, legends, and folk tales. Because of his uncritical and selective acceptance of ancient myths, he cannot be said to be doing history, either. Where myths can be favorably interpreted to fit his hypothesis, he does not fail to cite them. The contradictions of ancient myths regarding the origins of the cosmos of or people are trivialized. If a myth fits his hypotheses, he accepts it and interprets it to his liking. Where the myth doesn't fit, he ignores it. In short, he seems to make no distinction between myth, legends, and history. Myths may have to be interpreted, but Velikovsky treats them as presenting historical facts. If a myth conflicts with a scientific law of nature, the law must be revised:

> If, occasionally, historical evidence does not square with formulated laws, it should be remembered that a law is but a deduction from experience and experiment, and therefore laws must conform with historical facts, not facts with laws. (11)

One of the characteristics of a reasonable explanation is that it be a likely story. To be reasonable, it is not enough that an explanation simply be a *possible* account of phenomena; it has to be a likely account. To be likely, an account usually must be in accordance with current knowledge and beliefs, with the laws and principles of the field in which the explanation is made. An explanation of how two chemicals interact, for example, would be unreasonable if it violated basic principles in chemistry. Those principles, while not infallible, have been developed not lightly but after generations of testing, observations, refutations, more testing, more observations, and so on. To go against the established principles of a field puts a great burden of proof on the one who goes against those principles. This is true in all fields that have sets of established principles and laws. The novel theory, hypothesis, or explanation that is inconsistent with already established principles and accepted theories has the burden of proof. The proponent of the novel idea must provide very good reasons for rejecting established principles. This is not because the established views are consid-

ered infallible; it is because this is the only reasonable way to proceed. Even if the established theory is eventually shown to be false and the upstart theory eventually takes its place as current dogma, it would still have been unreasonable to have rejected the old theory and accepted the new one in the absence of any compelling reason to do so. As onetime Velikovsky lieutenant Leroy Ellenberger puts it: "The less one knows about science, the more plausible Velikovsky's scenario appears."

Velikovsky was bitterly opposed by the vast majority of the scientific community, but the opposition may have been elicited mainly because of his popularity with "the New York literati" (Sagan 1979: 83). It is doubtful that many scientists even read Velikovsky, or read very much of *Worlds in Collision*. A knowledgeable astronomer and physicist would recognize after a few pages that the work is pseudoscientific twaddle. But the New York literary world considered Velikovsky a genius on par with "Einstein, Newton, Darwin and Freud" (ibid.). To the scientific world it might be more accurate to say he was a genius on par with L. Ron Hubbard whose **Dianetics** was also published in 1950. A number of scientists threatened to boycott Macmillan's textbook division as a sign of their disgust that such tripe as Velikovsky's should be published with such fanfare, as if the author were a great scientist. According to Leroy Ellenberger, "when the heat was applied by professors who were returning Macmillan textbooks unopened in protest and declining to edit new textbooks Macmillan gave the book over to Doubleday, which had no textbook division."

Velikovsky may not have been a genius but he was certainly *in*genious. His books, like Hubbard's, have sold in the millions worldwide. His explanations of parallels among ancient myths are very entertaining, interesting, and apparently plausible. But his explanation of universal collective amnesia of these worlds in collision is amusing only because of its extraordinary improbability. Imagine we're on Earth 3,500 years ago when an object about the same size as our planet is coming at us from outer space. It whacks us a couple of times, spins our planet around so that its rotation stops and starts again, creates great heat and upheavals from within our planet, and yet the most anyone can remember about these catastrophes are things like "and the sun stood still" (Joshua 10: 12–13) and other stories of darkness, storms, upheavals, plagues, floods, snakes and bulls in the sky, and so on. No one in ancient times mentions an object the size of Earth nearly colliding with us. You'd think someone among these ancient peoples, who all loved to tell stories, would have told their grandchildren about it. Someone would have passed it on. But no one on Earth seems to remember such an event.

Velikovsky explains why our ancestors did not record these events in a chapter entitled "A Collective Amnesia." He reverts to the old Freudian notion of **repressed memory** and neurosis. These events were just too traumatic and horrible to bear; so all humans buried the memory of them deep in their subconscious minds. Our ancient myths are neurotic expressions of memories and **dreams** based on real experiences:

> The task I had to accomplish was not unlike that faced by a psychoanalyst who, out of disassociated memories and dreams, reconstructs a forgotten traumatic experience in the early life of an individual. In an analytical experiment on mankind, historical inscriptions and legendary motifs often play the same role as recollections (infantile memories) and dreams in the analysis of a personality. (12)

The typically unscientific theories and fanciful explanations of **psychoanalysis** seem even less credible when applied to the entire population of Earth, yet to the New York literati, in love as they were with all things Freudian, speculations such as these could pass for genius.

It is not surprising that when one thumbs through any recent scientific book on cosmology, no mention is made of Velikovsky or his theories. His disciples blame this treatment of their hero on a conspiracy in the scientific community to suppress ideas that oppose their own. Even now, more than 50 years later, after all of his major claims have been rejected or refuted, Velikovsky still has his disciples who claim he is not being given credit for getting at least some things right. However, it does not appear that he got anything of importance right. For example, there is no evidence on Earth of a catastrophe occurring around 1500 B.C.E. Leroy Ellenberger notes that

> the Terminal Cretaceous Event 65 million years ago, whatever it was, left unambiguous worldwide signatures of iridium and soot. The catastrophes Velikovsky conjectured within the past 3500 years left no similar signatures according to Greenland ice cores, bristlecone pine rings, Swedish clay varves, and ocean sediments. All provide accurately datable sequences covering the relevant period and preserve no signs of having experienced a Velikovskian catastrophe. [abob.libs.uga .edu/bobk/vlesson.html]

Current disciples think Velikovsky should get credit for anticipating catastrophism of the type that ended the reign of the dinosaurs some 65 million years ago. Critic David Morrison thinks otherwise:

> Velikovsky focuses narrowly on encounters between the Earth and planets—Mars

and Venus. While he refers to Venus being accompanied by debris, the dominant agents of his catastrophes are tidal, chemical, and electrical interactions between planets, not meteoritic impacts. Remarkably, Velikovsky did not even accept (let alone predict) that the lunar craters are the result of impacts—rather, he ascribed them to lava "bubbles" and to electric discharges. I see nothing in his vision that relates to our current understanding of interplanetary debris and the role of impacts in geological and biological evolution. I conclude that Velikovsky was fundamentally wrong in both his vision of planetary collisions (or near collisions) and in his failure to recognize the role of smaller impacts and collisions in solar system history.

If anything, says Morrison, "Velikovsky with his crazy ideas tainted catastrophism and discouraged young scientists from pursuing anything that might be associated even vaguely with him" (Morrison 2001: 70). Morrison polled 25 leading contemporary scientists who have played a significant role in the development of the "new catastrophism," and not one thought that Velikovsky had had any significant *positive* influence on "the acceptance of catastrophist ideas in Earth and planetary science over the past half-century." Nine thought he had had a negative influence (ibid.).

Morrison points out several other misleading claims about Velikovsky being right. For example, Velikovsky was right that Venus is hot but wrong in how he came to that conclusion. He thought it was because Venus is a recent planet violently ejected from Jupiter and having traveled close to the sun. Venus is hot because of the greenhouse effect, something Velikovsky never mentioned. As to the composition of the atmosphere of

Venus, Velikovsky thought it was hydrogen rich with hydrocarbon clouds. NASA put out an erroneous report in 1963 that said Mariner 2 had found evidence of hydrocarbon clouds. In 1973 it was determined that the clouds are made mainly of sulfuric acid particles. Velikovsky was also right about Jupiter issuing radio emissions, but wrong as to why. He thought it was because of the electrically charged atmosphere brought on by the turbulence created by the expulsion of Venus. The radio emissions, however, are not related to the atmosphere but to "Jupiter's strong magnetic field and the ions trapped within it" (ibid.: 65).

One of the few scientists to criticize Velikovsky's work on scientific grounds was Carl Sagan (Sagan 1979: 97), who was roundly criticized for committing fallacies, making errors, and being intentionally deceptive in his argumentation. Henry Bauer does not even mention Sagan in his lengthy entry on Velikovsky in the *Encyclopedia of the Paranormal* (1996), unless he is making an oblique reference to Sagan when he writes about "some sloppy or invalid technical discussions by critics purporting to disprove Velikovsky's ideas." Whether Velikovsky's critics were fair-minded or not, there can be no denying the scientific indifference and incompetence of Velikovsky. He seemed satisfied that his study of myths established events that science must explain, regardless of whether those events clashed with the beliefs of the vast majority of the scientific community. Still, it would have been interesting to hear what Velikovsky or his disciples would have said to astronomer Phil Plait's criticism that the orbits of our moon and the moons of Jupiter would now be much different if Venus had been ejected from Jupiter (Plait 2002). One wonders how the master would have responded to our current knowledge that Jupiter and

Venus "have entirely different compositions" (ibid.: 178). Finally, one wonders what Velikovsky would have said to the claim that had Venus come as close to Earth as the master claimed, it would have destroyed everything and killed every living thing on the planet (ibid.: 181).

See also **ancient astronauts** and Zecharia **Sitchin**.

Further reading: Bauer 1984, 1996b; Ellenberger 1995; Friedlander 1972, 1995; Gardner 1957; and Goldsmith 1977.

victim soul

A person who suffers pain or sickness for another person. This notion is clearly related to the belief that Jesus of Nazareth redeemed humankind by suffering and dying for our sins. Christ as scapegoat probably goes back to the ancient Jewish custom of letting a goat loose in the wilderness on Yom Kippur after the high priest had (symbolically) laid upon the goat all the sins of the people (Leviticus 16).

According to the Most Rev. Daniel P. Reilly, Bishop of Worcester (Massachusetts), the concept of the victim soul was popular in the 18th and 19th centuries. He has set up a committee to investigate the claims that a young girl in his diocese, Audrey Santo, is a victim soul. The girl allegedly made an agreement with the Virgin Mary to be a victim soul when asked while on a pilgrimage to Medjugorje in Bosnia-Herzegovina (formerly part of Yugoslavia). At the time the girl was 4 years old and in a comatose state due to an accident that had destroyed a good part of her brain a year earlier. Her mother, Linda Santo, had hoped for a miracle cure. Instead, she says, the Virgin Mary appeared to her daughter and talked to her about being a victim soul. Linda Santo claims that her daughter suffers so others can live. She has turned her lifeless daughter into a living relic. She has set up a shrine in her home and accepts visitors, who can view the comatose Audrey through a glass window.

Further reading: Nickell 1993, 1999.

vinyl vision

The ability to see groove patterns in vinyl recordings and correctly identify musical recordings without the benefit of identifying labels. Only one person is on record as having this amusing ability: Dr. Arthur B. Lintgen, who demonstrated his talent in the 1980s to none other than James **Randi**. Even though Randi promised to pay $10,000 to anyone who could demonstrate a paranormal ability, Lintgen's ability is merely *ab*normal, that is, rare. Hence, his award was little more than a few moments of fame.

The mass media and the public lost interest once it was disclosed that he used ordinary sense perception, his vast knowledge of recordings of orchestral music from Beethoven onward, and deductive inference from general rules about such music. Though Lintgen continues to have a core following of true believers in his psychic powers, based on such astounding feats as identifying Beethoven's *Fifth Symphony* from across a room without even looking at the record, most consider Lintgen a decent fellow for not abusing his power over the gullible. He admitted, for example, that Beethoven's *Fifth* was the most common recording he was asked to identify (Seckel 1987a). He was simply making an educated guess, not using psychic powers, when he identified the recording without looking at it.

Lintgen did not claim to read individual notes in the record grooves. "The trick is to examine the physical construction of the recording and look at the relative playing time of each one of the movements or separations on the recording" (ibid.):

All phonograph grooves vary minutely in their spacing and contour, depending on the dynamics and frequency of the music on them. Lintgen says that grooves containing soft passages look black or dark gray. As the music gets louder or more complicated, the grooves turn silvery. Percussive accents are marked by tiny "jagged tooth marks." The doctor correlates what he sees with what he knows about music, matching the patterns of the grooves with compositional forms.

According to Lintgen, a Beethoven symphony will have a slightly longer first movement relative to its second movement, while Mozart and Schubert would compose in such a fashion that each movement in many cases would have the same number of bars. Beethoven, however, had set out in a new direction and that changed the dynamics of the recording. In addition, if there was a sonorous slow beginning, one could look at the recording at that point and see a long undulating groove that would not contain the sharp spikes that would identify sharp percussion. (*Time,* January 4, 1982)

Lintgen was featured on the ABC-TV program *That's Incredible* in 1981. Before a live audience in the auditorium of Abington Hospital, near Philadelphia, Stimson Carrow, professor of music theory at Temple University, tested him. Dr. Lintgen correctly identified 20 out of 20 recordings just by studying the record grooves (Holland 1981).

He admitted that he used only his knowledge, experience, and reasoning power to accomplish this amazing feat. "I have a knowledge of musical structure and of the literature," he said. "And I can correlate this structure with what I see. Loud passages reflect light differently. . . . Record companies spread the grooves in forte passages; they have a more jagged, saw-tooth

look. I also know how the pressings of different labels look, so I can often figure out who is conducting" (Holland 1981). He can also occasionally figure out the nationality of the orchestra, an ability that amazed even Randi. In Randi's test of Lintgen, the doctor not only identified a recording correctly but announced that the orchestra was German. The recording, he said, had an upturned edge, a feature that was unique to the Deutsche Grammophon label. He also saw that there was a "lack of junk in between the grooves," from which he inferred that the recording was digital. He also knew that "Deutsche Grammophon, up to that time, had only recorded German orchestras for their digital recordings" (Seckel 1987a).

Lintgen discovered his unusual ability at a party in the mid-1970s. Some friends commented to him that he knows so much about music he could probably read the grooves of records. He tried it and found that as long as the recordings are of music that he knows—orchestral music from Beethoven to the present—he has a high rate of success.

vitalism

The **metaphysical** doctrine that living organisms possess a nonphysical inner force or **energy** that gives them the property of life. Vitalists believe that the laws of physics and chemistry alone cannot explain life functions and processes. Vitalism is opposed to mechanistic **materialism** and its thesis that life emerges from a complex combination of organic matter.

The vitalistic principle goes by many names: **chi** or qi (China), **prana** (India and **therapeutic touch**), ki (Japan), Wilhelm Reich's **orgone energy**, **Mesmer's** animal magnetism, Bergson's élan vital (vital force), and so on. American advocates much prefer the term "energy." Many kinds of **alterna-**

tive **health practices** or energy medicines are based on a belief that health is determined by the flow of this alleged energy.

See also **Ayurvedic medicine**, **reiki**, and **qi gong**.

Further reading: Park 1997.

W

Waldorf Schools

See **anthroposophy**.

Wallach, Joel D., M.S., D.V.M., N.D.

A veterinarian and **naturopath** who claims that all diseases are due to mineral deficiencies, that everyone who dies of natural causes dies because of mineral deficiencies, and that just about anyone can live more than 100 years if he or she takes daily supplements of colloidal minerals harvested from a pit in Utah. Wallach claims that minerals in foods and most supplements are "metallic" and not as effective as "plant-based" colloidal minerals, which is nonsense. Being colloidal has to do with the *size* of the mineral particles, not their effectiveness. Wallach learned all this from living on a farm, working with Marlin Perkins (of Mutual of Omaha's *Wild Kingdom* fame), doing necropsies on animals and humans, reading stories in *National Geographic* magazine, and reading the 1934 novel by James Hilton, *Lost Horizon*. He certainly didn't learn any of it from science texts.

Dr. Wallach makes his claims about minerals despite the fact that in 1993, a research team from the Centers for Disease Control and Prevention in Atlanta, Georgia, reported the results of a 13-year study on 10,758 Americans that failed to find any mortality benefits from vitamin and mineral supplements. The study found that even though supplement users smoke and drink less than nonusers, eat more fruits and vegetables than nonusers, and are more affluent than nonusers, they didn't live any longer than nonusers. The study also found no benefit from taking vitamin and mineral supplements for smokers, heavy drinkers, or those with chronic diseases.

The basic appeal of Dr. Wallach is the *hope* he gives to people who fear or are mistrustful of medical doctors and scientific knowledge. He gives hope to those who want to live for a really long time. He gives hope to those who are diagnosed with diseases for which current medical knowledge has no cure. He gives hope to those who are looking for a surefire way to avoid getting a terminal disease. And he gives hope to those who want to be healthy but who do not want to diet or exercise. All we have to do is drink a magic elixir of colloidal minerals and we'll be healthy. You can't just take your minerals in pill form, he warns us. You must take the colloidal variety in liquid form. Furthermore, this elixir must come from a pit in Utah, the only source approved by Dr. Wallach, and the only one, I suspect, in which he has a financial interest.

Dr. Wallach seems to be most famous for a widely circulated audiotape he calls "Dead Doctors Don't Lie." The label of the tape notes that Dr. Wallach was a Nobel Prize nominee. This is true. The Association of Eclectic Physicians nominated him for a Nobel Prize in medicine "for his notable and untiring work with deficiencies of the trace mineral selenium and its relationship to the congenital genesis of Cystic Fibrosis." The Association of Eclectic Physicians is a group of naturopaths founded in 1982 by two naturopathic physicians, Dr. Edward Alstat and Dr. Michael Ancharski. In his book *Let's Play Doctor* (coauthored with Ma Lan, M.D., M.S.), Wallach states that cystic fibrosis is preventable, is 100% curable in the early stages, can be managed very well in chronic cases, leading to a normal life expectancy (75 years). If these claims were true, he might have won the prize. He didn't win, but he gave a lot of false hope to parents of children with cystic fibrosis.

The basic danger of Dr. Wallach's theories is not that taking colloidal minerals will harm people, or even that many people will be wasting their money on a product they do not need. Many of his claims are not backed up with scientific **control studies**, but are anecdotal or fictional. The basic danger is that because he and other naturopaths exaggerate the role of minerals in good health, they may be totally ignored by the scientific community even if they happen to hit on some real connections between minerals and disease. Furthermore, there is the chance that legitimate scientific researchers may avoid this field for fear of being labeled a kook.

Dr. Wallach falsely claims that there are five cultures in the world that have average lifespans of between 120 and 140 years: the Tibetans in Western China, the Hunzas in Eastern Pakistan, the Russian Georgians and the Armenians, the Abkhasians, and the Azerbaijanis. He also mentions the people of the Vilcabamba in Ecuador, and those who live around Lake Titicaca in Peru and Bolivia. The secret of their longevity is "glacier milk," or water full of colloidal minerals. It is probably news to these people that they live so long. Dr. Wallach does not mention on what scientific data he bases his claims, but I am sure there are many anthropologists and tour book authors who would like to know about these Shangri-la havens.

The label on the "Dead Doctors Don't Lie" tape says, "Learn why the average life span of an MD is only 58 years." On his tape, Dr. Wallach claims "the average life span of an American is 75 years, but the average life span of an American doctor is only 58 years!" Maybe dead doctors don't lie, but this living one certainly stretches the truth. If he is telling the truth, it is not the whole truth and nothing but the truth. According to Kevin Kenward of the American Medical Association: "Based on over 210,000 records of deceased physicians, our data indicate the average life span of a physician is 70.8 years." One wonders where Dr. Wallach got his data. The only mention in his tape of data on physician deaths is in his description of a rather gruesome hobby of his: He collects obituaries of local physicians as he takes his mineral show from town to town. He may be somewhat selective as a collector, however.

On his tape, Dr. Wallach says:

> what I did was go back to school and become a physician. I finally got a license to kill (laughter), and they allowed me to use everything I had learned in veterinary school about nutrition on my human patients. And to no surprise to me, it worked. I spent 12 years up in Portland, Oregon, in general practice, and it was very fascinating.

Dr. Wallach is an N.D., a doctor of naturopathy, not an M.D., as his tape suggests. It is unlikely that most of the people in his audience know that naturopaths call themselves physicians and that there is a very big difference between an M.D. and an N.D. He also claims he did hundreds of autopsies on humans while working as a veterinarian in St. Louis. How does a veterinarian get to do human autopsies?

> Well, again, to make a long story short, over a period of some twelve years I did 17,500 autopsies on over 454 species of animals and 3,000 human beings who lived in close proximity to the zoos, and the thing I found out was this: every ani-

mal and every human being who dies of natural causes dies of a nutritional deficiency.

To accomplish this feat, he would have to do six autopsies a day, working 5 days a week for 12 years and taking only a 2-week vacation each year. He was allegedly performing all these autopsies in addition to his other duties, and presumably while he was writing essays and books as well. Maybe all those minerals gave him superhuman powers.

Dr. Wallach's "Dead Doctors Don't Lie" tape is both an attack on the medical profession and a panegyric for minerals. The attack is vicious and mostly unwarranted, which weakens his credibility about the wonders of minerals. He does not come across as an objective, impersonal scientist. He delights in ridiculing "Haavaad" University and cardiologists who die young from heart attacks, many of whom went into the field because of congenital heart defects. He reverts to name calling on several occasions, as well. Doctors, he says, routinely commit many practices that would be considered illegal in other fields. At one point he claims that the average M.D. makes over $200,000 a year in kickbacks. This ludicrous claim didn't even get a peep of skeptical bewilderment from his audience. He sounds like a bitter, rejected oddball who is getting even with the medical profession for ignoring him and his "research."

He cites a 1939 paper to demonstrate the need for minerals due to depletion in the soil. He then claims, without justification, that mineral supplements won't help us if they come from off the shelf of our local supermarket. The supplement must be "colloidal" and, as noted above, must come from a special pit in Utah. He suggests at one point in his tape that minerals in pill form aren't absorbed at all; they just pass right through the body and out into the sewer lines. But why would anyone trust someone who tells stories about people in China who lived to be over 250 years old or about a 137-year-old cigar-smoking woman? Not convinced of this man's trustworthiness? According to Dr. Wallach, for the past 20 years there have been cures for arthritis, diabetes, and ulcers. Veterinarians, who also discovered the cause of Alzheimer's disease years ago, discovered these cures. Tell that to the millions of people suffering from these diseases.

Ellen Coleman, a registered dietician and nutrition columnist, has another view of Wallach's product: "Colloidal mineral products have not been proven safe or effective. They are not better absorbed than regular mineral supplements" (www.hcrc.org/faqs/colloid.html). James Pontolillo, a research scientist, is concerned that colloidal mineral products may contain toxic organic compounds (www.quackwatch.com/01Quackery RelatedTopics/DSH/colloidalminerals.html). The National Nutritional Food Association says that some colloidal mineral products "contain aluminum or toxic minerals; others are high in sodium. Some do not contain detectable amounts of minerals listed on their labels. Finally, there is no evidence that colloidal minerals are more bioavailable than those found in other forms" (http://www.ncahf.org/nl/1997/1-2.html).

Nevertheless, Dr. Wallach has spawned a small industry of mineral sellers, including some **multi-level marketing** projects on the Internet.

Further reading: Barrett and Herbert 1994.

warlock

Literally, "one who breaks his word" (from the Old English w[ae]r-loga), that is, a deceiver. The word was used to designate

Satan and came to designate wizards, **sorcerers**, and male **witches**—those who practice black magic.

Watsonville, Our Lady of

On June 17, 1993, in Watsonville, California, Anita Contreras knelt down to pray for her children, and she saw a foot-high image of the Virgin Mary in the bark of an oak tree. Since then, thousands of pilgrims have flocked to Our Lady of Watsonville, hoping for a **miracle**.

Many Roman Catholics worship Mary as the Mother of God. Mexicans have been especially fond of her since her apparent apparition in 1531 to Cuauhtlatoatzin, a Nahuan peasant and Christian convert who took on the name of Juan Diego. (Watsonville is about 62% Mexican-American.)

The story of Our Lady of Guadalupe is a

Overlay of Our Lady of Guadalupe on the bark of tree in Watsonville.

bit more dramatic than that of Our Lady of Watsonville. Legend has it that Juan Diego was an ascetic mystic, who frequently walked barefoot the 14 miles from his village to church in Tenochtitlan (Mexico City). It was on these walks that he had several visions of the Virgin Mary. He allegedly brought to the bishop his cloak, on which an image of the Virgin had been painted. The cactus fiber cloak is the centerpiece of the Basilica of the Virgin of Guadalupe in Mexico City. Many believe that the painting is of heavenly origin. Skeptics believe it was done by a human artist and passed off as being of miraculous origin in order to win more converts to Christianity.

The name "Guadalupe" is Spanish and is a bit mysterious, since there was no town or shrine near Cuauhtitlan, Juan's village, by that name when the legend began. It is thought that the word derives from a Nahuatl word, *coatlaxopeuh,* which supposedly sounds like Guadalupe in Spanish and means something like "one who crushes the serpent." (The serpent can be identified with Satan or with the Aztec serpent-god Quetzalcoatl.) It is also possible that the legend has Juan saying that the Virgin was to be called Our Lady of Guadalupe because the one who invented it was Spanish. The creator of the name may have been intrigued by a statue of Our Lady of Guadalupe in Estremadura, Spain.

The improbability of the story of Juan Diego, his visions, and the miraculous painting has not deterred the faithful from belief. In fact, only a deep religious **faith** could account for the continued popularity of Virgin Mary sightings. The skeptic understands the desire to have a powerful ally in heaven, one who will protect and guide, console and love you no matter what troubles you have here on Earth. The skeptic also understands how easy it is to find confirmation for almost any belief, if one is very selective in one's thinking and perception. We understand how easy it is to see things that others do not see. Having visions also makes one feel special. Thus, it is not difficult to understand why many people see the Virgin Mary in the clouds, in a tortilla, in a dish of spaghetti, in patterns of light, or in the bark of a tree.

A shrine to Our Lady of Watsonville has been set up near the soccer fields and playgrounds of Pinto Lake County Park. Father Roman Bunda celebrated Mass at the site on the sixth anniversary of Contreras's discovery of the image in the bark. "For those who believe, no explanation is necessary," said Fr. Bunda. "For those who don't believe, no explanation is possible." He's right about the first part.

See also **pareidolia**.

werewolf

An animal from folklore believed to consume human flesh or blood, and who has the power to change from human to wolf and back again. (*Wer* is an Old English term for "man.") While there are no documented cases of any human turning into a wolf and back, there are documented cases of humans who *believed* they were werewolves. To suffer from such a delusion is known as **lycanthropy**.

Some have speculated that certain excessively hirsute individuals resemble wolves and that the legend of the werewolf may have a basis in the genetic disorder known as *hypertrichosis* or in some other endocrine disorder, such as adrenal virilism, basophilic adenoma of the pituitary, masculinizing ovarian tumors, or Stein-Leventhal syndrome.

Further reading: Noll 1992.

Wicca

A neo-pagan nature religion based on beliefs and rites said to be of ancient ori-

gin. Wicca claims a direct connection to the ancient Celtic tradition, which is thought to be more in tune with natural forces than Christianity and other modern religions of the West. However, rather than see Wiccans as members of a religion, it might be more accurate to see them as sharing a spiritual basis in nature and natural phenomena. For Wiccans have no written creed that the orthodox must adhere to. Nor do they build stone temples or churches to worship in. They practice their rituals in the great outdoors: in parks, gardens, forests, yards, or hillsides. According to a Wicca FAQ page,

> "Wicca" is the name of a contemporary Neo-Pagan religion, largely promulgated and popularized by the efforts of a retired British civil servant named Gerald Gardner [late 1940s]. In the last few decades, Wicca has spread in part due to its popularity among feminists and others seeking a more woman-positive, earth-based religion. Like most Neo-Pagan spiritualities, Wicca worships the sacred as immanent in nature, drawing much of its inspiration from the non-Christian and pre-Christian religions of Europe. "Neo-Pagan" simply means "new pagan" (derived from the Latin *paganus*, "country-dweller") and hearkens back to times before the spread of today's major monotheistic (one god) religions. A good general rule is that most Wiccans are Neo-Pagans but not all Pagans are Wiccans. [moonravens.topcities.com/intro1.htm]

A good general rule seems to be that there is no single set of beliefs or practices that constitutes Wicca, though a couple of beliefs seem to recur: *An' it harm none, do what you will.* This is known as "the Wiccan rede." Another common belief is the three-fold law, which states any good or evil that one does is repaid threefold. Both the rede and the threefold law seem to have been derived from Charles G. Leland's *Aradia or the Gospel of the Witches* (1889), the closest thing to a Wiccan sacred text.

Wiccans practice a number of rituals associated with such natural phenomena as the four seasons, the solstices, and the equinoxes. Their symbols are based on the connectedness of nature to human life. For example, they celebrate summer in a fertility rite known as Beltane. Rather than pray to a transcendent being, Wiccans seem more concerned with self-awakening, with arousing their connectedness to nature and nature gods, female as well as male. Their rituals seem to be metaphors for psychological processes. They sing, they dance, they chant. They burn candles and incense. They use herbs and **charms**. Often, Wiccans favor herbs to traditional medicines. In group rituals they express their desires to the community. However, they don't cast **spells**. They ask for blessings from north, south, east, and west. They meditate. They don't cook weird poisonous stews in cauldrons. They don't fly off on brooms. They don't pray for harm to their enemies.

Wiccans do share one thing in common with Christians, however. Both believe that the indifferent destructiveness of nature is essentially something good. We should be thankful for the blessings of nature (or God), including the pumiced humans at Pompei, the children swept away in flash floods, those sucked out of their homes by the tornado and thrown into the dark sky of the volcano, the millions who bake under an uncaring sun in parched lands, the innocent monsters deformed by uncaring biological laws, those devoured by great cracks in the earth, those drowned in hurricanes, the millions left homeless each year by indifferent forces ravaging an indifferent landscape. Only in their mythologies have Wiccan **magick** or Christian **prayer** stopped the flood, doused the lightning bolt, stilled the whirlwinds of tornado and hurricane, calmed the quaking earth, or put to sleep the tsunami.

The attractiveness of Wicca may be due to its friendliness toward women, its naturalistic view of sex, and its promise of power through magick. It is very popular among women, and it is tempting to say that Wicca is women's revenge for the centuries of misogyny and "femicide" or "gynicide" practiced by established religions such as Christianity. Wicca, like the Celtic religion, allows women full participation in the practice. Women are equals, if not superiors, of men. Women in Celtic mythology are unusual, to say the least. They are intelligent, powerful warriors, ruthless, sexually aggressive, and leaders of nations.

Finally, it should be noted that Wicca is not related to **Satan** worship. That practice is related to the persecution of "witches" by Christians, especially during the medieval and Spanish Inquisitions. The *Malleus Maleficarum (The Witch Hammer)* of Heinrich Kramer and Jakob Sprenger (1486) served as a guidebook for Inquisitors to help them identify and prosecute witches. It describes three necessary concomitants of witchcraft: the Devil, a witch, and the permission of Almighty God. The spirit of the witch-hunters lives on in the hearts of many devout Christians who continue to persecute Wiccans, among others, as devil worshippers. The modern inquisitors do not burn people to death. Rather, they try to abolish Halloween, school mascots, books that mention witches, and any sign, symbol, or number the Christians associate with Satan.

Further reading: Allen 2001; Cahill 1995; Sagan 1995.

wishful thinking

Interpreting facts, reports, events, and perceptions according to what one would *like* to be the case rather than according to actual evidence. If it is done intentionally and without regard for the truth, it is called misinterpretation, falsification, dissembling, disingenuous, or perversion of the truth.

See also **self-deception**.

witch

Someone in consort with **Satan**, the spirit who rebelled against God but whom God suffered to live. "Thou shalt not suffer a witch to live," according to the book of Exodus (xxii, 18). This and other Biblical admonitions and commands both defined the witch and prescribed his or her fate.

Today, the typical witch is generally portrayed as an old hag in a black robe, wearing a pointed black cap, and flying on a broomstick across a full moon. Children dress up as witches on Halloween, much to the dismay of certain pious Christians. Hollywood, on the other hand, conjures up images of sexy women with **paranormal** powers such as **psychokinesis**, **mind control**, hexing, and an array of other **occult** talents. Pious Christians who think pagans practice witchcraft sometimes identify "Pagan" or anti-Christian New Age religions with witches. Some pagans refer to themselves as "witches" and their groups as "covens." (Some male witches are very touchy about being called "**warlocks**.") Some of the members of these groups call themselves "**sorcerers**" and worship Satan, that is, they believe in Satan and perform rituals that they think will get them a share of Satan's supernatural occult powers. (Some are very touchy about being called "sorcerers.") Most New Age witches do not worship Satan, however, and are very touchy about the subject. They would rather be associated either with the occult and **magick**

or with attempts to reestablish a kind of nature religion, which their members associate with ancient pagan religions, such as the ancient Greek or the Celtic, especially **Druidism**. The neo-pagans also refer to both men and women witches as *witches*. One of the largest and most widespread of these nature religions is **Wicca**.

The witches of Christian mythology were known for their having sex with Satan and using their magical powers to do evil of all sorts. The culmination of the mythology of witchcraft came about from the 15th to the 18th centuries in the depiction of the witches' Sabbath. The Sabbath was a ritual mockery of the Mass. Witches were depicted as flying up chimneys at night on broomsticks or goats, heading for the Sabbath where the Devil (in the form of a feathered toad, a crow or raven, a black cat, or a he-goat) would perform a blasphemous version of the Mass. There would also be obscene dancing, a banquet, and the brewing of potions in a huge cauldron. The banquet might include some tasty children, carrion, and other delicacies. The witches' brew was apparently to be used to hurt or kill people or to mutilate cattle (de Givry 1971: 83). Those initiated into the satanic mysteries were all given a physical mark, such as a claw mark under the left eye. The Devil was depicted as a goat or satyr or some mythical beast with horns, claws, tail, and/or strange wings: a mockery of angel, man, and beast. One special feature of the Sabbath included the ritual kiss of the devil's ass (ibid.: 87), apparently a mockery of the traditional Christian act of submission of kneeling and kissing the hand or ring of a holy cleric. Numerous **testimonials** to having witnessed the witches' Sabbath are recorded. For example, a shepherdess, Anne Jacqueline Coste, reported in the middle of the 17th century that during the night of the feast of St. John the Baptist, she and her companions heard a dreadful uproar and

> looking on all sides to see whence could come these frightful howlings and these cries of all sorts of animals, they saw at the foot of the mountain the figures of cats, goats, serpents, dragons, and every kind of cruel, impure, and unclean animal, who were keeping their Sabbath and making horrible confusion, who were uttering words the most filthy and sacrilegious that can be imagined and filling the air with the most abominable blasphemies. (ibid.: 76)

Such stories had been told for centuries and were accepted by pious Christians without a hint of skepticism as to their veracity. Such tales were not considered delusions, but accurate histories.

Pierre de l'Ancre, in his book on **angels**, demons, and sorcerers published in 1610, claims he witnessed a Sabbath. Here is his description:

> Here behold the guests of the Assembly, having each one a demon beside her, and know that at this banquet are served no other meats than carrion, and the flesh of those that have been hanged, and the hearts of children not baptized, and other unclean animals strange to the custom and usage of Christian people, the whole savourless and without salt.

The claims made in books such as de l'Ancre's and the depictions of Sabbath activities in works of art over several hundreds of years were not taken as humorous fictions or psychological manifestations of troubled spirits. These notions, as absurd and preposterous as they might seem to us, were taken as gospel truth by millions of pious Christians.

The cruelties and delusions went on for centuries. Witch hunting was not abolished in England until 1682. The hunt spread to America, of course, and in 1692, in Salem, Massachusetts, 19 witches were hanged. (In 1711, the Massachusetts State Legislature exonerated all but six of the accused witches. In 1957 the state legislature passed a resolution exonerating Ann Pudeator "and certain other persons." They were all exonerated by the state legislature in 2001.)

The last judicial execution for witchcraft in Europe took place in Poland in 1793, when two old women were burned. A wizard, however, died as a result of an unofficial ordeal by water in England in 1865, and in 1900 two Irish peasants tried to roast a witch over her own fire (Smith 1952: 295).

Whatever the psychological basis for the creation of an anti-Church with witches and sorcerers joined with Satan to mock and desecrate the symbols and rituals of the Church, the practical result was a stronger, more powerful Church. No one knows how many witches, heretics, or sorcerers were tortured or burned at the stake by the pious, but the fear generated by the medieval and Spanish Inquisitions must have affected nearly all in Christendom. Being accused of being a witch was as good as being convicted. In truth, the Church ran a Reign of Terror the superior in many ways to those of Stalin or Hitler, whose terrors lasted only a few years and were restricted to limited territories. The Church's Terror lasted for several centuries and extended to all of Christendom. The Church's Terror was also aimed mainly at women. Thus, it is not strange that those religions today whose members call themselves witches or sorcerers should be anti-Christian, **pagan** and woman-centered, or satanic. It is not strange that these New Age religions exalt whatever the Church condemned (such as egoism and healthy sexuality in adults, whether homosexual or not) and condemn whatever the Church exalted (such as self-denial and the subservient role of women).

Witchcraft and sorcery may never be eradicated, and both are still practiced in many countries around the world.

Further reading: Carus 1996; Hicks 1991; Hill 1995; Sagan 1995.

wizard

Literally, "a wise person" (from the 15th-century Middle English *wysard*). The term came to refer to someone who claims to have supernatural knowledge or power, such as a **sorcerer** or one devoted to black magic. Today, the term is extended to refer to anyone who has a seemingly magical skill.

X

xenoglossy

The alleged speaking or writing in a language entirely unknown to the speaker. The probability of this happening is about zero.

Xenoglossy has been reported in cases of **automatic writing**, **past life regression**, and spirit **mediums**. These cases are reasonably dismissed as fraudulent, the uttering of gibberish, or, in the case of past life regression, the remembering of phrases or expressions heard in *this* life.

See also **automatic writing** and **glossolalia**.

Further reading: Thomason 1996.

Y

yeti

See **Bigfoot**.

yin and yang

According to traditional Chinese philosophy, yin and yang are the two primal cosmic principles of the universe. Yin is the passive, female principle and is associated with the dark and what is cold or moist. Yang is the active, masculine principle and is associated with light and what is warm or dry. According to legend, the Chinese emperor Fu Hsi claimed that the best state for everything in the universe is a state of harmony represented by a balance of yin and yang. Unsurprisingly, legend has it that, according to Fu Hsi, true harmony requires yang to be dominant. It's just the nature of things.

See also **acupuncture**, **chi**, *I Ching*, **koro**, and **macrobiotics**.

Z

Zener cards

Used by some **parapsychologists** to test **psychic** ability. There are five kinds of cards: a star, three vertical wavy lines, a plus sign, a circle, and a square. A deck of Zener cards consists of five of each symbol. The cards might be shuffled and the subject then tries to guess the order of the cards. Or a sender might look at a card and then try to **telepathically** communicate the perception to a receiver.

Zener cards.

Since there are 25 cards in the deck and five kinds of card, there is a one in five chance (or 20%) that any given card is on top of the deck or being viewed by a sender. A correct guess is called a "hit." Anything significantly higher than 20% hits in the long run would indicate that something other than chance is at work. In the short run, higher percentages are expected by chance. Thus, if you get nine out of 25 correct (36%), that would *not* be statistically significant. If you got 36% correct over 100 trials through the deck, that would be statistically significant and would indicate that something else besides chance is going on. Maybe you're psychic, maybe you are unconsciously picking up cues, or maybe you're cheating.

Zermatism

A **pseudoscience** invented by Stanislav Szukalski (1893–1987), a gifted Polish artist and immigrant to the United States, in a 39-volume work. Zermatism maintains that all human culture derived from Easter Island after the flood that destroyed all living creatures except those on **Noah's Ark**. All languages derive from a single source (the Protong) and all art is a variation on a few themes that can be distilled down to a single series of universal symbols. Zermatism explains the differences in races and cultures by claiming that they are due to the crossbreeding of species. The first

humans were nearly perfect but they mated with the Yeti, with abominable results.

zombie

A dead body with no soul, created by the black magic of voodoo **sorcerers**. Voodoo is a religion that originated in Africa and was brought to Haiti in the early 16th century by West African slaves. The slaves could not practice their religion openly and were forced to adopt in public the practices of the French Catholic settlers. Voodoo is still a popular religion in Haiti and in places where Haitians have emigrated, such as New Orleans. *Vodu* is an African word meaning "spirit" or "god." The black magic of voodoo sorcerers allegedly consists of various poisons (perhaps that of the puffer fish) that immobilize a person for days, as well as hallucinogens administered on revival. The result is a brain-damaged creature used by the sorcerers as slaves: the zombie. The zombie is not to be confused with the "zombie astral," whose soul (*ti-bon-ange*) is controlled by the sorcerer.

It is quite understandable that a religion practiced under slavery would emphasize evil spirits. It is a cruel irony that some in the religion would evolve to worship at evil's altar and engage in practices that not only enslave others but also keep the community in line from fear of being turned into a zombie slave.

Many people are skeptical of the existence of zombies; that is, they are skeptical that a dead person could be revived with or without retaining his or her "**soul**" or "self-consciousness" or "mind." Once you are dead, you are dead forever. For those who don't believe a person has a soul, death is not the separation of the body from the soul, but the end of life and consciousness. The voodoo zombie, if it exists at all, is not a dead person, but a living person who has been brain-damaged.

Further reading: Davis 1985, 1988.

Bibliography

Abanes, Richard. 1998. *End-Time Visions: The Road to Armageddon.* Four Walls Eight Windows.

Abell, George O. 1979. "The Moon and the Birthrate." *Skeptical Inquirer,* Summer.

———. 1986. "The Alleged Lunar Effect." In *Science Confronts the Paranormal,* Kendrick Frazier, editor. Prometheus Books.

Abell, George O., and Barry Singer, editors. 1981. *Science and the Paranormal.* Scribner.

Adams, James L. 1990. *Conceptual Blockbusting: A Guide to Better Ideas,* 3rd ed. Perseus Press.

Adler, Stephen J. 1993. "Debatable Device." *Wall Street Journal,* Feb. 3.

Alcock, James E. 1990. *Science and Supernature: A Critical Appraisal of Parapsychology.* Prometheus Books.

Alcock, James E. 1996a. "Channeling." In *The Encyclopedia of the Paranormal,* Gordon Stein, editor. Prometheus Books.

Alcock, James. 1996b. "Déjà Vu." In *The Encyclopedia of the Paranormal,* Gordon Stein, editor. Prometheus Books.

Alexander, Richard D. 1987. *The Biology of Moral Systems.* Aldine de Gruyter.

Allen, Charlotte. 2001. "The Scholars and the Goddess." *Atlantic Monthly,* Jan.

Allport, Gordon. 1954. *The Nature of Prejudice.* Addison-Wesley.

Amundsom, Ron. 1985. "The Hundredth Monkey Phenomenon." *Skeptical Inquirer,* Summer. Reprinted in *The Hundredth Monkey and Other Paradigms of the Paranormal,* Kendrick Frazier, editor. Prometheus Books, 1991.

———. 1987. "Watson and the Hundredth Monkey Phenomenon." *Skeptical Inquirer,* Spring.

Anderson, Robert B. 1996. "The Case of the Missing Link." *Pacific Discovery,* Spring.

Ankerberg, John, and John Weldon. 1996. *Encyclopedia of New Age Beliefs.* Harvest House Publishers.

Anscombe, Roderick. 1994. *The Secret Life of Laszlo, Count Dracula.* Hyperion.

Arnhart, Larry. 2001. "Evolution and the New Creationism—A Proposal for Compromise." *Skeptic* 8, no. 4.

Ashcraft, Mark H. 1994. *Human Memory and Cognition.* Addison-Wesley.

Ashmore, Malcolm. 1993. "The Theatre of the Blind: Starring a Promethean Prankster, a Phony Phenomenon, a Prism, a Pocket, and a Piece of Wood." *Social Studies of Science* 23.

Asimov, Isaac. 1988. "The Radiation That Wasn't." *Magazine of Fantasy and Science Fiction,* March.

Asimov, Isaac, Martin H. Greenberg, and Charles G. Waugh, editors. 1988. *Atlantis.* New American Library.

Asserinsky, E., and N. Kleitman. 1953. "Regularly Occurring Periods of Ocular Motility and Concomitant Phenomena During Sleep." *Science.*

Atack, Jon. 1990. *A Piece of Blue Sky: Scientology, Dianetics, and L. Ron Hubbard Exposed.* Carol Publishing Group.

Augstein, Hannah, editor. 1996. *Race: The Origins of an Idea, 1760–1850.* Thoemmen Press.

Babcock, Marguerite, and Christine McKay, editors. 1995. *Challenging Codependency. Feminist Critiques.* University of Toronto Press.

Baddeley, Alan D. 1998. *Human Memory: Theory and Practice.* Allyn & Bacon.

Baker, Robert. 1987–88. "The Aliens Among Us: Hypnotic Regression Revisited." *Skeptical Inquirer,* Winter.

Baker, Robert A. 1990. *They Call It Hypnosis.* Prometheus Books.

———. 1996a. *Hidden Memories: Voices and Visions from Within.* Prometheus Books.

Baker, Robert. 1996b. "Prophetic Dreams." In *The Encyclopedia of the Paranormal,* Gordon Stein, editor. Prometheus Books.

Barber, Paul. 1988. *Vampires, Burial, and Death: Folklore and Reality.* Yale University Press.

Barlow, Connie. 2001. *The Ghosts of Evolution.* Basic Books.

Bar-Natan, Dror, Alec Gindis, Aryeh Levitan, and Brendan McKay. 1997. "Report on New ELS Tests of Torah," www.answering-islam.org.uk/Religions/Numerics/report.html. May 29.

Barrett, Dierdre. 1996. *Trauma and Dreams.* Harvard University Press.

Barrett, Stephen, M.D. 1988. "Ayurvedic Mumbo-Jumbo," www.quackwatch.com/04ConsumerEducation/chopra.html.

Barrett, Stephen, M.D. 1991. "Homeopathy: Is It Medicine?" In *The Hundredth Monkey and Other Paradigms of the Paranormal—A Skeptical Inquirer Collection,* Kendrick Frazier, editor. Prometheus Books.

———. 1997. "Dubious Aspects of Osteopathy," www.quackwatch.com/04ConsumerEducation/QA/osteo.html.

———. 2000a. "Some Thoughts about Faith Healing," www.quackwatch.com/01QuackeryRelatedTopics/faith.html, July.

———. 2000b. "Therapeutic Touch," www.quackwatch.com/01QuackeryRelatedTopics/tt.html.

———. 2001a. "A Close Look at Naturopathy," www.quackwatch.com/01QuackeryRelatedTopics/Naturopathy/naturopathy.html.

———. 2001b. "Aromatherapy: Making Dollars Out of Scents," www.quackwatch.com/01QuackeryRelatedTopics/aroma.html.

Barrett, Stephen, and Kurt Butler, editors. 1992. *A Consumers Guide to Alternative Medicine: A Close Look at Homeopathy, Acupuncture, Faith-Healing, and Other Unconventional Treatments.* Prometheus Books.

Barrett, Stephen, and Victor Herbert. 1994. *The Vitamin Pushers: How the "Health Food" Industry Is Selling America a Bill of Goods.* Prometheus.

Barrett, Stephen, M.D., and Victor Herbert, M.D., J.D. 2001. "Questionable Cancer Therapies," www.quackwatch.com/01QuackeryRelatedTopics/cancer.html.

Barrett, Stephen, and William T. Jarvis, editors. 1993. *The Health Robbers: A Close Look at Quackery in America.* Prometheus Books.

Barry, Dave. 1997. "Altered States." *Miami Herald,* April 13.

Barsky, Arthur J., M.D. et al. 2002. "Nonspecific Medication Side Effects and the Nocebo Phenomenon." *Journal of the American Medical Association* 287, no. 5, Feb. 6.

Bartelt, Karen. 1999. "A Central Illinois Scientist Responds to the Black Box." *REALL News* 7, no. 12.

Bartholomew, Robert E. 1998. "The Medicalization of Exotic Deviance: A Sociological Perspective on Epidemic Koro." *Transcultural Psychiatry* 35, no. 1: 5–38, March.

Bartley, William Warren. 1978. *Werner Erhard: The Transformation of a Man, the Founding of est.* C. N. Potter.

Basil, Robert. 1991. "Graphology and Personality: 'Let the Buyer Beware.' " In *The Hundredth Monkey and Other Paradigms of the Paranormal,* Kendrick Frazier, editor. Prometheus Books, pp. 206–208.

Bauer, Henry H. 1984. *Beyond Velikovsky.* University of Illinois Press.

Bauer, Henry H. 1986. *The Enigma of Loch Ness: Making Sense of a Mystery.* Urbana: University of Illinois Press.

Bauer, Henry H. 1996a. "Cryptozoology." In *The Encyclopedia of the Paranormal,* Gordon Stein, editor. Prometheus Books.

Bauer, Henry H. 1996b. "Immanuel Velikovsky." In *The Encyclopedia of the Paranormal,* Gordon Stein, editor. Prometheus Books.

Behe, Michael. 1996. *Darwin's Black Box.* Free Press.

Belsky, Gary, and Thomas Gilovich. 2000. *Why Smart People Make Big Money Mistakes—And How to Correct Them: Lessons from the New Science of Behavioral Economics.* Fireside.

Belz-Merk, Dr. Martina. 2001. "Counseling and Help for People with Unusual Experiences at the Outpatient Clinic. Ambulanz of the Psychological Institute at the University of Freiburg," www.igpp.de/english/counsel/project.htm.

Bem, Daryl J., and Charles Honorton. 1994. "Does Psi Exist?" *Psychological Bulletin* 115, no. 1: 4–18.

Bem, Sandra Lipsitz. 1993. *The Lenses of Gender.* Yale University Press.

Benassi, Victor, and Barry Singer. 1980/81. "Fooling Some of the People All of the Time." *Skeptical Inquirer,* Winter.

Benski, Claude, et al. 1996. *The "Mars Effect."* Prometheus Books.

Benski, Claudio, et al. 1998. "Testing New Claims of Dermo-Optical Perception." *Skeptical Inquirer,* Jan./Feb.

Bergman, Jerry. 1999. "Censorship of Information on Origins." www.rae.org/censor.html.

Berman, David, editor. 1996. *Atheism in Britain,* 5 vols. Bristol, U.K.: Thoemmes Press.

Betz, Hans-Dieter. 1995. "Unconventional Water Detection: Field Test of the Dowsing Technique in Dry Zones." *Journal of Scientific Exploration.*

Beveridge, W. I. B. 1957. *The Art of Scientific Investigation.* Vintage Books.

Beyerstein, Barry. 1985. "The Myth of Alpha Consciousness." *Skeptical Inquirer,* Fall.

Beyerstein, Barry. 1997. "Why Bogus Therapies Seem to Work." *Skeptical Inquirer,* Sept./Oct.

Beyerstein, Barry. 1996a. "Altered States of Consciousness." In *The Encyclopedia of the Paranormal,* Gordon Stein, editor. Prometheus Books.

Beyerstein, Barry. 1996b. "Graphology." In *The Encyclopedia of the Paranormal,* Gordon Stein, editor. Prometheus Books.

Beyerstein, Barry, and Dayle F. Beyerstein, editors. 1991. *The Write Stuff—Evaluations of Graphology, the Study of Handwriting Analysis.* Prometheus Books.

Beyerstein, Dale. 1996. "Edgar Cayce." In *The Encyclopedia of the Paranormal,* Gordon Stein, editor. Prometheus Books.

Binns, Ronald. 1985. *The Loch Ness Mystery Solved.* Buffalo: Prometheus.

Blackmore, Susan J. 1982. *Beyond the Body: An Investigation of Out-of-Body Experiences.* London: Heinemann.

Blackmore, Susan. 1987. "The Elusive Open Mind: Ten Years of Negative Research in Parapsychology." *Skeptical Inquirer,* Spring.

Blackmore, Susan. 1991a. "Lucid Dreaming: Awake in Your Sleep?" *Skeptical Inquirer,* Summer.

Blackmore, Susan. 1991b. "Near-Death Experiences: In or Out of the Body?" *Skeptical Inquirer,* Fall.

Blackmore, Susan. 1992. "Psychic Experiences: Psychic Illusions." *Skeptical Inquirer,* Summer.

Blackmore, Susan J. 1993. *Dying to Live: Near-Death Experiences.* Prometheus Books.

Blackmore, Susan. 1998. "Abduction by Aliens or Sleep Paralysis?" *Skeptical Inquirer,* May/June.

Blum, Deborah. 1995. "Race: Many Biologists Argue for Discarding the Whole Concept." *Sacramento Bee,* Oct. 18.

Bolt, Christine. 1971. *Victorian Attitudes to Race.* London: Routledge and Kegan Paul.

Boston, Rob. 1996a. "Anthroposophy: Rudolf Steiner's 'Spiritual Science.' " *Church & State,* April.

Boston, Rob. 1996b. *The Most Dangerous Man in America? Pat Robertson and the Rise of the Christian Coalition.* Prometheus.

Bourke, Angela. 1999. *The Burning of Bridget Cleary.* Pimlico.

Brenneman, Richard J. 1990. *Deadly Blessings: Faith Healing on Trial.* Prometheus Books.

Brightman, Harvey J. n.d. "GSU Master Teacher Program: On Learning Styles," www.gsu.edu/ ~dschjb/wwwmbti.html.

British Columbia Civil Liberties Association. 1988. "The Use of Graphology as a Tool for Employee Hiring and Evaluation," www.bccla.org/ positions/privacy/88 graphology.html.

Brody, Jane E. 2002. "Cellphone: A Convenience, a Hazard or Both?" *New York Times,* Oct. 1.

Bronner, Stephen Eric. 2000. *A Rumor About the Jews: Reflections on Antisemitism and the Protocols of the Learned Elders of Zion.* St. Martin's Press.

Bronowski, Jacob. 1973. *The Ascent of Man.* Little, Brown.

Brooke, John Hedley. 1991. *Science and Religion: Some Historical Perspectives.* Cambridge University Press.

Browne, M. Neil, and Stuart M. Keeley. 1997. *Asking the Right Questions: A Guide to Critical Thinking.* Prentice Hall.

Bruer, John T. 1999. *The Myth of the First Three Years.* Free Press.

Brugger, Peter. 2001. "From Haunted Brain to Haunted Science: A Cognitive Neuroscience View of Paranormal and Pseudoscientific Thought." In *Hauntings and Poltergeists: Multidisciplinary Perspectives,* J. Houran and R. Lange, editors. McFarland.

Brugioni, Dino A. 1999. *Photo Fakery: The History and Techniques of Photographic Deception and Manipulation.* Brasseys.

Bulgatz, Joseph. 1992. *Ponzi Schemes, Invaders from Mars, and More Extraordinary Popular Delusions.* Harmony Books.

Bullard, Thomas E. 1996. "Ancient Astronauts." In *The Encyclopedia of the Paranormal,* Gordon Stein, editor. Prometheus Books.

Burl, Aubrey. 1976. *The Stone Circles of the British Isles.* Yale University Press.

Butler, Kurt. 1992. *A Consumer's Guide to "Alternative Medicine": A Close Look at Homeopathy, Acupuncture, Faith-Healing, and Other Unconventional Treatments.* Prometheus Books.

Byrnes, Gail, and I. W. Kelly. 1992. "Crisis Calls and Lunar Cycles: A Twenty-Year Review." *Psychological Reports* 71: 779–785.

Cahill, Thomas. 1995. *How the Irish Saved Civilization.* Nan A. Talese Publishing.

Cameron, Norman, and Joseph F. Rychlak. 1985. *Personality Development and Psychopathology: A Dynamic Approach,* 2nd ed. Houghton Mifflin College.

Camp, Gregory S. 1997. *Selling Fear: Conspiracy Theories and End-Times Paranoia.* Baker Book House.

Carlinsky, Joel. 1994. "William Reich—Epigones of Orgonomy: The Incredible History of William Reich and His Followers." *Skeptic* 2, no. 3, April.

Carr, Timothy Spencer. 1997. "Son of Originator of 'Alien Autopsy' Story Casts Doubt on Father's Credibility." *Skeptical Inquirer,* July/Aug.

Carroll, Robert Todd. 1974. *The Common-Sense Philosophy of Religion of Bishop Edward Stillingfleet (1635–1699).* Martinus Nijhoff.

Carroll, Robert Todd. 1996. "The Sound of One Mouth Blathering." Review of L. Ron Hubbard's *The Rediscovery of the Human Soul* (1996). skepdic.com/ refuge/hubbard.html, Oct. 22.

Carroll, Robert Todd. 1999. "Dowsing for Dollars: Fighting High-Tech Promises with Low-Tech Critical Thinking Skills," skepdic.com/essays/dowsingfordollars.html.

Carroll, Robert Todd. 2000. *Becoming a Critical Thinker: A Guide for the New Millennium.* Pearson Custom Publishing.

Carter, Ruth. 1999. *Amway Motivational Organizations: Behind the Smoke and Mirrors.* Backstreet Publishing.

Carus, Paul. 1996. *The History of the Devil and the Idea of Evil.* Reproduction of the original 1900 edition. Gramercy.

Cerone, Daniel. 1993. "Admitting 'Noah's Ark' Hoax." *Los Angeles Times,* Oct. 30.

Chittenden, Maurice. 1998. "Cherie's Magic Crystals Put the New Age in New Labour." *The Sunday Times,* London, July 19.

Chorvinsky, Mark. 1996. "The Makeup Man and the Monster: John Chambers and the Patterson Bigfoot Suit an Investigation." *Strange Magazine* 17, Summer.

Churchland, Patricia Smith. 1986. *Neurophilosophy: Toward a Unified Science of the Mind-Brain.* MIT Press.

Clark, Nancy, and Nick Gallo. 1993. "Do You Believe in Magic?—New Light on the New Age." *Family Circle,* Feb. 23.

Clark, Philip E., and Mary Jo Clark. 1984. "Therapeutic Touch: Is There a Scientific Basis for the Practice?" *Nursing Research* 33, Jan./Feb.

Clark, Ronald W. 1977. *Edison: The Man Who Made the Future.* G. P. Putnam's Sons.

Clark, Stephen. 1995. *How to Live Forever.* Routledge.

Cohn, Norman Rufus Colin. 1967. *Warrant for Genocide: The Myth of the Jewish World-Conspiracy and the Protocols of the Elders of Zion.* Harper & Row.

Cole, Richard. 1995. "U.S. Didn't Foresee Faults in Psychic Spies Program." Associated Press, *Sacramento Bee,* Nov. 29.

Collie, Ashley Jude. 1999. "Let the Force Be With You." *American Way,* March 15.

Collins, Paul S. 2002. *Banvard's Folly: Thirteen Tales of People Who Didn't Change the World.* Picador.

Colton, Michael. 1998. "You Need It Like . . . a Hole in the Head?" *Washington Post,* May 31.

Condon, Dr. Edward U. 1969. *Final Report of the Scientific Study of Unidentified Flying Objects Conducted by the University of Colorado under Contract to the United States Air Force.* E. P. Dutton.

Conn, Charles Paul. 1985. *The Possible Dream: A Candid Look at Amway.* Putnam.

Conway, Flo, and Jim Siegelman. 1997. *Snapping.* Sonoma, Calif.: Stillpoint Press.

Coons, P. M. 1986. "Child Abuse and Multiple Personality: Review of the Literature and Suggestions for Treatment." *International Journal of Child Abuse and Neglect* 10: 455–462.

Cooper, Claire. 1993. "Repressed-Memory Lawsuits Spur Backlash from Accused." *Sacramento Bee,* March 18.

Cooper, Joe. 1982. "Cottingley: At Last the Truth." *The Unexplained,* no. 117: 2338–2340.

Copi, Irving M. 1998. *Introduction to Logic,* 10th ed. Prentice Hall.

Coren, Stanley. 1997. *Sleep Thieves: An Eye-Opening Exploration into the Science and Mysteries of Sleep.* Free Press.

Coughlin, Paul T. 1999. *Secrets, Plots & Hidden Agendas: What You Don't Know About Conspiracy Theories.* Intervarsity Press.

Cromer, A. 1993. "Pathological Science: An Update." *Skeptical Inquirer,* Summer.

Crowley, Aleister. 1989. *The Confessions of Aleister Crowley: An Autohagiography.* Arkana.

Culver, Roger B., and Philip A. Ianna. 1988. *Astrology, True or False? A Scientific Evaluation.* Prometheus Books.

Cuneo, Michael W. 2001. *American Exorcism: Expelling Demons in the Land of Plenty.* Doubleday.

Daegling, David J., and Daniel O. Schmitt. 1999. "Bigfoot's Screen Test." *Skeptical Inquirer,* May/June.

Damasio, Antonio R. 1995. *Descartes' Error: Emotion, Reason, and the Human Brain.* Avon Books.

Damasio, Antonio R. 1999. *The Feeling of What Happens.* Harcourt.

Damer, T. Edward. 2001. *Attacking Faulty Reasoning: A Practical Guide to Fallacy-Free Arguments,* 4th ed. Wadsworth.

Darrow, Clarence. 1932. *The Story of My Life.* Scribner and Sons.

Darwin, Charles R. 1859/1958. *The Origin of Species.* New York: New American Library.

Darwin, Charles R. 1872/1998. *The Expression of the Emotions in Man and Animals,* 3rd ed. Oxford University Press.

Davidson, H. R. Ellis. 1989. *Myths and Symbols in Pagan Europe: Early Scandinavian and Celtic Religions.* Syracuse University Press.

Davies, Nigel. 1998. *The Ancient Kingdoms of Peru.* Penguin Books.

Davis, Wade. 1985. *The Serpent and the Rainbow.* Warner Books.

Davis, Wade. 1988. *Passage of Darkness: The Ethnobiology of the Haitian Zombie.* Chapel Hill: University of North Carolina Press.

Dawes, Robyn M. 1994. *House of Cards: Psychology and Psychotherapy Built on Myth.* Free Press.

Dawes, Robyn M. 2001. *Everyday Irrationality:*

How Pseudo Scientists, Lunatics, and the Rest of Us Systematically Fail to Think Rationally. Westview Press.

Dawkins, Richard. 1988. *The Blind Watchmaker: Why the Evidence of Evolution Reveals a Universe Without Design.* W. W. Norton.

Dawkins, Richard. 1995. *River Out of Eden: A Darwinian View of Life.* Basic Books.

Dawkins, Richard. 1996. *Climbing Mount Improbable.* W. W. Norton.

Dean, Geoffrey, Arthur Mather, and Ivan W. Kelly. 1996. "Astrology." In *The Encyclopedia of the Paranormal,* Gordon Stein, editor. Prometheus Books.

De Camp, L. Sprague. 1975. *Lost Continents: The Atlantis Theme.* Ballantine Books.

De Camp, L. Sprague. 1977. *The Ancient Engineers.* Ballantine Books.

Decker, Ronald. 1996. "Tarot." In *The Encyclopedia of the Paranormal,* Gordon Stein, editor. Prometheus Books.

Decker, Ronald, Michael Dummett, and Thierry Depaulis. 1996. *A Wicked Pack of Cards: The Origins of the Occult Tarot.* St. Martin's Press.

de Givry, Grillot. 1971. *Witchcraft, Magic & Alchemy.* Republication of the 1931 Houghton Mifflin Company edition. Dover Books.

Delaney, Frank. 1986. *The Celts.* Little, Brown.

Delgado, Jose, M.R., M.D. 1969. *Physical Control of the Mind: Toward a Psychocivilized Society.* Harper & Row.

Dembski, William. 1998. *Intelligent Design.* Cambridge University Press.

Dennett, Daniel Clement. 1978. *Brainstorms: Philosophical Essays on Mind and Psychology.* Bradford Books.

Dennett, Daniel Clement. 1984. *Elbow Room: The Varieties of Free Will Worth Wanting.* MIT Press.

Dennett, Daniel Clement. 1991. *Consciousness Explained,* illustrated by Paul Weiner. Little, Brown.

Dennett, Daniel Clement. 1995. *Darwin's Dangerous Idea: Evolution and the Meanings of Life.* Simon & Schuster.

Dennett, Daniel Clement. 1996. *Kinds of Minds: Toward an Understanding of Consciousness.* New York: Basic Books.

Dennett, Michael R. 1982. "Bigfoot Jokester Reveals Punchline—Finally." *Skeptical Inquirer,* Fall.

Dennett, Michael R. 1996. "Bigfoot." In *The Encyclopedia of the Paranormal,* Gordon Stein, editor. Prometheus Books.

de Rivera, Joseph. 1993. " 'Trauma Searches' Plant the Seed of Imagined Misery." *Sacramento Bee,* May 18.

Diaconis, Persi, and Frederick Mosteller. 1996. "Coincidences." In *The Encyclopedia of the Paranormal,* Gordon Stein, editor. Prometheus Books.

Diamond, Jared M. 1998. *Why Is Sex Fun? The Evolution of Human Sexuality.* Basic Books.

Dickson, D. H., and I. W. Kelly. 1985. "The 'Barnum Effect' in Personality Assessment: A Review of the Literature." *Psychological Reports* 57: 367–382.

Dineen, Tana. 1998a. *Manufacturing Victims: What the Psychology Industry Is Doing to People.* Montreal: Robert Davies Multimedia Publishing.

Dineen, Tana. 1998b. "Psychotherapy: The Snake Oil of the 90's?" *Skeptic* 6, no. 3.

Dobzhansky, Theodosius. 1982. *Genetics and the Origin of Species.* Columbia University Press.

Dodes, John E. 1997. "The Mysterious Placebo." *Skeptical Inquirer,* Jan./Feb.

Dolnick, Edward. 1998. *Madness on the Couch: Blaming the Victim in the Heyday of Psychoanalysis.* Simon & Schuster.

Donohue, Andrew. 2000. " 'Quick Fix' Concerns over Drug." *Sacramento Bee,* Dec. 23.

Dossey, Donald E. 1992. *Holiday Folklore, Phobias and Fun: Mythical Origins, Scientific Treatments and Superstitious "Cures."* Outcome Unlimited Press.

Dryer, Lisa M. L. 2001. "Ear Candling," www.quackwatch.com/01QuackeryRelatedTopics/candling.html.

Dudley, Underwood. 1998. "Numerology: Comes the Revolution." *Skeptical Inquirer,* Sept./Oct.

Dudley, William, editor. 1999. *UFOs.* At Issue—Opposing Viewpoint Series. Greenhaven Press.

Duncan, David Ewing. 1999. *Calendar: Humanity's Epic Struggle to Determine a True and Accurate Year.* Bard Books.

Duncan, Ronald, and Miranda Weston-Smith, editors. 1979. *The Encyclopedia of Delusions.* Wallaby Books.

Dundes, Allen, editor. 1992. *The Evil Eye: A Casebook.* University of Wisconsin Press.

Dunn, C., et al. 1995. "Sensing an Improvement: An Experimental Study to Evaluate the Use of Aromatherapy, Massage and Periods of Rest in an Intensive Care Unit." *Journal of Advanced Nursing* 21.

Easterbrook, P. J., et al. 1991. "Publication Bias in Clinical Research." *Lancet* 337: 867–872.

Eberle, Paul. 1993. *The Abuse of Innocence: The McMartin Preschool Trial.* Prometheus.

Edey, Maitland A., and Donald C. Johanson. 1990. *Blueprints: Solving the Mystery of Evolution.* Penguin.

Edis, Taner. 2001. " 'Intelligent Design' Meets Artificial Intelligence." *Skeptical Inquirer,* March/April.

Edwards, Diane D. 1988. "Cells Haywire in Electromagnetic Field?" *Science News* 133, n. 14, April 2.

Edwards, Iorwerth Eiddon Stephen. 1992. *The Pyramids of Egypt,* revised ed. Penguin USA.

Edwards, Paul. 1996a. "Karma." In *The Encyclope-*

dia of the Paranormal, Gordon Stein, editor. Prometheus Books.

Edwards, Paul. 1996b. *Reincarnation: A Critical Examination*. Prometheus Books.

Eisler, Robert. 1951. *Man into Wolf: An Anthropological Interpretation of Sadism, Masochism, and Lycanthropy*. London: Routledge and Paul.

Ekman, Paul, and Wallace V. Friesen. 1975. *Unmasking the Face: A Guide to Recognizing Emotions from Facial Clues*. Prentice-Hall.

Ekman, Paul, and Erika L. Rosenberg, eds. 1997. *What the Face Reveals: Basic and Applied Studies of Spontaneous Expression Using the Facial Action Coding System (FACS)*. Oxford University Press.

Ellenberger, Leroy. 1995. "An Antidote to Velikovskian Delusions." *Skeptic* 3, no. 4.

Ellis, Peter Berresford. 1998. *The Druids*. Wm. B. Eeerdmans.

Ellis, Richard. 1998. *Imagining Atlantis*. Knopf.

Ellis, Richard. 1994. *Monsters of the Sea: The History, Natural History, and Mythology of the Oceans' Most Fantastic Creatures*. Alfred A. Knopf.

Ellwood, Robert S. 1996. "Theosophy." In *The Encyclopedia of the Paranormal*, Gordon Stein, editor. Prometheus Books.

Enright, J. T. 1996. "Dowsers Lost in a Barn." *Naturwissenschaften* 83, no. 6: 275–277.

Enright, J. T. 1999. "Testing Dowsing—The Failure of the Munich Experiments." *Skeptical Inquirer*, Jan./Feb.

Enright, J. T. 1995. "Water Dowsing: The Scheunen Experiments." *Naturwissenschaften* 82: 360–369.

Evans, Bergen. 1957. *The Natural History of Nonsense*. Alfred A. Knopf.

Evans, Bergen. 1990. *Bias in Human Reasoning: Causes and Consequences*. Psychology Press.

Eysenck, H. J., and D. K. B. Nias. 1982. *Astrology: Science or Superstition?* St. Martin's Press.

Faltermayer, Charlotte. 1998. "The Best of Est? Werner Erhard's Legacy Lives on in a Kinder, Gentler and Lucrative Version of His Self-Help Seminars." *Time*, March 16.

Farady, Ann. 1985. *The Dream Game*. Washington Square Press, Pocket Books.

Feder, Kenneth L. 2002. *Frauds, Mysteries and Myths*, 4th ed. McGraw-Hill.

Feinberg, Todd E., and Martha J. Farah, editors. 1997. *Behavioral Neurology and Neuropsychology*. McGraw-Hill.

Ferris, Timothy. 1998. *The Whole Shebang: A State-of-the-Universe's Report*. Touchstone.

Fienberg, Howard. 2000. "Hands Off, Doctor," www.stats.org/spotlight/touch.htm, July 26.

Fiery, Ann. 1999. *The Book of Divination*. Chronicle Books.

Fikes, Jay Courtney. 1996. *Carlos Castaneda, Academic Opportunism and the Psychedelic Sixties*. Millennia Press.

Fingarette, Herbert. 1988. *Heavy Drinking: The Myth of Alcoholism as a Disease*. University of California Press.

Fingarette, Henry. 2000. *Self-Deception*. University of California Press.

Finucane, Ronald C. 1996. *Ghosts: Appearances of the Dead & Cultural Transformation*. Prometheus Books.

Fisher, Seymour, and Roger P. Greenberg, editors. 1997. *From Placebo to Panacea: Putting Psychiatric Drugs to the Test*. John Wiley and Sons.

Fitzpatrick, Robert L., and Joyce Reynolds. 1997. *False Profits: Seeking Financial and Spiritual Deliverance in Multi-Level Marketing and Pyramid Schemes*. Charlotte, N.C.: Herald Press.

Flanagan, Owen J. 2001. *Dreaming Souls: Sleep, Dreams and the Evolution of the Conscious Mind*. Oxford University Press.

Foreman, Judy. 1998. "New Therapy for Trauma Doubted." *Boston Globe*, Sept. 14.

Forer, B. R. 1949. "The Fallacy of Personal Validation: A Classroom Demonstration of Gullibility." *Journal of Abnormal Psychology* 44: 118–121.

Fraser, Steve, editor. 1995. *The Bell Curve Wars: Race, Intelligence, and the Future of America*. Basic Books.

Frazier, Kendrick. 1979–80. "Amityville Hokum: The Hoax and the Hype." *Skeptical Inquirer*, Winter.

Frazier, Kendrick, editor. 1986. *Science Confronts the Paranormal*. Prometheus Books.

Frazier, Kendrick, editor. 1991. *The Hundredth Monkey and Other Paradigms of the Paranormal*. Prometheus Books.

Frazier, Kendrick, editor. 1997. *The UFO Invasion: The Roswell Incident, Alien Abductions, and Government Coverups*. Prometheus.

Frazier, Kendrick, and James Randi. 1981. "Predictions After the Fact: Lessons of the Tamara Rand Hoax." *Skeptical Inquirer*, Fall.

Freud, Sigmund. 1928. *The Future of an Illusion*. Translated by W. D. Robson-Scott. Hogarth Press.

Freud, Sigmund. 1930. *Civilization and Its Discontents*. Translated by Joan Riviere. Hogarth Press.

Friedlander, Michael W. 1972. *The Conduct of Science*. Prentice-Hall.

Friedlander, Michael W. 1995. *At the Fringes of Science*. Westview Press.

Furnham, Adrian. 1991. "Write and Wrong: The Validity of Graphological Analysis." In *The Hun-*

dredth Monkey and Other Paradigms of the Paranormal, Kendrick Frazier, editor. Prometheus Books.

Gale, Richard. 2002. *The Blackwell Guide to Metaphysics*. Blackwell Publishers.

Gallo, Ernest. 1996. "Jung and the Paranormal." In *The Encyclopedia of the Paranormal,* Gordon Stein, editor. Prometheus Books.

Gallo, Ernest. 1994. "Synchronicity and the Archetypes." *Skeptical Inquirer,* Summer.

Gardner, Martin. 1957. *Fads and Fallacies in the Name of Science*. Dover.

Gardner, Martin. 1981. *Science: Good, Bad and Bogus*. Prometheus Books.

Gardner, Martin. 1983. *The Whys of a Philosophical Scrivener*. Quill.

Gardner, Martin. 1987. "Isness Is Her Business." *New York Review of Books,* April 9.

Gardner, Martin. 1988. *The New Age: Notes of a Fringe Watcher*. Prometheus Books.

Gardner, Martin. 1989. *How Not to Test a Psychic: Ten Years of Remarkable Experiments With Renowned Clairvoyant Pavel Stepanek*. Prometheus Books.

Gardner, Martin. 1991. "Water With Memory? The Dilution Affair." In *The Hundredth Monkey,* Kendrick Frazier, editor. Prometheus Books.

Gardner, Martin. 1992. *On the Wild Side*. Prometheus Books.

Gardner, Martin. 1995a. "Doug Henning and the Giggling Guru." *Skeptical Inquirer,* May/June.

Gardner, Martin. 1995b. *Urantia: The Great Cult Mystery*. Prometheus Books.

Gardner, Martin. 2000. *Did Adam and Eve Have Navels?: Discourses on Reflexology, Numerology, Urine Therapy, and Other Dubious Subjects*. W. W. Norton.

Gardner, Martin. 2001. "Distant Healing and Elisabeth Targ." *Skeptical Inquirer,* March/April.

Gaudiano, Brandon A., and James D. Herbert. 2000. "Can We Really Tap Our Problems Away? A Critical Analysis of Thought Field Therapy." *Skeptical Inquirer,* July/Aug.

Gauquelin, Michel. 1975. "Spheres of Influence." *Psychology Today* [Britain], no. 7, Oct.: 22–27; reprinted in *Philosophy of Science and the Occult*. Albany: State University of New York Press, 1990.

Gawande, Atul. 1999. "The Cancer-Cluster Myth." *The New Yorker,* Feb. 8.

Gee, Henry. 1996. "Box of Bones 'Clinches' Identity of Piltdown Paleontology Hoaxer." *Nature,* May 23.

Gelder, Ken. 1994. *Reading the Vampire*. London: Routledge.

Giere, Ronald. 1998. *Understanding Scientific Reasoning,* 4th ed. Holt Rinehart, Winston.

Gilovich, T., R. Vallone, and A. Tversky. 1985.

"The Hot Hand in Basketball: On the Misperception of Random Sequences." *Cognitive Psychology* 17: 295–314.

Gilovich, Thomas. 1993. *How We Know What Isn't So: The Fallibility of Human Reason in Everyday Life*. Free Press.

Gladwell, Malcolm. 2002. "The Naked Face." *The New Yorker,* Aug. 5.

Glymour, Clark, and Douglas Stalker. 1990. "Winning Through Pseudoscience." In *Philosophy of Science and the Occult,* 2nd ed. Patrick Grim, editor. Albany: State University of New York Press.

Glymour, Clark, and Douglas Stalker, editors. 1985. *Examining Holistic Medicine*. Prometheus.

Goeringer, Conrad. 1998. "Freemasons—From the 700 Club to Art Bell, an Object of Conspiracy Thinking." www.american/atheist.org/supplement/conspiracy.html, May.

Goerman, Robert A. 1980. "Alias Carlos Allende: The Mystery Man Behind the Philadelphia Experiment." *Fate,* Oct.

Gold, Mark S. 1995. *The Good News About Depression: Cures and Treatments in the New Age of Psychiatry*. Bantam Books.

Goldberg, Isaac. 1936. *The So-Called "Protocols of the Elders of Zion": A Definitive Exposure of One of the Most Malicious Lies in History*. Girard, Kans.: Haldeman-Julius Publications.

Goldsmith, Donald, editor. 1977. *Scientists Confront Velikovsky*. Cornell University Press.

Goodwin, Pamela J., et al. 2001. "The Effect of Group Psychosocial Support on Survival in Metastatic Breast Cancer." *New England Journal of Medicine* 345, no. 24, December 13.

Gordon, Henry. 1987. *Extrasensory Deception: ESP, Psychics, Shirley MacLaine, Ghosts, UFOs*. Buffalo, N.Y.: Prometheus Books.

Gordon, Henry. 1988. *Channeling into the New Age: The "Teachings" of Shirley MacLaine and Other Such Gurus*. Prometheus Books.

Gorman, Brian J. 1998. "Facilitated Communication in America: Eight Years and Counting." *Skeptic* 6, no. 3.

Gould, Stephen Jay. 1979. *Ever Since Darwin*. W. W. Norton.

Gould, Stephen Jay. 1982. "Piltdown Revisited." In *The Panda's Thumb*. W. W. Norton.

Gould, Stephen Jay. 1983. "Evolution as Fact and Theory." In *Hen's Teeth and Horse's Toes*. W. W. Norton.

Gould, Stephen Jay. 1987. *The Flamingo's Smile*. W. W. Norton.

Gould, Stephen J. 1993a. "American Polygeny and Craniometry Before Darwin: Blacks and Indians as Separate, Inferior Species." In *Racial Economy of Science*. Sandra Harding, editor. Indiana University Press.

Gould, Stephen Jay. 1993b. "Darwin and Paley Meet the Invisible Hand." In *Eight Little Piggies*. W. W. Norton.

Gould, Stephen Jay. 1996. *The Mismeasure of Man*. W. W. Norton.

Gould, Stephen Jay. 1997. "Evolution: The Pleasures of Pluralism." *New York Review of Books*. June 26.

Green, Gina, Ph.D. 1994. "Facilitated Communication: Mental Miracle or Sleight of Hand?" *Skeptic* 2, no. 3: 68–76.

Green, Saul. 2000a. "Can Alternative Treatments Induce Immune Surveillance Over Cancer in Humans?" www.hcrc.org/contrib/green/immunol.html.

Green, Saul. 2000b. "Can Any Cancer Treatment Strengthen the Immune System?" www.quackwatch.com/01QuackeryRelatedTopics/Cancer/immuneboost.html.

Grescoe, Taras. 2000. "Raël Love," www.salon.com/travel/feature/2000/03/08/raelians/index.html, March 8.

Griaule, Marcel. 1997. *Conversations With Ogotemmeli: An Introduction to Dogon Religious Ideas*. Reprint of 1948 edition. Oxford University Press.

Grim, Patrick, editor. 1990. *Philosophy of Science and the Occult*, 2nd ed. State University of New York Press.

Grof, Stanislav. 1975. *Realms of the Human Unconscious: Observations from LSD Research*. Viking Press.

Gross, Paul R., and Norman Levitt. 1997. *Higher Superstition: The Academic Left and Its Quarrels With Science*. Johns Hopkins University Press.

Guthrie, Stewart Elliott. 1995. *Faces in the Clouds: A New Theory of Religion*. Oxford University Press.

Gutiérrez-García, J. M., and F. Tusell. 1997. "Suicides and the Lunar Cycle." *Psychological Reports* 80: 243–250.

Hafernik, Rob. 1996. "Sitchin's Twelfth Planet," www.geocities.com/Area51/Corridor/8148/hafernik.html.

Hahneman, D., and A. Tversky. 1971. "Belief in the Law of Small Numbers." *Psychological Bulletin* 76: 105–110.

Haley, Jay. 1986. "Therapy—A New Phenomenon." In *The Power Tactics of Jesus Christ and Other Essays*. Rockville, Md.: Triangle Press.

Hallinan, Joseph T. 1997. "Money for Repressed Memories Repressed." *Sacramento Bee*, Jan. 12.

Halpern, Steven. 1998. "Fundamentally Sound," www.lightworks.com/MonthlyAspectarian/1998/January/0198-20.htm.

Hansel, C. E. M. 1980. *ESP and Parapsychology: A Critical Re-Evaluation*. Prometheus Books.

Hansel, C. E. M. 1989. *The Search for Psychic Power: ESP and Parapsychology Revisited*. Prometheus Books.

Harrington, Anne, editor. 1999. *The Placebo Effect: An Interdisciplinary Exploration*. Harvard University Press.

Harrington, Evan. 1996. "Conspiracy Theories and Paranoia: Notes from a Mind-Control Conference." *Skeptical Inquirer*. Sept./Oct.

Hart, Carol. 1999. "The Mysterious Placebo Effect." *Modern Drug Discovery*, July/Aug.

Haught, John F. 1995a. "Does Evolution Rule Out God's Existence?" www.aaas.org/spp/dser/evolution/perspectives/HAUGHT.shtml.

Haught, James A. 1995b. "True Believers and Utter Madness." *Free Inquiry*, Summer.

Haught, James A. 1996. *2000 Years of Disbelief: Famous People With the Courage to Doubt*. Prometheus Books.

Haught, John F. 1999. *God After Darwin: A Theology of Evolution*. Westview Press.

Hayes, Judith. 1996. *In God We Trust: But Which One?* Freedom from Religion Foundation.

Herm, Gerhard. 1977. *The Celts*. St. Martin's Press.

Heuvelmans, Bernard, Dr. 1955. *On the Track of Unknown Animals*, 3rd ed. London: Kegan Paul International Limited.

Hicks, Robert D. 1991. *In Pursuit of Satan: The Police and the Occult*. Prometheus Books.

Hill, Frances. 1995. *A Delusion of Satan: The Full Story of the Salem Witch Trials*. Doubleday.

Hines, Terence. 1990. *Pseudoscience and the Paranormal: A Critical Examination of the Evidence*. Prometheus Books.

Hines, Terence. 1991. "Biorhythm Theory: A Critical Review." In *Paranormal Borderlands of Science*, Kendrick Frazier, editor. Prometheus Books.

Hines, Terence. 1996. "Cattle Mutilations." In *The Encyclopedia of the Paranormal*, Gordon Stein, editor. Prometheus Books.

Hines, Terence M. 1998. "Comprehensive Review of Biorhythm Theory." *Psychological Reports* 83: 19–64.

Hitt, Jack. 1997. "Operation Moo." *Gentleman's Quarterly*, Feb.

Hobbes, Thomas. 1651. *Leviathan*. London: Printed for Andrew Crooke.

Hobson, J. Allan. 1988. *The Dreaming Brain*. Basic Books.

Hofstadter, Douglas. 1985. *Metamagical Themas: Questing for the Essence of Mind and Pattern*. Basic Books.

Hofstadter, Douglas R., and Daniel C. Dennett.

1981. *The Mind's I: Fantasies and Reflections on Self and Soul.* Basic Books.

Holland, Bernard. 1981. "A Man Who Sees What Others Hear." *New York Times,* Nov. 19.

Horowitz, K. A., D. C. Lewis, and E. L. Gasteiger. 1975. "Plant Primary Perception." *Science* 189: 478–480.

Hotz, Robert Lee. 1996. "Deciphering the Miracles of the Mind." *Los Angeles Times,* Oct. 13.

Houdini, Harry. 1924. *A Magician Among the Spirits.* Harper.

Houdini, Harry. 1981. *Miracle Mongers and Their Methods: A Complete Expose.* Prometheus Books.

Hover-Kramer, Dorothea. 1996. *Healing Touch: A Resource for Health Care Professionals.* Delmar Publishers.

Hoyningen-Huene, Paul. 1993. *Reconstructing Scientific Revolutions: Thomas S. Kuhn's Philosophy of Science,* trans. Alexander J. Levine. University of Chicago.

Hrobjartsson, Asbjorn, and Peter C. Gotzsche. 2001. "Is the Placebo Powerless? An Analysis of Clinical Trials Comparing Placebo with No Treatment." *New England Journal of Medicine* 344, no. 21, May 24.

Huff, Daryl. 1954. *How to Lie with Statistics.* W. W. Norton.

Hufford, David J. 1989. *The Terror That Comes in the Night: An Experience-Centered Study of Supernatural Assault Traditions.* University of Pennsylvania Press.

Hume, David. 1748. *An Inquiry Concerning Human Understanding.*

Humphrey, Nicholas. 2000. "The Power of Prayer." *Skeptical Inquirer,* May/June.

Huston, Peter. 1995. "China, Chi, and Chicanery—Examining Traditional Chinese Medicine and Chi Theory." *Skeptical Inquirer,* Sept./Oct.

Huston, Peter. 1997. *Scams from the Great Beyond: How to Make Easy Money Off of ESP, Astrology, UFOs, Crop Circles, Cattle Mutilations, Alien Abductions, Atlantis, Channeling, and Other New Age Nonsense.* Paladin Press.

Hutton, Ronald. 1993. *The Pagan Religions of the Ancient British Isles: Their Nature and Legacy.* Blackwell.

Huxley, Thomas Henry. 1992. *Agnosticism and Christianity and Other Essays.* Prometheus Books.

Hyman, Arthur, and James J. Walsh. 1973. *Philosophy in the Middle Ages,* 2nd ed. Hackett Publishing.

Hyman, Ray. 1977. " 'Cold Reading': How to Convince Strangers That You Know All About Them." *Zetetic* (now the *Skeptical Inquirer*), Spring/Summer.

Hyman, Ray. 1985. "The Ganzfeld Experiment: A Critical Appraisal." *Journal of Parapsychology* 49: 3–49.

Hyman, Ray. 1989. *The Elusive Quarry: A Scientific Appraisal of Psychical Research.* Prometheus Books.

Hyman, Ray. 1995. "Evaluation of Program on Anomalous Mental Phenomena." *Journal of Scientific Exploration* 10, no. 1.

Hyman, Ray. 1996a. "Dowsing." In *The Encyclopedia of the Paranormal,* Gordon Stein, editor. Prometheus Books.

Hyman, Ray. 1996b. "The Evidence for Psychic Functioning: Claims vs. Reality." *Skeptical Inquirer,* March/April.

Hyman, Ray. 1999. "The Mischief-Making of the Ideomotor Effect." *Scientific Review of Alternative Medicine* 3, no. 2: 34–43.

Hyman, Ray. 2002. "Guide to Cold Reading," www.skeptics.com.au/journal/coldread.htm.

Iacono, W. G., and D. T. Lykken. 1997. "The Validity of the Lie Detector: Two Surveys of Scientific Opinion." *Journal of Applied Psychology* 82, no. 3.

Imrie, Robert, and D. W. Ramey. 2000–2001. "The Evidence for Evidence-Based Medicine." *Scientific Review of Alternative Medicine,* Winter.

Jacobson, John W., James A. Mulick, and Allen A. Schwartz. 1995. "A History of Facilitated Communication: Science, Pseudoscience, and Antiscience: Science Working Group on Facilitated Communication." *American Psychologist* 50, no. 9: 750–765.

James, Peter, and Nick Thorpe. 1999. *Ancient Mysteries.* Ballantine Books.

Jamison, Kay R. 1999. *Night Falls Fast: Understanding Suicide.* Knopf.

Jansen, Karl. 2000. "Using Ketamine to Induce the Near-Death Experience: Mechanism of Action and Therapeutic Potential" and "The Ketamine Model of the Near Death Experience: A Central Role for the NMDA Receptor," www.lycaeum.org/drugs/Cyclohexamines/Ketamine/Ketamine_NDE_Model.html.

Jansen, Karl. 2001. *Ketamine: Dreams and Realities.* Multidisciplinary Association for Psychedelic Studies.

Jarvis, William. 1991. "Chiropractic: A Skeptical View." In *The Hundredth Monkey,* Kendrick Frazier, editor. Prometheus Books.

Jarvis, William T., Ph.D. 2000a. "Applied Kinesiology," www.ncahf.org/articles/a-b/ak.html.

Jarvis, William T., Ph.D. 2000b. "Reiki," www.ncahf.org/articles/o-r/reiki.html.

Jarvis, William T., Ph.D. 2001. "Some Notes on Cranial Manipulative Therapy," www.ncahf.org/articles/c-d/cranial.html.

Jerome, Lawrence E. 1977. *Astrology Disproved.* Prometheus Books.

Jerome, Lawrence E. 1996. *Crystal Power: The Ultimate Placebo Effect*. Prometheus Books.

Johanson, Donald C., and Maitland A. Edey. 1981. *Lucy: The Beginnings of Humankind*. Simon and Schuster.

Johnson, B. C. 1981. *The Atheist Debater's Handbook*. Prometheus Books.

Johnston, Moira. 1999. *Spectral Evidence: The Ramona Case: Incest, Memory, and Truth on Trial in Napa Valley*. Westview Press.

Jung, Carl G. 1973. *Letters*. Selected and edited by Gerhard Adler, in collaboration with Aniela Jaffé. Princeton University Press.

Kagan, Daniel, and Ian Summers. 1984. *Mute Evidence*. Bantam Books.

Kahane, Howard. 1997. *Logic and Contemporary Rhetoric: The Use of Reason in Everyday Life,* 8th ed. Wadsworth.

Kamin, Leon J. 1974. *Science and Politics of IQ*. Lawrence Erlbaum.

Kaminer, Wendy. 1992. *I'm Dysfunctional, You're Dysfunctional: The Recovery Movement and Other Self-Help Fashions*. Addison-Wesley.

Kandel, Eric R., and James H. Schwartz, editors. 2000. *Principles of Neural Science* 4th ed. McGraw-Hill.

Kasler, Dale. 1998. "Inside Business," *Sacramento Bee,* June 29.

Kava, Dr. Ruth. 1995. "Should There Be a Shark in Your Medicine Cabinet?" *Priorities* 7, no. 2.

Keen, Sam. 1973. "Interview with Oscar Ichazo." *Psychology Today,* July.

Keene, M. Lamar. 1997. *The Psychic Mafia*. Prometheus.

Kelly, I., and R. Martens. 1994. "Lunar Phase and Birthrate: An Update." *Psychological Reports* 75: 507–511.

Kelly, I. W. 1981. "Cosmobiology and Moon Madness," *Mercury* 10: 13–17.

Kelly, I. W. 1997. "Modern Astrology: A Critique." *Psychological Reports* 81: 1035–1066.

Kelly, I. W. 1998. "Why Astrology Doesn't Work." *Psychological Reports* 82: 527–546.

Kelly, I. W., G. A. Dean, and D. H. Saklofske. 1990. "Astrology, a Critical Review." In *Philosophy of Science and the Occult,* 2nd ed. Patrick Grim, editor. State University of New York Press.

Kelly, I. W., James Rotton, and Roger Culver. 1996. "The Moon Was Full and Nothing Happened: A Review of Studies on the Moon and Human Behavior and Human Belief." In *The Outer Edge,* J. Nickell, B. Karr, and T. Genoni, editors. CSICOP.

Kenney, James, et al. 1988. "Applied Kinesiology Unreliable for Assessing Nutrient Status." *Journal of the American Dietetic Association.*

Kincheloe, Joe L., et al., editors. 1997. *Measured Lies: The Bell Curve Examined*. St. Martin's Press.

King, Edward. 2002. "The Morgan Affair," www.masonicinfo.com/morgan.htm.

Kirsch, Irving, Ph.D., and Guy Sapirstein, Ph.D. 1998. "Listening to Prozac but Hearing Placebo: A Meta-Analysis of Antidepressant Medication." *Prevention & Treatment* 1, June.

Kistler Instrument Corporation. n.d. "The Piezoelectric Effect, Theory, Design and Usage," www.designinfo.com/kistler/ref/tech_theory_text.htm.

Kitcher, Phillip. 1983. *Abusing Science: The Case Against Creationism*. MIT Press.

Klass, Philip J. 1988. *UFO-Abductions: A Dangerous Game*. Prometheus Books.

Klass, Philip J. 1997. *The Real Roswell Crashed Saucer Coverup*. Prometheus Books.

Klebniov, Paul. 1991. "The Power of Positive Inspiration," *Forbes,* Dec. 9.

Klotz, Irving. 1980. "The N-Ray Affair." *Scientific American* 242, no. 5, May.

Klyce, Brig. n.d. "The Second Law of Thermodynamics," www.panspermia.org/seconlaw.htm.

Kmetz, J. M. 1977. "A Study of Primary Perception in Plants and Animal Life." *Journal of the American Society for Psychical Re-search* 71(2): 157–170.

Kmetz, John M. 1978. "Plant Perception." *Skeptical Inquirer,* Spring/Summer.

Kochunas, Brad. 1999. "Why Astrology Works," www.mountainastrologer.com/kochunas.html.

Koertge, Noretta, editor. 1998. *A House Built on Sand: Exposing Postmodernist Myths About Science*. Oxford University Press.

Koren G., and N. Klein. 1991. "Bias Against Negative Studies in Newspaper Reports of Medical Research." *Journal of the American Medical Association,* Oct. 2.

Korff, Kal K. 1995. *Spaceships of the Pleiades: The Billy Meier Story*. Prometheus Books.

Korff, Kal K. 1997. *The Roswell UFO Crash: What They Don't Want You to Know*. Prometheus.

Kossy, Donna J. 1994. *Kooks: A Guide to the Outer Limits of Human Belief*. Feral House.

Kottmeyer, Martin S. 1960. "Entirely Unpredisposed: The Cultural Background of UFO Abduction Reports." *Magonia,* Jan.

Kourany, Janet A. 1997. *Scientific Knowledge: Basic Issues in the Philosophy of Science,* 2nd ed. Wadsworth Publishing.

Kramnick, Isaac, and R. Laurence Moore. 1997. *The Godless Constitution: The Case Against Religious Correctness*. W. W. Norton.

Kroeber, A. L., Donald Collier, and Patrick H. Carmichael. 1999. *The Archaeology and Pottery of Nazca,*

Peru: Alfred L. Kroeber's 1926 Expedition. Altamira Press.

Kruger, Justin, and David Dunning. 1999. "Unskilled and Unaware of It: How Difficulties in Recognizing One's Own Incompetence Lead to Inflated Self-Assessments." *Journal of Personality and Social Psychology* 77, no. 6, Dec.: 1121–1134.

Kuhn, Thomas S. 1996. *The Structure of Scientific Revolutions,* 3rd ed. University of Chicago.

Kuhn, Thomas S. 2000. *The Road Since Structure: Philosophical Essays, 1970–1993, With an Autobiographical Interview,* Jim Conant and John Haugeland, editors. University of Chicago Press.

Kurtz, Paul. 1990. *Philosophical Essays in Pragmatic Naturalism.* Prometheus Books.

Kurtz, Paul. 1992. *The New Skepticism: Inquiry and Reliable Knowledge.* Prometheus Books.

Kurtz, Paul, editor. 1985. *A Skeptic's Handbook of Parapsychology.* Prometheus Books.

Kurtz, Paul. 1986. *The Transcendental Temptation: A Critique of Religion and the Paranormal.* Prometheus Books.

Kushe, Larry. 1995. *The Bermuda Triangle Mystery— Solved.* Prometheus Books. Re-print of the Warner Books, 1975 edition.

LaBerge, Stephen, Ph.D. 1990. "Lucid Dreaming: Psychophysiological Studies of Consciousness During REM Sleep." In *Sleep and Cognition,* R. R. Bootzen et al., editors. American Psychological Association.

Lamont, Peter, and Richard Wiseman. 1999. *Magic in Theory.* University of Hertfordshire Press.

Langmuir, Irving. 1989. "Pathological Science." Transcribed and edited by Robert N. Hall. *Physics Today* 42, Oct.

Langone, Michael. 1998. "Large Group Awareness Training Programs." *Cult Observer* 15, no. 1.

Langone, Michael D., editor. 1995. *Recovery from Cults: Help for Victims of Psychological and Spiritual Abuse.* W. W. Norton.

Lawson, Willow. 2000. "Cutting the Cranium: Skull Operation Is One of Oldest Surgeries, Say Archaeologists," ABCNews.com, Sept. 27.

LeBoeuf, Dave. n.d. "UFOs and the Extraterrestrial Hypothesis," www.the-spa.com/thirteen/ufo's/eth.htm.

Lefkowitz, Mary. 1996. *Not Out of Africa: How Afrocentrism Became an Excuse to Teach Myth as History.* Basic Books.

Lehner, Mark. 1997. *The Complete Pyramids.* Thames and Hudson.

Leibermann, M. 1994. "Growth Groups in the 1980's: Mental Health Implications." In *Handbook of Group Psychotherapy: An Empirical and Clinical Synthesis,* Addie Fuhriman and Gary M. Burlingame, editors. Wiley-Interscience.

Leikind, Bernard J., and William J. McCarthy. 1991. "An Investigation of Firewalking." In *The Hundredth Monkey and Other Paradigms of the Paranormal,* Kendrick Frazier, editor. Prometheus Books.

Lell, Martin. 1997. *Das Forum: Protokoll einter Gehirnwäsche. Der Psycho-Konzern Landmark Education.* München: Deutscher Taschenbuch Verlag GmbH & Co. KG.

Leonard, Dirk, M.A., and Peter Brugger, Ph.D. 1998. "Creative, Paranormal, and Delusional Thought: A Consequence of Right Hemisphere Semantic Activation?" *Neuropsychiatry, Neuropsychology, and Behavioral Neurology* 11, no. 4: 177–183.

Lerner, Sharon. 2001. "Magnets Un-plugged." *The Village Voice,* March 14–20.

Levins, Richard, and Richard Lewontin. 1985. "Lysenkoism." In *The Dialectical Biologist.* Harvard University Press.

Lewallen, Judy R. 1997. "The San Luis Valley Crystal Skull: A Transparent Mystery." *Skeptical Inquirer,* Sept./Oct.

Lifton, Robert Jay. 1989. *Thought Reform and the Psychology of Totalism: A Study of Brainwashing in China.* University of North Carolina Press.

Lilienfeld, Scott O. 1996. "EMDR Treatment: Less Than Meets the Eye?" *Skeptical Inquirer,* Jan./Feb.

Lilienfeld, Scott O., et al. 1999. "Dissociative Identity Disorder and the Sociocognitive Model: Recalling the Lessons of the Past." *Psychological Bulletin* 125, no. 5: 507–523.

Lin, Zixin, editor. 2000. *Qigong: Chinese Medicine or Pseudoscience?* Prometheus.

Linde K. N. Clausius, G. Ramirez, D. Melchart, F. Eitel, L. V. Hedges, et al. 1997. "Are the Clinical Effects of Homeopathy Placebo Effects? A Meta-Analysis of Placebo-Controlled Trials." *Lancet* 350: 834–843.

Lindskoog, Kathryn. 1993. *Fakes, Frauds & Other Malarkey: 301 Amazing Stories & How Not to Be Fooled.* Zondervan Publishing House.

Lingua Franca. 2000. *The Sokal Hoax: The Sham That Shook the Academy.* Bison Books.

Linton, Michael. 1999. "The Mozart Effect," www.firstthings.com/ftissues/ft9903/linton.html.

Lippard, Jim. 1994. "Sun Goes Down in Flames: The Jammal Ark Hoax." *Skeptic* 2, no. 3.

Lippard, Jim. 1996. "Charles Fort." In *The Encyclopedia of the Paranormal,* Gordon Stein, editor. Prometheus Books.

Lipstadt, Deborah. 1994. *Denying the Holocaust: The Growing Assault on Truth and Memory.* Plumsock MesoAmerican Studies.

Livingston, James D. 1997. *Driving Force: The Natural Magic of Magnets.* Harvard University Press.

Livingston, James D. 1998. "Magnetic Therapy: Plausible Attraction?" *Skeptical Inquirer,* July/Aug.

Locke, John. 1690. *Essay Concerning Human Understanding.*

Loftus, Elizabeth F. 1979. *Eyewitness Testimony.* Harvard University Press.

Loftus, Elizabeth F. 1980. *Memory, Surprising New Insights into How We Remember and Why We Forget.* Addison-Wesley.

Loftus, Elizabeth. 1994. *The Myth of Repressed Memory.* St. Martin's Press.

Loftus, Elizabeth, and Katherine Ketcham. 1991. *Witness for the Defense: The Accused, the Eyewitness, and the Expert Who Puts Memory on Trial.* St. Martin's Press.

Lohr J. M., D. F. Tolin, and S. O. Lilienfeld. 1998. "Efficacy of Eye Movement Desensitization and Reprocessing: Implications for Behavior Therapy." *Behavior Therapy* 29: 126–153.

Lonigan, Paul R. 1996. *The Druids: Priests of the Ancient Celts.* Greenwood Publishing Group.

Lutz, Peter, and Goran E. Nilsson. 1998. *The Brain Without Oxygen.* Chapman & Hall.

Lykken, David Thoreson. 1998. *A Tremor in the Blood: Uses and Abuses of the Lie Detector.* Plenum Press.

Lyons, Arthur, and Marcello Truzzi. 1991. *The Blue Sense: Psychic Detectives and Crime.* Mysterious Press.

Lyons, Janice, R.N. 1997. "Applied Kinesiology: A Christian Perspective," www.hcrc.org/contrib/lyons/kinesiol.html.

Magner, George. 1995. *Chiropractic: The Victim's Perspective.* Prometheus Books.

Malcolm, Norman. 1959. *Dreaming.* London: Routledge.

Marion, George. n.d. "Electronic Voice Phenomenon," www.ghostshop.com.

Marks, David F. 1989. "The Case Against the Paranormal." *Fate,* Jan.

Marks, David F., and John Colwell. 2000. "The Psychic Staring Effect an Artifact of Pseudo Randomization." *Skeptical Inquirer,* Sept./Oct.

Marks, David, and Richard Kammann. 1979. *The Psychology of the Psychic.* Prometheus Books.

Marlock, Dennis, and John Dowling. 1994. *License to Steal: Traveling Con Artists, Their Games, Their Rules—Your Money.* Paladin Press.

Marquis, Julie. 1996. "Erasing the Line Between Mental and Physical Ills." *Los Angeles Times,* Oct. 15.

Martens, R., I. Kelly, and D. H. Saklofske. 1988. "Lunar Phase and Birth Rate: A Fifty-Year Critical Review." *Psychological Reports* 63: 923–934.

Martens, R., and T. Trachet. 1998. *Making Sense of Astrology.* Prometheus Books.

Martin, Bruce. 1998. "Coincidences: Remarkable or Random?" *Skeptical Inquirer,* Sept./Oct.

Martin, Michael. 1990. *Atheism: A Philosophical Justification.* Philadelphia: Temple University Press.

Martin, S. J., I. W. Kelly, and D. H. Saklofske. 1992. "Suicide and Lunar Cycles: A Critical Review over 28 Years." *Psychological Reports* 71: 787–795.

Martin, Walter. 1985. *The Kindom of Cults,* revised and expanded. Bethany House.

Martin, Walter. 1989. *The New Age Cult.* Bethany House.

Masson, Jeffrey Moussaieff, trans. 1985. *The Complete Letters of Sigmund Freud to Wilhelm Fliess, 1887–1904.* Belknap Press of Harvard University Press.

Matheson, Terry. 1998. *Alien Abductions: Creating a Modern Phenomenon.* Prometheus Books.

Mayr, Ernst. 1989. *Toward a New Philosophy of Biology: Observations of an Evolutionist.* Harvard University Press.

Mayr, Ernst. 1998. *This Is Biology: The Science of the Living World.* Belknap Press.

McCoy, Bob. 1996. "Phrenology." In *The Encyclopedia of the Paranormal,* Gordon Stein, editor. Prometheus Books.

McCoy, Bob. 2000. *Quack! Tales of Medical Fraud from the Museum of Questionable Medical Devices.* Santa Monica Press.

McCrone, Walter. 1994. "Psychic ability? The Micro-PK Experiments," *New Scientist,* Nov. 26.

McCrone, Walter. 1999. *Judgment Day for the Shroud of Turin.* Prometheus Books.

McCutcheon, Lynn. 1996. "What's That I Smell? The Claims of Aromatherapy." *Skeptical Inquirer,* May/June.

McCutcheon, Lynn. 1997. "Taking a Bite out of Shark Cartilage." *Skeptical Inquirer,* Sept./Oct.

McDonald, John. 1998. "200% Probability and Beyond: The Compelling Nature of Extraordinary Claims in the Absence of Alternative Explanations." *Skeptical Inquirer,* Jan./Feb.

McGowan, Don. 1994. *What Is Wrong with Jung.* Prometheus Books.

McGuire, William, and R. F. C. Hull, editors. 1977. *C. G. Jung Speaking.* Princeton University Press.

McHugh, Paul R. n.d. "Multiple Personality Disorder. Dissociative Identity Disorder," www.psycom.net/mchugh.html.

McKay, Brendan. n.d. "In Search of Mathematical Miracles," http://cs.anu.edu.au/people/bdm/dilugim/index.html.

McNally, R. J. 1999. "Research on Eye Movement Densensitization and Reprocessing. EMDR as a Treatment for PTSD." *PTSD Research Quarterly* 10, no. 1: 1–7.

Meador, C. K. 1992. "Hex Death: Voodoo Magic or Persuasion?" *Southern Medical Journal* 85, no. 3: 244–247.

Mellen, Joe. 1966–67. "The Hole to Luck—Interview with Famous Self-Trepanner Dr. Bart Huges." *Transatlantic Review,* no. 23, Winter.

Michell, John. 1987. *Eccentric Lives & Peculiar Notions.* Citadel Press.

Mill, John Stuart. 1859. *On Liberty.* London.

Miller, Kenneth R. 1996. "Review of *Darwin's Black Box* by Michael Behe." *Creation/Evolution* 16: 36–40.

Milton, J., and R. Wiseman. 1999. "Does Psi Exist? Lack of Replication of an Anomalous Process of Information Transfer." *Psychological Bulletin* 125, no. 4: 387–391.

Montagu, Ashley. 1986. *Touching: The Human Significance of the Skin.* HarperCollins.

Montagu, Ashley. 1997. *Man's Most Dangerous Myth: The Fallacy of Race,* 6th ed. Altamira Press.

Montagu, Ashley. 1999. *Race and IQ: Expanded Edition.* Oxford University Press.

Moore, Brooke Noel, and Richard Parker. 2000. *Critical Thinking,* 6th ed. Mayfield Publishing.

Moore, Robert A. 1993. "The Impossible Voyage of Noah's Ark." *Creation/Evolution* 11: 1–43.

Morris, Ray Aldridge. 1990. *Multiple Personality: An Exercise in Deception.* Psychology Press.

Morrison, David. 2001. "Velikovsky at Fifty—Cultures in Collision on the Fringes of Science." *Skeptic* 9, no. 1.

Morse, Gardiner. 1999. "The Nocebo Effect." *Hippocrates* 13, no. 10, November.

Moser, Paul K., and J. D. Trout. 1995. *Contemporary Materialism: A Reader.* Routledge.

Moulder, J. E., et al. 1999. "Cell Phones and Cancer: What Is the Evidence for a Connection?" *Radiation Research* 151, no. 5, May.

Muscat, Joshua E., et al. 2000. "Handheld Cellular Telephone Use and Risk of Brain Cancer." *Journal of the American Medical Association* 284, Dec. 20.

Napier, John. 1972. *Bigfoot: The Yeti and Sasquatch in Myth and Reality.* E. P. Dutton and Co.

Nathan, Debbie, and Michael Snedeker. 1995. *Satan's Silence: Ritual Abuse and the Making of a Modern American Witch Hunt.* Basic Books.

NESS (New England Skeptical Society). "Illuminati." In *Encyclopedia of Skepticism and the Paranormal,* www.theness.com/encyc/illuminati-encyc.html.

Newberg, Andrew, M.D., Eugene G. D'Aquili, and Vince Rause. 2001. *Why God Won't Go Away: Brain Science and the Biology of Belief.* Ballantine Books.

Nickell, Joe. 1987. *Inquest on the Shroud of Turin.* Prometheus Books.

Nickell, Joe. 1992. "The Crop Circle Phenomenon: An Investigative Report." *Skeptical Inquirer,* Winter.

Nickell, Joe. 1993. *Looking for a Miracle: Weeping Icons, Relics, Stigmata, Visions and Healing Cures.* Prometheus Books.

Nickell, Joe. 1994a. "Pollens on the 'Shroud': A Study in Deception." *Skeptical Inquirer,* Summer.

Nickell, Joe, ed. 1994b. *Psychic Sleuths: ESP and Sensational Cases.* Prometheus Books.

Nickell, Joe. 1994c. *Camera Clues: A Handbook of Photographic Investigation.* University Press of Kentucky.

Nickell, Joe. 1996a. " 'Miraculous' Phenomena." In *The Encyclopedia of the Paranormal,* Gordon Stein, editor. Prometheus Books.

Nickell, Joe. 1996b. "Not-So-Spontaneous Human Combustion." *Skeptical Inquirer,* Nov./Dec.

Nickell, Joe. 1997. "Ghostly Photos." *Skeptical Inquirer,* July/Aug.

Nickell, Joe. 1999. "Miracles or Deception? The Pathetic Case of Audrey Santo." *Skeptical Inquirer,* Sept./Oct.

Nickell, Joe. 2000a. "Aura Photography: A Candid Shot." *Skeptical Inquirer,* May/June.

Nickell, Joe. 2000b. "Stigmata: In Imitation of Christ." *Skeptical Inquirer,* July/Aug.

Nickell, Joe, with John F. Fischer. 1991. *Secrets of the Supernatural: Investigating the World's Occult Mysteries.* Reprint edition. Prometheus Books.

Nienhuys, Jan Willem. 1997. "The Mars Effect in Retrospect." *Skeptical Inquirer,* Nov./Dec.

Nisbett, Matt. 2000. "The Best Case for ESP?" www.csicop.org/genx/ganzfeld.

Noll, Richard. 1997a. *The Aryan Christ: The Secret Life of Carl Jung.* Random House.

Noll, Richard. 1997b. *The Jung Cult: Origins of a Charismatic Movement.* Free Press.

Noll, Richard, editor. 1992. *Vampires, Werewolves, and Demons: Twentieth Century Reports in the Psychiatric Literature.* Brunner/Mazel.

Noonuccal, Oodgeroo. 1972. *Stradbroke Dreamtime.* Pymble, N.S.W., Australia: Angus & Robertson.

Norwood, Ann E., M.D., Robert J. Ursano, M.D., and Carol S. Fullerton, Ph.D. n.d. "Disaster Psychiatry: Principles and Practice." www.psych.org/pract_of_psych/principles_and_practice3201.cfm.

Nrretranders, Tor. 1999. *The User Illusion: Cutting Consciousness Down to Size.* Viking Press.

O'Connor, Eugene, trans. 1993. *The Essential Epicurus: Letters, Principal Doctrines, Vatican Sayings, and Fragments.* Prometheus Books.

Ofshe, Richard. 1980. "The Social Development of the Synanon Cult: The Managerial Strategy of Organizational Transformation." *Sociological Analysis* 41: 109–127.

Ofshe, Richard, and Ethan Watters. 1994. *Making Monsters: False Memories, Psychotherapy, and Sexual Hysteria.* Scribner's.

'O Hogain, Daithi. 1999. *The Sacred Isle: Pre-Christian Religions in Ireland.* Boydell & Brewer.

Ortega, Tony. 2000. "Witness for the Persecution." *New Times Los Angeles,* April 20–26.

Ortiz de Montellano, B. R. 1991. "Multicultural Pseudoscience: Spreading Scientific Illiteracy Among Minorities." *Skeptical Inquirer,* Fall.

Ortiz de Montellano, B. R. 1993. "Afrocentricity, Melanin and Pseudoscience." *Yearbook of Physical Anthropology* 36: 33–58.

Ortiz de Montellano, Bernard R. 1993. "The Dogon Revisited," www.ramtops.demon.co.uk/dogon.html.

Pagels, Heinz R. 1982. *The Cosmic Code: Quantum Physics as the Language of Nature.* Simon & Schuster.

Pain, Stephanie. 2000. "The Skull Doctors." *New Scientist,* Sept. 16.

Park, Michael Alan. 1986. "Palmistry or Hand-Jive?" In *Science Confronts the Paranormal.* Kendrick Frazier, editor. Prometheus Books.

Park, Robert L. 1997. "Alternative Medicine and the Laws of Physics." *Skeptical Inquirer,* Sept./Oct.

Park, Robert L. 2000. *Voodoo Science: The Road from Foolishness to Fraud.* Oxford University Press.

Paulos, John Allen. 1990. *Innumeracy: Mathematical Illiteracy and Its Consequences.* Vintage Books.

Paulos, John Allen. 1992. *Beyond Numeracy: Ruminations of a Numbers Man.* Vintage Books.

Paulos, John Allen. 1996. *A Mathematician Reads the Newspaper.* Anchor Books.

Paynter, Royston. n.d. "Zecharia Sitchin's Ancient Astronaut Theories—A Skeptical Archive," www.geocities.com/Area51/Corridor/8148/zindex.html.

Peebles, Curtis. 1994. *Watch the Skies!: A Chronicle of the Flying Saucer Myth.* Smithsonian Institution Press.

Peele, Stanton. 1989. *Diseasing of America.* Houghton Mifflin.

Pendergrast, Mark. 1996. *Victims of Memory: Sex-Abuse Accusations and Shattered Lives,* 2nd ed. Upper Access Book Publishers.

Pennock, Robert T. 1999. *Tower of Babel: The Evidence Against the New Creationism.* MIT Press.

Pergament, Eugene, and Morry Fiddler. 2001. "The Role of the p53 Gene," www.intouchlive.com/home/frames.htm?http://www.intouchlive.com/cancergenetics/p53.htm&3, March 1.

Perkins, David. 1982. *Altered Steaks: A Colloquium on the Cattle Mutilation Question, Participants: David Perkins, aka "Izzy Zane," Lewis MacAdams, Tom Clark.* Santa Barbara, Calif.: Am Here Books/Immediate Editions.

Persinger, Michael A. 1983. "Religious and Mystical Experiences as Artifacts of Temporal Lobe Function: A General Hy-pothesis." *Perceptual and Motor Skills* 57: 1255–1262.

Persinger, Michael. 1987. *Neuropsychological Bases of God Beliefs.* Praeger.

Persinger, Michael. n.d. "The Tectonic Strain Theory as an Explanation for UFO Phenomena: A Brief History and Summary 1970 to 1997," www.laurentian.ca/neurosci/tectonic.htm.

Peters, Fritz. 1976. *Gurdjieff: Containing Boyhood with Gurdjieff [and] Gurdjieff Remembered.* London: Wildwood House.

Pflock, Karl T. 2000. "What's Really Behind the Flying Saucers? A New Twist on Aztec." *The Anomalist,* no. 8, Spring.

Philp, Tom. 1996. "Homeopathic Remedies Besieged by Lawsuits." *Sacramento Bee,* Dec. 16.

Pickover, Clifford A. 1998. *Time: A Traveler's Guide.* Oxford University Press.

Pickover, Clifford A. 2001. *Dreaming the Future: The Fantastic Story of Prediction.* Prometheus.

Pincus, Jonathan, and Gary Tucker. 1985. *Behavioral Neurology,* 3rd ed. Oxford University Press.

Pinker, Steven. 1997. *How the Mind Works.* W. W. Norton.

Piper, Aug. 1997. *Hoax and Reality: The Bizarre World of Multiple Personality Disorder.* Jason Aronson.

Piper, Aug. 1998. "Multiple Personality Disorder: Witchcraft Survives in the Twentieth Century." *Skeptical Inquirer,* May/June.

Pipes, Daniel. 1997. *Conspiracy: How the Paranoid Style Flourishes and Where It Comes From.* Free Press.

Pittenger, David J. 1993. "Measuring the MBTI and Coming Up Short." *Journal of Career Planning & Placement.* Fall.

Plait, Philip. 2002. *Bad Astronomy: Misconceptions and Misuses Revealed, from Astrology to the Moon Landing "Hoax."* John Wiley & Sons.

Plimer, Ian. 1994. *Telling Lies for God.* Random House.

Polidoro, Massimo. 2002. "Ica Stones: Yabba-Dabba Do!" *Skeptical Inquirer* 26, no. 5, Sept./Oct.

Polidoro, Massimo. n.d. "Geller Caught Red-Handed," www.fi.muni.cz/sisyfos/geller.htm.

Pool, Robert. 1990. "Is There an EMF-Cancer Connection?" *Science* 249, no. 4973: 1096–1099, Sept. 7.

Pool, Robert. 1991. "EMF-Cancer Link Still Murky." *Nature* 349, no. 6310, Feb. 14.

Popkin, Richard H. 1967. "Skepticism." In *The*

Encyclopedia of Philosophy, vol. 7, pp. 449–461. Paul Edwards, editor. Macmillan.

Popkin, Richard H. 1979. *History of Skepticism from Erasmus to Spinoza.* University of California Press.

Popkin, Richard Henry. 1987. *Isaac La Peyrère (1596–1676): His Life, Work, and Influence.* Brill.

Popkin, Richard H., and Avrum Stroll. 2001. *Skeptical Philosophy for Everyone.* Prometheus Books.

Popper, Karl R. 1959. *The Logic of Scientific Discovery.* Harper Torchbooks.

Posner, Gary P. 1990. "God in the CCU? A Critique of the San Francisco Hospital Study on Intercessory Prayer and Healing." *Free Inquiry,* Spring.

Posner, Gary P. 1998. "Talking to the Living Loved Ones of the Dearly De-parted." *Skeptic* 6, no. 1.

Posner, Gary. 2000a. "Another Controversial Effort to Establish the Medical Efficacy of Intercessory Prayer." *Scientific Review of Alternative Medicine,* Spring/Summer.

Posner, Gary P. 2000b. "The Face Behind the 'Face' on Mars: A Skeptical Look at Richard C. Hoagland." *Skeptical Inquirer,* Nov./Dec.

Possel, Markus, and Ron Amundson. 1996. "Senior Researcher Comments on the Hundredth Monkey Phenomenon in Japan." *Skeptical Inquirer,* May/June.

Postman, Neil. 1992. *Technopoly.* Alfred A. Knopf.

Prenn, U. L. 1991. *Introduction to Ball Lightning: Rare Events.* Systems Co.

Pressman, Steven. 1993. *Outrageous Betrayal: The Dark Journey of Werner Erhard from est to Exile.* St. Martin's Press.

Prévost, Roger. 1999. *Nostradamus, le mythe et la réalité: Un histoiren au temps des astrologues.* Paris: Robert Laffont.

Prytz, John M. 1970. "The Hollow Earth Hoax." *Flying Saucers,* June.

Quenk, Naomi L. 1999. *Essentials of Myers-Briggs Type Indicator Assessment.* John Wiley & Sons.

Quigley, Christine. 1998. *Modern Mummies: The Preservation of the Human Body in the Twentieth Century.* McFarland & Co.

Rachels, James. 1971. "God and Human Attitudes." *Religious Studies* 7. Reprinted in *Philosophy and the Human Condition,* 2nd ed. Englewood Cliffs, N.J.: Prentice-Hall, pp. 509–518, 1989.

Radford, Benjamin. 2000. "Worlds in Collision: Applying Reality to the Paranormal." *Skeptical Inquirer,* Nov./Dec.

Radford, Edwin. 2000. *Encyclopedia of Superstitions.* Marlboro Books.

Radner, Daisie, and Michael Radner. 1982. *Science*

and Unreason. Belmont, Calif.: Wadsworth Publishing.

Ramachandran, V.S., M.D., Ph.D., and Sandra Blakeslee. 1998. *Phantoms in the Brain.* Quill.

Randi, James. 1982a. *Flim-Flam! Psychics, ESP, Unicorns, and Other Delusions.* Prometheus Books.

Randi, James. 1982b. *The Truth About Uri Geller.* Prometheus Books.

Randi, James. 1986. "The Columbus Poltergeist Case." In *Science Confronts the Paranormal,* Kendrick Frazier, editor. Prometheus Books.

Randi, James. 1989a. *The Faith Healers.* Updated edition. Prometheus Books.

Randi, James. 1989b. *The Magic World of the Amazing Randi.* Adams Media Corporation.

Randi, James. 1992. "Randi at CalTech: A Report from the Paranormal Trenches." *Skeptic* 1, no. 1, Spring.

Randi, James. 1995. *An Encyclopedia of Claims, Frauds, and Hoaxes of the Occult and Supernatural.* St. Martin's Press.

Raso, Jack. 1993. *Mystical Diets: Paranormal, Spiritual, and Occult Nutrition Practices.* Consumer Health Library. Prometheus Books.

Raso, Jack. 1994. *"Alternative" Healthcare: A Comprehensive Guide.* Prometheus Books.

Raso, Jack. 1995. "Mystical Medical Alternativism." *Skeptical Inquirer,* Sept./Oct.

Rau, Fred. 1992. "Snake Oil! Is That Additive Really a Negative?" *Road Rider,* Aug.

Rawcliffe, D. H. 1959. *Illusions and Delusions of the Supernatural and the Occult.* Dover Publications.

Rawcliffe, Donovan Hilton. 1988. *Occult and Supernatural Phenomena.* Dover Publications.

Raymo, Chet. 1998. *Skeptics and True Believers: The Exhilarating Connection Between Science and Religion.* Walker & Co.

Razdan, Rikki, and Alan Kielar. 1986. "Sonar and Photographic Searches for the Loch Ness Monster: A Reassessment." In *Science Confronts the Paranormal,* Kendrick Frazier, editor. Prometheus Books.

Ready, W. B. 1956. "Bridey Murphy: An Irishman's View." *Fantasy & Science Fiction* 11, no. 2: 81–88, Aug.

Redwood, Daniel, D.C. 1995. Interview with Deepak Chopra, www.healthy.net/asp/templates/interview.asp?PageType=Interview&Id=167.

Reed, Graham. 1988. *The Psychology of Anomalous Experience: A Cognitive Approach.* Prometheus Books.

Reid, Brian. 2002. "The Nocebo Effect—Placebo's Evil Twin." *Washington Post,* April 30.

Reilly, David, Morag A. Taylor, Neil G. M. Beattie, Jim H. Campbell, Charles McSharry, Tom C. Aitchison, Roger Carter, and Robin D. Stevenson. 1994. "Is Evidence for Homeopathy Reproducible?" *The Lancet* 344, Dec. 10.

Restak, Richard M., and David Grubin. 2001. *The Secret Life of the Brain.* Joseph Henry Press.

Richards, Bill. 1993. "Elusive Threat: Electric Utilities Brace for Cancer Lawsuits Though Risk Is Unclear; Companies Spend on Cutting Electromagnetic Fields as Lawyers Smell Blood." *Wall Street Journal,* Feb. 5.

Ridpath, I. 1978. "Investigating the Sirius Mystery." *Skeptical Inquirer,* Fall.

Roberts, J. 1972. *Mythology of the Secret Societies.* Macmillan.

Robertson, Blair Anthony. 1999. "Speed-Reading Between the Lines." *Sacramento Bee,* Oct. 21.

Rojecewicz, Peter M. 1987. "The Men in Black Experience and Tradition: Analogues with the Traditional Devil Hypothesis." *Journal of American Folklore* 100, April/June.

Rommel, Kenneth M., Jr., Project Director. 1980. "Operation Animal Mutilation: Report of the District Attorney, First Judicial District, State of New Mexico." June.

Rosa L., E. Rosa, L. Sarner, and S. Barrett. 1998. "A Close Look at Therapeutic Touch." *Journal of the American Medical Association* 279: 1005–1010.

Rosen, Gerald M., Ph.D., and Jeffrey Lohr, Ph.D. 1997. "Can Eye Movements Cure Mental Ailments?" www.hcrc.org/ncahf/newslett/n120-1.html#emdr.

Rosenberg, Howard L. 1974. "Exorcizing the Devil's Triangle." *Sealift,* no. 6, June.

Ross, Colin A. 1996. *Dissociative Identity Disorder: Diagnosis, Clinical Features, and Treatment of Multiple Personality.* John Wiley & Sons.

Rotton, James. 1997. "Moonshine." *Skeptical Inquirer,* May/June.

Rowland, Ian. 2002. *The Full Facts Book of Cold Reading,* 3rd ed. Ian Rowland Limited.

Rubin, Robert T., editor. *Extraordinary Disorders of Human Behavior.* Plenum Press, 1982.

Ruscio, John. 1998. "The Perils of Post-Hockery." *Skeptical Inquirer,* Nov./Dec.

Russell, Bertrand Arthur. 1977. *Why I Am Not a Christian, and Other Essays on Religion and Related Subjects.* Simon and Schuster.

Ryle, Gilbert. 1949. *The Concept of Mind.* Barnes and Noble.

Ryle, Gilbert, and Rene Meyer, eds. 1993. *Aspects of Mind.* Blackwell.

Ryman, Daniele. 1993. *Aromatherapy: The Complete Guide to Plant and Flower Essences for Health and Beauty.* Bantam Doubleday Dell.

Sacks, Oliver W. 1974. *Awakenings.* Doubleday.

Sacks, Oliver W. 1984. *A Leg to Stand On.* Summit Books.

Sacks, Oliver W. 1985. *The Man Who Mistook His Wife for a Hat and Other Clinical Tales.* Summit Books.

Sacks, Oliver W. 1989. *Seeing Voices: A Journey into the World of the Deaf.* University of California Press.

Sacks, Oliver W. 1995. *An Anthropologist on Mars: Seven Paradoxical Tales.* Knopf.

Sagan, Carl. 1972. "UFO's: The Extraterrestrial and Other Hypotheses." In *UFO's: A Scientific Debate,* Carl Sagan and Thornton Page, editors. Cornell University Press.

Sagan, Carl. 1979. *Broca's Brain.* Random House.

Sagan, Carl. 1994. "Carlos." *Parade Magazine,* Dec. 4.

Sagan, Carl. 1995. *The Demon-Haunted World: Science as a Candle in the Dark.* Random House.

Sagan, Leonard A. 1992. "EMF Danger: Fact or Fiction?" *Safety & Health* 145, no. 1, Jan.

Saler, Benson, Charles A. Ziegler, and Charles B. Moore. 1997. *UFO Crash at Roswell: The Genesis of a Modern Myth.* Smithsonian Institution Press.

Sampson, Wallace, and Lewis Vaughn, eds. 2000. *Science Meets Alternative Medicine: What the Evidence Says About Unconventional Treatments.* Prometheus Books.

Sargant, William. 1957. *Battle for the Mind: A Physiology of Conversion and Brainwashing.* Greenwood Press.

Schacter, Daniel L. 1996. *Searching for Memory: The Brain, the Mind, and the Past.* Basic Books.

Schacter, Daniel L. 2001. *The Seven Sins of Memory: How the Mind Forgets and Remembers.* Houghton Mifflin.

Schacter, Daniel L., editor. 1997. *Memory Distortion: How Minds, Brains, and Societies Reconstruct the Past.* Harvard University Press.

Schadewald, Robert. 1986. "Creationist Pseudoscience." In *Science Confronts the Paranormal,* Kendrick Frazier, editor. Prometheus Books.

Schaeffer, Robert. 1986. *The UFO Verdict.* Prometheus Books.

Schaeffer, Robert. 1996. "Unidentified Flying Objects. UFOs." In *The Encyclopedia of the Paranormal,* Gordon Stein, editor. Prometheus Books.

Schafersman, Steven D. 1998. "Unraveling the Shroud of Turin." *Approfondimento Sindone,* year 2, vol. 2.

Scheflin, Alan W., and Edward M. Opton. 1978. *The Mind Manipulators: A Non-Fiction Account.* Paddington Press.

Schick, Theodore, Jr., and Lewis Vaughn. 1998. *How to Think About Weird Things,* 2nd ed. Mayfield Publishing.

Schnabel, Jim. 1994. *Round in Circles: Physicists, Poltergeists, Pranksters and the Secret History of the Cropwatchers*. London: Penguin.

Schultz, Ted. 1989. "Voices from Beyond: The Age-Old Mystery of Channeling." In *The Fringes of Reason*. Harmony Books.

Schwarcz, Joe, Ph.D. 2000. "Colorful Nonsense: Dinshah Ghadiali and His Spectro-Chrome Device," http://www.quackwatch.com/01QuackeryRelatedTopics/spectro.html.

Seckel, Al. 1987a. "The Man Who Could Read the Grooves." *Los Angeles Times*, Oct. 19.

Seckel, Al. 1987b. "Sensing Just How to Help the Police." *Los Angeles Times*, Nov. 16.

Seebach, Linda. 2000. "Science Offers Little Support for 'Energy Medicine' Ideas." *Rocky Mountain News*, March 19.

Seely, D. R., S. M. Quigley, and A. W. Langman. 1996. "Ear Candles—Efficacy and Safety." *Laryngoscope* 106, no. 10: 1226–1229.

Sefton, Dru. 1998. "A Spirited Debate." *San Diego Union-Tribune*, July 10.

Segel, B. W. 1995. *A Lie and a Libel: The History of the Protocols of the Elders of Zion*, Richard S. Levy, translator and editor. University of Nebraska Press.

Selby, Carla, and Bela Scheiber. 1996. "Science or Pseudoscience? Pentagon Grant Funds Alternative Health Study." *Skeptical Inquirer*, July/Aug.

Selby, Carla, and Bela Scheiber. 2000. *Therapeutic Touch*. Prometheus.

Seligmann, Kurt. 1997. *The History of Magic and the Occult*. Reprint of 1948 edition. Random House.

Shane, Howard C. 1994. *Facilitated Communication: The Clinical and Social Phenomenon*. Singular Publishers Group.

Shapiro, Arthur K., and Elaine Shapiro. 1997. *The Powerful Placebo: From Ancient Priest to Modern Physician*. Johns Hopkins University Press.

Sheaffer, Robert. 1989. *UFO Sightings: The Evidence*. Prometheus Books.

Shermer, Michael. 1997. *Why People Believe Weird Things: Pseudoscience, Superstition, and Other Confusions of Our Time*. W. H. Freeman.

Shermer, Michael. 1998. "A Mind Out of Body." *Skeptic* 6, no. 3.

Shermer, Michael. 2001a. *The Borderlands of Science: Where Sense Meets Nonsense*. Oxford University Press.

Shermer, Michael. 2001b. "Deconstructing the Dead 'Crossing Over' to Expose the Tricks of Popular Spirit Mediums." *Scientific American*, Aug. 1.

Shermer, Michael, and Arthur Benjamin. 1992. "Deviations: A Skeptical Investigation at Edgar Cayce's Association for Research and Enlightenment." *Skeptic* 1, no. 3, Fall.

Shermer, Michael, and Alex Grobman. 2000. *Denying History: Who Says the Holocaust Never Happened and Why Do They Say It?* University of California Press.

Sherwood, John C. 1998. "Gray Barker: My Friend, the Myth-Maker." *Skeptical Inquirer*, May/June.

Shneour, Elie. 1990. "Lying About Polygraph Tests." *Skeptical Inquirer*, Spring.

Shuker, Karl P. N. 1997. *From Flying Toads to Snakes with Wings: From the Pages of Fate Magazine*. Llewellyn Publications.

Siegel, J. M., et al. 1999. "Sleep in the Platypus." *Neuroscience* 91: 391–400.

Siegel, Ronald K. 1992. *Fire in the Brain: Clinical Tales of Hallucination*. Dutton.

Singer, Barry, and Victor A. Benassi. 1986. "Fooling Some of the People All of the Time." In *Science Confronts the Paranormal*, Kendrick Frazier, editor. Prometheus Books.

Singer, Margaret Thaler, and Janja Lalich. 1995. *Cults in Our Midst*. Jossey-Bass.

Singer, Margaret Thaler, and Janja Lalich. 1996. *Crazy Therapies*. Jossey-Bass.

Singer, Peter. 1993. *Practical Ethics*. Cambridge University Press.

Singh, Nagendra Kumar. 1997. *Divine Prostitution*. New Delhi: A. P. H. Publishing Corporation.

Skerret, P. J. 1996. "DHEA: Ignore the Hype." *HealthNews*, a newsletter from the publishers of the *New England Journal of Medicine*, Nov. 19.

Skolnick, Andrew A. 1991. "The Maharishi Caper: Or How to Hoodwink Top Medical Journals." *ScienceWriters: The Newsletter of the National Association of Science Writers*, Fall.

Skrabanek, Peter, and James McCormick. 1990. *Follies and Fallacies in Medicine*. Prometheus Books.

Slade, Peter D., and Richard P. Bentall. 1988. *Sensory Deception: A Scientific Analysis of Hallucination*. Johns Hopkins University Press.

Smith, George H. 1979. *Atheism: The Case Against God*. Prometheus Books.

Smith, Homer. 1952. *Man and His Gods*. Little, Brown.

Smith, Ralph. 1984. *At Your Own Risk: The Case Against Chiropractors*. Simon and Schuster.

Smith, Rodney K. 1984. *Multilevel Marketing: A Lawyer Looks at Amway, Shaklee, and Other Direct Sales Organizations*. Baker Book House.

Sokal, Alan. 1996. "A Physicist Experiments with Cultural Studies." *Lingua Franca*, May/June.

Sokal, Alan, and Jean Bricmont. 1998. *Fashion-*

able Nonsense: Postmodern Intellectuals' Abuse of Science. St. Martin's Press.

Spanos, Nicholas. 1987–88. "Past-Life Hypnotic Regression: A Critical View." *Skeptical Inquirer,* Winter.

Spanos, Nicholas P. 1996. *Multiple Identities and False Memories: A Sociocognitive Perspective*. American Psychological Association.

Spanos, Nicholas P., and John F. Chaves, editors. 1989. *Hypnosis: The Cognitive-Behavioral Perspective*. Prome-theus Books.

Spinoza, Baruch de. 1670. *Theologico-Political Treatise*.

Squatriglia, Chuck. 2001. "Mystery Unlocked? A Scientist Says He's Solved a Monster Controversy—The 'Beast' in Loch Ness Is Merely an Illusion Created by Earthquakes." *San Francisco Chronicle,* June 27.

Stalker, D., and C. Glymour. 1989. *Examining Holistic Medicine*. Prometheus Books.

Stanovich, Keith E. 1992. *How to Think Straight About Psychology,* 3rd ed. HarperCollins.

Steele, K. M., K. E., Bass, and M. D. Crook. 1999. "The Mystery of the Mozart Effect: Failure to Replicate." *Psychological Science* 10: 366–369.

Stein, Gordon, ed. 1993. *Encyclopedia of Hoaxes*. Prometheus Books.

Stein, Gordon, ed. 1996a. *The Encyclopedia of the Paranormal*. Prometheus Books.

Stein, Gordon. 1996b. "Spiritualism." In *The Encyclopedia of the Paranormal,* Gordon Stein, editor. Prometheus Books.

Steiner, Robert A. 1996. "Fortunetelling." In *The Encyclopedia of the Paranormal,* Gordon Stein, editor. Prometheus Books.

Stenger, Victor J. 1990. *Physics and Psychics: The Search for a World Beyond the Senses*. Prometheus Books.

Stenger, Victor J. 1995a. *Not by Design: The Origin of the Universe*. Prometheus Books.

Stenger, Victor J. 1995b. *The Unconscious Quantum: Metaphysics in Modern Physics and Cosmology*. Prometheus Books.

Stenger, Victor. 1997a. "Intelligent Design: Humans, Cockroaches, and the Laws of Physics," www.talkorigins.org/faqs/cosmo.html.

Stenger, Victor J. 1997b. "Quantum Quackery." *Skeptical Inquirer,* Jan./Feb.

Stenger, Victor J. 1999. "Bioenergetic Fields." *Scientific Review of Alternative Medicine* 3, no. 1, Spring/Summer.

Stenger, Victor. 2000. "The Emperor's New Designer Clothes," spot.colorado.edu/~vstenger/Briefs/Emperor.html.

Stenhoff, Mark. 2000. *Ball Lightning: An Unsolved Problem in Atmospheric Physics*. Kluwer Academic Publishers.

Stern, Jess. 1967. *Edgar Cayce: The Sleeping Prophet.* Doubleday and Co.

Stern, Madeleine B. 1982. *A Phrenological Dictionary of Nineteenth-Century Americans*. Greenwood Publishing.

Sternberg, Esther M., and Philip W. Gold. 1997. "The Mind-Body Interaction in Disease." *Scientific American,* Jan.

Stevens, Phillips, Jr. 1996. "Evil Eye." In *The Encyclopedia of the Paranormal,* Gordon Stein, editor. Prometheus Books.

Stewart, J. 1977. "Cattle Mutilations." *Zetetic* (now *Skeptical Inquirer*), vol. 1, no. 2, Spring/Summer.

Storr, Anthony. 1996. *Feet of Clay: Saints, Sinners, and Madmen: A Study of Gurus*. Free Press.

Story, Ronald. 1976. *The Space-Gods Revealed: A Close Look at the Theories of Erich von Däniken,* 2nd ed. Barnes & Noble.

Story, Ronald. 1980. *Guardians of the Universe*. St. Martin's Press.

Strassman, Rick. 2001. *DMT: The Spirit Molecule: A Doctor's Revolutionary Research into the Biology of Near-Death and Mystical Experiences*. Inner Traditions International Ltd.

Sulloway, Frank. 1993. *Freud: Biologist of Mind*. Harvard University Press.

Sutherland, Stuart. 1979. "The Myth of Mind Control." In *The Encyclopedia of Delusions,* Ronald Duncan and Miranda Weston-Smith, editors. Wallaby Books.

Sykes, C. J. 1992. *A Nation of Victims: The Decay of the American Character*. St. Martin's Press.

Symons, Donald. 1981. *Evolution of Human Sexuality*. Oxford University Press.

Talbot, Margaret. 2000. "The Placebo Prescription." *New York Times Magazine,* Jan. 9.

Tanner, Amy. 1994. *Studies in Spiritism.* Prometheus Books.

Tavris, Carol. 1992. *The Mismeasure of Woman*. Simon & Schuster.

Tavris, Carol. 1993. "Hysteria and the Incest-Survivor Machine." *Sacramento Bee,* Forum section, Jan. 17.

Taylor, Shelly E. 1989. *Positive Illusions: Creative Self-Deception and the Healthy Mind*. Basic Books.

Thiriart, P. 1991. "Acceptance of Personality Test Results." *Skeptical Inquirer,* Winter.

Thomas, David E. 1997. "Hidden Messages and the Bible Code." *Skeptical In-quirer,* Nov./Dec.

Thomas, David E. 1998. "The Aztec UFO Symposium: How This Saucer Story Started as a Con Game." *Skeptical Inquirer,* Sept./Oct.

Thomason, Dr. Sarah Gray. 1996. "Xenoglossy." In *The Encyclopedia of the Paranormal,* Gordon Stein, editor. Prometheus Books.

Thompson, Tony. 1994. "The Hidden Persuaders." *Time Out,* June 22–29.

Thorburn, W. M. 1918. "The Myth of Occam's Razor." *Mind* 27: 345–353.

Thornton, E. M. 1976. *Hypnotism, Hysteria, and Epilepsy: An Historical Synthesis.* London: Heinemann.

Thouless, R. H., and B. P. Weisner. 1942. "The Present Position of Experimental Research into Telepathy and Related Phenomena." *Proceedings of the Society for Psychical Research* 47.

Toronto, Richard. n.d. "The Shaver Mystery," www.parascope.com/nb/articles/shaverMystery.htm.

Torrey, E. Fuller. 1992. *Freudian Fraud: The Malignant Effect of Freud's Theory on American Thought and Culture.* HarperCollins.

Trafford, Abigail. 1995. "Truth About Human Error in Hospitals." *Sacramento Bee,* p. B7, March 21.

Trimble, Russell. 1996. "Alchemy." In *The Encyclopedia of the Paranormal,* Gordon Stein, editor. Prometheus Books.

True, George Nava. n.d. "The Facts About Faith Healing," www.netasia.net/users/truehealth/Psychic%20Surgery.htm.

True, George Nava. n.d. "Psychic Dentistry," www.netasia.net/users/truehealth/Psychic%20Dentistry.htm.

Trull, D. n.d. "A Really Light Lunch—The Prana Diet Plan," www.parascope.com/en/lightlch.htm.

Tversky, A., and D. Hahneman. 1971. "Belief in the Law of Small Numbers." *Psychological Bulletin* 76: 105–110.

Valenstein, Elliot S. 1973. *Brain Control: A Critical Examination of Brain Stimulation and Psychosurgery.* Wiley.

Vallee, Jacques F. 1994. "Anatomy of a Hoax: The Philadelphia Experiment 50 Years Later." *Journal of Scientific Exploration,* Spring.

van Beek, Walter E. A. 1992. "Dogon Restudied: A Field Evaluation of the Work of Marcel Griaule." *Current Anthropology* 32.

Vankin, Jonathan, and John Whalen. 1998. *The Seventy Greatest Conspiracies of All Time: History's Biggest Mysteries, Coverups, and Cabals.* Citadel.

Victor, Jeffrey S. 1993. *Satanic Panic: The Creation of a Contemporary Legend.* Open Court.

Vistica, Gregory. 1995. "Psychics and Spooks: How Spoon-Benders Fought the Cold War." *Newsweek,* Dec. 11.

Vogt, Evon, and Ray Hyman. 2000. *Water Witching U.S.A.,* 2nd ed. University of Chicago Press.

Volkey, J., and J. Read. 1985. "Subliminal Messages: Between the Devil and the Media." *American Psychologist* 40, no. 11.

Wagner, M. W., and M. Monnet. 1979. "Attitudes of College Professors Toward Extra-Sensory Perception." *Zetetic Scholar* 5: 7–16.

Wakefield, Hollida, and Ralph Underwager. 1994. *Return of the Furies: An Investigation into Recovered Memory Therapy.* Peru, Ill.: Open Court.

Walker, Clarence E. 2001. *We Can't Go Home Again: An Argument About Afrocentrism.* Oxford University Press.

Walker, Tom, and Judith O'Reilly. 1999. "Three Deaths Linked to 'Living on Air' Cult." *Sunday Times* (London), Sept. 26.

Wall, Ernst L. 1995. *The Physics of Tachyons.* Hadronic Press.

Walsh, James. 1998. *You Can't Cheat an Honest Man: How Ponzi Schemes and Pyramid Frauds Work . . . and Why They're More Common Than Ever.* Merritt Publishing.

Washington, Peter. 1996. *Madame Blavatsky's Baboon: A History of the Mystics, Mediums, and Misfits Who Brought Spiritualism to America.* Schocken Books.

Watkins, Arleen J., and William S. Bickel. 1991. "A Study of the Kirlian Effect." In *The Hundredth Monkey and Other Paradigms of the Paranormal,* Kendrick Frazier, editor. Prometheus Books.

Watters, Ethan, and Richard Ofshe. 1999. *Therapy's Delusions: The Myth of the Unconscious and the Exploitation of Today's Walking Worried.* Simon and Schuster.

Weinstein, Harvey. 1990. *Psychiatry and the CIA: Victims of Mind Control.* Washington, D.C.: American Psychiatric Press.

Wheeler, Thomas J., Ph.D. n.d. "Deepak Chopra and Maharishi Ayurvedic Medicine," www.hcrc.org/contrib/wheeler/chopra.html.

Wicklein, Robert C., and Jay W. Rojewski. n.d. "The Relationship Between Psychological Type and Professional Orientation Among Technology Education Teachers," scholar.lib.vt.edu/ejournals/JTE/jte-v7n1/wicklein.jte-v7n1.html.

Willey, David. n.d. "Firewalking: Myth vs. Physics," www.pitt.edu/~dwilley/fire.html.

Williams, Barry. 1998. "Cold Water on a Hot Topic." *The Skeptic* (Australia) 18, no. 4, Dec.

Williams, Barry. n.d. "Pyramids, Pyramyths & Pyramidiots," www.skeptics.com.au/journal/paramyth.htm.

Williams, George C. 1996. *Adaptation and Natural Selection: A Critique of Some Current Evolutionary Thought.* Princeton University Press.

Williams, Susan M. 1989. "Holistic Nursing." In *Examining Holistic Medicine,* Douglas Stalker, editor. Prometheus Books.

Williams, William F., editor. 2000. *Encyclopedia of Pseudoscience*. Facts-on-File.

Willis, Paul. n.d. Interview with Jasmuheen, "Can People Live on Nothing but Air?" Correx Files, www.abc.net.au/science/correx/archives/jasmuheen .htm.

Wilson, Ian. 1988. *The Bleeding Mind: An Investigation into the Mysterious Phenomenon of Stigmata*. London: Weidenfeld and Nicholson.

Wirth-Pattullo, V., and K. W. Hayes. 1994. "Inter-rater Reliability of Craniosacral Rate Measurements and Their Relationship With Subjects' and Examiners' Heart and Respiratory Rate Measurements." *Physical Therapy*, Oct.

Wiseman, Richard. 1997. *Deception & Self-Deception: Investigating Psychics*. Prometheus Books.

Wiseman, Richard, and Robert L. Morris. 1995. *Guidelines for Testing Psychic Claimants*. Prometheus Books.

Wiseman, Richard, and Matthew Smith. 1998. "Can Animals Detect When Their Owners Are Returning Home?" *British Journal of Psychology* 89: 453.

Wiseman, Richard, et al. 1996. "Psychic Crime Detectives: A New Test for Measuring Their Successes and Failures." *Skeptical Inquirer*, Jan./Feb.

Witztum, D., E. Rips, and Y. Rosenberg. 1994. "On Equidistant Letter Sequences in the Book of Genesis." *Statistical Science* 9.

Wolf, Lucien. 1921. *The Myth of the Jewish Menace in World Affairs; or, The Truth About the Forged Protocols of the Elders of Zion*. Macmillan.

Woodruff, Dianne L., CMA, Ph.D. n.d. "A Defense of Craniosacral Therapy," www.vicpain .com/therapy.htm#CRANIO.

Worrall, Russell S. 1986. "Iridology: Diagnosis or Delusion?" In *Science Confronts the Paranormal*, Kendrick Frazier, editor. Prometheus Books.

Wright, Lawrence. 1995. *Remembering Satan*. Vintage Books.

Wynn, Charles M., and Arthur W. Wiggins. 2001. *Quantum Leaps in the Wrong Direction*. Joseph Henry Press.

Zelicoff, Alan P. 2001. "Polygraphs and the National Labs: Dangerous Ruse Undermines National Security." *Skeptical Inquirer*, July/Aug.

Zusne, Leonard, and Warren Jones, editors. 1990. *Anomalistic Psychology: A Study of Magical Thinking*. Lawrence Erlbaum Association.

Picture Credits

Page 11: drawing by J. Ruhe; page 22: photo by R. Carroll; page 34: scanned image courtesy of Mike Horodniak; page 42: digigraph by R. Carroll; page 55: copyright 2002 Dave Rubert Photography; page 59: printed with permission of Tom Anheyer; page 65: photo courtesy of Ken Feder; page 84: image courtesy of Dr. Peter Friesen, Plattsburgh State University; page 89: photo courtesy of Peter Sorensen; page 107: courtesy of the Clendening History of Medicine Library, University of Kansas Medical Center; page 114: photo by Leslie A. Carroll; page 128: photo by R. Carroll; page 133 top: photo courtesy of NASA; page 133 bottom: photos courtesy of NASA; page 154: digitally altered photo (using the magic of Paint Shop Pro) by R. Carroll; page 189: digigraph by R. Carroll; page 199: photo by Leslie A. Carroll; page 203: courtesy of NASA; page 221: courtesy of the Clendening History of Medicine Library, University of Kansas Medical Center; page 223: courtesy of the Clendening History of Medicine Library, University of Kansas Medical Center; page 249: photo courtesy of NASA/GSFC/MITI/ ERSDAC/JAROS, and U.S./Japan ASTER Science Team; VE Record ID: 9468; page 263: digigraph by R. Carroll; page 275: photo by R. Carroll; page 286: courtesy of Library of Congress, Prints & Photographs Division (reproduction number LC-USZ62-100747); page 287: photo courtesy of Bob McCoy; page 288: courtesy of the Clendening History of Medicine Library, University of Kansas Medical Center; page 312: photo by L. A. Carroll; digital ghosting by R. Carroll; page 353: digigraph by R. Carroll; page 406: digigraph by R. Carroll.

Name Index

Subject Index